STRATA MECHANICS IN COAL MINING

STRATA MECHANICS IN COAL MINING

By
M.L. JEREMIC
Laurentian University, Sudbury, Ontario

A.A.BALKEMA/ROTTERDAM/BOSTON/1985

ISBN 90 6191 508 2 cloth edition
ISBN 90 6191 556 2 paper edition
© 1985 A.A.Balkema, P.O.Box 1675, 3000 BR Rotterdam, Netherlands
Distributed in USA & Canada by: A.A.Balkema Publishers, P.O.Box 230, Accord, MA 02018
Printed in the Netherlands

Contents

PREFACE   VII

SYMBOLS   IX

1. GENERAL CONSIDERATIONS   1
   1.1 Development of strata mechanics   1
   1.2 Fundamentals of strata mechanics   4
   1.3 Philosophy of strata mechanics   9
   1.4 Rationalization of coal strata   15

2. COAL BEARING STRATA   21
   2.1 Geology of coal strata   21
   2.2 Sedimentology of coal strata   27
   2.3 Anisotropy of coal strata   39
   2.4 Tectogenesis of coal strata   45

3. COAL SEAM FEATURES   59
   3.1 Formation of coal   59
   3.2 Petrography of coal   66
   3.3 Coal rank   74
   3.4 Cleats in coal   83

4. ROOF AND FLOOR STRATA   97
   4.1 Weak roof strata   97
   4.2 Competent roof strata   106
   4.3 Structural defects of roof strata   116
   4.4 Behaviour of subcoal strata   125

5. PRIMITIVE STRATA PRESSURE   137
   5.1 Geological stresses   137
   5.2 Lateral tectonic stress   145
   5.3 Stress state of soft coal seam   157
   5.4 Hydrodynamic stresses   163

6. DEFORMATION AND FAILURE OF COAL STRUCTURE	173
6.1 Coal and gas outbursts	173
6.2 Yielding fracturing of coal	189
6.3 Shear displacement of coal	205
6.4 Extrusion deformations	218
7. ROOM AND PILLAR MINING	233
7.1 Principal mining systems	233
7.2 Stability of stope pillar structures	244
7.3 Strata mechanics	261
7.4 Caving mechanics	282
8. LONGWALL MINING	297
8.1 Longwall mining systems	297
8.2 Stability of panel-stope structures	309
8.3 Strata mechanics	324
8.4 Caving mechanics	335
9. SLICE MINING	354
9.1 Principal mining systems	354
9.2 Stability of slice-face structure	367
9.3 Strata mechanics	378
9.4 Caving mechanics	393
10. LONG PILLAR MINING	404
10.1 Principal mining systems	404
10.2 Stability of stope-pillar structure	413
10.3 Strata mechanics	421
10.4 Caving mechanics	430
11. SUBLEVEL CAVING	441
11.1 Principal mining systems	441
11.2 Stability of stope-pillar structure	453
11.3 Strata mechanics	462
11.4 Caving mechanics	480
12. COAL PILLAR STRUCTURE	493
12.1 Mini coal pillars	493
12.2 Time dependent deformations of pillars	504
12.3 Strength of coal pillars	521
12.4 Protective pillars	530
13. REFERENCES	548
SUBJECT INDEX	564

Preface

Development of coal mining throughout the world has increased in the last decade due to the need for energy, and this trend will continue, at least, until the end of this century. Increased production requires enlarging underground mine structures and rapid coal extraction, in which coal mine strata mechanics play an essential role.

Coal mine strata mechanics first interested me in the late 1940's during practical work as a student and later as a graduate mining engineer in St. Barbara Coal Mine, Yugoslavia. I was fascinated in observing the mechanics of stress concentration and relaxation manifested by deformation and failure of coal pillars and roof falls in roadways, as well as effective strata control by almost illiterate miners as an integral part of mine operations. Miners were aware of particular acoustic emission effects before roof falls during retreat long pillar mining; realizing that they could be indicated by a mouse's hearing system and its reaction by run. These warnings of roof falls were primitive, but effective, because accidents under these circumstances never occurred. This earlier practical knowledge was of fundamental importance for later sophisticated research in strata mechanics and mine stability. Generally, in the last two decades the subject of rock mechanics in coal mining has been extensively investigated and discussed, as shown by the abundance of publications, where the authors more or less considered the strata and mining conditions of the region in which they live and work. This volume is no exception, because a large number of example are from the western coal fields of North America. At present there are several different philosophies of coal mine strata mechanics, arising from the different backgrounds of the specialists involved in this particular field, for example, highly theoretical concepts versus very practical concepts in solving problems. Also, the conceptual approach to coal mine strata mechanics by a mining engineer does not necessarily coincide with that of a civil engineer. Many underground civil engineering structures are integral and functional systems of static rock masses in which the primary concerns are stresses induced by the size and shape of openings. The determination of such stresses requires a good engineering background and knowledge of the theory of mechanics. Mining engineering structures are temporary or semipermanent and are loosely integrated in the rock strata, and are dynamic rather than static. The mining stresses of primary interest are those with different play during coal extraction due to uncaved or caved roofs, stable or unstable pillars, and others, for whose determination ingenuity and mining experience is necessary in addition to theoretical knowledge.

This volume, *Strata Mechanics in Coal Mining* is offered as a contribution to the spread of knowledge and understanding in this important engineering field. It is written by a mining engineer for professionals in the coal mining industry and students of mining engineering. It includes some of the author's own philosophy of coal mine stability.

The author acknowledges the assistance given by supervising personnel in coal mines of Western Canada and Western Australia. Also, he greatly appreciates the assistance given by mining students of the University of Alberta, Western Australian School of Mines and Laurential University, Sudbury, Ontario.

It is a pleasure to record the financial support given by Department of Energy Resources of the Province of Alberta with particular thanks to Mr. R. D. McDonald, executive director and by Laurential University, Sudbury, Ontario, also with particular thanks to Dr. D. E. Goldsack, Dean of Faculty of Science and Engineering. The author is sincerely grateful to Mr. R. A. King, President of Crownest Resources for the research grant on strata mechanics and his encouragement to publish this volume.

I wish to express my deepest appreciation to my colleagues from the Laurentian University: Dr. R. A. Cameron, Dr. R. S. James, Dr. P. H. Lindon, also my colleagues from the industry Mr. P. Oliver, Superintendent Rock Mechanics, INCO Metals Company, Mr. M. Masson, Superintendent Mine Engineering Services, Falconbridge Nickel Mines Limited and finally, Mr. C. B. Graham, Senior Roof Control Engineer of the Ministry of Labour, Province of Ontario for reading the manuscript and the suggestions which considerably improved the quality of this publication.

Sudbury, Ontario
January 1, 1983

Dr. M. L. Jeremic, P. Eng., P. Geol.,
Professor of Rock Mechanics and
Mining Engineering

Symbols

The symbols are described where they are represented. Some of the more commonly introduced symbols are listed below:

A   Used for areas

A_e Excavated area

A_p Cross-sectional area of a pillar

A_t Tributary area

a   Subsidence factor dependant on the type of mining

B   Width of the draw area

b   Width of the opening behind support

C   Convergence, factor of pillar stability

C_c Percent of coal left in pillars (1–R)

c   Cohesion, caving height, coefficient of velocity of strata displacement, initial yielding stress

c_J Cohesion of joint

D   Used for diameters, length of draw area, flexural rigidity

d   Average thickness of rock stratum, constant

d_f Density of fissures

E   Young's modulus

E_d Dynamic modulus of elasticity

F   Used for forces, including Protodyakonov factor of hardness

F_a Active force

F_p Passive force

f   Length of longer semi-axis of ellipse

f_c Unconfined yielding strength

G   Shear modulus, weight of rock

g   Acceleration of gravity

H   Used for heights

h   Mine depth, thickness of overburden

h_o Height of the gob

h_p Penetration depth

I   Moment of inertia, index of roof span stability

i   The aspirity angularity

J   Coefficient of coal fracturing

K   Used for different factors as defined locally, including curvature functions

K_t Lateral thrust

k   Used for different coefficients as defined locally, including strength constant and coal porosity

k_n Foundation modulus of the pillars

L   Used for lengths

L_p Length of the prototype

L_n Length of the model

l   Used for lengths

M   Cohesion index

n   Poisson's number

N   Used variously for forces

N_m Modified bearing capacity factor

n   Coefficient of stress concentration

P   Used for pressures

P_a Active earth pressure

P_p Passive earth pressure

p   Gas and water pressure, hydraulic pressure, restraint at free boundary

p_k Pressure of hydraulic fracturing

Q	Weight of solid coal	δ	Angle of caving and draw, elastic constant
Q_m	Coal discharge		
q	Surcharge	δ_i	Beam or plate deflection
q_o	Uniformly distributed load	ε	Normal strain
q_{ult}	Bearing capacity	ε_d	Permitted strain
R	Used for different purposes as defined locally, including gas constant and recovery factor	η	Used variously for viscosity, roof sag or beam deflection
R_c	Cohesion indices	η_o	Elastic deflection
r	Used for radius, including length of influence of subsidence	η_v	Viscous deflection
		Θ	Used variously for angles
S	Stability coefficient, support spacing along the face	κ	Strength factor of coal cube, compaction factor
S_p	Shaft pillar safety factor	μ	Coefficient of rock coherency, compaction factor
s	Length of arch, resultant displacement		
T	Temperature	ν	Poisson's ratio
t	Used variously for time, thickness of coal seam and workable thickness of coal seam	ν_d	Dynamic Poisson's ratio
		π	Dimensionless product
		ρ	Radius of curvature, scaling factor
U	Energy	σ	Used for stresses
U_c	Energy contraction factor	$\sigma_1, \sigma_2, \sigma_3$	Principal stresses
u	Horizontal deflection and displacement, mass velocity of the particles	σ_V, σ_H	Vertical and horizontal stresses
		σ_r, σ_t	Radial and tangential stresses
		σ_N	Normal stress
V	Used for various velocities, volume of rock or coal	$\sigma_x, \sigma_y, \sigma_z$	Normal stress components
V_n	Velocity of coal discharge	σ_c	Compressive stress
v	Vertical deflection and displacement, ground surface subsidence	σ_T	Tensile stress
		σ_p	Strength of pillar
		σ_1	Fracture strength
v_{max}	Maximum ground surface subsidence	σ^*	Ratio of uniaxial compressive strength of proto type rock and model material
W	Least width of various coal pillars		
W_R	Weight of rock	τ	Used for shear stresses
w	Width of various mine openings	τ_o	Shear strength
w_e	Fracture energy	τ_r	Residual shear strength
X	Length of subsided ground	τ_β	Shear stress along plane of failure
x	Used for various lengths of mine structures	φ	Angle between the direction of σ_1 and the plane of a joint
β	Used for various angles of inclination, including elastic constant and strength coefficient	ψ	Angle of limit of subsidence
		α	Used variously for angles, including elastic constant and strength coefficient
γ	Unit weight, elastic constant, shear strain	ϕ	Variously used as internal and surfacial friction angles as defined locally
γ_c	Unit weight of coal	ϕ_{eff}	Effective friction angle
Δ	Phase differences	ϕ_r	Residual angle of friction
		ϕ_J	Friction angle for a joint

A dot over a symbol (e.g. $\dot{\varepsilon}$) could be considered differentiation with respect to time.

CHAPTER 1

General considerations

These introductory general considerations of coal mine strata mechanics represent more or less the practical point of view on this subject which involves the unquantifiable state of nature. Simplification of the concepts and solutions in strata mechanics are necessary because natural conditions do not allow exact and precise definitions.

The topics for general consideration are in accordance with the conceptual structure of this volume. They reflect considerations given in other headings on strata mechanics.

1.1 DEVELOPMENT OF STRATA MECHANICS

The development of strata mechanics coincides with the development of coal mining, and should be particularly relevant for high technology.

1.1.1. *Coal mining and strata mechanics*

From archeological data on ancient mining it is obvious that layout and shape of underground excavations are in accordance with the stress state in the surrounding rocks. This requirement rule can be equally applied to coal mining which was mentioned by the Greeks in the Fourth Century B.C., and with the remark that the Chinese were already familiar with the use of coal.[1] Aristotle described coal as combustible matter: 'Those bodies which have more earth than smoke are called coal-like solids.'

In underground coal workings preserved from the Middle Ages before the Industrial Revolution, the mining method was one of so-called bell-pits, whose arching shape offered stress distribution and stability suitable for extracting very shallow coal deposits. Each bell-pit consisted of one individual heading (Fig. 1.1.1) but were extremely numerous within one coal field. The main problem of mine production was the stability of excavations at shallow depths which rarely exceeded 10 m. The individual bell-pits were linked by spider-like headings and were commonly used until the seventeenth century.[2] In this period the need for coal was very limited because its use was only for domestic heating and blacksmithing. Nevertheless coal was actively traded in England as early as the thirteenth century. However, many cities in Europe objected to the use of coal, because smoke and dust produced during burning, often adversely affected the environment.[3]

1

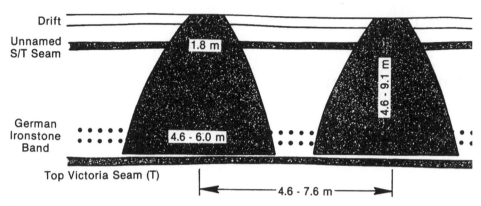

Fig. 1.1.1. The bell-pits for coal and ironstone mining at Sproats, Northumberland (The National Coal Board, England).

Advances in coal mining came with relatively improved strata stability after the Industrial Revolution. The demand for coal for use as a fuel for steam engines was great. This demand was met by introducing shaft mining and utilizing steam engines for hoisting large quantities of coal, mine dewatering, and other purposes. Subsequently coal mining engineering gradually advanced until the present technological revolution which introduced continuous mechanized mining with automation and computer programming. The new mining technology allowed for the change from small underground mines to large mines with productions of up to 2 million tonnes per year. This change was followed by introducing advanced analyses of strata mechanics particularly continuum theory in rock mechanics which came into use and also when improved underground instrumentation permitted new concepts of stress interpretation. Without adequate analyses of mine stability the efficiency of new technology would be jeopardized.

The recent energy crisis has focused further attention on the potential use of coal for the development of power, heat, gas, and liquid fuel. Such requirements anticipate a new breakthrough in underground mining technology which will increase coal production up to 6 million tonnes per year, with a single production face yielding 6000 to 8000 tonnes of coal per day. A model of such technology could be the coal mine at Lecie Colliery, Poland, which is fully automated with a production capacity of 25000 tonnes per day,[4] or the prototype of a computerized coal mine designed by the National Coal Board of Great Britain. Without doubt, the achievement of satisfactory mine stability requires the integration of strata mechanics in mine design, because strata in such mines cannot be controlled by human factors.

At present several concepts and proposals for new coal mining technologies have been proposed. They are irrelevant to practical mining because they do not have strata mechanics analyses and related derivations for ground control. For example, the concept of manless underground mining with remote control from the ground surface completely ignores the response of the strata to such a type of coal excavation: ground control for such a type of mining technology may be so expensive, that it will prohibit coal production.

Finally, the knowledge of strata mechanics became an important factor for successful design and stability control of underground coal mines and should be utilized much more than at present, particularly in future large coal mines with rapid extraction to make them safe and profitable operations.

1.1.2. *Principles of strata mechanics*

Strata mechanics are an integral part of rock mechanics, which is the theoretical and applied science of the mechanical behaviour of rock. This science has recently developed several theoretical solutions for problems in the design of stable underground structures. Some of these theoretical analyses involve complex solutions which are not in accordance with actual mine situations because they have many errors rooted in the absence of mining knowledge and practical experience.

In contrast to this approach some practical solutions, known as 'rules-of-thumb,' have been developed throughout the history of mining. For example, investigations of ancient mining workings in Central Bosnia, Yugoslavia indicated that the width of a stable self-supporting roof span of an underground excavation does not exceed 1/6 of the overburden depth, with the majority still intact.[5] These investigations indicated that shape of rock excavations had been accommodated to structural discontinuity of rock strata (Fig. 1.1.2). Yet, solutions by 'rules-of-thumb' which prescribe the size of openings, pillars, and others, regardless of the mechanical behaviour of the strata and the mining conditions can also produce many misjudgments, which can cause the collapse of mine structures, serious injuries and even deaths.

It is obvious that in analyses of strata mechanics it is necessary to give equal consideration to theory and practical investigations. The optimal solution of a problem should be an integration of theoretical findings with practical aspects. A combination of the results of the two opposite fields should be rational and effective to endure

Fig. 1.1.2. Photographs of adit from Roman era (Central Bosnia, Yugoslavia).

solutions which produce adequate mine stability. To achieve this it is necessary to have both good theoretical understanding and also practical appreciation as expressed by the saying 'The practice without theory is blind, the theory without practice is dead'. The beginning of theoretical consideration of strata mechanics can be related to the introduction of soil mechanics in 1943 by Dr. K. Terzaghi. Of particular interest for coal mining has been the approach to arching theories based on the increase in stress.[6]

Strata mechanics have been investigated as an integral part of rock mechanics since the end of World War II and represented mainly by the theoretical principles of both statics (beam and plate theories) and strengths of materials (rock properties and behaviour). At that time the methods for laboratory testing of rock and coal materials were extensively developed. However, the main setback of these analyses is their approximation of rock materials as solids without structural defects. Development of the concept of rock mass structure and implementing continuum mechanics for theoretical analyses (regardless of the necessity make a certain number of simplifications) produced reasonable results. These analyses can indicate zones of stress concentration and relaxation as well as the patterns of principal stresses and their role in failure mechanics. Following the principles of logic of application, it is necessary that theoretical results must come under the scrutiny of practice before any further consideration is given. For example, the analytical results should be checked against the real mining situation to which the theoretical model corresponds. During this exercise, all the theoretical data which are doubtful and contradictory to mining practice should be eliminated from the model solution.

The knowledge of generalized strata mechanics in interaction with underground coal mining operations represents a valid parameter for the design of stable structures. No longer can be sharp pencil and ruler used to design a mine structure by directly copying of the geometry of the existing layouts. In the near future we should expect a further refinement of strata mechanics relating to the design of mine stability, in which theoretical solutions would be based on data obtained during in situ instrumentation rather than on laboratory testing and the development of dynamic rather than static models. This concept of strata mechanics analyses is essential for the design of structures in sophisticated and mechanized underground coal mining with rapid excavation at great mine depths, where pillarless mining would probably be practiced.

1.2. FUNDAMENTALS OF STRATA MECHANICS

At the present time, there are many hypotheses advanced to explain theoretically the strata mechanics in underground mining. For practical reasons, strata mechanics in this volume is considered in terms of two principal concepts: stress and strain and the mechanics of failure as is briefly discussed below.

1.2.1. *Principles of stress and strain*

The majority of approaches for evaluation of strata mechanics uses the concepts of stress and strain where continuum media are postulated. Each medium is subject to volume and surface forces which are continuous functions of the space coordinates. It can be shown that the equations of equilibrium are not sufficient for the definite determination of the stress state within the solid body. It becomes obvious that both the movement of material points within the solid body have to be examined and information related to the constitution of the body must be avoidable in order that the solution of the equilibrium equations becomes unique.[7]

The classic approach to the problem by the linear theory of elasticity is probably the most idealized concept of stress and strain regarding strata material. It is argued by many rock mechanics scientists that for strata mechanics analyses a theoretical concept should be used which is much closer to the behaviour and properties of sediments. The most common deviation from the classic approach is that the rock is not considered to be an isotropic body, but an elastically homogenous and anisotropic one. It should be pointed out that sedimentary strata appear to fit this premise best.

S. Boshkov and M. T. Wane commented that the general linear relationship between stress and strain is expressed by equations which contain at most 21 independent coefficients. The algebra necessary to manipulate the relationships for different coordinate systems becomes quite tedious and, following the minimum effort principle, it is advisable to retreat to a more favourable position.[8]

With three mutually perpendicular planes of symmetry, the number of coefficients connecting stress and strain are reduced to nine. If, further, the material is considered to have an axis of symmetry, the number of coefficients is reduced to five. At this point, the problem is idealized almost to the classical theory of linear elasticity.

It should be pointed out that the majority of theoretical analyses of stress-strain theory applied to rock mechanics or strata mechanics neglect an important independent variable of the system. The fact is that stress and strain are not only a function of the coordinates in space but also of time. Under these circumstances, the analysis of strata mechanics must take in account the time-dependent elastic and non-elastic effects.

For elastic stress, when the rock is suddenly loaded, it does not respond with a resulting strain immediately but approaches its final value exponentially: similarly when the rock is unloaded, the strain returns exponentially.

For yield stress, when a rock is loaded, the resulting strain also approaches asymptotically its final value. This is called transient creep. For long-time loading conditions it is called steady state creep, where a rock exhibits slow continuous deformation due to constant load. This mechanics is in the area of rheology, where the time effects could be represented phenomenologically by a combination of simple elements.[8]

S. Boshkov and M. T. Wane also commented that rheological models are used only for descriptive purposes. When they are applied to three dimensional problems usually a statement, in the form of the generalized Hook's Law, must be made. In essence, a linear connection is postulated between the time derivatives. Again, the problem is limited to solutions similar to those in the classic linear theory of elasticity.[8]

At least conceptually the mechanics of the relation between stress and strain for instant loading as well as the time dependent loading could be brought to a common nominator which is the classic theory of linear elasticity.

Granting this conceptual point of view, the most logical approach for strata mechanics would be the application of the classical theory of linear elasticity for an orthogonal body. Under these circumstances theoretical mathematical studies as well as digital model investigations should be used to provide rational solutions of stresses

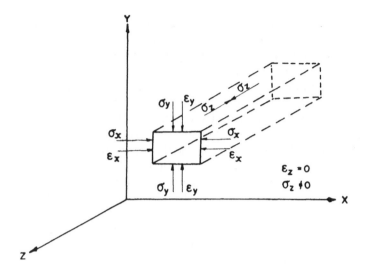

Fig. 1.2.1. Plane strain
representation.

and strains around structures. The degree of restriction of these solutions depends
on the complexity of the mining configurations and approximation of strata discontinuity. In practice the analyses are limited to simplified mine layouts, with
assumptions that sections of excavations are identical to plane strain, with infinite axial
extent. The philosophy of the plane strain solution is in accordance with the
geometrical representation of underground mine structures by cross-sections, longitudinal sections and level planes. For example, imagine a long roadway in a rock strata,
where stress is applied in all directions, but strain exists in only two planar directions
because the third strain is zero since deformation in its direction cannot occur. This
corresponds to the usual mining geometry within a strata cross-section which is a
convenient aspect of design, support installation and other factors (Fig. 1.2.1).
Conditions of plane strain could be analyzed in terms of deformation of elements in the
rock mass by normal strains in the $X - Y$ coordinates. Shear strain also could be
analyzed in this same plane.

It is obvious that analyses of strata mechanics are based on many assumptions: a
strata considers the continuity of the medium and the compatibility of the stress;
geological structural continuities and discontinuities are disregarded; mining excavations have smooth regular surfaces which do not effect the mechanical properties of
the immediate rock strata, the simplification of mine layout geometries. Although it is
assumed that mine excavations are placed in the continuum under stress, this
equilibrium is generally upset by triggering a variety of displacement and deformations.

1.2.2. *Criteria of failure*

Underground mining experience sometimes does not predict strata failure due to
stresses imposed by subsequent mining operations.[10] The unpredicted strata failures
often result in fatalities and they should be of primary concern to mine designers and
operators. To introduce a measure of failure prevention and strata control, it is
necessary to know the relationship between cause (acting stresses) and effects (failure

mechanics) in underground coal mining. The studies of failure mechanics have been carried out by many rock mechanics scientists, considering rock as a brittle material. However, with the aspect of mine stability the failure mechanics of strata could be fundamentally described by two concepts:

1. *Failures due to quasi-static loading* are defined by criteria derived merely from simple mathematical assumptions, and they are rather expressions of physical hypotheses. For strata bound deposits, fracture formed by tensile and compressive stresses are of equal importance.

Observations in underground mines show that the initial failure of a mine roof is not necessarily caused by the maximum shear stress concentration, but rather is due to tensile stresses and insufficient tensile strength of the rock strata. The geometry of tensile failure is easily observable and a stress pattern is easily recognizable, particularly in the case when the lower portion of a roof is in tension and the rock simply pulls apart (in the same manner as the fracture of an overloaded beam).

Strata fracture is of a more complex nature and conceptually there are four fundamental theories which are of interest for strata mechanics, as briefly given below:

a) Brandtzaeg's Theory, where fracturing is parallel to major principal compressive stress. It is postulated that lateral tensile stress is induced by the compression. This theory does not give quantitative results which can be used in mathematical analyses, and usually is neglected in rock mechanics. However, it has practical value for evaluation of cracks on the walls of roadways, coal ribs and coal faces (Fig. 1.2.2). Such fractures are naturally formed near a face surface. In the strata, away from the face surface, lateral confining stress tends to build up to prevent this type of tension failure. Other types of fracture than these may form instead, as has been discussed further.[5]

b) Griffith's Theory postulates that an overstressed brittle material will fracture due to tensile stresses set up in cracks which appear along grain boundaries. He derived equations for the principal stresses required to cause fracture, and for the tensile strength of the material and the orientation of the most severely stressed cracks. However, his theory does not predict how the crack will propagate, and if the crack grows it may not follow the plane of the initial flow. The modified Griffith theory states that cracks will partially close under the influence of compressive stresses and that frictional forces will develop across the cracks surface.[11]

c) Coulumb's Theory postulated that shear resistance of a cohesive soil or rock varies with the magnitude of normal stress on the fracture planes as well as the shearing properties of the material (friction and cohesion). This fundamental theory was further extended by the Navier-Coulumb criterion of rock failure, which assumes that the fracture takes place as a result of shear movement along a plane inclined at an angle (β) to the minor principal stress direction. Finally it should be pointed out that Coulumb's criterion is equivalent to Mohr's linear envelope to Mohr's circle as discussed below.

d) Mohr's Theory postulated that of all planes having the same magnitude of normal stress, the one along which failure is most likely to occur is that which has the maximum shearing stress. The maximum shear stress occurs at a plane of 45° to the direction of major principal stress. However, normal stress induces frictional forces on potential shear surfaces (in accordance with Coulumb's law), so that actual shearing takes place along planes which are inclined at an angle $45°-\phi/2$ to the direction of the major principal stress (ϕ = the angle of rock internal friction). Mohr introduced

Fig. 1.2.2. Axial failure of coal rib (Western Collieries, Australia).

graphical determinations of shear stress (magnitude and direction) at a point in a stressed rock.[11, 12]

Very brief introduction of fracture criteria in brittle rock have led to a certain number modifications by rock mechanics scientists, which are of more theoretical importance, than a fundamental influence on the mechanics of strata failure. Several aspects of failure mechanics of coal strata are now discussed under headings which consider deformation and failure of mine structures.

2. Failure due to dynamic loading in coal mining is formed within a zone with an intense stress distribution where the loss of equilibrium leads to shock bump. The main cause of the formation of such zones is the redistribution of internal forces under the influence of mining excavations, which causes appreciable departures from a hydrostatic state of stress within the zone.[13]

The arise of shear and tensile stresses together with their associate deformations represents a potential source of fracturing. A similar view of the formation of a mechanical system in the Earth's crust, leading to loss of equilibrium and earth tremors, was suggested by some other authors.[14]

A qualitative or quantitative expressions of the forces causing loss of equilibrium and related shock bumps could be defined by analyses of the stress-strain state in the surrounding rocks of mine openings. The failure mechanics is basically represented by the principal normal stress and maximum tangential stresses. This is due to the change of load when the shear reaches its maximum. Under this circumstance, shear cracks appear in the solid rock, followed by elastic vibrations relative to some state of equilibrium. A sharp displacement of the side rocks pushes forward the coal face and vibrations can act as dynamic loads on the edge of the seam.[13] The face zone collapses instantaneously where fracture is accompanied by releases of the elastic energy of bulk and shear deformations, which is converted to the energy of the field of displacements of fracture.

For the analysis of failure mechanics, dynamic models are used which are based on the theoretical and experimental characteristics of shock waves. The waves convey information on both seismic energy fraction and the types and directions of forces at the foci of bumps.

Finally it should be pointed out that in some cases the failure criteria of geological material were defined on the basis of theoretical analyses which contradict mining facts. The incongruity of these examples is rooted in many errors of judgment, which have been caused by a lack of mining knowledge both in terms of education and practical experience.

1.3. PHILOSOPHY OF STRATA MECHANICS

The degree of understanding of strata mechanics depends on a thorough understanding of natural factors, which are the salient features of all mining engineering enterprises. With this point of view it could not be expected that a precise definition of the conditions and final solutions to strata stability could be achieved. The simplification and rationalization of valid solutions for strata stability is a philosophical concept rather than an engineering one. The careful and expertly formed judgments of strata mechanics, by selecting important and relevant factors and ignoring the unimportant and irrelevant is, rather a matter of the human mind than an array of technical criteria.

The philosophical approach for the analyses of strata mechanics and related mine stability should be the integration of the theoretical analyses, model analyses and analogy analyses. Apparent pitfalls have been seen in cases in which the researcher may be tempted to use one method of analysis to the exclusion of others. Under these circumstances, the results could be ridiculous, especially when coupled with an unquestioned faith in theoretical solutions which cannot be supported by the testimony of experience.

1.3.1 *Theoretical considerations*

The strata loaded by underground mine structures may undergo displacements and deformations or, if overloaded they may become cracked, fractured and collapsed. The effect of loads on rock strata depends on the physical property of such materials, which is the important factor for their behaviour under stress. It should be pointed out that sedimentary rocks are anisotropic rather than isotropic. Thus it is possible to assume,

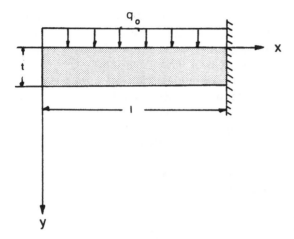

Fig. 1.3.1. Model of cantilever for calculation of the normal stress.

under the same conditions of non-isotropy, that the preferential orientation of major openings may have a greater effect on mine stability than the choice of preferred shapes.

The philosophical consideration of theoretical analyses distinctly recognizes particularities of structure of sedimentary strata which are to be contrasted with igneous or massive rocks. The theory regarding excavation openings in massive rocks tends to adhere to the classical approach of elasticity, homogeneity and isotropy of material and the stress distributions concerning openings in an infinite plate of solid.[15] The theory regarding the creation of openings in sedimentary strata tends to recognize the difficulties of nonhomogeneity due to transverse anisotropy by adopting the theoretical considerations for plates, beams and cantilevers.[16, 17]

The many theoretical analyses considering structures either of massive rock mass or sedimentary strata are based on the Theory of Elasticity for which the solutions seldom contain the material parameters. Reflection upon this result intuitively raises some apprehensions. An example of roof strata loaded by cantilever structure is now given where the analysis of normal stresses has been carried out without taking into account rock material parameters, as briefly listed below:[18]

a) The mathematical approach was to derive formulas for component stresses and deformations for the approximated roof slab as an elastic cantilever uniformly loaded by its own weight (Fig. 1.3.1). The formulas for the stresses were derived for the general case of an orthotropic elastic cantilever in plain strain conditions. The stress function was taken in the form of a fifth-degree polynomial.

b) Through mathematical derivation and integration a system of equations were obtained, and by solving them, the coefficients for an isotropic cantilever were found. Furthermore, by substituting the values of these coefficients in the formulas for stresses and making certain transformations for convenience of computerized calculations, equations for the normal and tangential stresses were derived.

c) The result of this theoretical exercise is a simple equation, which contains a certain number of assumptions, for conditions of equilibrium as given below:

$$(2-4)\sigma_T = 6q_o \frac{l^2}{t^2} \tag{1}$$

where

σ_T = tensile stress;
q_o = unit load;
l = length of rock cantilever;
t = thickness of rock cantilever.

The philosophical evaluation of this theoretical exercise, can be given by a consideration of this final equation.

a) The theoretical findings of the problem are combined with practical underground findings for a particular coal seam (Seam No. 5, Chertinskaya Mine, Longwall face No. 95). The suggested equation rationalizes the practical aspects because it ensures the appearance of longwall face performance.

b) However, to represent the actual longwall situation a factor between 2 and 4 is introduced which makes it possible to obtain a 'practically correct' solution. This results in for example a critical length of roof span 10 to 20 m.

c) The absence of a parameter for the physical properties of rocks which compromise a cantilever beam, under the assumption that they are of equal strength and behaviour has never been taken into account.

There is an opinion that theoretical analyses should be used to learn the mechanism of strata loading by mine structures rather than in ill-integration with geological and mining parameters obtained during underground excavations, for representing practical solutions.[16]

1.3.2. *Digital model concept*

Digital modeling became an important tool for the analysis of strata mechanics loaded by mine structure. The mathematical models incorporate the characteristics of the physical properties of geological materials. This basic premise may depict such material as elastic, plastic, viscous and any combination, orthotropic obeying the behavioral laws of linearities, non-linearities and time dependent deformations. Whenever one of these hypotheses is granted it leads to indisputable analytic solutions.

The introduction of the digital computer allows many repetitive computations to be performed on large quantities of data. Under these circumstances, it is possible to simplify complex problems into a series of discrete units which have known solution. These discrete units are analytically independent, but are connected to form the complex problem and a solution should be obtained which satisfies the defined boundary conditions. The analyses of mine stability by discrete units are principally carried out with finite elements, finite differences and numerical boundary methods. In all cases each element or block is defined by a prescribed relationship between stress and strain, or load and deformation. The combination of these units simulates the underground mine structure. The several techniques of digital model analyses are briefly described as follows:

1. Finite element method is used for stress analysis and other tensor field problems, for the case of the division of a mine structure into a series of elements which are connected at nodal points to form a framework. The method produces continuous variations of stress and strain, within elements but not between elements. A relationship between the forces and displacement is defined by a series of equations which depend upon the material parameters, the geometry of mine structure and the degree to which the strain forces or displacement are applied to the element nodes produces stresses and strains within the element which are transformed into displacement at the nodes. The advantages of this method are that different parameters can be assigned to each material type, so that the mine structure is more completely defined. More complex rock mass behaviour can be modeled by using special elements (e.g. for joints)[19] or complex material behaviour (non-linear elastic, elasto-plastic, visco-plastic or no-tension analysis).[20]

2. The finite differences method is used for investigations of a large deformation and the prediction of stress distribution in mine structure. These models are capable of representing regular and irregular boundaries and a wide range of material responses, as well as various applied loading conditions and applied fractions. This modeling technique has been used to simulate blocky or jointed rock mass as a discrete system of rigid blocks. The blocks are assumed to be rigid but interact with one another at the edges and corners, where the force and displacement conditions are satisfied. This method produces an explicit finite difference which enables evaluation of collapse mechanisms to develop intrinsically as the analysis proceeds in real time and arbitrarily large displacements can occur. Cundall has developed a computer model, where the system can be solved by analyzing the forces, displacements, velocities and accelerations of blocks for incremental time steps until equilibrium is attained.[21] His model technique can solve a class of low stress, if large displacements are to be expected. If it is assumed that the blocks are delineated by continuities, then contact exists at a few points and thus the analysis of these contact forces has practical validity. The finite difference method is of particular importance for stress analyses of pillars under conditions of time dependent deformations (creep).

3. The boundary element method is based on the division of the boundary surfaces into discrete elements from which the stress and displacements at any point of the rock strata can be calculated. These modeling techniques reduce the number of elements in comparison to the finite elements finite differences programs which divide the whole area of influence of the mine structure into discrete elements and approximations are made to each element, is time consuming and extremely costly to run on the computer. However, the simplicity of boundary element programs of a reduced number of elements, have disadvantages because they are not powerful enough for modeling complex geometries and material properties as for example with the final element and final difference techniques. The boundary element program for a two-dimensional formulation (plane strain) has been developed by S. L. Crouch, which utilizes the Displacement Discontinuity Method.[22] This program analyses multiple mine structures of linearly elastic bedded mineral deposits in an orthotropic elastic half space. The displacement discontinuity method employs an 'influence function' to construct a system of algebraic equations involving boundary displacements and stresses. This modeling technique, however, is applicable under the assumption that the mineral deposit beds have negligable width in comparison to their length and lay parallel and not too close to the surface. This limitation suggests that the Displacement Discontinuity Method might be applicable to the analyses of displacement and stresses for pillar-stope structures of lateral mining systems (room-and-pillar, shortwall and longwall) in regions where the surface topography is flat.

The application of solutions obtained by digital models, and their utility must rest on the investigator's judgment, which should be based on his complex knowledge.

1.3.3   *Physical model concept*

The physical models could be represented by two principle groups: firstly, the photoelastic modeling developed in other engineering disciplines from structural element critical stress analyses was borrowed for application in mine stability investigations; secondly, the models of equivalent materials used in the solutions of structure deformation and failure, which could be visualized and monitored, could be an important asset in strata mechanics analyses, particularly due to the restriction of digital model analyses.

1. Photoelastic modeling is postulated on the changes of stresses which existed prior to formation of the structure. The stress concentration due to excavation can be related to the failure of rocks around the excavation, which is an important factor from the view point of the estimation of the stability of the structure. The stress analyses around and between openings are carried out by photoelastic stress analysis techniques. This method uses the certain optical properties of most transparent materials (plastic or glass). The photoelastic technique is applied for two-dimensional modeling in the linear elastic range, and should be a simple method which is based on elementary theory. In this case, the photoelastic model study is similar to numerical analyses which takes into account all the complicated structural restraining conditions and verifies the structure in the linear elastic range according to elastic theory.[23] For example, if a stressed photoelastic model is viewed in polarised light, interference fringe patterns in color can be observed, which is represented in terms of the direction of the principal stresses and the magnitude of the maximum shear stress around the excavation.[24] It is generally accepted that there is a great limitation of the photoelastic model analyses for practical underground solutions, because there is a large difference between the linear elastic material and the heterogenous rock mass, particularly in coal mining. However, for mine stability analysis, particularly in the case of pillar structure it could be applied if the limitations are known and the parameters which do not agree with the actual situation are eliminated.

2. Models of equivalent materials are devices which relate to the physical system in a such manner, that it could be used to predict deformation and failure of prototypes. The problem of design of physical models with accurate simulation of a prototype, calls for a basic understanding of this mechanical phenomena. This implies the necessity of knowing all of the pertinent variables (scale, geometry, property, loading and others), and the ability to satisfy kinematic and dynamic similarities between model and prototype. Using dimensional analyses, any physical phenomenon could be described by the following equation:

$$\pi_1 = f(\pi_2, \pi_3, \pi_4, \dots, \pi_n) \tag{1}$$

where the π represents a complete set of dimensionless products formed from the variables.[16] A solution could be effected for any term as a function of all the others. If complete similarity is to exist between the model and the prototype, their correspond-

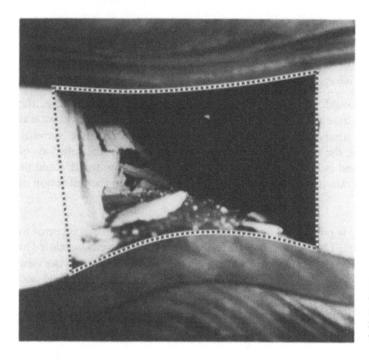

Fig. 1.3.2. Physical model of the deformation of roadway (the roof and floor strata-rubber mats; the coal ribs-plaster).

ing terms must be equal. The minimum number of terms has been shown to be equal to the number of independent variables minus the number of fundamental dimensions. Thus, if the variables are present in the description of a phenomenon and they may be measured in terms of mass, length and time dimensions, the number of terms must be seven, and must be picked in such a manner that they are independent of each other. It is obvious that this places mechanical limitations on the construction of a suitable model. The modeling by equivalent materials was introduced at the end of the last century, and it has been applied for analysis of mine pillar strength as a function of the shape. This type of investigation is still in use particularly to define deformation and failure mechanics of pillars (Fig. 1.3.2). Further development of modeling of equivalent materials has been particularly developed in civil engineering for investigation of the stability of permanent rock structures.[25] With the introduction of digital modeling techniques, where solutions can be achieved more rapidly and economically, a certain number of authors consider physical modeling of lesser importance. This consideration is not correct because the modeling of equivalent materials offers qualitative data of the deformations and failure of mine structure, what cannot be obtained by digital models. In our opinion, physical modeling of rock structures has an important role for analyses of mine stability and such modeling should further contribute to the solution of problems of practical mining.

Physical modeling might have difficulties in the case of stress studies. For example, the photoelastic models have inherent limitations which are rooted in the basic premises that underlie the classical theory of elasticity, where the stress distribution and concentrations are considered to be idealized solids. However, the models of equivalent materials have difficulties for stress analyses, because of the differing physical and strength properties of the model and prototype materials. If the model and prototype are constructed of the same material, than an apparently inconsequential solution exists which states that prototype may be its own model. A certain number of examples exist which describe distorted models, in which some of the conditions dictated by the theory of modeling are violated to a sufficient degree to require correction of data.

1.3.4. *Analogy considerations*

The general theory of models does not preclude the use of any coal mine in operation as a model for another that is still in the planning stage. The success of this analysis depends on the approach of interpolation of data from a producing mine to a mine in the planning stage, and on a more profound thinking effort than the one that goes into the direct copying of geometry.

A good example of the successful analogy analysis and the creative use of data from a mine in production to design a new mine is the hydraulic coal mine operated by B. C. Coal Limited in Western Canada. For example, all previous knowledge gained through sublevel conventional mining regarding strata mechanics have been incorporated into the present hydraulic mining of the soft thick Balmer seam. The particularities in strata behaviour in interaction with entries driven to the footwall of the seam, roof stability and cavability due to blasting the top coal down using long drillholes, and other underground phenomena were essential for the design of ground control and mine stability. It is obvious that previous mining experience regardless of the technology of obtaining coal might be a key factor in the philosophy of control of strata stability.

There are numerous examples, where 'the laboratory of nature' has been essential for the solution of mine stability problems. For example, Miike Colliery (Mitsui Coal Mining Co.) in Japan practices coal mining under the sea (5 million tonnes of clean coal per year). The bottom of the sea compromises quarternary sediments composed by clay minerals. Previous practical knowledge on the mechanics of the closing of fissures in these sediments and their rehealing by clay minerals has been incorporated in the control of ground subsidence due to longwall mining. Under these circumstances the sea water flows in the mine although subsidence fissure has been eliminated.

It is a point of interest, that analogy analyses should not be abused, because blind adherence to the 'rules of thumb' is seldom a successful solution. Hard mining practice has shown that many underground designs with the concept of 'modus operandi' required extensive 'practical research' when they came in operation to avoid their closedown. At the present time the criteria for analogy analyses are sufficiently developed that can be successfully used. One example, using experience from an adjacent mine by the acquisition of practical know how through the study of former mistakes. However, excessive extrapolation without regard to variables of nature, permits the mistaken adoption of a successful pattern of underground excavations from one area to a problem in another.

1.4. RATIONALIZATION OF COAL STRATA

Rock excavations, particularly in underground mining have shown that the quantitative approximation of strata is difficult because of the uncertainty of nature. For this reason all rational analyses of coal strata should be considered as limited, with the degree of limitation being a function of the strata complexity as well as the scope of the investigations.

From the aspects of strata stability and coal mineability, the rationalization of coal measures should be carried out by at least two principal evaluations.

1.4.1.   *Investigation of coal strata*

The evaluation of strata mechanics and stability during the mine design stage is one of the most important parameters. Adverse natural effects are an integral part of any mining enterprise which can effect strata stability and related economics as well as safety. In some coal mines the uncertainty of strata mechanics is due to insufficient investigations rather than to natural factors. To decrease the factor of uncertainty and the related risk of future mine operations, it is necessary to employ more sophisticated methods in obtaining information on the physical, mechanical, and geological universe in which the coal seam is buried.

Approximations of strata behaviour are based on two principal physical methods of access to coal deposits during exploration.

1. Exploration drilling is the most common and sometimes the only method of investigation of coal-bearing strata. The exploration is carried out by drilling either with or without coring, but always with geophysical borehole logging. The degree of uncertainty of strata mechanics will depend on the spacing of the exploration drillholes and the number of drillholes cored. Geotechnical information is obtained simultaneously with geological information, as briefly described below:

Geological interpretations are based on the geological and geophysical logging of

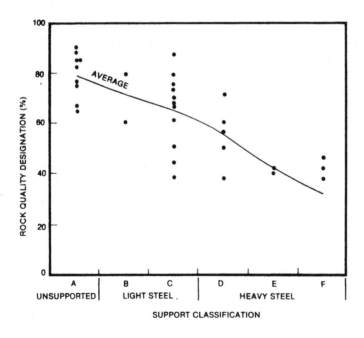

Fig. 1.4.1. Comparison of estimated **RQD** from pilot borehole. Support for the Straight Creek Tunnel (after Coon).

the boreholes, and visual (including the use of hand lenses and miscropes) examination of drill cuttings, drill core, and the sludge recovered from the drilling fluid. The continuity and persistance of geological structures between drillholes is estimated and interpretated by the geologist.

Initial geotechnical indications are based on the percent of recovery, the degree of fragmentation, and the hardness of core samples. The most common evaluation is by the rock quality designation parameter (RQD), which is defined from analyses of drill cores, which can be correlated with the support requirements (Fig. 1.4.1).[26] The procedures of geotechnical investigations of drillholes have been extensively and repeatedly described in journals and books, thus they do not have to be repeated here.[27, 28] However, some particular geotechnical aspects related to coal are considered in the sections on the characteristics of coal strength.

Exploration drillholes can also be used to obtain information on the hydrogeology of the strata including permeable and impermeable beds, static and dynamic water heads, and water flow rates as well as the temperature and gas characteristics, such as thermal anomalies and activities, gas pressure and flow, and gas absorption.

2. Underground excavations which block out some part of the ground are rarer but more useful. To achieve satisfactory results entry excavations should be in several directions with the opening up of an area sufficient for visual observations and instrumentations. Underground instrumentations can be classified in four main groups as follows:

a) Borehole stress determination, for which exist a wide number of devices as for example: U.S.B.M. borehole deformation gauges; C.S.I.R. doorstopper triaxial strain cell, photoelastic stressmeter (photoelastic disc and glass plug); vibrating wire stressmeter, hydraulic fracturing. These instruments are placed mostly in hard rock strata rather than in coal seams due to its cleated nature.

b) Pressure cells and load cells of various manufacturers, which are used in conjunction with mine support and the bearing capacity of mine roofs and floors.

c) Borehole extensometers of the various types and manufacturers with single and multiple wires. They are used for measurements of strata displacement. This method of instrumentation is most common in underground coal mining.

d) Geophysical surveying by seismic or sonic techniques which are primarily used for locating stress concentrations which can lead to either rock bursts or coal and gas outbursts.

Descriptions of procedures for in situ instrumentation is not necessary here because such procedures are well described in numerous manuals and classifications as well as in publications and books which relate to these topics.[29, 30] However, some peripheral consideration is given in this volume for their specific relations to coal strata mechanics.

3. Laboratory investigations of the drill core samples and rock as well as coal samples taken from a pilot underground mine, should be considered as the final approximation to strata behaviour. Generally several groups of laboratory investigations should be considered:

a) Evaluations of elasticity modulii (static and dynamic);
b) Evaluations of compressive and tensile strengths (uniaxial and triaxial);
c) Evaluations of direct shear strengths (impact rock and rock joints).

The laboratory investigations of rock and coal samples as well as in situ investigation of rock and coal strengths are described not only in numerous publications but also in several volumes of books,[31, 32] some laboratory investigations of rock and coal samples are briefly given at an appropriate place in this volume.

Strata mechanics should be evaluated during the mine design phase because ignorance of these phenomena can cause great difficulties in mine stability. For example in one underground mine in Great Britain due to mechanization, hydraulic props replaced friction props. However, the hydraulic supports failed, because their greater load capacity caused excessive rupture in the mine roof. The hydraulic props were subsequently removed and the old supports brought back, providing a more expensive and less productive system, but one better suited to local ground conditions. Another example is a coal mine in the U.S.S.R. where the main roadways were excavated in shale and mudstone with greater mine depths, and with the increased pressure were completely closed after four months, due to accelerated floor heave.[33] The solution to this problem was found by locating the roadways in de-stressed zones. This delayed production added to the costs of their excavation.[34] It is obvious in both examples that if investigations of rocks properties and bearing capacities as well as instrumentation of strata convergence and lateral displacement had been carried out before actual mining, such adverse consequences would have been avoided.

1.4.2. *Geometrization of coal seams*

For rationalization of coal strata the key factor is correctly extrapolated seam morphology. The four parameters briefly listed below define the geometry of coal seams and related mine stability.

1. Thickness of seam greatly influences mineability and strata stability in coal mines. The following thicknesses are classified according to mining conditions:

a) Very thin seams with thicknesses up to 0.5 m are seldom mined. However, recently a seam of coking coal 0.35 m thick sandwiched between sandstone strata was extracted by longwall mining (coal getting by plough) without face support.

b) Thin seams with thicknesses from 0.5 to 1.5 m are extensively extracted primarily by continuous mechanized mining technology: both room-and-pillar and longwall mining. The mine stability depends greatly on the roof cavability and the floor-bearing capacity.

c) Medium-thick seams with thicknesses from 1.5 to 3.5 m are extracted by many mining methods. Which depends not only on the thickness but also on the angle of inclination of the seam. The behaviour of these seams and adjacent strata and the related ground control have been studied all over the world, so that their strata mechanics are well defined and known.

d) Thick seams with thickness from 3.5 to 25.0 m are mined by full-face or multiple-face mining methods. They exhibit unstable mining conditions primarily due to the large mass of coal excavated and the strata intensively displaced. The mining of these seams particularly if they are inclined, is treated cautiously in North America, because of limited knowledge of their strata mechanics and corresponding ground control.

e) Very thick seams with thickness greater than 25 m usually are not mined underground. Stabilities of their structures are considered in open-pit mining.

This preceding classification of seam thicknesses suggests that mining methods vary with the coal seam thickness. In fact the application of similar extraction systems to seams of various thicknesses will produce different overall mine stabilities and related economics.

2. Gradient of seam also greatly influences the mineability and strata stability.

a) Flat coal seams are most favourable for continuous mechanized mining if they are not of excessive thickness. The strata mechanics and strata control of these seams due to geometrical simplicity and good underground conditions has been well defined by sophisticated methods.

b) Gently dipping seams are inclined up to 25°, an angle limited by the capability of continuous miners, which are also limited to seams up to 3.5 m thick. Mining methods are the same as in flat seams, but mining conditions are a little more difficult as is the strata control engineering.

c) Dipping seams are inclined from 25° to 45°, an upper limit defined by the initiation of the rolling of coal lumps under the influence of gravity. Mining methods such as longwall are employed in West Germany, where seams up to 55° are mined, but sublevel mining is a more efficient method, particularly for thicker seams. Strata mechanics are not well defined, particularly for hydraulic mining.

d) Steeply dipping seams are inclined from 45° to 75°, and are presently mined by retreat up-dip or down-dip by long-pillars. Hydraulic mining systems regardless of the seam thickness are particularly effective, even the drilling-and-reaming method for very thin seams. Strata mechanics and corresponding strata control are the weakest points of all mining methods.

e) Vertical seams are mineable by the same mining methods and technologies as steeply dipping seams, with equal problems in strata control.

It is generally agreed that whether a seam is flat or vertical or is 25 cm or 25 m thick, there is no fundamental reason why it cannot be mined. In fact, there is always some way, some known way, to mine coal under all those conditions by applying existing technology. The problem is in applying the particular technology so that strata control is adequate. Like most things that can be engineered, it is only a question of the cost. This highly important factor is often ignored during consideration of strata control methods.

3. Strike and dip of seam are important parameters for the rationalization of coal strata, because they strongly affect coal mineability and mine stability. Two extreme cases can be considered:

a) Regular strike and dip belongs to seams or portion of seams which were not directly affected by folding. Under this circumstance the extraction panels or blocks are approximated by a plate, which the strike line intersects by a horizontal plane. The great majority of mining engineering considerations and strata mechanics is related to this structural type.

b) Irregular strike and dip belongs to seams or portion of seams which are folded and very often refolded. They cannot be approximated as panels of plate geometry, because the coal seam could be of a very complicated structure discussed elsewhere in this volume.

The degree of deformation of an approximated coal plate is a limiting factor for employing conventional mining systems where geometrical distortion is combined with mechanical stability deterioration. In terms of strata control and cost, a deformed coal seam should be mined by hydraulic mining if other factors are in favour of this technology.

4. A continuity of seam inferred from exploration drillholes should be always considered with reservation if facts do not exist to prove differently. In particular, the continuity of the seam should be questioned in folded and faulted coal fields. For example, in Figure 1.4.2 three possible interpretations of the coal seam extension are shown. Each case shows a different interpretation of the data: from a slight change in dip to a major offset of the seam. If there exists a discontinuous structure of the seam, particularly as shown in the last drawing, the designer has to face two possibilities: either develop two separate mine structures avoiding the interference of the fault zone as much as possible, or develop one unit by making connections between the two tectonic plates by rock tunneling, which could be time consuming and expensive because of excavation in weak (faulted) ground. However, in terms of strata mechanics there is a fundamental difference between these two possibilities, because it is a question of mining either a single or multiple structure. Furthermore, application of a room-and-pillar system in a multiple structure could cause grave difficulties in controlling underground stability. Under these circumstances it is necessary to rationalize each tectonic block separately.

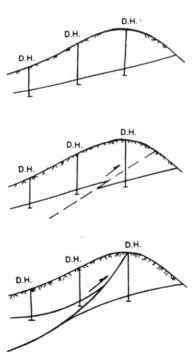

Fig. 1.4.2. Different inerpretations of coal seam structural continuity.

CHAPTER 2

Coal bearing strata

This is a very broad topic which cannot be considered in a single heading such as coal bearing strata. However, it is necessary to represent principal geological and structural factors in an orderly manner and give brief comments regarding their influence on coal mine design and stability. These factors are listed below:

1. The geology of coal bearing strata is represented from a geological point of view rather than engineering considerations, to provide the necessary introduction of the subject into the topic of coal mine mechanics.

2. Sedimentology of coal bearing strata is represented in similar manner but with the addition that stability factors are controlled by sedimentary features.

3. Diagenesis of coal bearing strata is an important element in mine stability, mechanics of deformation and failure of mine roof and floor strata.

4. The tectogenesis of coal bearing strata and the part it plays in the failure mechanics is considered seriously.

Mechanical stability or instability of coal bearing strata might depend to a large extent on the depositional environment, subsequent burial history and orogenic movement of particular parts of the earth crust.

2.1. GEOLOGY OF COAL STRATA

It is difficult to generalize geological features of individual coal basins on a world wide scale, particularly because of great variations in their geological history. For this reason the geology of coal deposits for one basin, the Plains region of Western Canada is presented. This area offers an instructive example or model of coal deposit geology.[1]

2.1.1. *Coal strata*

The coal strata of the Plains region extends southwards from Lesser Slave Lake and is bounded on the west by the Foothills coal belt. The Plains coal strata occurs in massive laterally persistent zones associated with the Upper Cretaceous Oldmen and Foremost formation; which make up the Belly River group and Horseshoe Canyon formation (Fig. 2.1.1).

The geological formation of the Plains region dips very gently toward the southwest. Consequently, from a mining point of view, the coal strata could be considered

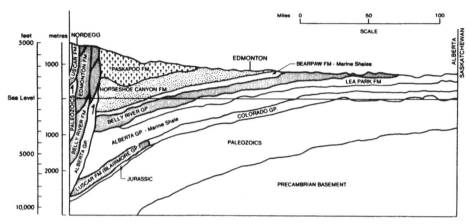

Fig. 2.1.1. A geological cross-section of Alberta showing the coal bearing strata.

horizontal. However, the lateral continuity of the coal strata has been broken locally by sedimentary processes and the effects of glaciation.[1]

The depth of the coal seams, in the area considered, vary from several metres to a depth of over 650 m.

The coal strata in the Plains region is covered by glacial till, the thickness of which varies from several metres to several hundred metres. Discussion of the coal bearing strata has been based on the data from the Ardley area[2], located in the central part of the Plains region. The geological factors which relate to underground mining are briefly described below:

a) The coal strata exhibits large lateral extensions, their internal structure and thicknesses being variable. Individual coal seams are also present. The thickness of siltstone, shale and sandstone and their position in the stratigraphic profile may vary as well (Fig. 2.1.2).

b) The coal bearing strata consist of three coal zones:
– Upper Ardley (inconsistent coal distribution);
– Lower Ardley 'B' (consistent coal distribution);
– Lower Ardley 'A' (local coal distribution).
It may be concluded that only the Lower 'B' zone can satisfy a large scale mine project.

c) Inconsistent coal distribution in the Upper Ardley zone and local coal distribution in the Lower Ardley 'A' zone suggests that they are not suitable for multiple seam mining but rather single seam mining.

d) Presence of a mineable coal seam in Lower Ardley zone 'A' and zone 'B', could be, from a mining point of view, approximated as one thick mineable coal seam which has a centrally located clay parting 6 to 9 m thick.

Using conventional mining methods and a 50 per cent recovery factor the amount of recoverable coal in Ardley area has been estimated to be 15 to 20 billion tonnes. Such enormous potential can offer excellent sites for large scale underground mines. For example, near Red Deer, Pigeon Lake, south-west of Wabamun Lake, and east of Entwistle.

The general conditions of the coal bearing strata in most of these areas should favour

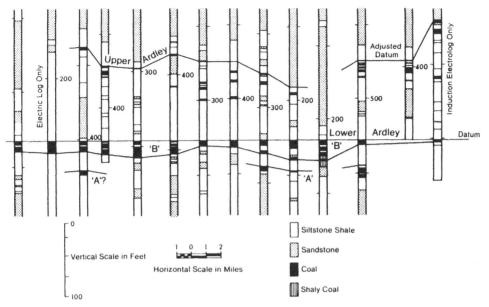

Fig. 2.1.2. The north half-section of the coal bearing strata of the Ardley area (after Holter et al.).

normal single thick seam mining. However, the strength characteristics and behaviour of individual rock types of the rock strata will be a key factor in underground mine design. Generally, the rock strata of the Plains region belongs to the classification of soft rock with a tendency to deform plastically and cave in readily if external support is not installed.[3,4]

2.1.2. *Coal zone*

When dealing with the practical aspects of mining, the most important part of the coal bearing strata is the coal zone which corresponds to the production horizons (Upper Ardley zone 'B' and Lower Ardley zone 'A'). The morphology and internal structure of the coal seam zones is an important parameter for underground mine design:[1]

   a) Due to the mechanism of sedimentary processes in the coal basin, the coal had a specific frequency of vertical deposition and of lateral deposition. The unproductive profiles are represented by pinching out, splitting or fingering of the coal. Due to this phenomena individual coal seams in the coal zone have specific lateral extension variable thicknesses and changing thicknesses of the partings within the coal seam.[5] Generally speaking, the variation of the interseam strata in the coal zone corresponds to a variation of individual coal seams and could be appreciable if the coal seam zone is continuous (Fig. 2.1.3).

   b) The common feature of this type of sedimentation is the general balance in the accumulation of vegetal matter. For example, if one seam in the coal zone is pinching out another below or above may increase in thickness. Under these conditions the workability of the coal field could be almost unlimited. In this case it is of paramount importance to design a layout of the main entry and development entries to satisfy

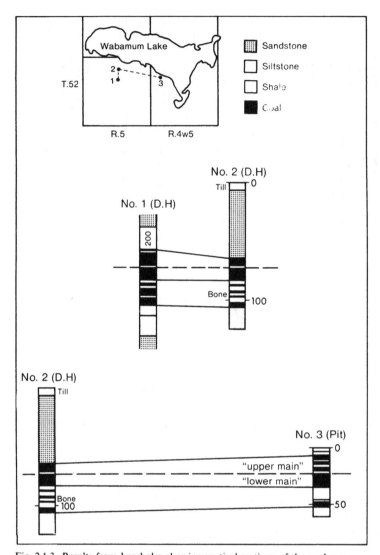

Fig. 2.1.3. Results from boreholes showing vertical sections of the coal seam zone, at Wabamun.

conditions of coal extraction where there is a possible lateral change over from one seam to another.

c) The coal seam zone having a heterogenous internal structure of different layers with variable thickness is going to influence the design of coal extraction and coal seam mineability. For example, large lateral mine development working faces cannot be located within certain seams as changes in the seam thickness may be encountered.

From the mine design point of view the coal seam zone deserves particular attention. It will be important to establish a relationship between the workable coal seams, coal seam zones and the overall mine stability. For example, the position of the water bearing horizon, cavibility of the immediate overlying rock strata and their relative compaction at shallow depth of mining directly depend on the character of the coal zone.

2.1.3. *Coal seam*

The coal seam profile is delineated by geological parameters and does not necessarily correspond to a mineable coal seam. The principal features of a coal seam profile should be the analysis of coal seam mineability and strata stability. For example, several factors which have been considered in the evaluation of the coal seam profile in the Plains region, are listed below.[1]

a) The workable seams are of a particular lensoid shape where the floor exhibits a gently undulating morphology. The resulting gradual changes of coal seam thicknesses should not affect the coal extraction operations. Floor undulation may influence the layout of the mine development. The orientation of the individual development blocks will depend on the axis of orientation of the gently rolling floor.

b) The internal structure of the coal seam seldom exhibits homogeneity due to the presence of partings. The influence of the partings on the coal seam mineability may be outlined as:

– The increasing thickness of clay partings decreases the coal quality;[4]

– The partings of clay bands may disrupt the operation of the coal face cutting machine;

– The partings in the roof may give rise to local roof falls and restrict the mobility of the self-advancing coal face supports.

c) The criteria for the mineable coal seam thicknesses are different for underground mining than for open pit mining. For example, Figure 2.1.4 represents five mineable coal seams for open pit mining that requires a coal seam thickness of 0.3 m or more.[5] However, there is only one coal seam that may be considered suitable for underground mining; requiring a coal seam of 1.52 m or more. As illustrated in Figure 2.1.4, for a coal seam with an average thickness of 3.1 m the working section would be about 2.6 m leaving the shale coal to enhance the competence of the roof. Leaving 0.15 m of coal bottoms would result in improving the bearing capacity of the floor.

Generally, for practical mining purposes, the Plains region may be considered to be flat and therefore applicable to Lateral Mining.

The thickness of individual coal seams may be classified with respect to mineability as follows:

a) Thin coal seams less than 1.5 m are presently uneconomic for underground mining.

b) Medium thick coal seams 1.5 to 3.5 m; the upper limit being described by the coal cutting machine. This range affords efficient mining operations.

c) Thick coal seams 3.5 to 15.0 m; this category also includes the coal seams divided by thick partings. However, in both cases slice mining may be considered.

It has been estimated that the majority of the workable coal reserves are of the medium thick coal seam category.[6]

The coal seams of the Plains region of Alberta exhibit strong geological stratification features, with slight or no orogenic deformation. There are three mining methods which are applicable in such a geological setting. These are:

– Room-and-pillar mining,

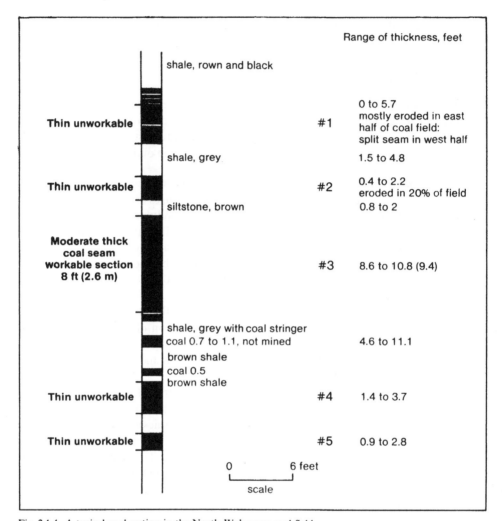

Fig. 2.1.4. A typical coal section in the North Wabamun coal field.

– Shortwall mining,
– Longwall mining.

There are, however, certain geological constraints which must be considered relative to these mining methods. In summary, the constraints are:

a) Coal seams less than 2 m thick preclude the conventional in-seam technique of leaving unmined shaley coal partings in the roof and floor to enhance the competence of the mine opening. Under these circumstances, a longer span of soft roof must be exposed which means potential instability. This is of particular importance in the case of the wide webs and the headgate curve extension encountered in shortwall mining.

b) The undulation of the floor (particularly high frequency undulation), will require a shield type of support if longwall mining is to be applied. Generally speaking, the soft roof strata will not permit longwall faces to be used to the extent that they are in

Europe. The length of the faces under such conditions may have an upper boundary of about 100 m.

c) Some of the coal fields in the Plains region contain small fault throws. This may mean that longwall mining cannot be applied. In this case, the most effective mining method might be continuous mining room-and-pillar system. This method has the inherent flexibility of copying with such a situation.

It is the author's opinion that the geological elements of the coal bearing strata are the most important factors in:

– Choice of mining method,
– Design of strata control and
– Mechanization.

It is for this reason that the geology of the coal strata must be given close consideration, particularly in the phase of mining studies in the test mines, where strata mechanics and related mine stability should be evaluated together as at Canada's Lethbridge coal mine.

2.2. SEDIMENTOLOGY OF COAL STRATA

There are differences in the overall nature of sedimentary mechanics in basins or parts of basins. These variations in depositional environments can be related to the tectonic setting of the basins. The mechanical properties and behaviour of rock strata primarily depend on the lithology, sedimentary structure and stratification discontinuities.

2.2.1. *Lithology of rock strata*

Each coal basin contains major rock types related to the coal bearing strata. Their lithology will depend on the rock petrology of the surrounding land and the character of the depositional basin. Within the aspect of practical mining three groups of the sedimentary rocks can be delineated. Igneous rocks which might occur occasionally in some coal basins are excluded (Sydney Basin, Australia).

1. Carbonate sediments, of interest to coal mining are represented primarily by rocks which were deposited in brakish marine environments (delta) or calcium rich environments (skeletons of preserved animals contain calcium from dissolution of the underlying limestone). Three basic rock types are briefly described below:

a) Limestone's original fine-grained texture, is recrystallized in a coarse grained fabric the possible cause being orogenic movement. The compressive and tensile strengths of the intact limestone (rock specimens) depend on the degree of cementation and content of the clay minerals. The investigated strength of limestone of the coal bearing strata of Zenica Colliery (Yugoslavia) showed uniaxial strengths ranging between 30 and 74 MN/m^2, with excellent elasto-brittle properties.[7] The tensile strength is considerably lower, between 3.2 and 5.5 MN/m^2. I could be concluded that the intact strength of limestone is sufficient for the stability of the mine structure, and is confirmed by underground stability analyses. For example, the limestone strata of the mine floor 'upper coal seam' exhibited high bearing capacity. When overloading conditions came into effect, the limestone mine floor sometimes responded with limited

Fig. 2.2.1. The weathering of dolostone beds with weakening contributed by joints and stratification planes (B.C., Canada).

displacement, but more often with sudden and violent failure (rock bursts).

b) Dolostone has similar fabric and strength properties to limestone. Although the intact strength and behaviour of rock samples is similar to limestone, the strata strength of dolostone beds (also a limestone) can be reduced due to weathering. The amount of reduction of strength depends on the porosity and permeability of dolostone. However, the joints and bedding could facilitate weathering (Fig. 2.2.1) which usually causes heterogenous strength degradation of rock strata in shallow mines. This phenomenon might have a very serious influence on mine stability because the rock strata looses self-support and load bearing abilities.

c) Shaly limestone is a transitional rock type with its intact strength lower than limestone and higher than shale. The shaly limestone can be physically integrated with thinly carbonaceous and clay laminas which intercalate with each other. However, the strength properties of shaly limestone samples are particularly degradated upon swelling. These mechanical properties and behaviours obviously have an influence on mine stability and its self-supporting capacity.

2. Siliceous sediments are the most common rock strata of coal deposits all over the world. Petrological studies of individual siliceous rock types within the same coal basin showed small differences in mineralogical composition, but great differences in grain sizes and fabrics. The lateral and vertical continuity of individual siliceous rock types as well as their repetition depend on the depositional history of the coal basin and cyclic changes in climate. Several basic rock types are illustrated as follows:

a) Conglomerate sediments have no continuity of lateral development. For example, lateral development is present in the Northern Coalfield, but absent in the Southern Coalfield of the Sydney Basin (Australia). The conglomerates are usually thickly bedded sediments, with massive resistance to weathering and their outcrops are of cliff morphology. Their strength is particularly increased if the cement of the silica pebbles is of siliceous composition, for example the uniaxial compressive strength of such rock samples are over 120 MN/m² (Mount Head coal field, Western Canada). If conglomerate forms the main roof of a coal seam, the mine stability and safety could be dangereously effected due to the resistance of conglomerates to caving. For example, a roof span of 3000 m² was observed before it started to cave in a mined-out area.

b) Sandstones exhibit a wide range of grain size and cement composition. They are massive and bedded competent rocks. For example, laboratory testing of the uniaxial compressive strength of rock samples of the coal bearing strata of Southern Coalfield showed that with variation in grain size a large range of strengths existed (37 to 98 MN/m²). The sandstones exhibit linear elasticity and have a high elastic modulus, which are in a range between 3.5×10^4 to 5.5×10^5 MN/m². Sandstone rock strata offers very supportive and high bearing structures, but this stability can be adversely effected by sudden and violent failure (rock burst). For this reason in some mines hydraulic fracturing must be carried out to control the accumulated strain and its violent release. The main setback in mine stability, similar to that of the conglomerate, is the lack of its tendency to cave during coal extraction. This phenomenon is very observable on numerous weathered outcrops of washed out coal seams in the Rocky Mountain coal field (Fig. 2.2.2). Also, rock fragmentation during caving is in large blocks which do not facilitate progressive caving and the further de-stressing of the strata which surround the working faces.

c) Siltstone is a lithological unit transitional to sandstone (silty sandstone) and shale (shaly siltstone). The definition of the basic lithological type and its hybrids depends on

Fig. 2.2.2. An erroded coal seam with laminated fine-grained sandstone (roof) and massive medium-grained sandstone (floor), at the Mount Head Basin.

the grain size, content of quartz and clay, and finally on the degree of lamination. These elements are constricted to the variations in compressive and tensile strengths of rock samples. It is often the case that some coal bearing strata are made up of interbedded siltstone and its transitional rock types, where interbedding is extended to sandstone and shale bed. The evaluation of the mechanical stability and roof cavibility of siltstone beds should primarily be done on the basis of underground mine investigations. Uniaxial compressive strengths of siltstone samples from the Rocky Mountain coal field were in the range of 37 to 70 MN/m². The tensile strength was very low; and ranged between 3.5 to 5.5 MN/m².

d) Shales are represented by fine grained fabrics and appreciable clay content. The term shale in coal mines often applies to a range of argillaceous rocks with a wide variation in strength and elastic modulii. For example, shales of the Sydney Coal Basin (Australia) have an axial compressive strength of 30 to 55 MN/m², a tensile strength of 2.5 to 4.0 MN/m² and a Young's modulus of elasticity of $(3.0–6.0) \times 10^3$ MN/m²; factors which are given particular consideration in roof support.[8] The main considerations in the self-supporting abilities of a roof support and the bearing capacity of a mine floor are the swelling indices and the slake durability capacities. With respect to the mine roof stability of the shale strata, moisture is a major consideration because it results in a swelling pressure, which could exceed several times the strength of the rock.

3. Clay sediments are present in rock strata where lithification is not complete, for example; the Moscow Coal Basin (U.S.S.R.). The Collie Basin (Western Australia), and the Plains region (Western Canada). This particular feature of rock strata is of importance in the stability of the mine structure.

These lithological features of coal bearing strata are typical for complex and broad swampy flood plains bordering flowing rivers. The ratios of coarse grained elastic sediments and fine grained sediments vary locally for each coal basin. Thus, local yielding of rock strata is nonuniform in both horizontal planes (less yield) and vertical planes (more yield).[9] Superimposed on this variability in the sediments are the inherent variations of rock properties, which have been observed during investigations of rock samples and rock strata in the coal bearing strata of the Plains region. Practical underground mining in this area proposes three lithological types:

a) Clay shales have fine bedded textures with coal stringers (remnants of plant roots) and slicken-sides. These rocks are made up of clay of various mineralogical compositions. During compressive loading of core samples the axial strain rapidly increases and the strength and elasticity significantly decreases. The stress-strain curves and rock shear flow failure suggest viscous-plastic behaviour for these rocks, corresponding to underground mining situations where the predominant deformation of these rocks is extrusion. The compressive strength of these rocks is low, from 0.33 to 12.49 MN/m² and is due to the variable diagenesis.

b) Mudstone-siltstones also have fine bedded textures. The silt-to-clay ratio of this rock type varies widely, resulting in an equally wide range of strengths. Core samples, tested in compression have shown that axial strain rapidly decreases before failure, with increases of strength and elasticity. However, rock strength increases with the decrease of clay content. Underground mines experience sudden roof falls of mudstone or siltstone strata. The compressive strength of these rocks range from 0.79 to 20.98 MN/m².

Fig. 2.2.3. A strata of a
clay rock complex at
Wabamun Open Pit Mine.

c) Clay-sandstones have mostly clay material as cement and on exceptional occasion have a carbonate cement. In the Plain region sandstones adjacent to workable coal seams have mainly clay cements, which allow them to weaken when they are exposed in mine openings (Fig. 2.2.3). Compression tests have revealed that axial strain decreases rapidly during loading, but immediately increases before failure strain. Failure results in shear planes at angles of 15° to 20° to the loading direction (core axis). Shear movement along the failure surface exhibits residual strength. This mode of deformation and failure is typical of elasto-plastic behaviour and corresponds to the caving behaviour of sandstone during coal extraction. Sandstone has the greatest strength of all the rock units, ranging from 2.3 to 30.0 MN/m^2.

From this representation it is obvious that the lithology directly influences the rock strength and its behaviour. Rock lithology is dominant in strata control by coal pillars and artificial supports. For example, clay shale and mudstone might have very low compressive strengths but they do not have the ability to be reinforced by internal support (rock bolts) so that strata stability must be maintained only by external support. Contrary to this hard and strong sandstone and siltstone are excellent environments for reinforcing by internal support.

2.2.2. *Clay mineralogy*

Mineralogical composition is variable for the various rock lithologies which make up the two general rock groups related to stability: mineralogically stable rock types are common for the majority of coal basins all over world; and, mineralogically unstable rocks are characteristic only for a limited number of beds of coal bearing strata or coal basins as a whole. Unstability of the rock beds is caused by the presence of clay minerals, the degree of their instability being dependant on the content and mineralogy of the clay. The Alberta Research Council has carried out a series of investigations on clay mineralogy and on the amount of clay in the coal-bearing strata of the Plains region[9, 10]. This research has shown that the strength and elastic properties of these rocks depend not only on the amount of clay but also on the type of clay present. These features influence both the strength and behaviour of fine-grained and coarse-grained sediments equally.

The fine-grained sediments have clay as a matrix, interspersed with variable amounts of silt. Common characteristics of fine-grained rocks containing clay can be summarized as follows:

1. The clay in a rock matrix directly effects the strength and behaviour of the rock. Laboratory investigations of sandstone showed that samples with high clay content are considerably weaker than those with small amounts of clay.[9] Failure of rocks with high clay content under uniaxial loading usually occurs along the cemented grains.[11] Sandstones with calcite cement have a higher uniaxial compressive strength (Fig. 2.2.4).[12] Exemplifying the effect of clay, sandstone and conglomerate instability has been observed in Sydney Basin (Australia).[13] Sandstone and conglomerate are quite variable in the degree of instability, which probably relates to the amount of clay material present in the rocks cement matrix. Irregular slabbing of the mine roof effects mine stability. Particular instability is observed in conglomerates if the clasts are composed by soft, puggy mudstone pebbles and cobbles where moisture is introduced. The lithic grains and matrix tend to decompose and the fabric becomes less cohesive.[13]

2. The clay minerals which are present in rock masses can be considered very important. Various types of clay minerals influence mine stability differently. For example, rocks which contain montmorillonite can cause considerable support problems in underground mining because they are subject to rapid disintegration in the presence of moisture. However, the rocks which have kaolinite and illite as dominant clay minerals are more stable. In the Plains region the predominant clay mineral is montmorillonite, as illustrated in Table 2.2.1, and the mine roof and floor heaves due to the intensive disintegration and swelling of this clay mineral. To improve roof conditions, it is necessary to leave a coal layer of up to 0.6 m thickness to avoid exposing the rock strata to atmospheric moisture (Fig. 2.2.5).

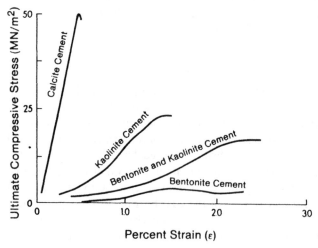

Fig. 2.2.4. Sandstone strength and behaviour depending on the cement material.

Table 2.2.1. Relative content of clay minerals in clay-shale rocks

Rock description	Montmorillonite	Kaolinite	Illite
Very fine-grained dessication cracks, coaly stringers	70	10	20
Semi-competent rock with coal stringers	35	15	50
Dense, dark gray with some silt	85	5	5

Quartz and plagioclase feldspar are always present in the rock samples

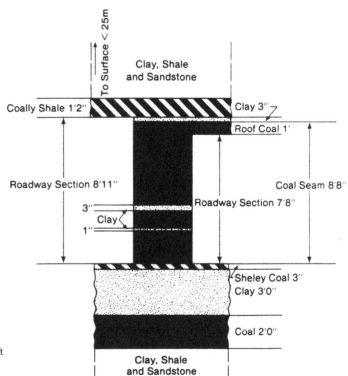

Fig. 2.2.5. A coal layer left in a mine roof to protect weak clay rock strata.

3. *The natural moisture in rock strata* depends on the porosity and permeability of the rock. There is a direct relationship between moisture content and rock strength. With an increase of moisture the strength is decreased. Studies in the Medicine Hat area of the Plains region show that coal beds which have a very high moisture content, has a correspondingly high coal porosity and permeability due to cleat. The overseam strata always has a higher moisture content (up to 2.5 times) than the subcoal strata of the

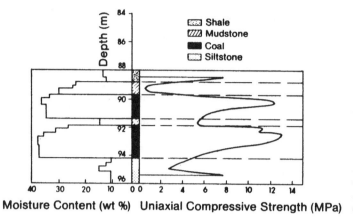

Fig. 2.2.6. A profile of a coal zone with a strength and moisture distribution (Medicine Hat area).

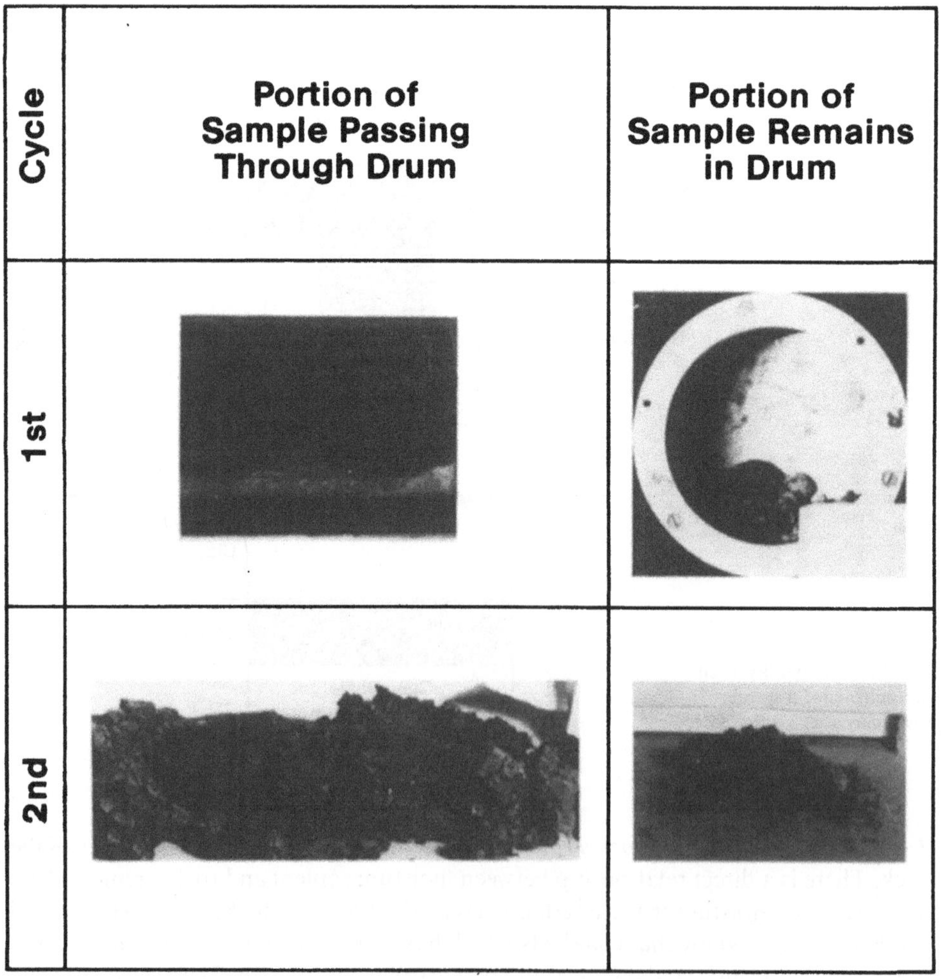

Cycle	Portion of Sample Passing Through Drum	Portion of Sample Remains in Drum
1st		
2nd		

Fig. 2.2.7. The intensity of decomposition of clay-shale rock in relation to bentonite content.

workable seams (Fig. 2.2.6). The differences in moisture in this case are due to cleat development in the roof strata and not to the mineralogical composition of the rock. Under these circumstances the strength of the roof strata is decreased up to 5 times compared to the floor strata of the same mineralogical composition.

4. Slake durability investigated on clay-shale samples from the Medicine Hat area showed that the main factor which determines the intensity of decomposition of a rock is the ratio of silt to bentonite. For example, bentonite shales were completely disintegrated during slake durability tests, which is opposite to silty shales (Fig. 2.2.7). Table 2.2.2 presents the mechanical and slaking properties of clay-shale rocks also from the Medicine Hat area, from which it could be concluded that the wearing capacity of all rock types is high, particularly where bentonite is present. This might effect a mine floor to the extent that the application of mine mechanization for coal production would be in question.

5. Swelling of rocks from same area showed, that these rocks can cause considerable difficulties in maintaining a mine opening. For example in laboratory investigations, uniaxial swelling of clay-shale rocks with a high silt content could swell over 9 per cent within seven days (Fig. 2.2.8).[12]
Rock samples with high bentonite content could not be prepared for testing because they rapidly swelled and disintegrated when exposed to atmospheric moisture. With respect to rock squeezing and pressure on mine support, the confined swelling tests should be performed on clay-shale rock samples. From the intensity of the axial and lateral rock expansion adverse effects on the stiffness of mine support could be inferred.

Clay mineralogy is a very important element for the analyses of underground mine stability. Unfortunately, the progress in this direction is insufficient because the application of geotechnical indices defined in the laboratory cannot readily be transferred to the actual mine situation where an integrated system of ground control and mechanization exist.

Table 2.2.2. Mechanical and slaking properties of clay-shale rocks

Rock type	Per cent of rock remnant after slaking		Slake characteristics
	1st cycle	2nd cycle	
Very fine-grained, rock with bentonite	20.55	4.51	Complete disintegration
Very fine-grained rocks with some silt	57.70	26.55	Major slaking into thick mud
Very dense rocks with some silt	60.02	18.12	Major slaking into mud

Compressive strength is from $0.32 \, MN/m^2$ to $10.98 \, MN/m^2$; Elasto-plastic to plastic behaviour with shear flow failure

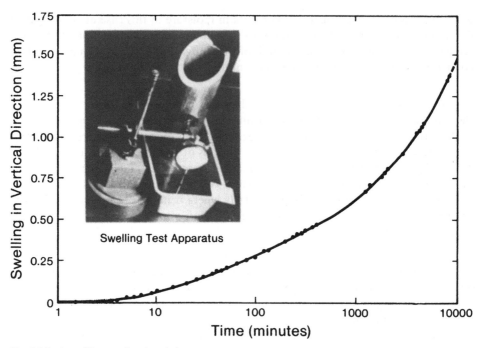

Fig. 2.2.8. A swelling test for clay-shale.

2.2.3.   *Rock diagenesis*

The effect of diagenesis on the mechanical properties and behaviour of the strata is significant. The coal deposits in Western Canada are a practical example of strata which have undergone a range of metamorphic processes.

1. Plains region coal bearing strata which have undergone mild diagenetic effects, maintain some of the original features of sedimentation. The geology of this area is discussed in great detail at the beginning of this chapter, and only the diagenetic effects of the mechanical stability of the rocks will be briefly considered here.[11]
    Numerous tests on mine samples and drill cores done by geotechnical consulting firms and by researchers at the University of Alberta show the following rock strength characteristics.
    a) The large majority of the rock samples tested exhibited very low uniaxial compressive strength (range 0.30 to 30.00 MN/m²).
    b) Laboratory results suggest that strength increases with increased depth of burial. It is obvious that there is a relationship between diagenetic effects and rock strengths.
    c) On the basis of rock strength three depth zones can be delineated. Strata buried less than 75 m show rock strengths up to 6 MN/m² (soil-like rock). Burial of 75 to 150 m produced very weak rock with strengths between 3 and 15 MN/m². Finally strata buried more than 150 m (the majority of rocks) have strengths up to 30 MN/m² (Fig. 2.2.9).

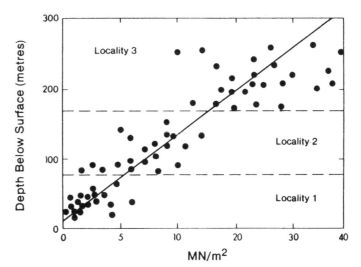

Fig. 2.2.9. The uniaxial compressive strength of rock in relation to the depth of burial.

The low intensity of diagenesis is caused by particular geological events:

a) The coal-bearing strata were buried in shallow parts of the Cretaceous-Tertiary geosyncline at depths estimated at not more than 1000 m. The overburden pressure was not sufficient to cause appreciable rock induration.

b) Movement was vertical only (uplift) during the initial stages of the Laramide orogeny and without lateral movement (folding and thrusting). Thus the coal-bearing strata lie flat or nearly so with maximum gradients of 1.5 per cent, almost in the original attitude of sedimentation, and with no diagenetic effects from tectonic pressures.

Limited diagenetic effects are produced merely by the overburden loading which caused compaction, decrease in pore space, and lower moisture content. Geophysical investigations, using a sonic-pulse transmitter and an oscilloscope resister mechanism, provided the velocities of compressional and shear waves from which elastic parameters were calculated (Alberta Research Council).[9] The elastic parameters of the strata suggest:

a) That rock compaction and elastic modulii increase with depth of burial, similar to the laboratory results represented by static loading (Fig. 2.2.10).

b) That the rate of rock compaction and magnitude of Young's modulus differ between very fine grained sediments (shale) and coarse-grained sediments (sandstone).

Diagenesis and compaction deserve particular consideration for underground mine design because the strata will respond differently to lateral mining at various depths. The difference in the behaviour of the strata in lateral mining results from changes in lithology, which may depend directly on the diagenetic effects.

2. The Rocky Mountain Belt consists of rock strata which have undergone a strong diagenetic effect. From the cross-section illustrated in Figure 2.1.1 it is obvious that Luscar (Blairmore) sediments which consists of deeply buried coal bearing strata in the area of the Rocky Mountain belt (over 2,500 m) only occur near the ground surface, if they are brought up by orogenic movement. There is evidence that the degree of rock diagenesis depends on the tectonic shearing stresses under intensive confining

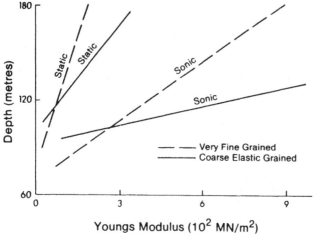

Fig. 2.2.10. Elastic modulii of rock strata showing different rock lithologies and depths of burial.

(lithostatic) pressure rather than on the dead load of the overburden (geostatic pressure). Sediments of the Plains region show that as a result of the depth or the weight of overburden there is a particular tendency to bring the fine-grained sediment grains closer together. The sediments of the Rocky Mountain belt exhibit along with very intensive compaction, changes in internal fabric and mineral composition as discussed below:

a) The original rocks which were subjected to powerful orogenic pressure were deformed by enormous shearing stresses, which produced differential movement between individual rock grains. These phenomena have been observed on the samples of coarse-grained sandstone (salt and pepper texture) where quartz grains are elongated in the direction of the main tectonic shear stresses.

b) The moisture content in the rock is possibly governed by the presence of clay minerals. For example, in the Plains region the low degree of diagenesis may account for the presence of deposited clay particles in a dispersed state. The effect of diagenesis was insufficient to squeeze out the water providing a possible explanation for the layer of molecularbound water which exists around each clay particle.[9] High diagenetic effects in the Rockey Mountain coal belt eliminated such a possibility and clay minerals with molecular water are not present.

c) The orogenically disturbed coal basins (Mount Head, Western Canada) consist of some recrystallized and metamorphosed sediments (Fig. 2.2.11).

d) The sandstones from these two regions exhibit two distinct stress-strain relationships, which was observed during uniaxial compressive loading of rock samples. The sandstone from the Plains region has very low strength (7.98 MN/m²) and it is characterized by elasto-plastic behaviour until failure and thereafter by plastic behaviour with a tendency to flow (Fig. 2.2.12). However, the sandstone from the Rocky Mountains is of high strength (108 MN/m²) with linear elastic behaviour having failure and violent strain energy release (Fig. 2.2.12).

This very brief representation of rock diagenesis from an engineering point of view, without any doubt suggests that this factor is an important element in the analyses of

Fig. 2.2.11. Metamorphosed
shale beds exhibit a slaty
character (black bands) at the
Cascade Coal Basin.

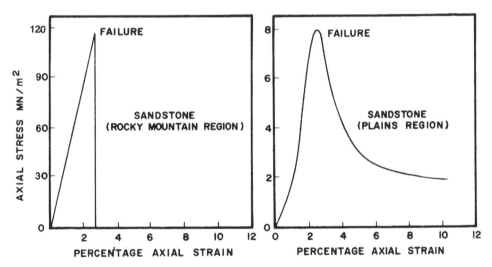

Fig. 2.2.12. Typical stress-strain diagrams of sandstones from the Plains region and the Rocky Mountain region.

the mechanical stability of coal bearing strata and on the choice of ground control in underground mines.

It should be noted that the final stratification processes of coal bearing strata are modeled not only by sedimentary mechanics but also by the presence of orogenic pressure, which is to a greater extent discussed under future headings.

2.3. ANISOTROPY OF COAL STRATA

The anisotropy of coal bearing strata will be considered with respect to sedimentation mechanics and stratification structure and will be discussed in several separate sub-heading.

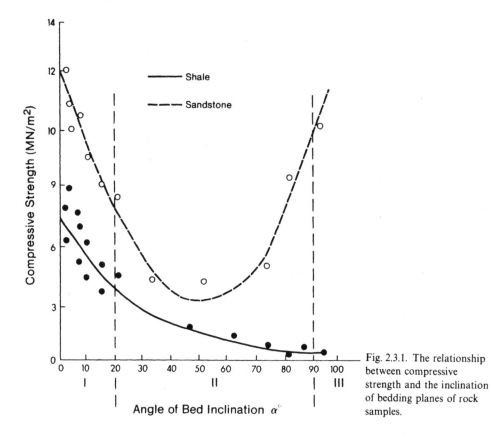

Fig. 2.3.1. The relationship between compressive strength and the inclination of bedding planes of rock samples.

2.3.1.   *Sedimentary structure*

Sedimentary structure may be well preserved under condition of moderate diagenesis. However, in cases of severe orogenic movement the original sedimentary structure might be completely modified and destroyed.

Numerous analyses of the interaction between sedimentary structure and mine stability have been carried out relative to underground mining. These analyses have been based on the laboratory testing of core samples and on geophysical investigations in the field, with particular interest in the following:

a) The relation between the angle of inclination of the strata and its uniaxial compressive strength is a common anisotropy parameter of sedimentary rocks. Numerous investigators have obtained a more or less similar function for dependence of compressive strength on the angle of inclination of sedimentary beds. In our case the strength decreases with the increase in the angle of the bedding and reaches minimum at inclinations of approximate 45°, after which the strength starts to increase reaching a maximum at an angle of inclination of 90°. However, in the case of some weak rocks or coal seams this postulate does not hold true. For example, a clay-shale rock can have a maximum compressive strength when the load direction is normal to the beds and it reaches a minimum when the load is parallel to the beds (Fig. 2.3.1).

b) The rock anisotropy can be influenced by certain mineral compositions. On the

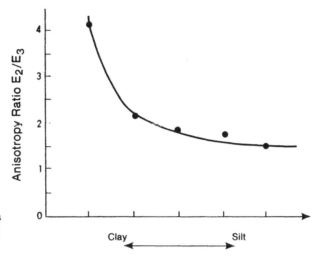

Fig. 2.3.2. The anisotropy ratio as a function of clay and silt content in the rock strata.

basis of in the field sonic method investigations of elastic constants of rock strata it has been concluded that the dynamic elastic anisotropy ratio E_2/E_3 ($E_2 =$ Young's modulus normal to bedding; $E_3 =$ Young's modulus parallel to bedding) is increased with increasing clay contents and is lower with higher silt content (Fig. 2.3.2). It is obvious that coal-bearing strata are highly anisotropic with sharp variations between lithological units.[14]

c) Rock anisotropy has also been clearly exhibited in the tension tests. For example, a set samples of weak rock tested by the Brazilian discs method (with natural moisture content) showed that the tensile strength of fine-grained sediments varies due to load orientation. In the samples with a load direction perpendicular to sedimentation the rock exhibited a tensile strength from 0.48 to 1.12 MN/m^2. With load direction parallel to bedding, the rock exhibited a lower strength between 0.16 to 0.23 MN/m^2. However, the investigation of metamorphised sandstone showed the same functional relationship between tensile strength and angle of inclination of bedding as in case of compressive strength for hard rock beds.[15]

It should be pointed out that one more factor influences strata anisotropy. The rock can exhibit a wide range of bedding intervals, by virtue of their fluvial depositional enviroments. The coal bearing strata and density of rock laminations might be represented by three basic types:

a) Thick bedded sediments, as in the case of some limestone or sandstone rocks, could be members of coal bearing strata. If the thickness of individual beds is two or three times greater than the heights of the mine opening, then the rock anisotropy with respect to sedimentation could be considered irrelevant. Under these circumstances it can be assumed that the rocks take on the mechanics of failure and deformation of homogeneous massive rocks. These rock beds sometimes present difficulties because they are not liable to caving and when they do cave then they fragment in large blocks, which further reduces caving.

b) Medium bedded sediments which are adjacent to coal seams are the most common case in underground mining. They facilitate caving and ground control

Fig. 2.3.3. Laminated shale beds without self-supporting abilities (false roof) covered with medium bedded siltstone (immediate roof).

because they usually have good rock fragmentation and flushing characteristics. However, this optimism is válid only for rock lithologies which are not liable to the adverse effects of slaking and swelling. The thicker the bedding the more stable is the roof of the mine opening.

c) Laminated or thinly bedded rock sediments usually represent rock types of the shale-siltstone complex. The reinforcement and stability of laminated strata depends a great deal on the absence or presence of shear displacement along bedding planes and on the thicknesses of such thinly bedded rock complexes (Fig. 2.3.3.).[16]

Of particular importance in underground mining is the location of the individual lithologic anisotropical rock types with respect to the mine opening. Laminated rock as a mine roof should be considered as a false mine roof, because there is no self sustaining property until medium or thick beds above the coal seam are reached. The strata above the false roof is called the immediate roof and further above this is the main roof.

2.3.2. *Sedimentation continuity*

The continuous planes of weakness in the rock strata are bounded sedimentary planes. The degree of these structural continuities depend mainly on three factors: [15, 17]

a) Direct shear strength (along bedding planes) might influence rock competence more than joint planes which are widely spaced. Two types of deformations along bedding planes are predominant: first, bed separation (rock dilatation) where resistance depends on the cohesion; and second, bed displacement (shearing) where resistance depends on the angles of friction. The intensity of the bed separation in the mine openings depends on the rock types as well as the loading magnitude and the bounding character of the beds. In Western Canadian coal mines it has been established that bed separation is greater at heterogenous rock contacts. For example, roof spans of roadways exhibited dilatation of 15 cm at the contact of siltstone and shale. Direct shear laboratory tests on rock samples from the Medicine Hat area have shown that the shear strength of siltstone is approximately double that of the shale at

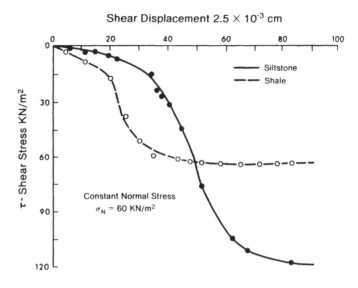

Fig. 2.3.4. A diagram of shear strength versus shear displacement (Medicine Hat area).

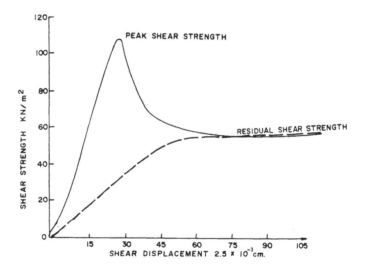

Fig. 2.3.5. Peak and residual strength of siltstone from Grande Cache Mine.

constant normal loads. Underground observations also suggest that at the same loading conditions the shear displacement along bedding planes is different for different rock types (Fig. 2.3.4).

b) The orogenic shearing of bedding planes can considerably further increase the weakening of the sedimentary continuity. The degree of weakening along lateral strata extensions usually vary due to different intensities of orogenic shearing forces. Shear strength deterioration along bedding planes can be evaluated from laboratory tests of samples on a metric direct shear box. Laboratory tests for shear strength of siltstone samples from Grande Cache coal mine (Fig. 2.3.5) have shown that they coincide with the classic representation of peak shear strength and the residual strength of a high cohesion intercept. Further discussion of this phenomenon is omitted, because it is very extensively described in geotechnical literature.

c) The presence of high moisture or water along bedding planes can considerably reduce their strength. For example, the shale from Lethbridge (Petro-Canada test mine) showed that shale samples with natural moisture have higher cohesion ($c = 0.6 \, MN/m^2$) and higher angles of friction ($\phi = 26°$).

The most suitable roof control for rocks that are weakened by sedimentary continuity is tensioned roof bolts. These have the effect of resisting shear movement along bedding planes and creating a thick beam.

2.3.3.   *Sedimentation discontinuities*

The anisotropy of rock strata can be effected by cross bedding to some degree, this is particularly present in the false or immediate roof of coal seams. The presence of a carbonaceous or micaceous film on these planes promotes localized slabbing of such a rock, and causes "poor" roof conditions. Roof control can be achieved by the use of roof bolts with straps and mesh to suspend the loose material.

Sedimentation discontinuities can be caused by diverse influences of a sedimentary environment such as paleo channels. An excellent example of this particular sedimentation discontinuity is described for the Bullieseam mines (NSW, Australia) by W. A. Williams and P. A. Gray as follows:[18]

a) Where fluvial channels, subsequently filled with sandstone, have cut down to, or nearly to, the seam roof through the primary laminite or mudstone, difficult roof conditions often occur. The erosional interface is poorly cohesive and may be further weakened by differential compaction since the compaction ratio of mud into mudstone is far greater than that of sand into sandstone.

b) Apart from the discontinuities occurring in these environments which have a detrimental effect on roof stability, a recent intensive study of a gateroad driven for a future longwall panel in Corrimal Colliery has revealed a complex lithological distribution. The presence of a weaker sandstone, marginal to the main sandstone facies had not been suspected from the mapping of the seam roof in the workings. This weak marginal sandstone required more support than the main channel sandstone.

c) The height of the mudstone-sandstone interface above the roof may be important in two ways. During development, the choice of bolt anchorage horizons may be influenced by the thickness of the mudstone. Where seam height allows, it is usually preferable to use bolts which are long enough to anchor into the sandstone.

The influence of these particular sedimentary discontinuities on mine operation is exhibited by the thickness of the mudstone-laminite facies. It follows the requirements of shortwall or longwall systems, as shown in Figure 2.3.6. Strong bridging sandstone impedes gob formation, resulting in loading of the area. This may be tolerable where sandstone forms the roof at the face if the inherent strength is sufficient to resist the stresses (stage 1). However, when the roof at the face is a thin stratum of brittle, weak mudstone, severe roof breakage ahead of the supports may occur (stage 2). Such difficulties will be much greater with a shortwall extraction system than with longwall. Stage 3 represents a reasonably good caving condition although inferior to that associated with a very thick shale roof.

Finally, it should be concluded that anisotropy of coal bearing strata either in

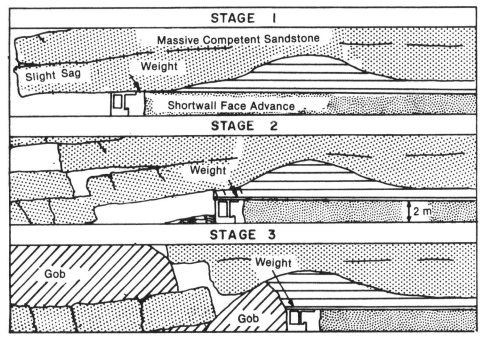

Fig. 2.3.6. Behaviour of a mine roof in paleo-channels.

transverse or lateral direction is a variable which in large part depends on the environment of sedimentation conditions.

2.4. TECTOGENESIS OF COAL STRATA

A large number of coal deposits, particularly those with a higher rank have been intensively deformed during tectogenetic movement of the earth's crust.

To get a better understanding of mine stability as a function of the existence of deformation and structural defects of coal bearing strata it is necessary to discuss very briefly the relationship between geological structures and orogenic stresses, within the frame work of recent global theory of geological mechanics.

The character of deformation of the coal bearing strata is exhibited primarily by folding, refolding and thrusting. The underground coal mines might be located within a folded structure (syncline) or part of this structure (limb), and will be briefly discussed later.

Finally, most important are the geological structural defects of coal bearing strata which are exhibited by faults, joints and cracks. In this heading, consideration is given only to geological stress fields and the mechanics of their formation.

2.4.1. *Orogenic movement of coal strata*

The origin of deformation and failure of coal bearing strata can be explained by

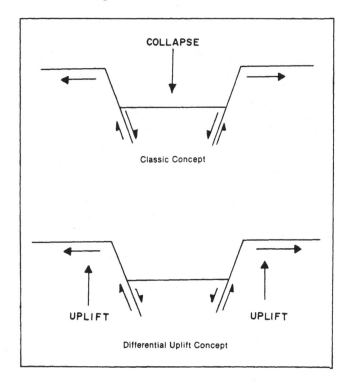

Fig. 2.4.1. The concepts of graben formations.

structural geological processes. Generally speaking two principal types of orogenic movement could be suggested:[19]

1. Geological extension structures are the product of graben formations in the earth's crust. At the present time there are two concepts of graben origination, the collapse concept, and the differential uplift concept. There are differences between these two graben concepts. A collapse structure has the collapsed block displaced downwards. The angle of the boundary faults dip opposite to each other (Fig. 2.4.1). The apparent mechanics of this structure might be a contravention of Archimedes Principle because the density of the rocks increases below the displaced block towards depth in the earth's crust. In the case of the differential uplift concept, the graben was formed by uplifting blocks on their sides (Fig. 2.4.1). The intensity of uplift of each block was different, resulting from unequal internal expansion of the earth's crust. The subsided block was not a product of the collapse structure, actually it was stationary, whereas the sideblocks were moved against gravity. A structure of this origin represents faulting with a steep dip and large displacement. Due to tension mechanics, the fracturing of coal bearing strata was by an echelon failure pattern.

2. Geological compressive structures are the result of tension in the earth's crust, where a lateral component causes a compression in the uplifted block. This type of geological structure relates to mountain building processes, where a threefold stress pattern exists as illustrated in Figure 2.4.2. These stresses produce fold on thrust structures, which are

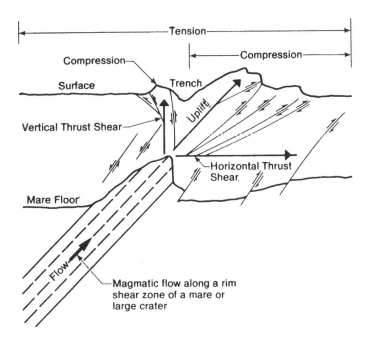

Fig. 2.4.2. Stress-strain pattern due to uplift by magma flow (after Mauritsen).

formed by continuous and slow strain build up in the earth's crust with a sudden release in the form of earth-quakes. The type of fold structure may govern the choice of underground mining methods and overall mine design. However, the main effect of compression on underground mine stability is the number of ruptures produced, which have a particular geometry and pattern as well as the direction of displacement of the coal bearing strata.

Further consideration of tectogenesis of coal bearing strata is directed toward the aspects of folding deformation (lateral stresses) and rupturing (gravitational and lateral tectonic stress), which is briefly represented in the following subheadings.

2.4.2. *Folding deformations*

Folding of coal strata is deformation which could be observed on a regional scale to a micro-size scale. The influence of folded structures on strata mechanics and related mine stability is particularly important, and one must keep in mind the virgin stress distribution of geological structures. Folding of the almost exclusively sedimentary rocks that compromise the coal strata is a yielding process and results from differential forces.

The folded strata exhibits wide ranges of morphological variations and can be related to the degree of deformation. This phenomenon can be divided into three main groups as discussed below:

1. Open folds are the simplest structures which result from orthogonal forces acting on essentially homogenous material. More complex, assymetrical structures can result because these forces change as the yielding process takes place because of minor and

Fig. 2.4.3. Open fold of a syncline with outcroping fault (Mount Head Basin).

major nonhomogeneities and anisotropic aspects of the rock masses, or because through geological time more than one series of forces may have affected a particular strata. A common example of multiple deformation is the faulting of the already existing open fold (Fig. 2.4.3).

Fold geometry represented by anticlines, synclines and limbs is important when relating structures to mines. Usually the thickness of the coal strata (particularly coal seams) is increased in the area of the anticline and the syncline and decreased at the limbs. A particular fold geometry will have an influence on the choice of the mining methods as well as on the strata control, which in the recent past was very often done by stowing.

2. Tide folds have been represented briefly by considering the fold belts of the Canadian Rockies.[20] These folds exhibit major changes along strike and dip. The main characteristic of the tide folds are that they have both anticline and syncline compression with an asymmetrical geometry in cross-section. The limbs of the folds dip in the same direction but at different angles. The steepest limbs are closed as the result of tectonic stress. Tide folds of isoclinal morphology can be observed from hand specimen size (Fig. 2.4.4), to large folds which have amplitude greater than 1,200 m and wavelengths approximately 2,600 m with strike extension for many kilometres.

With regard to mining and strata stability, these fold structures facilitate open pit mining (anticline) but they do not favour underground mining. However, the application of the hydraulic mining technology to this type of fold structure has proven satisfactory if the enlarging of the coal seam thickness is not out of proportion.

3. Overfolds represent the most complex fold geometry, because they are overturned with the inclined axial plane. The overturned limb has been rotated more than 90° to attain its present attitude. Large overfolds, the dimensions of which range in size from panels to underground mine units are of lesser influence on mine stability. Smaller

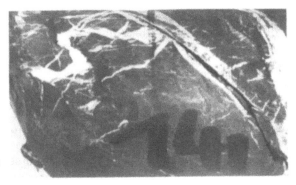

Fig. 2.4.4. Tide macrofolds exhibited in core samples (Cascade Basin).

Fig. 2.4.5. Overfold structure (Grande Cache).

overfolds (Fig. 2.4.5), however, are more significant when dealing with mine layouts and stability, because they change the geometry of coal strata and corresponding thickness of coal seams abruptly. Because overfolds cannot be detected during the exploration and the mine development phase, they can cause difficulties in day to day operations.[21] The greatest difficulties in underground mining arises when overfolds are crushed and displaced by thrust faults. Further ground instability due to the overfold structure is generated by branch trusting (it might slice fold in several sections) by tear fault deformation and other structural defects. The geometry and mechanical origin of these is discussed in the next subheading.

2.4.3.   *Rupturing deformations*

The mechanics of rupturing of the coal bearing strata is related to gravity faulting, thrusting, overthrusting, and folding. Under this heading critical stresses of rock rupturing are discussed, these have been limited to upper crustal tectonics excluding deformations at elevated temperatures (related to the high rank of coal).

The criteria of rupture are mainly inferred from studies of core samples of sandstone, which have been used as a model material. Data representative of sandstone has been extended to siltstone and shale, which fits to a certain degree, but cannot be generalized due to the limitations relevant to their different properties. The cleat relates to soft strata, because this type of fracturing is not characteristic of hard sandstone rock.

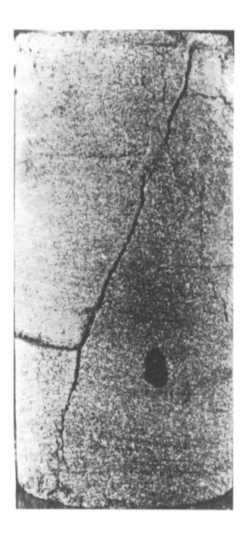

Fig. 2.4.6. Model of gravity fault with a hanging wall throw.

1. Gravity faults are formed by horizontal extensions of the earth's crust and the movement of tectonic blocks.[19] It is the most common structural defect of coal bearing strata where thrusting and folding are not present. Using sandstone as a model material (Fig. 2.4.6), the following rupture criteria are suggested:

a) Fracture geometry is usually represented by echeloned shear planes with a certain strike and rather steep dip. The hanging wall strata is displaced downwards along the shear plane and is known as the throw. Displacement could range from several microns to several thousand metres. Generally speaking the echeloned faults have a cascade geometry and they are called step faults. However, step faults might run in two directions intersecting each other by forming tectonic blocks which are delineated by the zone of the throw. This is the case when gravity faulting of older tectogenetic events were in one direction and gravity faulting of younger tectonic events were in the other (perpendicular to the first one).

b) The origin of the stress field which caused normal faulting is explained under the subheading of orogenic movement. It might be a product of the stress field formed by the movements of large crustal blocks during uplift, where orientation of the principal stresses are illustrated in Figure 2.4.7. The geometry and orientation of shear failure depends on distribution of the principal stresses. It should be pointed out that single faults do not exist in nature but are replaced by multiple faults, as illustrated in Figure 2.4.8.

c) The mechanics of fracturing is governed by the orientation and magnitude of the principal stresses. The stress on the shear plane is postulated below:

$$\tau = \tfrac{1}{2}(\sigma_1 - \sigma_3)\sin 2\beta \tag{1}$$

where

β = angle of failure;
σ_1 = major principal stress;
σ_3 = minor principal stress;
τ = shear stress.

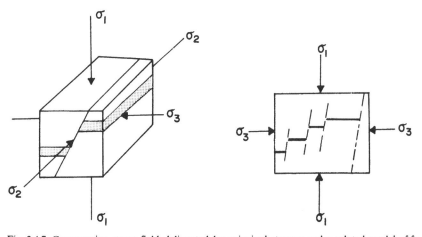

Fig. 2.4.7. Compressive stress field delineated by principal stresses and a related model of fault fracturing.

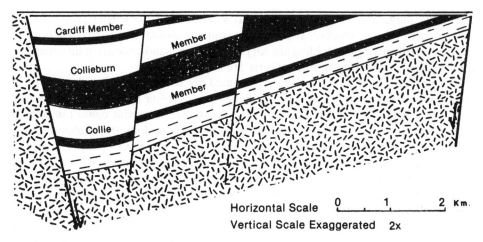

Fig. 2.4.8. Diagrammatic cross-section of the Cardiff sub-basin (Western Australia).

The mechanics of fracturing is an important parameter for assessing the direction and magnitude of the shear stresses as well as the principal stresses.

Normal or gravity faulting is extended through complete sections of coal bearing strata regardless of petrological anisotropy. The angle of inclination of the fault plane deviates along the strata section because of changes in compression magnitude as well as changes in the property and behaviour of individual lithological units.

2. Minidislocations are common for orogenically deformed strata. The fracturing is more pronounced within the coal seams than in the rock beds. The presence and density of minidislocations in rock beds depends a great deal on the strata lithology.

a) Fracture geometry can be observed on individual cracks, which are randomly oriented in intensively folded areas. The orientation and inclination of cracks is specific to bedding planes. The minidislocations occur in single and pair patterns. The intersected cracks occurring in pairs in some cases can cause a rock weakening and will break along these planes. A limited number of minidislocations may be healed or filled by calcite to form white veinlets and stringers (Fig. 2.4.9).

b) The origin of the stress field can be inferred on the basis of the relationship between the tectonic stresses and the fracture mechanics. These types of tectonic stresses are usually related to uplift in the earth's crust. It might be inferred that the magnitude of the major principal stress (σ_1) relates to the uplift stress (Fig. 2.4.10), and that the magnitude of the minor principal stress (σ_3) depends on the constraint provided by the rock mass during compressive loading.[22] This particular stress field in the later phase of tectonic movement is dissipated by the fracturing.

c) The mechanics of fracturing of minidislocations have been described by J. Hucka and B. Das, what is summarized as follows.[23] The cracks formed and to shear stresses, should have been orientated to the resultant effect of the major (σ_1) and minor (σ_3) principal stresses in accordance with Mohr's theory. Mohr's theory predicts formation of dislocations symmetrical to the major principal stress. The angle of inclination of

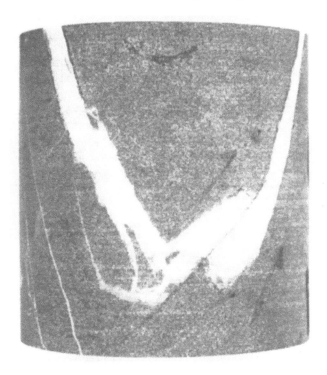

Fig. 2.4.9. Brittle shear
fracture filled by calcite
and deformed by a younger
tectonic movement.

cracks subtends with the major principal stress and can be expressed by Mohr's
criteria:

$$\beta = 45 - \frac{\phi}{2} \qquad (2)$$

where
β = the angle between crack plane and the major
principal stress;
ϕ = the angle of internal friction of the rock.

There is a definite relationship between the angle of joint orientation to the major
principal stress and the angle of friction of the rock mass. This is discussed briefly
below:

$\phi = 0$ and $\beta = 45°$: no substance with $\phi = 0$ is known, so $\beta = 45°$ is the ideal
upper limit of inclination. This suggests that cracks are
oriented with major principal stress at 45°;

$\phi = 20$ and $\beta = 35°$
$\phi = 30$ and $\beta = 30°$
$\phi = 50$ and $\beta = 20°$ } common values for rocks;
$\phi = 60$ and $\beta = 15°$

$\phi = 90$ and $\beta = 0°$: no substance with infinite friction is known, hence joints
cannot be parallel to the major principal stress.

The orientation of minidislocations to the major principal stress might be utilized to

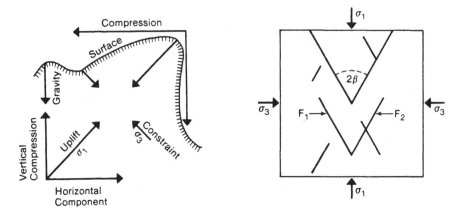

Fig. 2.4.10. Formation of a compressive stress field (uplift movement) and a related model of brittle fracturing.

define the direction of the major tectonic stress, whose remnants may still be present in the ground.

3. Thrust faults probably originated during the main structural phase of the tectonic movement. This fracture type belongs to the very slow process of folding of the rock strata and is related to large-scale thrust movements. Again, using sandstone as a model material, the following elements of rupture are suggested:

a) Fracture geometry is represented by the shear planes of crushed grains and particles. The shear movement occurs as a result of asperities on the shear planes riding over one another, causing dilatation or volume increase just before failure. The geometry of the asperities of the majority of sandstone core samples is in a zig-zag pattern, with the angle between the shorter and the longer cracks being approximately 45° to the bedding planes. However, with later tectonic events the asperities and fracture orientations as well as bedding planes are, to a certain degree, displaced and distorted (Fig. 2.4.11).

b) Origin of the stress field can be inferred from a number of structural cross-sections in the Mountain coal basins. The proposed stress field is well known from many fold belts around the world, where the horizontal component played the main role in strata deformation.[24] Thrusting is expressed by the riding of the upper tectonic block over the lower one, which is then followed by a stress release along the thrust plane and on internal stress redistribution in the individual displaced tectonic blocks (Fig. 2.4.12).

c) Mechanics of fracturing can be derived from the phenomena that cohesion between solids is destroyed at a very slow rate of folding (yielding fractures due to the machanics of permanent load and geological time). This fracture follows a definite plane of shearing, whose orientation is a function of the direction of the lateral tectonic pressure. The principal stresses are compressive, and the motion of the upper block suggests, without any doubt, that the lateral tectonic stress (σ_1) was greater than the overburden pressure (σ_3). It is obvious that the upwards shearing motion has been

Fig. 2.4.11. Trust faulting at an angle along bedding planes.

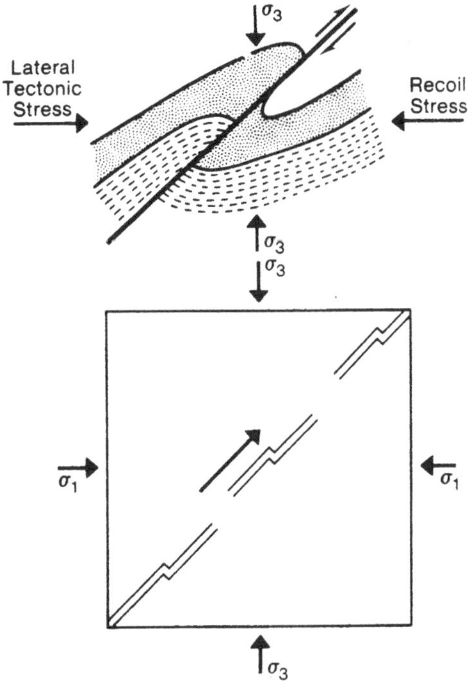

Fig. 2.4.12. Formation of a trust fault due to lateral tectonic stress and a related model of rupture.

favoured with an increase in σ_1 (lateral tectonic stress) and the decrease of σ_3 (overburden pressure). The general trend of fracturing tends to be less than 45° to the acting major principal stress, which is an indication that the frictional component of shear resistance was positive. This can be expressed by Mohr's equation

$$\tau = \sigma_o + \sigma_N \tan \phi \tag{3}$$

where

σ_o = intrinsic shear resistance;
σ_N = normal stress;
ϕ = friction angle.

This type of fracturing is similarly present in the siltstone, mudstone and shale of the coal bearing strata. The main difference between the fracturing in sandstone and the other rocks is a smaller dilatation, (probably dependent on rock porosity), and in the greater angle between zig-zag fractures (90° and 130°), found in sandstone.

4. Shear slips occurred during thrusting processes but are younger than the major flow-fracturing (which represents the main period of thrust faulting), and can be easily detected in core samples.[20]

a) Fracture geometry is related to shear displacement planes with limited dilatation, because of the almost flat asperities or lack of asperities. The fracture geometry of the rock primarily depends on the frictional forces resulting from this intimate contact.

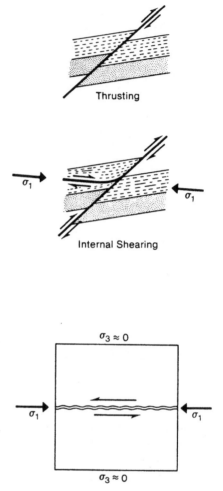

Fig. 2.4.13. Shear slips along distorted bedding planes.

Fig. 2.4.14. Internal shearing of the thrust sheet and a related model of slippage.

Fracture initiation tends to be in the area of minimum shear resistance, e.g., close to or along the bedding planes (Fig. 2.4.13).

b) Origin of the stress field can be described by the cross-section model of S. A. Mouritsen (Fig. 2.4.14) where the wedge-shaped block of sheared rock delineated by the main thrust plane and newly sheared plane most likely represents the relaxation of lateral tectonic stress due to shearing motion upwards, where the vertical resistance was small or zero.[19] Geologically this case is quite possible, which suggests that the inferred criteria of the rock mass fracturing might be similar to this structural interpretation.

c) Mechanics of fracturing is governed by the mobilization of the shear stress in the upper block. The stress distribution is governed by high lateral stress (in the direction of motion), which ceases due to the resisting stress on the front of the block and the vertical relief towards the thrust plane and toward the ground surface. Under these circumstances the major principal stress (horizontal) is high and the minor principal stress (vertical) is small or even negative (tension). The fracture will occur when the shear resistance of the rock is overcome.[25] The angle (β) between the major principal stress and the fracture, in the majority of cases, is between $0°$ and $15°$, and on a local scale the fracture often coincides with bedding planes. The model of rock fracturing is illustrated in Figure 2.4.14. The criterion of failure in this case can be based on the assumption that shear stress depends on the normal stress as well as the friction forces, and can be expressed as:

$$\tau = \sigma_N \tan \phi \tag{4}$$

The equation states clearly that conditions for shear displacement are governed by the residual strength of the bedding planes of the strata.[26] This type of shearing is well exhibited in all rock units, particularly in the siltstone and shale.

5. Tensile fracturing possibly represents the youngest type found in the coal strata. It is a common type and preferentially takes place within a zone which has already been displaced by an earlier tectonic movement. This type of rupturing in the sandstone strata has been delineated as follows:[20]

a) Fracture geometry is characterized by individual tensile fractures which are steep or vertical ($\beta = 0° - 10°$). This type of fracturing is extensively propagated. It has fault characteristics rather than the characteristics of joints with bridges. It seems most likely that the crack once initiated, continuously ran at a high velocity until it either emerges from a free boundary or it passes into a compressive stress zone and ceases. Tensile fractures are not quite straight lines; rather they have a deviating geometry (Fig. 2.4.15).

b) Origin of the stress field might be related to the displacement of the upper block by shearing, because the relation of this type of fracturing to the main thrusting period is not identified. The formation of the horizontal tensile (σ_3) stress field might be assumed to be a rock relaxing phenomenon, when shear displacement ceases (Fig. 2.4.16). The relatively small magnitude of vertical compressive stress might be related to the fact that the displaced block was also relaxed in a vertical direction.

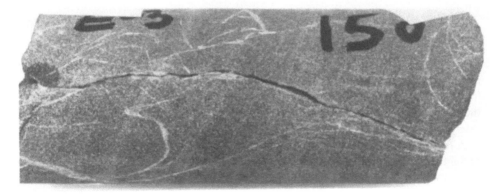

Fig. 2.4.15. Tensile failure perpendicular to bedding planes.

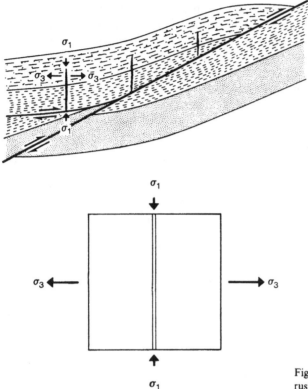

Fig. 2.4.16. Tensile stress coused by thrust sheet restrain.

c) Mechanics of fracturing can be inferred from the fracture geometry, where the breaks in the sandstone took place by the pulling apart of the individual quartz grains. The coarse grains were not split by tensile stress, probably because the acting tensile stress was lower than their tensile strength. It is obvious from the fracture position that the tensile stress was equal to the tensile strength of the cement material, i.e.,

$$\sigma_3 = \sigma_T \qquad (5)$$

where

σ_T = tensile strength.

and that the vertical compressive stress (σ_1) was of the same magnitude, as illustrated in Figure 2.4.16. The stress relation indicates that the frictional component of shear resistance is zero, and that the postulated state of pure shear stress would lead to tensile fracture. It should be noted that tensile fractures are elongated along the axial plane of the folds, and are characterized by large vertical extension and limited longitudinal extension.

A further discussion of the tectogenesis of the coal strata is extended in chapter 4 under the heading of the 'Structural defects of the rock strata'. In this particular heading the knowledge of strata rupturing is utilized to define geotechnical elements and indices for the stability analyses of the mine structure.

CHAPTER 3

Coal seam features

Coal seam features are the product of particular events that can be related to their depositional setting such as peat accumulation. The coal genesis and subsequent deformation of these strata is by tectonic events. It is important to recognize that in the four groups listed below, the coal seam features are classified according to their origin:

1. Seam structures in coal formations whose origins are related directly to depositional mechanics, morphology of sedimentary basins and syngenetic fabrics.

2. Coal petrography as the most important element in the formation of transverse anisotropy of coal seams.

3. The rank of coal exhibits, a known relationship to its strength and behaviour, which may vary depending upon changes in the compositions of the coal caused by deposition and/or orogenesis.

4. Cleats in coal, regardless of their origin and distribution patterns, have an appreciable influence on the mechanical properties of coal, particularly their shear strength.

Each of these individual topics is discussed in the following text. Practical applications to coal mine strata mechanics and stability is also presented.

3.1. FORMATION OF COAL

The subject of coal formation has been intensively studied in the science of sedimentology and is represented in various volumes all over world.[1] The aim of this brief discussion is to relate certain features affecting formation to coal seam mineability and stability.

3.1.1. *Depositional mechanics*

The formation of coal is essentially influenced by two groups of factors: paleo-biological and paleo-geographical. Only the latter is considered in this text.

1. The subsidence of the paleo-basin is an important factor in the chemical and physical modeling of coal seams. For example, the slow and continuous rise of the groundwater table protects swamps against river flood waters and inundations by sea water. However, if the groundwater table rises too high (rapid subsidence) the swamps

Fig. 3.1.1. A very thick coal seam exposed in an open pit mine (South Blackwater, CSR Company, E. Australia).

will drown and limnic sediments will be deposited. However, if the subsidence is too slow then the deposited vegetal matter will rot and the already formed peat will be destroyed. This mechanism of ground oscillation can be related to the relief energy of the hinterland, which results either in a restricted supply of fluviatile sediments (continuous peat formation) or an unrestricted supply (interrupted peat formation). Depositional mechanics have a major influence on the formation of sedimentary geometry and the syngenetic structures of coal seams.

2. Depositional milieux is one of the most important factors which influences or controls coal seam thicknesses and multiple seam structure. Two specific kinds of paleo-geographic sites can be identified in which vegetal matter is depended on to form coal seams.

The first type belongs to sea coast environments where deltaic swamps are developed and so-called paralic coal deposits are formed. These basins which contained brakish and semi-brakish water are the sites of coal deposits characterized by structureless vitrain which has a high sulphur (2 to 3 per cent) and nitrogen (2 per cent) content and only a small amount of mineral matter (< 6 per cent). The formation of this 'paralic milieux' is facilitated by marine regression with deposition in narrow belts (Pennsylvanian coal deposits) or broad basins (Ruhr coal deposits). In the first case the thickness of the coal strata are limited to a few coal seams in succession. In the second case, the coal strata are thick and contain numerous coal seams. For example, the Ruhr Basin has a thickness up to 4,000 m and consists of 40 workable coal seams. In both cases the thickness of individual coal seams is small (0.15 to 1.5 m).

The second type belongs to platform – Continental basins, where the deposition of vegetal matter took place in forest swamps and along the shores of large lakes, as well as in ground depressions to form 'limnic coals'. In these paleo-graphic conditions coal seams are usually thick but their number in vertical succession is limited (French Central Plateau, Carboniferous deposits of Saar and others). The thickness of the coal seams are particularly large where two depressions intersect each other. Under these circumstances coal seams nearly 100 m thick have formed as is evidenced by those in Queensland, Australia (Fig. 3.1.1). However, these type of inland coal deposits are restricted laterally. These deposits contain vitrain with cubical fabric and have a low-sulphur content (< 1 per cent) and high abundance of mineral matter (> 12 per cent). Inland depressions formed due to orogenesis (Western Canada) or glacial erosion where channels containing ice-melted water which formed a favourable milieux for the deposition of vegetal matter (Western Australia). However, some paleo-depressions were formed by the solution of thick beds of limestone, gypsum or salt (South Africa and Brazil). Coal from this type of deposit has a high sulphur content, (11 per cent S at Rasa Coal Mine Yugoslavia) and high nitrogene content (3 to 4 per cent).[1]

3. Climate is an equally important factor in coal formation, particularly with respect to the syngenetic structure of seams. It is a known fact that a warm and humid climate is most favourable for forest growth and swamp development. For example, in most warm climatic zones in the Upper Carboniferous, very rich coal deposits with many broad bands of bright coal were formed. In the Southern hemisphere and in Siberia, large coal deposits were formed in humid but colder climates (Permian deposits). In the case of a humid climate where annual rainfall is greater than the total yearly evaporation, boghead coal cannot be formed. This discussion illustrates then, that climate is a factor which affects the structural types of deposits and hence their mechanical stability.

3.1.2. *Sedimentary morphology*

The mining methods and strata control techniques used to extract a coal deposit depend upon the morphology of the coal seams. Unfortunately, only geological criteria have been used to characterize the seams. Lack of standardized engineering criteria have resulted in different techniques being used to mine the seams and maintain mine stability for similar coal seam morphologies. Listed and discussed below are four morphologically distinct coal seam types, all of which are of primary sedimentary origin.

1. Regular morphology: in this case the seams have an somewhat idealized tabular sheet-like geometry. However, from practical knowledge it is known that such a simple idealized model does not correspond to an entire coal seam but only to certain portions of it (Fig. 3.1.2). For example, from data obtained during exploration drilling and

Fig. 3.1.2. Coal seam of regular thickness in open pit mine (Griffin Coal Mine, W. Australia).

underground coal operations it is known that Number 4 coal seam, at Grande Cache has a thickness of between 2 and 12m. This variation in thickness is spread over a large distance and so does not influence underground mining. In this example the changes of seam thickness within a single underground mine (up to 2km length) are small, the accepted average thickness of the coal seam is 6 m. A coal seam with 'regular' morphology offers a relatively good understanding of strata mechanics and the related ground control measures to be employed.

2. Irregular morphologies are represented by seams which have abrupt changes in thickness. The irregularity of the seams is due to the nature of the sedimentary environment and exhibits a 'roll' to 'horseback' geometry. Rolling coal seams may have been formed by streams flowing into swamps where the vegetation gave rise to the coal being formed. These streams would bring fluvitale sediments into the basins or swamps and channels of clastic sediment deposits which would become buried under vegetal matter as the formation of the plant material progressed.[2] Although these types of circumstances lead to the formation of coal seams with 'irregular' morphology, it is correct to say that most often seams of this type form due to deformation related to orogenic events. This morphological type in interaction with coal extraction, particularly in the case of hydraulic mining, is a most unstable mine environment where little is known about strata mechanics.

3. Composite coal seams perhaps better known as 'dirty coal seams' are common phenomena in many coal fields. They originated mainly during periods of alternately slow and rapid subsidence within the basin of deposition. The seams contain bands of coal and intervening layers of clastic rocks. The mineability of dirty coal seams depends not only upon the rank of the coal and seam thickness but also the proportion of interlayered rock. If the geological thickness of a coal seam is greater than the workable thickness, a dirty coal section can be eliminated from the mining profile as illustrated in Figure 3.1.3. In this case the workable thickness can be defined as transitional between clean and dirty coal. The stability of coal faces in mechanized mining is decreased due to large displacement (softer coal layers) or constrained displacement ('jammed' hard coal layers). However, working faces of dirty coal seams are of lesser influence on strata mechanics in conventional mining than in the case of continuous mining technology.[3]

4. Split coal seams or fingered seams exhibit the sedimentary history of coal deposition particularly well. This morphology is formed by sedimentation mechanics similar to those which form composite or dirty coal seams. In this case however the periodic nature of the subsidence of the basin is less pronounced and occurs over a longer time interval. The clay or sand brought in by water from surrounding lands is swept out over vegetal matter. The deposit of sediment grows thinner as it extends away from the eroded land mass and for some distance from the edge of the basin the deposition of plant material persists without interruption. Under these circumstances continuity of rock and coal beds does not exist in either the vertical or horizontal dimension.[2] The composite and split structures of coal seams or coal zones require the particular layout and ground control approaches as discussed in slice mining methods.

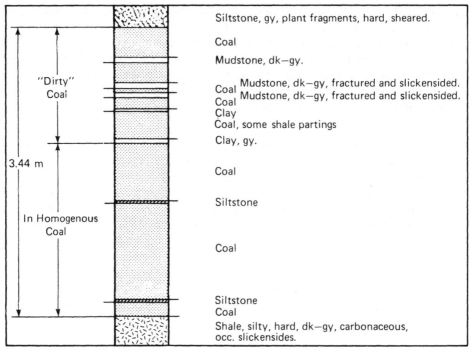

Fig. 3.1.3. Profile of composite coal seam (Number 10 seam, Grande Cache, W. Canada).

3.1.3. *Syngenetic structures*

The investigations of syngenetic structures in coal seams have focused on their physical features. As a parameter for classification of these internal structures the phenomena of transverse anisotropy of coal seams has been used. By making certain simplifications three types of structures can be recognized.

1. Homogeneous internal structure characterizes a coal seam which has uniform properties and behaviour along its profile. Homogeneous transverse anisotropy in coal seams has been studied in Siberian coal deposits which originated in cooler climates. They are typically dull coals with massive textures, in which bedding planes are rather minor in abundance.[3] Similar behaviour of workable seam profiles have been observed in underground mines in Western Australia, where the coal seams exhibit uniform hardness and similar response to the attack of a continuous miner (Fig. 3.1.4). Underground phenomenological studies of coal faces with homogeneous transverse anisotropy suggest limited displacement along stratification planes resulting in satisfactory stability of the mine structure. Strata mechanics in interaction with the mine excavation could be analyzed by assuming that the coal is an elastic and isotropic material.

2. Semi-homogeneous internal structure relate to the majority of coal seams of coking rank. For example, investigations of a Pittsburg coal seam of 1.6 m thickness, indicated

Fig. 3.1.4. Homogeneous coal face (Collie Coalfield, W. Australia).

the existence of two distinct horizons, both of which were assumed to be orthotropic, elastic material.[4] One horizon had a modulus of elasticity of 3.78×10^3 MN/m^2 and a Poisson's ratio of 0.45, and the other 2.76×10^3 MN/m^2 and 0.42 respectively, and with a uniform shear modulus for both of 5.16×10^3 MN/m^2. Cook (1978) pointed out that from an engineering point of view the variability of mechanical properties of the two coal horizons along a transverse seam profile could be ignored, because the differences in their properties are small compared to the observed scatter of data. The coal could be approximated as an elastic and orthotropic material. This structural type exhibits satisfactory stability of a mine structure and so strata mechanics could be analyzed for different mining configurations.

3. Heterogeneous internal structure primarily belongs to the younger coking coal deposits of Jurasic-Cretaceous age, which originated in very warm climates. The main feature of transverse anisotropy in these coal seams is the presence of broad layers of alternately dull and bright coal of extremely different strength properties and behaviour. The investigation of this type of coal deposit from Western Canada showed the following mechanical characteristics of the seam profile.[5]

a) The major content of soft and very soft coal (up to 75 per cent) in the seam profile is indicated by low coal recovery during exploration drilling. The recovery is less than 50 per cent and the coal is broken and pulverized (Fig. 3.1.5). This evaluation of coal seam softness coincides with run-off-mine coal of 40 per cent fines. In these mines, the mechanical instability of mine structures is so severe that pillar mining should not be used.

b) The sectional strength of the coal seam profile, from low strength to high strength is directly proportional to pillar elastic constants (Fig. 3.1.6). Coal under initial loading, with a uniform behaviour, relates to the compaction of cracks. When compaction is

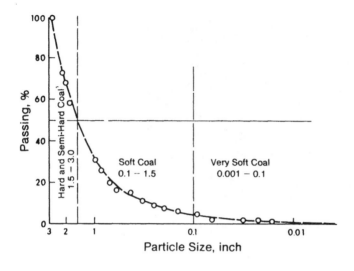

Fig. 3.1.5. Strength heterogeneity exhibited by particle size distribution of recovered coal from borehole.

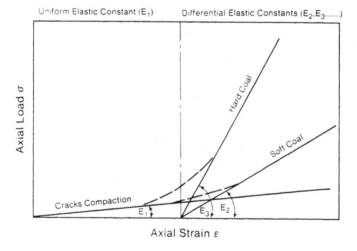

Fig. 3.1.6. Heterogeneity of elastic properties of coal seam profile.

completed, the pillar starts to exhibit differential properties for each coal section along the seam profile. Under these circumstances, differential or sectional pillar behaviour is developed and this leads to mine instability in a short period of time.

The heterogeneous internal structure is particularly well exhibited in thick coal seams. The enormous differences in strength properties of individual coal layers results in particular types of structural instability. These are discussed in subsequent sections of this book, particularly under coal slip displacements.

3.2.   PETROGRAPHY OF COAL

Coal is not a mineral, as people often assume, because it is composed of a number of constituents which can be identified by the marked eye or by the microscope.

Lithotypes, considered in mining technology, can be seen by the unaided eye, while macerals which are considered important in coal processing technology, can only be seen through the microscope. Severals aspects of the relationship between transverse anisotropy of coal seams and coal petrology are of considerable interest.

Fig. 3.2.1. Vitrain of friable and cubical fabrics.

Fig. 3.2.2. Banded coal fabrics with clarain (less cleated) and vitrain (more cleated).

3.2.1.   Lithology of the coal profile

Of interest in mining operations are only individual lithotypes which are recognizable as banded constituents of coal. The principal individual coal lithotypes in Western Canadian coal fields are briefly described below.[5]

1. Vitrain: is a bright, brilliant, vitreous and brittle coal. In Western Canadian seams, vitrain exhibits a cubical structure (Fig. 3.2.1). It can be usually be broken by hand into small, sharp-edged cubes. Vitrain usually occurs in beds up to 1.5 m thick and it is mainly responsible for the differential strength of a coal seam profile. In Central European bituminous coal seams however, vitrain occurs in non-laminated bands which are disseminated through the seam. In this environment, vitrain does not influence the homogeneous strength of a seam profile.

2. Clarain: occurs in bright and semi-bright coal of less brilliant luster than vitrain and has poor cleat manifestation. Clarain occurs in laminated bands with thicknesses up to 20 cm. The occurrence of vitrain and clarain together is common since both are formed by overbank flooding in relatively shallow water (Fig. 3.2.2). Clarain bands promote the inelastic deformation of mine structures when confinement is removed, particularly in the case of orogenically sheared bands.

3. Fusain: is a bright coal of fibrous structure which is very friable and easily crumbles to dust. Physical and chemical similarity of fusain to wood charcoal suggests that it was

Fig. 3.2.3. Fusain grains enlarged 400 times, with shearing fabrics.

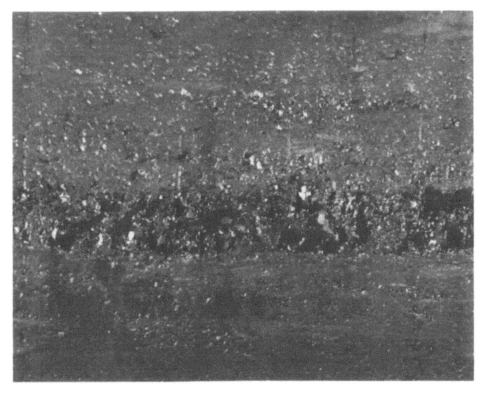

Fig. 3.2.4. Durain of fine grained fabrics interbedded with cleated vitrain.

formed under a shallow water cover with frequent exposure to air. Also a forest fire had probably swept across the swamp peat deposits. Fusain bands are lensoid in shape and are 1 to 10 cm in thickness. Fusain bands are soft and often pulverized (down to micron size) by orogenic shearing (Fig. 3.2.3). For this reason fusain usually represents coal layers of minimal strength.

4. Durain: this lithotype was formed under a deep water cover and in a basin which exhibits a variable subsidence rate. Durain occurs in finely bedded layers with thicknesses up to 1 m. It is a hard, dull, dark gray, or black coal and represents agglomeration of carbonized plant remains (Fig. 3.2.4). Durain contains very few cleats and cracks and unlike vitrain is not affected by orogenic movements. In terms of its properties and behaviour, durain resembles hard Carboniferous coals.

5. Black and dull coal – Cannel and Boghead Coal: this coal exhibits conchoidal fracture, the cleat is absent. Usually it occurs as layers at the top or bottom of coal seams, and as such is not mineable. It has a woody or hollow sound when struck. Thickness of individual layers seldom exceeds 0.75 m; it burns readily with a bright smoky flame and is a hard and brittle coal.

The existence of different coal lithotypes is also related to different plant communities. The most important fact with respect to rock mechanics in coal mining is that each lithotype has particular mechanical properties and behaviours which might cause instability of coal seams if they are represented in broad bands.[6]

3.2.2. *Hardness of coal lithotypes*

The hardness of individual coal lithotypes has been investigated on the coal faces in underground and open pit mines, in exploration diamond drillholes, and in the laboratory on individual coal samples. For example, laboratory testing included investigation of 32 coal samples using the INSTRON testing system TT-D. The most representative results of sectional strength of individual layers, particularly for thick and composite coal seams has been obtained from diamond drilling, coring, and geophysical logging of the exploration boreholes.[7] On the basis of these types of investigations a simple strength categorization of coal layers has been developed. The criteria for the determination of hardness of individual coal lithotypes have been based on the following data:

1. Hardness of coal beds inferred from geophysical borehole surveying (gamma-gamma log).

2. Enlarging a hole diameter, inferred from caliper deflection.

3. Geological coal core logging; lithology, brightness and cleavage of coal recorded as definitive indicators.

4. Laboratory testing of coal samples under uniaxial compressive load.

Research on the relationship between coal density (defined by gamma-gamma logs) and individual coal lithotypes have been carried out references, because in many instances during exploration drilling the only physical evidence that coal is present is the geophysical log data as coal samples are often not recovered.

3.2.3.   *Density seam profile*

In Table 3.2.1 and Figure 3.2.5, the relationship between a coal lithotype and its density is illustrated. As indicated in Table 3.2.1, there is also a clear correlation between coal lithotypes and its uniaxial compressive strength. In underground mining operations, the stability of an entire coal seam consisting of several coal lithotypes is of much more significance than that of individual coal horizons. Hence the following discussion will focus on 'compound lithotypes'. On the basis of data from geophysical gamma-gamma

Table 3.2.1. Classification of lithotypes by hardness

Physical-mechanical characteristic	Hard coal	Semi-hard coal	Soft coal	Very soft coal
Lithology	Durain, cannel & boghead	Clarain, or composite of clarain, durain and vitrain	Vitrain	Fusain
Brightness	Dull	Semi-bright	Bright	Lustrous
Cleavage	Traces, or none	Irregular, erratic	Intensive, regular	Sheared, open fractures
Enlarged hole	0.3 to 0.5 D caliper deflection; walls of hole stable	1.0 to 1.5 D caliper deflection; hole caves with time	2.0 to 2.5 D caliper deflection; holes cave often	2.5 to 3.5 D caliper deflection; holes cave instantly
Density	Dense, structureless	Less dense, cracked	Low density, well cracked	Porous, with very low density
Uniaxial compressive strength	20 to 40 MN/m^2	7 to 18 MN/m^2	1 to 7 MN/m^2	0 to 1.0 MN/m^2

D = drillhole diameter

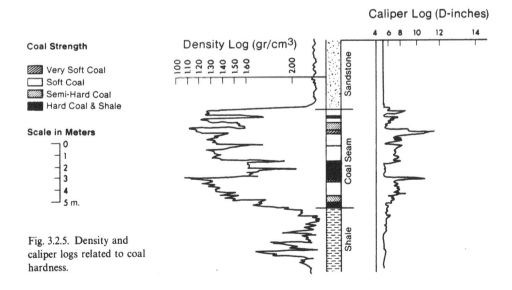

Fig. 3.2.5. Density and caliper logs related to coal hardness.

logs and the relationship of coal density to coal lithotypes, three principal density levels of transverse anisotropy of coal seams are suggested:

1. Mono-lithotype profile can be the product of any individual lithotype. The two examples described below are for coal lithotypes with extremely different density and hence mechanical characteristics.

   In Table 3.2.1 a classification of coal lithotypes defined by hardness is presented. The reader is reminded that the listed mechanical characteristics in this table for each coal type are of a general nature and that seams from an individual coal field often exhibit additional specific features unique of that region.[8]

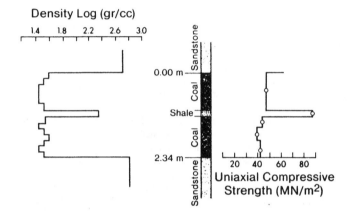

Fig. 3.2.6. Density log and strength profile of a mono-lithotype hard coal.

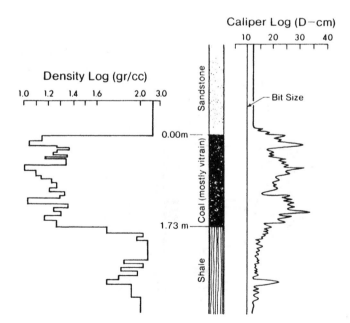

Fig. 3.2.7. Density and caliper logs of a mono-lithotype soft coal.

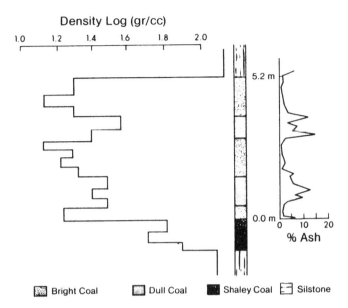

Fig. 3.2.8. Density log and ash content profiles of multi-lithotype coal seams.

a) A hard coal profile is composed mainly of durain which has a high density (1.4 to 1.6 g/cm³) and a high uniaxial compressive strength ($< 40\,MN/m^2$–Fig. 3.2.6). For such a transverse anisotropy the mechanical stability of coal faces is very high.

b) A soft coal profile is largely represented by vitrain whose original low strength and density is further reduced by orogenic shearing (Fig. 3.2.7). This coal is rather porous and has a low density (1.0 to 1.3 g/cm³). Coal faces with a high proportion of this lithotype are normally quite unstable.

The number of coal seams which are mono-lithotypes is rather limited, and for this reason the remainder of this section is given to a discussion of seams which contain several coal lithotypes.

2. Multi-lithotypes profile: in terms of the stability of the coal seam (face) and strata control, two principal cases should be considered.

a) The broad-banded lithotype (thickness of layer is 10 to 100 cm), profile of Cretaceous coking coal, consists of alternating layers of hard and soft coals of different densities and strengths (Fig. 3.2.8). Coal faces with this distribution of lithotypes are typically quite unstable.

b) The narrow-banded lithotype (thickness of layer 1 to 10 cm) profile of Carboniferous coking coal, also consists of intercalated layers of hard and soft coals but the physical properties of the entire coal seam in terms of density and hardness are like the monolithotype profile for hard coal. The very regular alternation of hard and soft coals is such that significant variations in the mechanical stability of the coal face are minor even in geological environments where thrust faulting is observed.

3. The erratic lithotype profile represents coal seams which contain irregular thicknesses and distributions of either hard or soft coal beds.

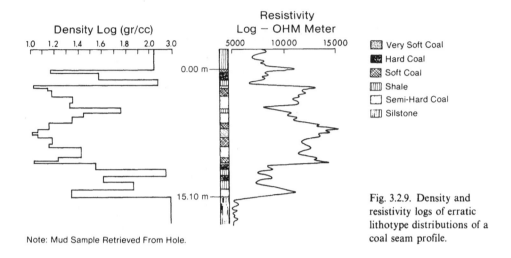

Note: Mud Sample Retrieved From Hole.

Fig. 3.2.9. Density and resistivity logs of erratic lithotype distributions of a coal seam profile.

    Further, erratic variations in the density and corresponding hardness of a profile is particularly evident when the coal seam contains a significant proportion of clastic (sedimentary) rock material (Fig. 3.2.9). Coal seams with erratic petrographic characteristics such as these are particularly unstable in underground mining operations if the seam has greater than 25 per cent elastic rock (e.g. sandstone, shale etc.). This instability is due to the very large differences in the elastic constants between the coal beds and the silicate rock material.

    Within multi-lithotype coal seams that have been deformed, it is common to observe that the black or less bright bands of coal (durain, clarain) are commonly folded and that the brighter harder bands are crushed. The differences in the way each type of coal reacts to similar tectonic stresses is attributed to the differences in their physical properties (e.g. density, strength etc.). Dr. Das has noted that the material forming slip planes in coal seams must have been ductile and that tectonic forces moved and strongly folded the coal above this plane without affecting the material beneath it.[9]

    The preceeding discussion indicates that knowledge of the density profile of a coal seam is an important parameter in evaluating the stability of a coal face. As with other physical phenomena of traverse anisotropy in mineral deposits, it is difficult to interpret and transfer the geological-structure data into quantitative terms applicable to all deposits. For this reason each major deposit should be examined individually to evaluate problems related to coal face stability. In the Western Canadian coal fields where geophysical logging is a common exploration technique, the data collected should be expanded to allow for the evaluation of coal face stability. At the present time there is insufficient research in this field particularly with respect to the concept of expressing or considering geophysical anomalies as potential mine design parameters.

3.3.   COAL RANK

Every step in the gradual transition from organic material to various coals can be traced from growing vegetation into peat, from peat into lignite, and so on into bituminous coal and into anthracite. Where the anthracitic coals have been affected by strong metamorphism, graphite is a common product.

3.3.1. *Coal rank formation*

The chemistry of 'intramolecular combustion' is only mentioned here, the main consideration is given to factors which might relate to the mechanical stability of coal seams which is an important element of strata mechanics.[10] Two groups of factors determine the rank of a coal. They are briefly discussed below:

1. Coalification events represent the processes which cause the transformation of vegetal matter to various types of coals. The important ones are listed below:

a) Biochemical processes of coalification result from the·alteration of plant material into peat and are due to microbial and associated chemical processes. Under these circumstances humic substances are formed which increase oxygen supply and temperature and cause the formation of an alkaline environment.

b) Diagenetic processes result in the conversion of peat into lignite (soft brown coal). This is due to the compaction and decrease in porosity of the peat caused by physical pressures resulting from the increase of overburden and by chemical processes which result in a greater than 60 per cent increase in carbon content of the organic material. The mining of lignite deposits is mainly done by large open pit mines, which are usually coupled with power plants. The stability of underground lignite mines is not well defined because of complex strata mechanics due to very thick seams, low strength of surrounding rocks and usually the presence of water-bearing horizons. The same general processes of diagenesis are known to convert dull, dark brown 'hydrogel' into much higher rank brilliant black 'bitumogel' which is a bituminous coal. Underground mines of this coal rank may be more stable due to the increased strengths of the coal and adjacent strata and the decrease in the thickness of coal seams (peat/coal ratio 3/1) as a result of the diagenetic processes.

c) Metamorphism is a key factor in the formation of high rank coals. The fundamental characteristic of these processes is that mobile products are given off (gas and crude oil) leaving solid residual products in the form of coal or keregon. Abrupt changes in the hydrocarbons present and an increase in the degree of bituminization are the main factors observed in the transition of sub-bituminous to bituminous coal. The rapid decrease in the content of hydrogen and volatile matter, and the increase of carbon and methane gas are the main factors observed in the transition from bituminous coal to coking coal, anthracite and meta-anthracite.

2. Coalification is the main factor required for formation of various coal ranks.[11] Different coalification relates to the different mechanical stabilities of coal seams. The principal causes of coalification are briefly listed below:

a) Overburden depth and related strata pressure retards chemical reactions and promotes 'physico-structural coalification'. However, the main role of the overburden depth is that it controls temperature (the geothermal gradient is $3°$ to $4°/100$ m) and static pressure (1 MN/m^2 per 45 m). For example at very low temperatures little happens to buried organic material, even over long periods of geological time. In the Moskow Basin (U.S.S.R.), the lower Carboniferous lignite deposit has never sunk to depth greater than that which corresponds to a temperature of 20° to 25° since its deposition. A similar case exists with regards to Permian sub-bituminous coal in the Collie Basin, Western Australia and the Plains region, Western Canada. These sub-bituminous and lignite coals of Carboniferous to Tertiary age[12] are commonly interlayered with rock strata which are not nearly so competent or lithofied as the coal units themselves. As a result the main problems in mine stability relate not to the coal seams but rather the adjacent rock strata. It is obvious that lack of overburden pressure resulted in insufficient compaction of the coal seams and rock strata as well as insufficient dehydration and recrystallization. These parameters influence the stability of coal faces because with increase of moisture content there is a decrease in the coal elastic constants and related self-supporting abilities.

b) Temperature should be considered as one of the key factors because it promotes chemical coalification which directly relates to coal rank. With a rise in temperature there is rise in the rank of a coal. Increase in temperature can be related to several factors. For example, as the subsidence of coal bearing strata increases, the temperature to which the coal is exposed increases, and bituminous coals are known to originate at depths between 2,000 and 4,000 m, equivalent to temperatures ranging from 60° to 160°C. Additional heat from large intrusive magmatic bodies are responsible for the formation of anthracite deposits. In such cases the required temperature is near 300°C, which is difficult to reach at the estimated depths of burial by normal geothermal heat alone. An example of such an anthracite deposit is the German Westphalian D deposit of Lower Cretaceous age. Finally, the promotion of coalification can result from frictional heat, due to tangential tectonic stresses where the temperature may rise from 600° to 800°C. Accelerated, strong tectonic shearing stress can cause an increase in coal rank at lower temperatures and this is probably the case for bituminous seams of the Lower Cretaceous age[12] in the Rocky Mountain Coal Belt, Coal Valley.

c) Time is also an important element which affects the degree of coalification of an organic deposit. The time factor is a critical parameter provided that the depth of burial and temperature of a coal deposit is correct for a sufficient coalification process to proceed. Under these circumstances an increase in time results in an increase in coal rank. For this reason the majority of coal seams of higher rank belong to older geological formations. The degree of coalification can be less if subsidence was rapid and the 'cooking' time short. Degree of coalification can be higher for short period of time if additional heat is contributed by a magmatic intrusion or friction due to tectonic movement. In these latter cases, the high rank coal is not sufficiently 'aged'. These coal seams often are weak and higher than normal moisture contents which contributes to their mechanical instability.

d) Tectonic pressure or fold pressure do not influence coalification except locally along faults and overthrusts. The promotion of coalification due to fold pressure is restricted to the place, where rock strata have been buried to very great depths and heated to high temperatures before folding commenced (Ruhr Basin). In most fold belts coalification has largely gone to completion prior to tectonic deformation. The vitrinite in coal seams where the coalification path is due only to overburden pressure anisotropy, is commonly orientated parallel to the bedding planes. In cases where tectonic deformation has occurred, the vitrinite layers are in an oblique position to bedding planes or have disintegrated. From the attitude of the vitrinite seams with respect to the bedding plane the direction of tectonic principal stresses can be determined.[12]

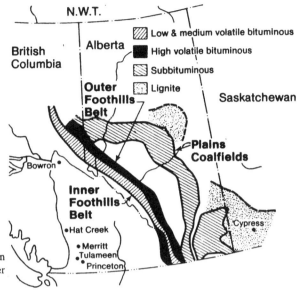

Fig. 3.3.1. Coal rank distribution in Alberta (thickness of coal seams greater than 1.5 m).

3.3.2. *Strength of coal due to rank*

Coal rank is known to vary both vertically within a coal seam succession and laterally on a regional scale (e.g.) 'coal belts'. A good example of the latter type of distribution is exhibited in Western Canada, where there exists a distinct correlation between physiographic features of an area (Fig. 3.3.1) and the coal rank found in formations in that region. They are listed below for Western Canada:

Inner Foothills Belt (Rocky Mountains Region) contains semi-anthracite (limited to Cascade Basin), low-volatile bituminous coal (Luscar formation – northern part) and medium-volatile bituminous coal (Kootney formation–southern part).

Outer Foothills Belt (Foothill Region) contains high-volatile bituminous coal.

Inner Plains Region (Plains coal field) contains sub-bituminous coal.

Outer Plains Region (extends into Saskatchewan) contains lignite.

Underground and laboratory investigations have established that the diagenetic and metamorphic processes which have contributed to the formation of the different coal ranks in Western Canada also have influenced coal strength. These are important for mine stability, coal preparation, coal fragmentation and other technological parameters.

Very extensive exploration by drilling of coal deposits in Western Canada in the last decade offers enough data so that at a very early stage of mine development it is possible to evaluate coal strength for a particular rank of coal as discussed below:

1. Excessive strength-anthracite rank could be evaluated on the basis of following exploration diamond drilling parameters:

Core recovery	< 90 per cent
Core breakage	In cylinders with height equal or greater than two core diameters (Fig. 3.3.2)
Fabric	Amorphous or with a very small contrast between bands

Fig. 3.3.2. Coal of excessive strength of semi-anthracite rank (Cascade Basin).

2. Moderate strength-high volatile bituminous and sub-bituminous rank could also be estimated on the basis of criteria used in (1):

Core recovery   > 75 per cent
Core breakage   In cylinders with cracks and with height less than two core
                                  diameters (Fig. 3.3.3)
Fabric   ·   Clearly banded

3. Low strength-medium volatile and low-volatile bituminous coal:

Core recovery   < 50 per cent
Core breakage   Broken cylinders and discs (Fig. 3.3.4)
Fabric   Great contrast between bands

4. No strength-sheared low volatile bituminous coal and weathered high volatile bituminous or sub-bituminous coal:

Core recovery   None, only sludge recovered with drilling fluid
Core breakage   Fines, down to 320 mesh (Fig. 3.3.5)
Fabric   Porous; with an increase of moisture there is exponential
                                  decrease of elastic constants (Fig. 3.3.6)

The classification of coal rank by strength criteria can not be generalized for all coal deposits. For example, coal of the same rank might have all four strength categories as given below:

– Uncleated anthracite-excessive strength,
– Cleated anthracite-moderate strength,
– Entonic anthracite-low strength,
– Porous anthracite-no strength (high moisture content).

Fig. 3.3.3. Coal of moderate
strength of high-volatile
bituminous rank (Coal Valley).

Fig. 3.3.4. Coal of low strength of medium-volatile bituminous coal (Crows Nest).

Fig. 3.3.5. Coal of no
strength of low volatile
bituminous coal—
(Stark-Key Mine).

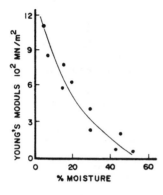

Fig. 3.3.6. Relationship between moisture content and modulii of elasticity (Stark-Key Mine).

As a matter of fact the classification of coal strength might be independent of coal rank because of the effects of such parameters as coal 'aging', post-tectonic movement, weathering etc., which seriously modify the initial value of this parameter.

The core recovery and core breakage of the coal could be considered as satisfactory parameters for evaluation of coal strength and this has been confirmed by laboratory testing. It is also apparent that the present geotechnical determinations and classifications of hardness of rock core samples which places emphasis on the number and spacing of joints, is not adequate, and in many cases irrelevant. This is particularly correct in relation to the determination of the rank of a coal from core samples.

3.3.3.   *Mechanical stability of coal rank*

The mechanical stability of coal seams and their behaviour under load is of primary interest for the analyses of strata mechanics and ground control. As mentioned before, it is impossible to generalize coal properties and behaviour with coal rank and most likely these mechanical phenomena should be considered for each coal province individually. For example such consideration has been given to Western Canadian coal fields where the properties and behaviour of individual coal ranks have been classified from exploration drilling data. These field data and observations were supplemented by core sample testing in the rock mechanics laboratory. Large diameter (76 mm) core samples of various lengths (dependent on coal strength) were cutted and polished with the ends parallel to 0.25 mm. A total of 48 coal samples were classified according to coal rank and related strengths. Only two coal samples of low strength coal were prepared, and only one sample of the minimal strength, because the rest of the 21 coal samples of these two types broke during their preparation. On the basis of laboratory testing the three following criteria have been used to characterize the mechanical stability of coal seams of different coal ranks: coal strength, stress-strain diagrams and character of failure.[14] Data of this type for each major rank of coal is presented in Table 3.3.1.

The definition and classification of the coal ranks and their relationship to mechanical properties and behaviour of coal seams has importance because it might offer a qualitative estimation of the strength and mechanical stability of coal seams in a

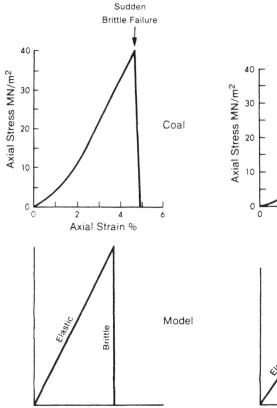

Fig. 3.3.7. Stress-strain diagram of coal of excessive strength.

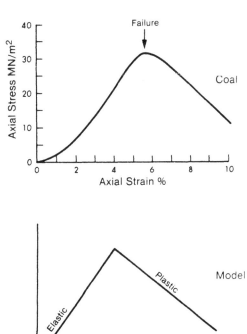

Fig. 3.3.8. Stress-strain diagram of coal of moderate strength.

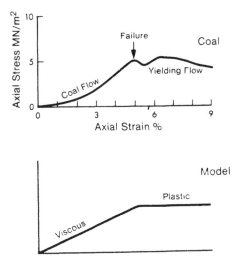

Fig. 3.3.9. Stress-strain diagram of coal of low strength.

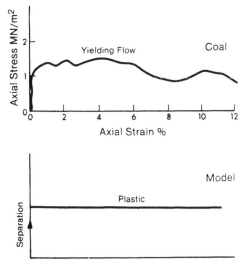

Fig. 3.3.10. Stress-strain diagram of coal of almost no strenth.

Table 3.3.1. Mechanical stability of coal seams of different rank

Coal rank	Density of coal (gamma-gamma log)	Field observations — Borehole phenomena	Laboratory investigations — Compressive strength MN/m²	Stress-strain relationship	Failure
Semi-anthracite	Very dense coal: 1.5–1.8 g/cm³	Underground noise effect (similar to the tander-storm), seismic effect (easy to record): large amount of gas release (bubbling flow of water at collar). After outburst, the hole is in collapsed state and its further drilling is abandoned	20–40	Elasto-brittle (Fig. 3.3.7)	Violent release of strain energy
High volatile bituminous, sub-bituminous coal	Dense coal 1.3–1.5 g/cm³	The borehole experiences enlarging its diameter (recorded by Caliper log) and larger content of the coal in drilling fluid (coal falling down in hole has been fragmented and lifted to surface by fluid circulation). No difficulties during exploration drilling	7–18	Elasto-plastic (Fig. 3.3.8)	Yielding with gradual release of strain energy
Low and medium volatile bituminous coal	Porous coal 1.1–1.3 g/cm³	The borehole experiences closure deformation which can be very intensive if drilling was stopped in such a layer for a short period of time (for example 1 h). In this case, further drilling cannot be commenced because of closure of hole filled by soft coal	1–17	Visco-plastic (Fig. 3.3.9)	Shear flow extrusion
Low volatile bituminous coal	Very porous coal: 1.0–1.1 g/cm³	The borehole experiences blockage difficulties, because when coal layer is intersected the shear forces are activated, causing differential displacement. The hole under these conditions can be lost before the coal seam intersection has been drilled out	0–1	Plastic-friction (Fig. 3.3.10)	Fracturing, desintengration

mine structure.[13] In this section simple criteria have been suggested for the evaluation of coal seam rank and stability on the basis of geophysical borehole logging (gamma-gamma density log) and visual observations of borehole stability. These data have been supplemented with laboratory testing to determine the coal core strength, behaviour and failure. At the present there is insufficient research on the application of geophysical logging techniques relating to the evaluation of coal structure stability. This work should be significantly expanded. Particular emphasize should be placed on exploring the use of data from geophysical anomalies as mine design parameters.[12]

Finally, determination of the mechanical stability of a coal seam profile during the evaluation stage of a resource potential, is of the paramount importance. For example, the rock mechanics factors might be unfavourable for underground mining, regardless of optimal coal reserves and quality. Of equal importance however is that a knowledge of the mechanical characteristics of the coal seams in an optimal mining operation, can be used to avoid at the earliest stages of development costly mine design problems and/or errors.

3.4. CLEATS IN COAL

The presence of cleats or structural defects in coals seams is promoted by a variety of factors which include the depositional mechanics, coal petrography, coal rank, orogenic movement, and mining activities. These same elements are also the factors which control the distribution, density, and extent of the joints in coal seams.

The cleats are mainly exhibited by the major joint systems elongated along and across the face of the coal seams, and also be weaker and more randomly oriented fracture planes with orientations different to the strike and dip of seams.

The presence and orientation of the cleats in coal seams influences the choice of mine layouts, the direction of coal extraction, the application of coal excavation technologies etc. For this reason, cleat development can be a factor which affects coal mine stability, particularly if the cleat is equally well developed in the strata of the mine roof. For mine stability analyses, it is necessary to know the nature of the coal seams, the fabric of the cleated coal, density of cleat material along the seam profile, and finally the shear strength of the cleated joints. All of these topics are important to mine stability and are discussed individually with illustrations, in the following pages.

3.4.1. *Origin of the cleat*

The origin of the cleat is described in great deal by many authors, [15, 16, 17] however determination of the origin of jointing in local seams within mine structures is not an easy task because of the complexity of structural features and the variations of mining activities. Generally speaking three types of coal cleats can be identified; each is briefly described below:[15]

1. Endogenous cleavage is mainly oriented perpendicular to the bedding planes of coal seams. The individual joint planes tend to divide the coal beds into a large number of thin tabular fragments. They are formed by internal forces which arise due to the drying and shrinkage of organic material as well as its compaction and release of volatile

Fig. 3.4.1. Outcrop of cleated coal (Wabamun Open Pit Mine).

Fig. 3.4.2. Exogenic cleat formed by folding (Number 4 seam, Grande Cache).

matter. It is difficult to predict the orientation of these planes conclusively, but they form along preferred planes at random, where the net effect of the resultant stresses (due to the above parameters) and the resistance to separation (strength) is most favourable for their formation. It is certainly correct that chances for identification and definition of the orientation of endogenous cleat are much greater in coal seams which have not been effected by tectonic stresses, than in the case of tectonically disturbed seams. For example, since the coal seams in the Plains Region of Western Canada are horizontal in attitude (non deformed), the cleat in these seams can be considered to be of endogenous origin. These coal seams exhibit two intersecting sets of vertical joints whose orientations are given below:

a) Cleat in the N-E direction, which is perpendicular to the Rocky Mountain Range. This is the principal cleat and is clearly visible.

b) Cleat in the SW direction, which is parallel with the Rocky Mountain Range. This is a secondary cleat (Butt cleat).

In this example it is obvious that the orientation of the endogenous stress field was influenced by the neighbouring mountain building process regardless of the distance that the seams are from the orogenic event or their shallow depth. Such shallow coal seams are often affected by glaciation which tends to enlarge the dimensions of existing joint planes in length and width (Fig. 3.4.1).

2. Exogenic cleavage is formed by external forces related to tectonic events (Fig. 3.4.2). The mechanics of formation of exogenic cracks depends on the character of faulting of the coal bearing strata.[16] This cleat is oriented in the direction of the major principal stress and usually consists of two sets of parallel joints, inclined to each other. The investigations of cleat development and the orientation of coal seams in the Rocky Mountain Belt show the following:

– Original-endogenous cleats were destroyed by the formation of tectonic cleavage.

– The orientation of tectonic joints is governed by the geometry of folding and fault structures.

Strongly deformed coal seams (due to tectonic events) exhibit erratic and irregular cleat distribution. This can cause deterioration in the strength of a coal seam profile at certain points, for example, at the crest of a bed where fracturing is at a maximum (Fig. 3.4.3), or along the whole seam profile. Investigations of coal seams in Western Canada suggest that there is a proportional relationship between intensity of tectonic stresses, degree of cleat development, and the deterioration of coal strength.[17]

3. Induced cleavage does not exist in virgin coal seams because it originates during mining operations due to load transfer onto mine structures. The formation and frequency of induced cleat depends on the mine layout and type of mining technology. This cleat type cannot be easily distinguishable from endogenous and exogenic jointing. The major criteria used to distinguish induced or mining joints is their larger propagation and greater opening particularly for lower rank coals which are susceptible to weathering (penetration of oxygen and moisture is favourable through these cracks). The joints are formed mainly in the outer zone of the coal faces where restrain (tensile stress) is in effect (Fig. 3.4.4). This is in contrast to the inner constrained zone (compression) where virgin coal fractures are closed and induced fracturing is very limited.

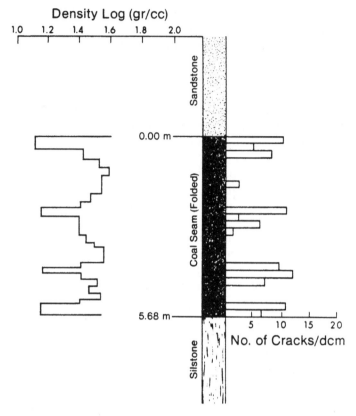

Fig. 3.4.3. Density
variations due to orogenic
processes.

The exposure of cleavage on coal pillar faces might depend on their origin. For example, endogenous cleat is usually exhibited on both coal pillar faces, but exogenic and induced cleat is on only one coal face, which is parallel to the main principal stress. The intersection of the planes which define the coal faces and cleat is of great importance in the analysis of coal mine stability. It is essential that the orientation of the working face of a coal seam be such that its intersection with cleat cause minimal adverse effects on the stability of the mining face.

3.4.2.   *Cleat fabric of coal seams*

For underground mining studies the origin of cleat in coal is certainly less important than the pattern of fractures in the coal seams which affect the stability of the coal face.[18] The relationship between the pattern of fractures or joints and the bearing capacity of a coal structure becomes increasing important as the thickness and pitch of a coal seam increases. Generally speaking, there are a large number of types of structural patterns of the cracks which might have different orientations to the strike and dip of coal seams (Fig. 3.4.5). With the engineering point of view, using to a certain degree some standards from rock engineering, five patterns of cracks are proposed.

Fig. 3.4.4. Induced cleavage by mining operation (Canmore Mine).

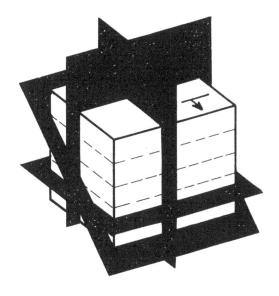

Fig. 3.4.5. Model of the cracks orientation to coal seam structure (Number 11 seam, Grande Cache).

Fig. 3.4.6. Cubical crack pattern of coal seam
(Lethbridge Mine).

1. Cubical cleat fabric is well exhibited in coal seams where endogenous cleat is present. This fabric is common in geologically young (not 'aged' seams) coals and coals of lower rank. The cubical fabric (Fig. 3.4.6) has particular influence on the stability of the coal pillar. In this case a model of a joint system develops an orthogonal anisotropy to the pillar faces. The influence of this structural pattern on pillar stability depends on several factors as listed below:

a) The uniformity or heterogeneity of cubical fabrics. For example, the pillars are more stable with heterogenous cubical fabrics.

b) The size of coal cubics is directly related to pillar stability. For example, coal from the Rocky Mountain Belt has cubical fabrics in sizes below 2.5 cm and has a very low strength. With an increase of the size of the cubical, the strength of the pillar increases exponentially.

c) The mode of pillar failure is not rapid, rather it is a progressive failure. The failure starts with one or a limited number of cubicals and the proceeds with more of them failing simultaneously.

The observed deformation and failure of pillars with cubical fabrics suggest gradual failing mechanics with a rapid increase of deformations. It has been established that when the percentage of deformations is large (greater than 25 per cent), the pillars still support the load. Even at total pillar collapse, it still supports the load which is directly superimposed above it.

Fig. 3.4.7. Pattern of
laminated fabrics of durain
and vitrain layers.

2. Laminated cleat fabric is primarily exhibited in coal seams which have alternate beds of hard and soft coal. In this case laminated crack zones are errected by exogenic processes, where hard coal layers were stationary and soft layers were sheared over them.[19] Under these circumstances hard coal layers are without or with limited cleat development of endogenous origin, but soft layers are extremely well cleated due to coal fracturing during shearing (Fig. 3.4.7). Underground phenomenological studies of

Fig. 3.4.8. Pattern of discontinuity crack pattern.

the pillars showed that the laminated fabrics of coal governs the stress concentration and its orientation.

3. Discontinuous cleat fabric of coal seams can be related to endogenous and exogeneous processes of crack formation. For this particular cleat fabric, it should be pointed out that discontinuous fracturing is along planes of stratification and at vertical or steep angles to these planes. Such a structure exhibits joints which are bridged by solid coal (Fig. 3.4.8). Discontinuous joint patterns exhibit wide variations in spacing and intensity of fracture propagations. The influence of this discontinuous pattern on the coal pillar stability is exhibited by the facilitation of shear failures along ribs.[20] The shear stress concentration is below the shear strength of solid coal (coal bridge), so that shearing takes place only along discontinuous joints. Under these

Fig. 3.4.9. Pattern of
continuous crack pattern.

circumstances small coal blocks are formed which are delineated by joint surfaces, bedding surfaces, and outer pillar surfaces. Visual observation of yielding pillar deformation suggests more fragmentation fracturing, rather than possible coal slabbing deformation.

4. Continuous cleat fabric is mainly related in origin to tectonic and mining stresses which propagate existing joints and destroy coal bridges between them (Fig. 3.4.9). Continuous joints of this type exhibit a variable geometry. Outer pillar zone weakening is by slabbing in the case of smooth continuous surfaces and by block falling in the case of rough continuous surfaces. Maximum pillar weakening is achieved in cases where shear displacement occurs along bedding or small thrust planes.

5. Blocky cleavage fabric is primarily caused by tectonic events, where shearing along bedding planes and at low angles to bedding planes is the primary pattern of fracturing. Under these circumstances the coal is fractured to form rhomboidal blocks (Fig. 3.4.10). In this case the coal face is unstable even before excavation is started. Research on the deformational behaviour of coal with this unique cleavage pattern has been carried out through investigations in Western Canada. In this region, the coal deposits have been intensively deformed within the Cordillerian thrust belt. This orogenic event deformed coal bearing strata which prior to tectonism was already structurally complex and erratic in its distribution. This combination of primary and secondary processes is responsible for locking the coal seams into folded and faulted structures as well as the

Fig. 3.4.10. Pattern of blocky type cracks.

deterioration of their original strength, fragmentation of individual coal beds or seams, and their displacement along adjacent beds. Due to tectonism of this type coal masses in some deposits behave much like viscous bodies where extrusion deformations are predominant.

3.4.3.   *Strength of coal joints*

The evaluation of the strength of coal joints has been carried out for coal types in various Western Canadian coal fields. In this case the testing program has been concerned primarily with shear strength characteristics of plane weaknesses encountered in the sloughing deformation of coal pillars.

The samples for shear testing were cut from coal blocks by a circular saw. In order to fit the dimensions of the shear box precisely, they were cast in DEVCON B Liquid Steel Epoxy. The shear box apparatus used was a high capacity type built at the Civil Engineering Department, The University of Alberta, where the tests were carried out. The normal load was produced by a weight lever arm while the shear load was generated by a chain driven gear box. The rate of shearing was held constant for all tests at 0.125 cm/min and the maximum shear displacement was unlimited for all samples. The samples were incrementally loaded. During a test the shear load and corresponding shear displacement were recorded directly on X-Y recorder.[21]

The tests have been carried out on coal sample exhibiting three types of crack defects: joints with coal bridges, joints without coal bridges, and open joints, which are typical for deformation of the outer zone of coal pillars. Individual specimens were tested several times (usually 2 to 3 times) under various normal loads. The results from the tests for each crack unit are represented separately by graphs. For each test the peak value of shear resistance has been chosen as the highest point reached during shearing. The peak shear stresses are plotted against corresponding normal stress σ_N. All peak values are plotted in a normal stress-shear stress diagram.

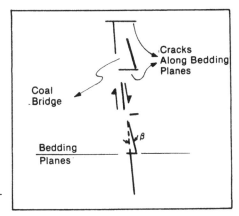

Fig. 3.4.11. Model of discontinuous cracks of potential shear failure.

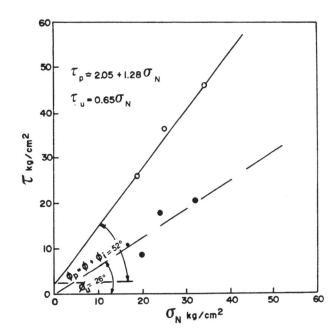

Fig. 3.4.12. Shear strength of discontinuities parallel to coal face.

Testing results are given separately for each engineering classified crack pattern, which are more or less parallel to the coal faces.

1. Joints with coal bridges are the most common category in many coal basins throughout the world. The model of such a joint is illustrated in Figure 3.4.11. Shear failure takes place along joints whose angle of inclination β varies in the seam profile and results in the formation of a steep surface transverse to the coal face.[22] The testing data for coal samples exhibiting only endogeneous cleat showed that the relationship between shear stress and displacement clearly exhibited the "peak-ultimate" type of

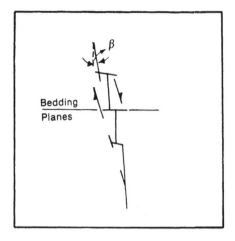

Fig. 3.4.13. Model of continuous cracks of potential shear failure.

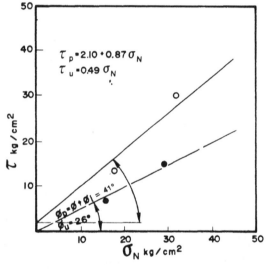

Fig. 3.4.14. Shear strength of continuities parallel to coal face.

behaviour. From the diagram in which normal stress versus ultimate shear stress is plotted (Fig. 3.4.12), it is noted that there is an appreciable change in the value of the frictional angle of the "ultimate strength" envelope, from approximately 26° to roughly 52° respectively. It is obvious that the "peak" shear strength of a coal with a discontinuous crack pattern is significantly influenced by the existence of coal bridges which have to be over-ridden or sheared off as sliding commences, and that the frictional component (sliding of two surfaces) is probably of secondary importance.

2. Joints without coal bridges are most common in the Western Canadian coal fields as a result of tectonic deformation. A model of a continuous cleat pattern is illustrated in Figure 3.4.13. The shearing along continuous joints results in the formation of coal

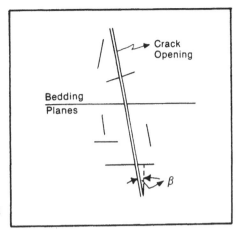

Fig. 3.4.15. Model of propagated crack of potential shear failure.

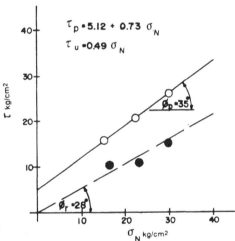

Fig. 3.4.16. Shear strength of open joints parallel to coal face.

slabs with a steep failure geometry, which is governed by the value of the angle of inclination β. However, the steepness of the path varies considerably for individual coal faces, due to variations in pattern spacing, inclination, and the connection of joints. The testing data of coal samples which also exhibited exogenic cleat, showed that they have a lower shear resistance for 20 to 25 per cent of the discontinuity joints (Fig. 3.4.14), which is possibly due to the crushing of interlocking joint surfaces during shearing (over-ridding). The geometrical shear component is more important than the frictional component.

3. Open joints are developed during mining operations and are due to induced mining stresses. Generally speaking they are propagated along coal faces at a steep angle or they are vertical (Fig. 3.4.15). In this case there does not exist a rough geometry of asperities between joints, because planes of stratification have very little influence on their formation. The shear strength of these joints is further decreased as illustrated in Figure 3.4.16. This type of joint pattern has a major influence on underground mine stability. For example, instability increases with progressive slabbing along these joints, which are usually developed in an echelon pattern.

In summary, it is the shear strength of coal joints which control the stability of coal faces and coal pillars and for these reasons research on this topic deserves adequate attention.

CHAPTER 4

Roof and floor strata

The mechanics of deformation and failure of the roof and floor strata varies widely due to rock types and mining technology. Roof and floor strata stability is governed by the lithology of rocks and their structural defects, as listed below:

1. Weak rock strata stability became an increasingly important element of mine stability, because it hinders mechanized and rapid mining techniques.

2. Competent rock strata exhibit satisfactory stability, but in the case of excessive strength it could be subject of violent and sudden failure.

3. Structural defects of roof strata are of particular importance for strata mechanics, roof control and block caving geometry.

4. Competent subcoal strata is essential for overall mine stability as well as for interaction with coal pillars and external support.

Each of these topics is discussed individually, giving certain attention to practical mining applications.

4.1. WEAK ROOF STRATA

The lithology of weak roof strata plays an important role, particularly with the aspect of reinforcement of the mine roof and its external support. Weak roof strata in interaction with underground mining excavations, exhibit appreciable displacement and rapid failure after coal removal. Excavation in this case does not favour mine stability. However, weak strata exhibits optimal caving and flushing characteristics in the gob area, contrary to the first case, this is favourable for mine stability.

It is interesting to note that rock materials with uniaxial compressive strengths less the $5\,MN/m^2$ cannot be truly classified 'as rocks' and should be considered as a soil. Within the aspect of mechanics of deformation and failure weak rock might be considered as a transition between solid and soil.

4.1.1. *Roof displacement*

The analyses of roof displacement of weak rock were approximated from the concept that a mine roof represents structures of the beams or plates, which are loaded vertically by their own weight and the weight of overlying strata. Under these circumstances, the roof strata displaces vertically and horizontally due to internal tensile stress. The

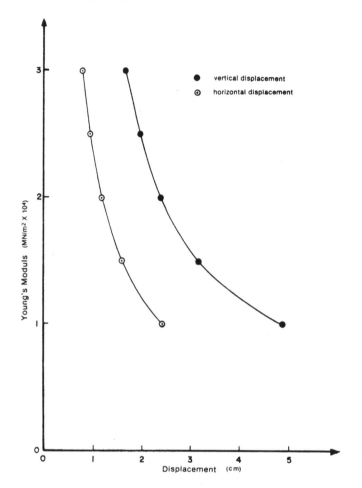

Fig. 4.1.1. Modulus versus maximum displacement (boundary element displacement analysis).

relationships between vertical and horizontal displacements are a function of the modulus of elasticity, and have been studied by digital model analyses (displacement discontinuity method developed by Crouch), as illustrated in Figure 4.1.1. From this study it is obvious that magnitude of displacement increases with an increase of rock weakening (lower Young's modulus), which has been confirmed by underground monitoring of vertical roof displacement at Star-Key Mine (near Edmonton, Western Canada). Unfortunately, the mathematical model lacks versatility because intensity of displacement cannot be modeled with respect to time. For example, monitoring showed that hourly displacement of very weak rock (clay-shale) is approximately the same as daily displacement of moderately weak rock (clay-sandstone). For a roadway 3 m wide and 1.8 m high, the maximum vertical displacement (before failure) was 7.8 cm and horizontal displacement was 1.6 cm, suggesting a ratio of 5 for vertical versus horizontal displacement. Weak rocks have no self-supporting roof spans and displacement analyses are only possible for supported mine openings. Representations of this mechanical phenomena, discussed below, are from two case histories in the Leghbridge area:

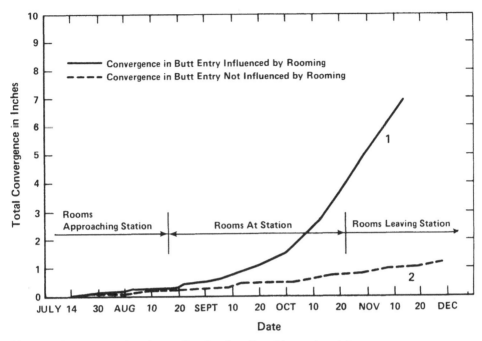

Fig. 4.1.2. Convergence of roadways affected and unaffected by nearby mining.

1. Galt Number 8 Mine has been mined-out at shallow depth up to 80 m, for several decades up until the 1970's. This mine had been the subject of investigations by CANMET personnel in the 50's and 60's. Roof displacements were studied for two particular cases of support as briefly given below:

a) Roof supported by two leg timber sets were monitored in the 50's for roadways of approximately 2.8 m width and 1.6 m height.[2] The monitoring was related to the convergence measurements. This instrumented data showed some roadways were affected by coal extraction and others were not as illustrated in Figure 4.1.2. The rate of convergence did not exceed 2.5 mm per day for the roadway unaffected by coal extraction. The roadway affected by nearby mining showed an increased rate of convergence as coal extraction approached the observation station. It was observed that increased movement in the roadway started when coal extraction was very close to, or at, the monitoring station, although that convergence accelerated when coal extraction passed the station. It was obvious, that a weak roof readily falls in a mined-out area, shortly after coal removal. Under these circumstances the roof strata was de-stressed simultaneously with coal extraction, and load transfer ahead of the extraction face was limited. Increased roadway convergence behind the extraction line was due to strata movement from gob settlement rather than the transfer of mining stresses.

b) Roof support by rock bolts has been investigated on two occasions. In the 50's the strata reinforcement had been carried out by expansion roof bolts with metal roof plates. These were 1.1 to 1.4 m in length, installed across a 2.8 m wide crosscut. The roof bolts were installed at angles up to 45° over timbered ribs with four in a row at 0.55 m spacing. Timber support was then removed and after a period of 18 months of favourable support a second set of roof bolts were installed in a main roadway. After a duration of about one year convergence of the bolted roof span did not exceed 2.5 cm. Roof bolting in the 50's by shur-grip bolts of 1.25 m length was not successful, either for support of entry immediately after coal removal or entry which had already been driven 10 to 15 m. The mine roof defined as well stratified thinly laminated thickness 15 to 20 cm, with horizontal beds of very soft shale, after removal of timber failed immediately along a line of bolts, and later the rest of the roof strata caved overriding the bolts.

2. Petro-Canada experimental mine–depth up to 250 m, was opened in 1980 to obtain more knowledge about displacement and support of weak strata. The knowledge obtained depended a great deal on the type of support installed, and is briefly discussed below:

a) Patterns of vertical bolt installation have been investigated by application of three fundamental types of rock bolts: resin anchorage, high strength mechanical and split set stabilizer. The generalized pull test results of these rock bolt types are illustrated in Figure 4.1.3. The pull test at Lethbridge Mine Number 6 on 20-mm 1.8 m long threated rebars, grouted with five 2.8cm diameter catridges of Celtite polyester resin, showed that the bolt was capable of resisting 55 to 83 MN/m² before a steady yield was indicated. This type of rock bolting was considered satisfactory for reinforcement of a mine roof. Regardless of the rock bolt capabilities, when a roof beam was in tension, the roof displacement became appreciable, particularly in the region of bed separation and roof sagging (Fig. 4.1.4). It was obvious that under these circumstances, the rock bolts could do nothing to resist convergence except hold the weak rock together, this eventually failed in the same manner as the 1960's CANMET experiments.

b) Patterns of inclined rock bolts over the ribs and vertical bolts in a roof span area, similar to the experiment carried out in the 50's by CANMET personnel, showed satisfactory roof support. Under these circumstances the experiment could be extended for roof support of width up to 6 m, as illustrated in Figure 4.1.5. For a weak rock strata

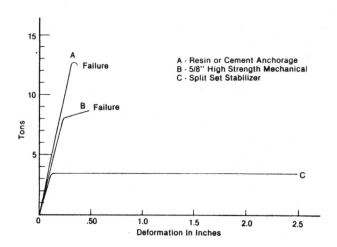

Fig. 4.1.3. Generalized diagrams of the pull tests of rock bolts.

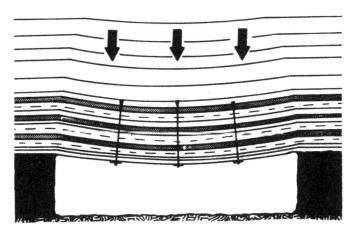

Fig. 4.1.4. Displacement of a rock beam reinforced by vertical rock bolts.

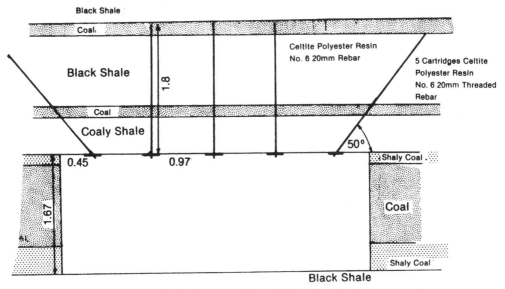

Fig. 4.1.5. Bolting pattern of inclined and vertical rock bolts with polyester resin grouting.

rock bolting with fast setting polyester resin grout is essential for strata reinforcement. The measured convergence showed maximum vertical displacement up to one third more than in the first case for a roof span of half the length.

c) Roof truss support showed that an upward truss prevents rock beam buckling and displacement (Fig. 4.1.6). In these conditions horizontal compression prevents lower layers from pulling apart, because the truss roads are pulled into high tension during installation and takes a tremendous force from the roof to stretch them farther. As long as anchorage of roof trusses hold steady over coal pillars the roof will overcome the strength of the trusses and stretch the steel before roof deflection begins. Maximum convergences below trusses have been reported below 1 cm.

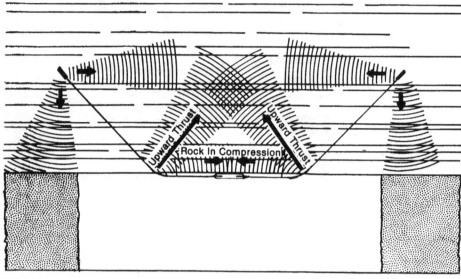

Fig. 4.1.6. Roof truss support against displacement of a 'rock beam', with stress diversion into the coal pillars.

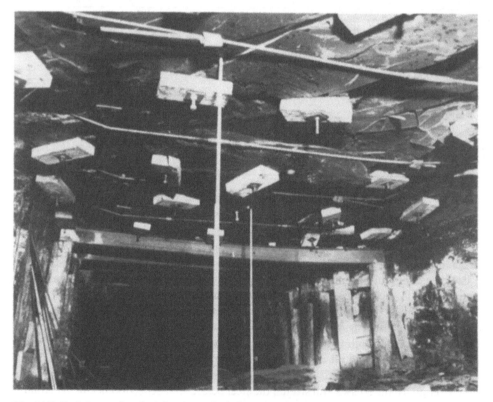

Fig. 4.1.7. Techniques of roof reinforcement, including I-beam frame support.

d) Steel frame I beams and wooden props exhibited maximum convergence, as can be visualized from photography (Fig. 4.1.7). The maximum vertical displacements were recorded over 5 cm for a period of approximately six months. It was obvious that there had been bed separation above the support and a dead load rested on the beam.

The comprehensive representation of roof displacement and convergence aims to explain strata mechanics of weak rock giving consideration to their reinforcement. The old mining technology with cyclic operations done by manual work, and roadways of width up to 3 m have been satisfied by timber support. Continuous mechanized mining requires elimination of external support by timber, because increase machine mobility and intensive roadway advancing cannot be followed by this type of support. Under these circumstances roof bolting is required, and the most likely solution would be roof truss support with polyster resin grouting.

4.1.2. *Roof span failure*

Roof span failure of weak rock strata has been investigated in several coal mine throughout world. Generally speaking, roof falls could be related to two stress mechanisms which have been investigated in Star-Key Mine. The underground phenomenological study suggestions are briefly listed below:

a) Tensile failure occurs in the roof of the extraction stopes, where the length of the

Fig. 4.1.8. Diletation (separation) and shearing of roof layers.

Fig. 4.1.9. Uncaved part of a mine roof after the timber was withdrown from the room.

roof span is up to 12 m. The observed mechanics of roof collapse follows a classic example of beam failure in extension, as briefly listed below:
– Dilatation of the beds;
– Shear displacement of the beds;
– Sag of the beds;
– Pulling apart of rock beam at the middle;
– Breaking of the beam at the rib edges;
– Roof collapse.
The underground observations suggest that strata diletation starts where bounding between layers is minimum, as between the left coal layer and the weak bed of sandstone (Fig. 4.1.8), and upwards by diletation for other weak rock beds.[3]

b) Shear failures have been observed in the roof of roadways, with a span length up to 3 m. The observed phenomena of roof falls suggest the following:
– Instead of pulling apart in the middle of a roof span the beam is broken by a shearing force at the rib edges;
– Rock collapses as a continuous beam and does not initially break in the middle;
– The roof fall is sudden and rock is well fragmented due to the impact force.
In the Star-Key Mine, protection against roof collapse has two legs (entry) or three legs (room) timber supports. However, the most efficient roof reinforcement is exhibited where partial roof falls occur in the mined-out rooms, and one part of the mine roof in the vicinity of the rib still stands up (Fig. 4.1.9). Similar phenomena have been observed in some roadways which stand up without any support for 20 years. Underground and laboratory investigations showed that the optimal self-supporting ability of a roof span

is reached if the thickness of the left coal layer is greater than 0.3 m and its uniaxial compressive strength is greater than 25.0 MN/m².

4.1.3. Roof caving

Roof caving phenomena have been studied for flat coal seams, because at present there are very limited numbers of pitching coal seams in weak rock strata. Representative roof caving was also found from underground investigations at the Star-Key Mine. Roof falls after coal removal ranged in time from a couple of hours to several days. The lag time before a roof caved depended on the length of roof span and variation in rock lithology. A fair statement would be, that a roof cannot stand up without support, because it has no ability to sustain its own load.

The main characteristic of roof caving in the extraction area is as follows:

a) It is complete, forming a cupolated (cup shaped) roof cave,

b) No remnants of the uncaved roof in gob area remain, this is similar to the formation of cantilever beams,

c) Flushing characteristics of the rock material are well exhibited during caving processes,

d) The rock fragmentation and compaction in the gob area is more than satisfactory, as illustrated in Figure 4.1.9.

It is obvious that listed caving characteristics are not optimal for strata control in room-and-pillar and shortwall mining, because of the existence of exposure of some

Fig. 4.1.10. Mine roof reinforcement by leaving coal layers (Western Collieries, Australia).

portion of the roof without support for a certain time before the support is installed. The described characteristics of roof cavibility could be optimal for longwall mining with a self-advancing shield support. The readiness of roof caving with good flushing characteristics immediately after support advance, might be beneficial for de-stressing the coal face and removing small builds up of abutment stresses.

Finally, it should be pointed out that mining in an incompetent rock environment, could cause great difficulties in strata control, particularly if water accumulation exists above mineable coal seams as it does in Western Collieries (Australia) and Moskow Basin (U.S.S.R.).[4] For example, successful underground mining in Western Collieries has been done only in thick coal seams in which a coal layer 0.5 m thick was left as support of weak strata and the board-and-pillar system was applied. In this case the coal pillars were left intact to protect roof strata against fissuring and caving (Fig. 4.1.10). For example in the 60's the Muja Coal Seam was very successfully mined until an exploration drillhole intersected a water bearing horizon after which the mine was flooded. The recovery of the coal seam became impossible by underground mining, and so its recovery started in the late 70's by an open pit operation.

4.2.   COMPETENT ROOF STRATA

Competent roof strata should sustain load and stand unsupported for a certain period of time after coal removal. The majority of underground coal mines have competent rock strata as a mine roof. Deformation of the mine roof for various mine structures depends on the stress state of the strata and could range widely from yielding to violent and sudden failure. A similar statement could also be applied to roof caving which ranges from good to very poor.

The rock lithologies and their internal structures above the coal seam are significant factors in mine stability, particularly with respect to their strength properties. For a determination of the transverse anisotropy of rock strata profiles there is a very useful tool called a borehole hard rock penetrometer. This unit, developed in Poland, is synchronized for automatic recording of rock hardness, and contributes to the evaluation of roof span stability relating to the strength of individual rock units.[5]

Due to the difference in mechanical strength, properties of individual rock layers above a coal seam and their different response to mine excavation three particular rock sections should be delineated:[6]

– A false roof, consists of thinly bedded rock and weak rock usually intercalated by coal bands. The thickness of a false roof has an influence on the choice of mine support and corresponding mining systems;

– An immediate roof, consists of competent strata which could be either above the false roof or if a false roof is absent it is directly exposed. With the aspect of mine support and strata cavibility, this roof is of primary importance, and following discussion mainly considers this roof sequence;

– A main roof is above an immediate roof and is usually represented by massive and strong strata that does not break or bend until all of the coal is excavated by the applied mining system. This sequence of roof strata controls vertical extensions of roof caving due to coal removal, as well as surface subsidence if the intact coal pillars are left behind.

4.2.1. *Roof displacement*

The roof movement of competent rock strata follows the same mechanics of vertical and horizontal displacement as discussed for weak rock beds, where intensity of deformations depends on the rock modulus of elasticity. However, the majority of calculations of 'rock beam' deformation, based on mechanics of solids, do not fit the actual displacements in the mine. The most difficult task of underground mining is to predict roof falls on the basis of the critical displacement of strata. For example, some detecting systems set for critical displacement of mine roofs in several South European coal mines, either failed to give warning before roof falls or gave warning when the roof did not fall. The reasons for this are that there are wide variations of elastic constants of rock strata due to the nonhomogeneous mineralogy and the heterogeneity of geological structural defects.

Coal mines without any roof detoriation and without requirements for roof support are very rare. For this reason a competent roof span should be considered as interacting with an artificial support such as primary rock bolts. Under these circumstances the primary interest are pull out tests of rock bolts (relationship between load and a deflection of bolt), which indicate reinforcement capacity and roof span stability. The roof displacement could be governed by the type of applied rock bolts, mine layout and the lithology of the rock strata. This phenomena is described in great deal with respect to design implications by Dr. S. Peng, and further consideration of it is not necessary.[7]

The mechanics of displacement of hard roof strata (reinforced by rock bolts) in interaction with weak coal pillars and ribs, and in a state of yielding deformations is of particular interest. These phenomena have been studied in great deal on Number 4 coal seam at McIntyre Mining Company, where coal exhibits very low strength and inelastic behaviour. The early studies of roof convergence have been directed toward evaluation of roof sagging deformations as well as vertical and horizontal displacements of roof strata by introducing bed separation and flexural concepts. Roof convergences in mine entries (5 m width; 2.5 m height) of 0.5 m have not been rare. It is obvious that such a deformation could hardly be related to displacement due to roof sag, this has been confirmed by physical model investigations. The model simulated rock beams, which corresponded to the massive sandstone roofs of mine entries. The extrapolated data of flexural deformation from physical modeling to rock beams are as follows:
- Elastic limit of deflection up to 0.5cm
- Viscous limit of deflection up to 10.0cm
- Beam flexural failure at 10.0cm

After this experiment, which suggested a convergence at failure five times lower than recorded underground, the solution to this phenomena had been labeled as an interaction between strong and hard rock plates with the roof strata and yielding pillars in a state of viscous deformation. Following this, monitoring of subsurface subsidence of the main roof of Number 4 seam from Number 10 Mine indicated that roof strata gradually moved downwards without losing continuity, at a rate which corresponded to the collapse of coal pillars. It was noticed at this time that mechanics of roof convergence could be used to indicate yieldability of coal pillars and their state of stability. Further studies suggested that displacement of roof plates which rested on yielding pillars have similarities with the mechanics of ice flotation.

Fig. 4.2.1. The break of timber caps due to vertical displacement and horizontal slips of roof strata.

The predominant movement is a vertical displacement of the roof strata and has varying rates. For example at Terrace Cliff Mine, the rate of vertical movement accelerates after coal removal, resulting in breakage of newly installed timber. The observations of failure geometries of the timber support suggest that sag deflection of the roof strata is not present and that instead there are sharp vertical movements (sink) and limited horizontal slips of the exposed roof layer from the yielding pillars to the entry, e.g. in a direction of relative lateral movement of the pillar (Fig. 4.2.1). The intensity of the lateral slip depends on the contact pressure between stratification planes of the roof strata and the coal pillars, as discussed below:

– Shear stress between bedding planes is activated by normal stress and equals the supporting capacity of yielding pillars;

– Lateral thrust is due to the passive pressure of the yielding pillar;

– Lateral tectonic stress.

Due to roof strata reinforcement by rock bolts, it could be assumed that there are rock plates up to 3 m thickness, and that slips between thin beds do not exist. Drilling in the roof strata indicated that this thick roof plate was an integral part of the large rock block which was delineated in the lower plane within the extraction area of the Number 4 coal seam and in the upper plane by the separation of the rest of the strata 30 m above and in the vicinity of Number 10 coal seam. From the virgin parts of the coal zone and in a longitudinal direction a rock block can be delineated by tensile and shear failure (Fig. 4.2.2). The approximated diagram of vertical displacement composed from Dr. H. Bielenstein's data[8] showed rotation of the roof block downwards to the main extraction area and upwards above the virgin coal area.

APPROXIMATED VERTICAL DISPLACEMENT

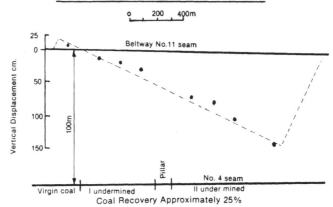

ROTATED IMPACT ROOF BEAM

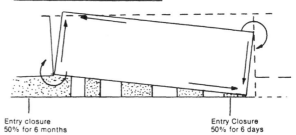

Fig. 4.2.2. The relation of the mechanics of vertical displacement to the rock beam model and the related convergence.

The interaction between the roof block movement and the yielding rate of the coal pillars which induced particular character displacement is exhibited mainly in a longitudinal direction (slip from entry to pillar) but could also be contractual roof displacement (slip from pillar to entry) in a perpendicular direction. The magnitude of vertical displacements should be the same in both cases as they are a function of the distance from the virgin coal and the size of the mined-out area.

4.2.2. *Roof span failure*

Competent roof strata could exhibit several fundamental types of roof span failure, each of them are individually discussed below:

1. Tensile failure is similar to the roof failure of weak strata. The mechanics of failure of the roof strata are usually considered to approximate a rock beam with clamped edges which fails in the middle under load deflections at critical rock tensile strengths (Fig. 4.2.3). Modes of such a roof beam deformation have been observed in underground mines and they could be analyzed on the basis of the mechanism of beam stresses and failure.

It is interesting to note that the state of the acting stresses depends on the length of

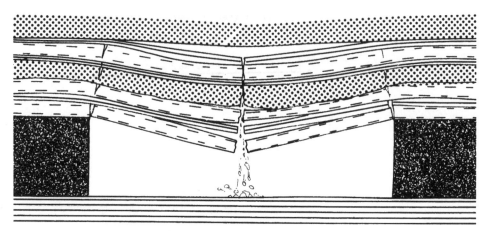

Fig. 4.2.3. A model of a roof fall due to a tensile failure of the rock beam.

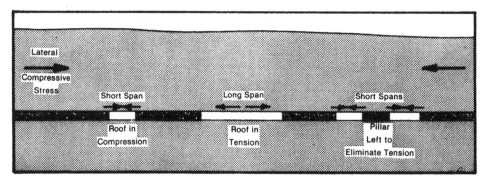

Fig. 4.2.4. The influence of the length of the roof span on the state of stress in a rock beam in a lateral compressive stress field.

the rock beam. For example, if coal strata is in compression due to geological stresses, then roof span should also be in compression. This is correct only for a short roof span, because long roof spans will induce tension regardless of the existence of lateral compressive stresses due to their own weight. To control such mechanics of roof beam deformation and to obtain moderate compression which facilitate strata stability it is necessary to define an optimal length of roof span which is delineated by adequate distance between the coal pillars (Fig. 4.2.4).

2. *Steep shear failure* is also similar to roof failure in weak rock strata. It would be fair to state that shear failure with cupola rock falls is most common for soft rock and is not characteristic of competent rock strata except in the case of interaction with lateral tectonic stresses. Under these circumstances shear failure are developed just at the rib edge. The rock falls as a solid beam without first breaking in the middle (Fig. 4.2.5).

3. *Low angle shear failure* is discussed in great detail under the subheading of lateral tectonic stresses in the next heading, and so their discussion is omitted at this place.

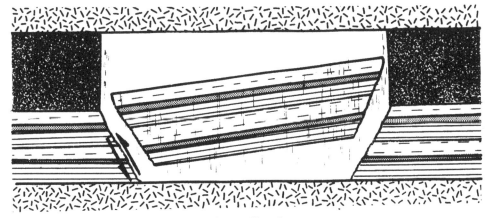

Fig. 4.2.5. A model of a roof fall due to shearing at pillar edges.

4. Violent failure is of particular interest in the case of competent roof strata. The roof stability in all three types of failure listed above could be successfully controlled by internal or external support, which is unfortunately not case with violent and sudden failure.

There are a certain number of rock mechanics specialists in several countries, particularly in Japan and Poland, who devote great effort to defining failure mechanisms and elements for strata stabilization, for example hydraulic fracturing. Investigations in the Upper Silesian coal fields showed that strata pressure and related rock bursts increase with mine depth particularly in thick bedded massive sandstone which exhibits dynamic fracture in the first phase of deformation, and typical roof bumps (shock)[9] in the second phase. The Polish specialists investigated the approximate location of violent fracture of the roof strata by the digital model (finite element analyses), for longwall faces at depths of 400 m, where immediate sandstone roofs were considered as elastic bodies subjected to a plain state of strain. The probable position of the dynamic fracture over the longwall face is shown in Figure 4.2.6. This fracture has a steep angle of about 70° and it begins about 2 m ahead of the longwall face in the immediate roof area. Beyond the area of this zone the shear stress concentration decreases, consequently, the probability of dynamic breakage of roof strata is reduced.

On the basis of the knowledge of stress distributions in roof strata and the rock strength a critical hydraulic pressure can be determined which initiates rock fracturing, and is given by the following equation:

$$p_K = (1 - v)(\sigma_H + \sigma_T) \tag{1}$$

where

p_K = pressure initiating hydraulic fracturing;
σ_T = tensile strength of rock;
σ_H = major horizontal stress;
v = Poisson's ratio.

The drillholes for hydraulic fracturing of roof strata should be located in the zone where the local concentration of tensile stresses favours an easy formation of fissure net. For

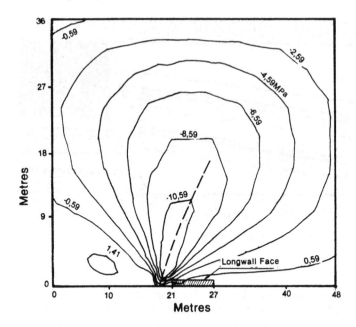

Fig. 4.2.6. Location of a dynamic fracture of the main roof in the plane of a shear stress concentration (after Kidybinski).

example, for the digital model in question indicates such a zone 10 to 20 m ahead, and 30 m above the longwall face.

Sikora presented a practical example for de-stressing of roof strata (Dymitrow Colliery) by hydraulic fracturing, which is briefly represented below:[9]

a) Because of longwall mining with hydraulic stowing and the rock strength characteristics the roof strata is very prone to rock bursts. This was confirmed by very high seismic activity which persisted far ahead of the longwall face.

b) The zone for hydraulic fracturing was selected under the edges at about 300 m ahead of the longwall face. Two sets of boreholes were drilled at distance of 20 m from the roadways. The boreholes were about 17 m over the seam and were sealed by heads at a height of 10 m over the seam (Fig. 4.2.7). Water infusion was carried out between shifts for safety reasons.

c) De-stressing by hydraulic fracturing has been monitored by seismological and seismoacoustic responses. For example, before hydraulic fracturing the number of bumps per month were between 107 and 109 and energy release was $(4.9–9.2)\ 10^6$J. However, after hydraulic fracturing the number of bumps were decreased to between 19–52 per month, and energy release was $(1.9–4.1)\ 10^6$J.

d) About 150 m^3 of water at effective pressures up to 35 MN/m^2 were injected into the fractured zone.

This example, obviously suggested that in special cases of roof control technology other means such as hydraulic fracturing beside external and internal support has to be taken into account. Hydraulic fracturing of strata with excessive strength induces roof caving and better rock fragmentation as well as preventing rock bursts.

Finally it should be pointed out that in both cases (virgin rock and artificially fractured rock) the failure mechanics is reflected by the elasto-brittle character of the rock environment, and exhibits sudden and violent release of strain energy. The

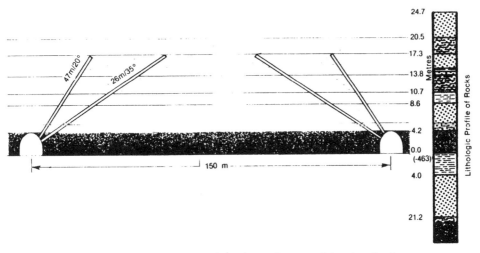

Fig. 4.2.7. Position of boreholes for hydraulic fracturing at the longwall face Number 2.

differences are mostly in the magnitude and amount of energy release, as given in the example above.

4.2.3.   *Roof caving*

The consideration of competent roof strata cavibility has been studied for decades, particularly for cases of mechanized longwall mining.

The state of the roof, when supported at the face, depends on the stress level and resulting deformation, as discussed in previous paragraph. Whittaker in a certain number of his publications defined strata mechanics in longwall mining with particular attention to roof caving mechanics.[10, 11] He established, that more than 30 m ahead of the face, the strata is in a virgin state of stress. As the face advances the vertical stresses increase, and a few metres from the face the crack is initiated in the seam and in the immediate roof. When crack propagation pressure decreases, there is a great deal of deformation, which will bring the roof to cave when coal and external supports are removed. Generally speaking the mechanics of caving are influenced by mining parameters such as depth, layout of the mine workings, mining methods, method of treatment of the gob area as well as the thickness and angle of inclination of the coal seams. Further discussion to these phenomena is given for each principal system of mining.

Roof cavibility will now be limited to roof lithology and strength, as briefly described below:

a) Thick strong roof strata, is characterized by limited fractures and rock burst phenomena. The mining problem is to prevent rock bursting and to get rapid caving behind extracted areas, and in case of longwall mining to have an adequate load-bearing capacity for stability. For example, in coal mines of Western Canada with room-and-pillar mining systems, the roof caving of hard and massive sandstones is incomplete and less than two heights of the mine face. It is a general rule that a roof

Fig. 4.2.8. Incomplete roof caving of the massive sandstone (fragmentation in large blocks).

Fig. 4.2.9. Complete roof caving of bedded strong rocks with irregular cave planes, due to the differential properties of individual layers.

Fig. 4.2.10. Caving
geometry and caved rock
fragmentation of the thin
and friable roof.

caves in sequences with large rock fragments (Fig. 4.2.8). Delayed and partial caving of
mine stopes is common. Sudden falls of large areas of the roof have been observed when
mining excavations intersect fault planes, here the roof fails due to its own dead weight.
These sudden roof falls have resulted in fatal accidents. Incomplete roof caving has a
grave consequence on strata de-stressing, and transfers additional stresses onto already
yielding coal pillars.

   b) Laminated strong roof strata are usually related to thin bedded sandstone and
siltstone strata. The underground observations in the Zenica Colliery (Yugoslavia),
showed complete caving of mine roof during long-pillar mining. The caved roof was
extended about two heights of the coal face and the cave had a trapezoidal shape. Roof
caving was facilitated by diletation and shearing of individual layers, which were by this
type of deformation transferred from self-supporting beams acting as dead load. The
cave plane is usually irregular due to the differential break of individual rock beds (Fig.

4.2.9). The rock fragmentation is satisfactory, and strata pressure is relieved after caving. This type of mine roof is most desired for underground mining regardless of which method is implemented.

c) A thin bedded and friable roof is prone to caving and fracturing in small rock blocks and are additionally disintegrated by impact during fall. These types of mine roofs have similar characteristics of cavibility as the weak rock described in the previous subheading. Thin bedded roof after caving also exhibits cupola geometry, the height of which is usually greater than two heights of the working coal face (Fig. 4.2.10). The self-supporting capacity of the roof span is limited, particularly in the case of very thinly bedded shaly rock strata. Under these circumstances, the roof caves immediately after coal removal if the artificial support is not installed.

Finally, it should be pointed out that deformation and failure of mine roof of competent rock strata in interaction with the mining technology of coal extraction and chosen mining configurations have been investigated by numerical model analyses by a number of authors. The research data is based on acceptance of the roof span being an elastic body, for which principles of solid mechanics are applicable. Sometimes contradictory to the existence of structural defects within the rock strata, displacement can take place along the tension cracks or shear fractures. Digital modeling does not take this into account. Under these circumstances the prediction of displacement as well as failure mechanics could be misleading and could result in wrong evaluations of roof support and rock cavibility characteristics.

4.3.   STRUCTURAL DEFECTS OF ROOF STRATA

The stability of a mine roof could also depend on the structural defects of the strata. The origin of structural defects is discussed under the heading tectogenesis of coal measure. Structural defects could be categorized into three principal groups, on the basis of continuity, attitude, frequency and other factors, as briefly given below.

4.3.1.   *Continuous defects*

These types of structural defects are represented mainly by faults (normal, thrust and tear) and might have a localized influence on coal strata stability. Although, roof

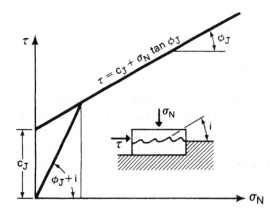

Fig. 4.3.1. A generalized diagram of the shear strength of discontinuities.

conditions more or less deteriorate at the vicinity where additional, usually external, reinforcement is required.

In the case of faults the mine roof stability does not depend on the rock type, but on the shear strength of continuities. There is a relationship between orogenic mechanics and the strength of structural defects. The strength primarily depends on the rupturing criteria of rock strata and the stress pattern, which reflect apparent cohesion of the continuities (c_j) and the angle of inclination of asperities (i),[12] as illustrated in Figure 4.3.1. These types of structural defect should be considered dangerous if they are not detected in advance, because they relate to large block falls, particularly in the flat and gently dipping parts of coal seams.[13] The influence of each fault type on the mining is briefly summarized.

1. Normal faults have an influence on roof stability as well as on other aspects of underground mining. For example, the longwall mining in the Ruhr Basin (West Germany) could be locally disrupted by normal faults, because of the discontention of continuity of the extraction panel. Under these circumstances the advancing coal face has to be stopped and the rock has to be excavated, to make a connection with the new level of the seam. The bridging of displacement levels of coal panels is in weak ground, and intensive material and labour application for support is required.

Influence of normal faults on roof stability has been investigated in various coal basins all over world. For example, the investigations of interaction between faults and mine roof failure at New South Wales (Great seam) showed three characteristic phenomena:[14]

a) The existence of two swarms of hundreds of minor normal faults with small throws (3 to 300 mm), deteriorated the strength of the roof strata which facilitated roof falls.

b) The faults with a throw greater than 300 mm, caused a severely failed roof condition.

c) The faults classified as normal showed that normal dip-slip faults do not appear to be associated with severe roof failure, but that oblique slip and strike-slip faults appear to be associated with a roof classed as severely failed.

d) Particularly, severe failure conditions were observed where fault slips intersect each other, regardless of magnitude or fault type (Fig. 4.3.2).

For this particular basin, the effect of normal faults on roof conditions has been qualified using a concept called the 'zone of influence', a rectangular area around a fault which has a potentially deleterious effects. For example, the 'zone of influence' on oblique slip faults may extend up to 3.0 m on each side of it in headings and up to 1.14 m in the cut throughs. The fault 'zone of influence' could have design implications for mine openings and possibly represent the maximum length of a cantilevered roof beam likely to fail during mining.[14]

The influence of faults on coal seam mineability, mine stability and strata mechanics have been investigated by many. Particular attention is given to strata mechanics in the case where faults represent stress accumulated zones with sudden release by rock burst phenomena.[15, 16] The brief consideration of this phenomena is given under the heading of 'virgin strata pressure'.

2. Thrust faults represent displacement, where one thrusted sheet, rides over another one. The thrust faults are commonly oriented along the strike and dip of coal bearing

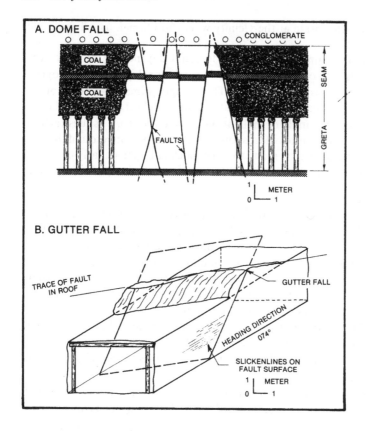

A. DOME FALL

CONGLOMERATE

COAL

COAL

FAULTS

SEAM

GRETA

1 | METER
0 |___ 1

B. GUTTER FALL

TRACE OF FAULT
IN ROOF

GUTTER FALL

HEADING DIRECTION
074°

SLICKENLINES ON
FAULT SURFACE

1 | METER
0 |___ 1

Fig. 4.3.2. Interaction
between faults and roof
failure (after Shepher &
Fisher).

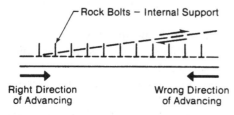

Rock Bolts – Internal Support

Right Direction
of Advancing

Wrong Direction
of Advancing

Fig. 4.3.3. A roof wedge formed by a thrust fault.

strata. The upward extension intersects the contacts between rock strata and the coal seams at a low angles, forming a wedge type structure.[13]

Strength of the joints depends primarily upon the intensity of orogenic movement. This type of rupture should be accepted as very dangerous because the roof falls in large wedges.

The direction of the headings should be such that the thrust fault will be detected in advance, and that the wedge of the faulted roof block can be controlled by external support (Fig. 4.3.3).

The dead load of the rock wedge is delineated by the thrust plane along the dip, and by nearby bridging joints which are normal to the dip. These should be of great concern during the excavation of the coal. The best solution for mine stability is to orient the mine workings oblique to the thrust planes.

3. Tear faults are located along fold axis with strike-dip displacements. The underground mine of the Rocky Mountain fold belt have shown tensile fractures in the echelons which are elongated in the direction of the strike of the strata. There are open fractures where the openings increase with time. Usually, these types of fault have a limited lateral extension (up to 100 m) and larger vertical extensions (over 250 m). They are often water feeders in the mine.

The shear strength of these faults is non-existent, they are open fractures without filling where friction and cohesion equals zero.

These types of faults adversely effect mine development if entries are driven in their direction. This phenomena is particularly prevalent in shallow mines where lateral compressive tectonic stress is released to a certain degree, so that roof spans are in a tensile stress field. Under these circumstances, the roof falls might be so severe, that all mine layouts have to be changed in a direction oblique to the fault trends.

Continuous structural irregularities play important roles in mine stability, and influence each fault type differently because of their origin, intensity of displacement and the density of small faults which follow the larger fault structure.

4.3.2. *Discontinuous defects*

This type of structural defect is represented mainly by joints and cleats, which reduce the mechanical strength of the strata. The degree to which they adversely influence stability depends on their magnitude, orientation, frequency, type of bridging and eventually healing or filling. Mining experience suggest that joints and cleats are ubiquitous and may have a more widespread influence on roof or rib stability. The rock lithology is the main factor which controls the magnitude and density of joints. For example, joints and cleats in shaly roof strata are densely spaced and discontinuous whereas joints and cleats in massive sandstone are rare or widely spaced and do not contribute significantly either to roof stability or cavibility. With regards to mining three types of structural discontinuities are listed below:

1. Filled joints have been studied in the sandstone strata of the Rocky Mountain Belt and their influence on mine roof stability have been designed. They usually occur in sets of intersected joints and vertical or steeply inclined stratification planes, and they are usually filled with calcite cement. The joints have a wide spacing (over 50 cm) and are partially detectable during exploration drilling. The extension of the joints is limited laterally and vertically (3 to 30 cm) so that between them rock bridges exist in both directions.

The joints, in general, have been shown to have an integrated strength by shear resistance along filling (lower-friction angle up to 27°), and a residual shear resistance between rock surfaces (higher-friction angle over 45°), as illustrated in Figure 4.3.4. The strength of the joint decreases as the thickness of the filling is increased.[17]

This type of joint should be considered the most stable in relation to roof falls and rock caving. For example, in relation to this type of joint roof falls in entries are observed to be less than 5 per cent of the total. In the pillar-stope structure, the roof suffers caving due to bridging joints, whereas rock bridges offer high shear resistance. When roof caving is initiated, the breakage of the rock strata is in large blocks due to the wide spacing of the joints. The rock is further fragmented when it fails down to the

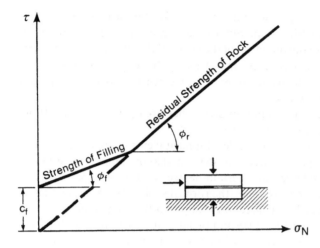

Fig. 4.3.4. A generalized diagram of the shear strength of a filled joint-rock structure.

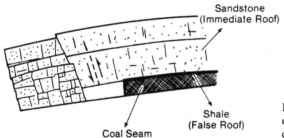

Sandstone (Immediate Roof)

Shale (False Roof)

Coal Seam

Fig. 4.3.5. Delayed roof caving due to the overall strength of coarse sandstone strata containing filled joints.

stope floor (Fig. 4.3.5). In sandstone this resistance to caving may cause a large area of the roof to stand up for weeks until a fault, not exposed by coal extraction operations, suddenly causes the roof to collapse. The fact that siltstone and shale strata are more susceptible to roof caving and better rock fragmentation, is probably related to lithological and other structural phenomena, rather than to filled joints.

2. Cleats in the rock strata have been studied for decades, particularly their directional effect on roof stability. There is a relationship between directional weakness of rock strata and the direction of the advancing coal faces. It is important to know the direction of dip of the cleat toward the coal face.[18] For example, it is postulated that cleat planes parallel to coal faces facilitate caving. However, this effect could be decreased to a certain degree if dip of the cleat planes is opposite to the direction of the maximum shear stress concentrations (Fig. 4.3.6).

Generally speaking, the joint planes are approximately at right angles to stratification planes and together with those, divide the rock strata into blocks which attain equilibrium in the excavated area through friction along faces of discontinuities. If such blocks are exposed in the roadways and their intersection is without adequate support then they could suddenly slip down and cause dangerous roof falls.

Fig. 4.3.6. Cleat and stratification planes dividing rock strata into blocks, which are in the opposite direction to maximum shear stress planes.

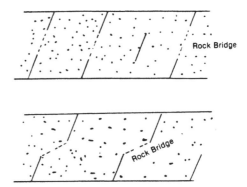

Fig. 4.3.7. A discontinuity pattern of massive sandstone beds.

The principal cleat direction of the roof strata should be oriented to the extraction face at an angle of 15°, with the dip parallel to the maximum shear stress orientation. In poor roof conditions, some authors recommend orientation of up to 45°, to the advancing face.

3. Bridged joints are primarily developed with hard rock, for example, massive sandstone strata, where some of them are being healed. The key factor in the self-supporting ability of roof spans are the pattern and density of joints and rock bridges (Fig. 4.3.7).

Generally speaking this type of joint has very little influence on the strength deterioration of the mine roof. For example, a shortwall mining method in Australia was very successfully applied due to the existence of an immediate roof of massive sandstone with a poorly developed jointing system. Under these circumstances the wide roof span exposed ahead of the support stood up until a continuous miner completed the coal cut of 3 m. Poor jointing prevented roof caving in the mined-out area, even in cases when joint orientations were parallel to working faces and in the direction of maximum shear stress.

It would be fair to say that the influence of joints on roof stability has to be considered in conjunction with the rock types. The lithological environment of the roof strata is the main factor which facilitates certain types of joints as well as their characteristic development.

4.3.3. *Stratification slips*

Shear displacements are extensive in coal bearing strata, are prevalent along the bedding planes and their vicinity, and within the coal seams. Displacement of mine structures due to mining stresses are facilitated along orogenically sheared planes.

The density of the shear planes and intensity of displacement along slips depend on the rock types:

a) Sandstone – slips are over 5 cm apart; displacements are along shearing planes and are restricted due to roughness of the joints;

b) Siltstone – slips are typically from 2.5 to 5.0 cm apart; displacement is not restricted by the roughness of the joints;

c) Shale – slips are less than 2.5 cm apart; displacements along bedding planes are followed by roof falls (Fig. 4.3.8).

The strengths of the orogenically sheared planes are residual because they lost cohesion due to displacement along bedding planes. The shear strength depends on the frictional (ϕ – sliding of two surfaces) and geometrical (i – interlocking of surface irregularities) components, as illustrated in Figure 4.3.9. In the case of sheared slips, the

Fig. 4.3.8. Carbonaceous shale sheared along planes of stratification (Mount Head Basin).

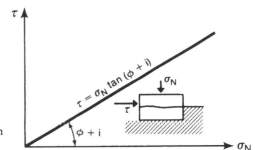

Fig. 4.3.9. A general diagram of the shear strength of sheared stratification planes.

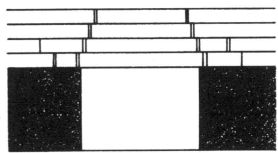

Fig. 4.3.10. The mechanics of roof faults of thin bedded strata containing stratification joints.

angle of inclination of the asperities to the rock surface (i) depends on the rock type[19].

Stratification continuities are well exhibited in shaley type rocks (shaley coal to shaley siltstone) and are in the majority of cases thinly bedded. The bed diletation and displacement is governed by stratification defects which can be further deteriorated by the effect of humidity on the exposed mine roof. Under these circumstances the roof overstressing is due to swelling pressure and roof falls are eminent. A mine roof can undergo considerable deformation and failure along bedding planes near the exposed surface without disturbing the roof bolts. After the roof falls, the bolts are isolated from the roof surface and are left dangling. Three types of geometry of roof falls have been observed: first, a 'dome' type nearly circular in plane view with a flat upper surface in cross-section where stress is controlled by stratification planes (Fig. 4.3.10); secondly, an 'arch' type linear or serpentine, with snake-like appearance in plane view; third, 'irregular' types of falls from sloughing to minor.

It would be fair to say that these types of roof falls originated from a particular rock lithology, rather than a geological structural defect.

4.3.4. *The structural quality of roof spans*

The evaluation of the strength of rock masses on the basis of geological structural defects is an important element in the analysis of mine stability, particularly in the evaluation of roof span quality. Particular effects to study underground coal mining quantitative classifications of roof strata have been done in Poland,[19] as further discussed.

K. Pawlowicz[20] introduced that the evaluation of roof span stability was upon the compressive strengths of rocks and the average frequency of bedding planes. The assessment of roof stability was given by the index I, which was determined from the following equation:

$$I = \frac{\sigma_c \cdot d}{100} \qquad (1)$$

where

σ_c = average compressive strength determined on the basis of rock samples in the laboratory;

d = average thickness of the layers of the strata (cm).

The same author also took the joints into consideration, but their density in relation to the bed thickness of siltstone and sandstone was expressed as constant which was less than the 1.1–1.2 thickness of the intersected layer. Also, this author classified, in relation to his evaluation, five types of rock strata with respect to roof span stability as follows:

Index 0–18, the roof falls immediately after exposure or has a small delay.

Index 18–35, the roof is difficult to maintain, is brittle and breaks up, and has cavities, fractures and fissures. It hangs on the supports, and is dangerous, and easy falls down.

Index 35–60, the fractures roof falls locally and is poor to good.

Index 60–130, is a good to very good roof, and creates excellent working conditions, is breakable and prone to caving.

Index 130, a strong roof, it ensures excellent working conditions, but exhibits uncompleted caving which is a cause of difficult and dangerous mining conditions.

Improvement of this classification by A. Bilinski et al.,[21] determines the 'I' index as follows:

$$I = 0.016 \sigma_M \cdot d_f \qquad (2)$$

where

σ_M = the compressive strength of the rocks of the strata;

d_f = the average distance between the fissures (cm).

The value σ_M is determined below

$$\sigma_M = \sigma_c \cdot K_1 \cdot K_2 \cdot K_3 \qquad (3)$$

where

K_1 = the mechanical scale factor which is constant for each rock type (0.33 for sandstone; 0.42 for mudstone; 0.50 for siltstone);

K_2 = a factor of strength degradation due to loading time which should be determined from test curves for each rock type;

K_3 = a factor of strength degradation due to moisture, which should also be determined from test curves for each rock type.

These authors also accepted rock classifications suggested by K. Pawlowicz, except that his category with Index 130 was divided in two sub-categories. First, Index 130–250 is a very strong roof, where the longwall mining system requires appropriate procedures of forced caving. Second, Index 250, rocks of very excessive strength do not permit longwall mining without great difficulties.

It would be fair to say, that these evaluations of the roof span strength depend on the compressive strength of deterioration due to geological structural defects. The evaluations might satisfy local mining conditions, for example Upper Silesian coal fields, where roof strata are well known on the basis of a century of hard mining experience.

The determination of roof span quality due to structural defects should not be applied to individual classifications which suite particular local conditions. The most realistic solution is developed by independent rock classifications which suite the local conditions of coal deposits.

4.4. BEHAVIOUR OF SUBCOAL STRATA

The competency of the mine floor depends on the strength of the rock stratum, as well as on the differential strength of the subcoal strata which is composed of multiple lithological units. The overall stability of subcoal strata is equally influenced by the distribution and thickness of single lithological layers. Several topics of particular importance are discussed with respect to mine stability.

4.4.1. *Deformation and failure of the mine floor*

Underground phenomenological investigations suggest four types of failure resulting from interaction between coal pillar-panel structures and mine floors.[22] Deformation and failure of each individual type are briefly described below:

1. Floor heave (*shear flow*) deformations are particularly well exhibited for thicker subcoal strata of low strength with yielding behaviour, for example clay-shale and silt-sandstone rocks. The presence of slickensides within strata and higher moisture content will promote floor heave even under low pillar loading conditions because under these circumstances there is very little resistance to the initiation of shear flow.[23]

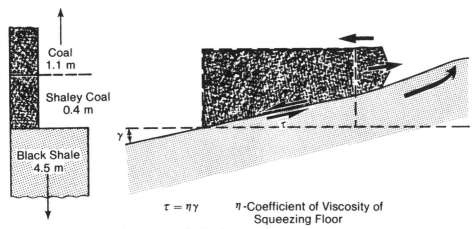

Fig. 4.4.1. Model of soft floor heave and pillar flow.

This type of deformation is exhibited by strata flow beneath the coal pillar with movement outwards and upwards towards the excavated area along the path of least shear resistance (Fig. 4.4.1). Underground studies suggest that the intensity and acceleration of floor heave also depend on the strength and behaviour of coal pillars. For example, in the Lethbridge area (Galt seam), where coal has a higher strength and rigidity, the subcoal strata of black shale (up to 5 m thick) exhibited limited heave over a long period of time. In the Grande Cache area (Number 4 seam), where coal has very low strength and yielding property, the subcoal strata of coal-shale section (up to 6 m thick) exhibited intensive floor heave in a short period of time (total closure had been observed for a period of between 6 days and 6 months).[24] Floor heave at the very shallow Star-Key Mine exhibited moderate uprise over long periods of time, but it had been followed by fissure development both in the coal rib and on the shale floor (Fig. 4.4.2).

Floor heave is a serious problem in underground mining, which necessitates the fragment brushing of roadways to keep them open. Brushing accelerates the problem by decreasing the bearing capacity of the floor on the ribs. The underground observations in the Grande Cache and Lethbridge areas suggest that the intensity of

Fig. 4.4.2. Brushed layer of heaved floor (Star-Key Mine).

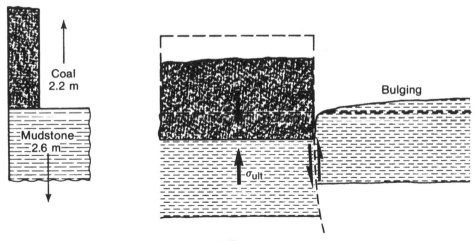

Fig. 4.4.3. Model of soft floor punching by coal pillar.

floor heave depends on the stress magnitude and competence of the mine floor. It is a combined effect of:

a) An increase in floor stress level from the start line of the face;

b) Build up of failure in the floor causing an apparent swelling;

c) Swelling due to water if clay rock is present.

The prevention of floor heave could sometimes be carried out by floor bolting or dwelling. In longwall mining gateside pack can also move the abutment stresses away from the roadway and reduce heave.

2. Floor punching (vertical shear) by coal pillars is promoted by the thicker and weaker subcoal strata, and is particularly facilitated by an increase in moisture content.[23]

The penetration of pillars into subcoal strata exhibit punching failures which are a product of vertical shear stress in the top stratum and large deformations in the weak strata below (Fig. 4.4.3). In these cases, the cracks are visible at the intersection of the pillar-coal floor seam, and ridges of displaced material built up along these junctions and parallel to the pillar.

Underground observations suggest that pillar punching in the mine floor is due to progressive loading from nearby mining, rather than being due to a constant load versus time mechanism which is the case with floor heave (creep deformation). This is in agreement with investigations of the floor bearing capacity in the Lethrbridge area which was monitored by prop penetration into the coal seam floor and measures as a function of load transferred from nearby coal extraction areas. Props with dynamometers installed in the butt entries (Galt Number 8 Mine) showed that there was very little increase in prop loading until the room extraction had advanced to within 12 m of the dynamometer. A noticeable increase of prop loading then occurred in three typical increments of 8, 13, and 14 tonnes, respectively (Fig. 4.4.4). It was then observed that the loads decreased sharply and this was atrributed to the penetration of the dynamometer props into the subcoal strata.[25]

Coal pillar deformation follows the mechanics of mine floor punching and the

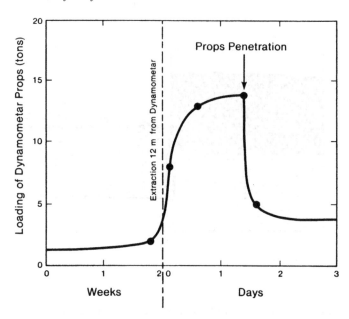

Fig. 4.4.4. Dynamometer prop loading and penetration during load transfer from nearby mining (Lethbridge).

development of vertical shear stresses. For example, in Olga Number 1 Mine (West Virginia) in the head entry vertical shear failures were observed along the intersection line of the chain pillar rib and the roof. Consequently, the wooden posts set along and/or close to the pillar ribs were frequently crushed or punched into floor sometimes as deep as 20 cm. In this case, the pillar exhibited coal falls in large slabs along vertical shear failures.[26]

The loss of the shear strength of the rock in the direction of loading is the main factor contributing to penetration deformations, and is facilitated by the following factors:
   a) The fabric of clay effects the transverse properties of the strata;
   b) The thickness of the immediate mine floor stratum;
   c) The extension of vertical fissures.

The prevention of penetration floor deformation should be a design consideration, because strata reinforcement in this case is either nonexistent or rather limited.

3. Floor displacement (horizontal shear) deformation could be correlated to displacement along stratification planes as in the case of a normal stress (abutment load) activating a horizontal shear stress, which then overcomes shear strength along the bedding planes of a subcoal strata.[27]

Horizontal shear displacement is promoted in laminated subcoal strata when individual beds are of different lithological composition and strength. This is opposite to the thick and weak subcoal strata of uniform lithology which relates to floor heave and penetration.

Shear failure along bedding planes can be developed either in subcoal strata or both subcoal and roof strata. Shear movement of the strata is facilitated if clay-shale partings have low residual shear resistance (Fig. 4.4.5), particularly if the mine floor is saturated by water. Laboratory testing has shown a significant decrease of shear strength of coal-

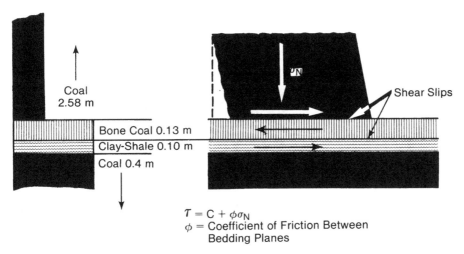

$\tau = C + \phi\sigma_N$
ϕ = Coefficient of Friction Between
Bedding Planes

Fig. 4.4.5. Model of horizontal pillar-floor displacement.

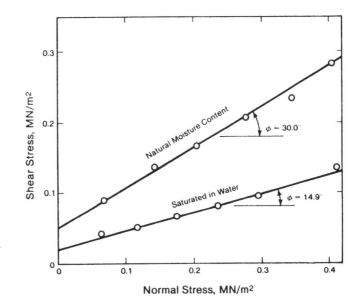

Fig. 4.4.6. Deterioration of shear strength of shale due to water saturation (Lethbridge).

shale rocks after saturation (Fig. 4.4.6). If clay partings do not exist then the preferential planes for shear movement are stratification contacts between the bone coal and the coal seam. At the very shallow Star-Key Mine (< 30 m) underground observations showed such displacements in mine floor, those were of a limited range (3.9 cm). It had been observed that the stability of coal pillars was influenced more by their restraint than by the limited floor movement. The preferential direction of coal pillar and floor displacement was approximately perpendicular to the Rocky Mountain Range.

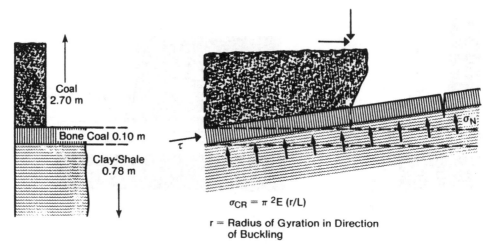

$$\sigma_{CR} = \pi^2 E\,(r/L)$$

r = Radius of Gyration in Direction
of Buckling

Fig. 4.4.7. Model of floor bending and pillar fracturing.

This type of failure in the mine floor was present if the following elements were implemented:

a) When the normal stress mobilized the shear stress which was then greater than the shear strength of the stratification planes;

b) When slickensides can be observed along bedding planes, to be the cause of deterioration of the original shear strength;

c) When water content was correlated to lithology and the strength of the floor rock.

With the aspect of mine stability, the adverse effects of shear failure could be decreased but not eliminated. Coal pillar areas should be large enough, so that the stresses are transmitting and the mobilizing shear stresses are lower than the shear strength.

4. *Floor bending* (*shear fractures*) of the mine floor occurs in laminated subcoal strata which have alternating firm and soft layers. The bending deformation is facilitated by the existence of lateral tectonic stress, which confines the firm floor rock beam, and the existence of greater shear strength between the firm and the soft layers. The floor bending is more readily observable in wider mining openings.

The mechanics of floor bending can be explained by the load transmission of coal pillars onto the firm stratum, resulting in two phenomena. First, the firm layer takes and accumulated all tangential stress. Second, the soft layer beneath does not have the ability to accumulate stress and so transfers radial stress upwards.[28] Under these stress conditions the firm layer, as the elastic member of the subcoal strata, will be exposed to bending forces which will cause buckling and breaking of the mine floor (Fig. 4.4.7).

Bending of the firm mine floor is exhibited in the rooms of Star-Key Mine where the mine floor is a relatively thin bone coal layer.

The stress concentration in subcoal strata is a maximum at the intersection of coal pillars and the mine floor because the stress release from the firm (elastic) member effects maximum pillar fracturing the least at this point.

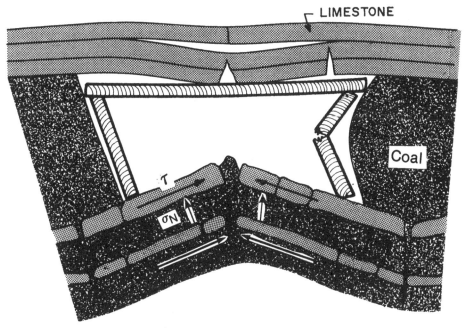

Fig. 4.4.8. Rock burst of the mine floor.

Such mechanics of floor failure and deformation have been observed in some South European coal mines where stress release is by violent fracturing of the firm layer of the mine floor (tangential rock burst) and by extrusion of a soft layer from beneath a firm one. If the profile of the subcoal strata is alternating with firm and soft layers then violent fracturing of each elastic member could be triggered in different time sequences. Under these circumstances coal pillars can be fractured and reach a yielding state with limited or with no supporting capacity (Fig. 4.4.8).

These are two main factors effecting this type of floor deformation:

a) Transfer of high abutment stresses in the mine floor strata because of insufficient stress relief of roof strata during coal excavation;

b) The existence of elastic members, with high modulii of elasticity, that can store and violently release strain energy.

The most successful results have been achieved by floor fracturing and wetting, but rock bursting has not been totally eliminated.

4.4.2. *Bearing capacity of the mine floor*

The investigations of the bearing capacity of the mine floor became an integral part of the study of mine stability. An excellent example of this is the investigations in Ellalong Coal Basin in Australia, which was done by CSIRO geotechnical specialists.[29] They conducted a series of in situ plate bearing tests to determine the relative suitability of mine floors for longwall mining. Their considerations and investigations are briefly represented below:

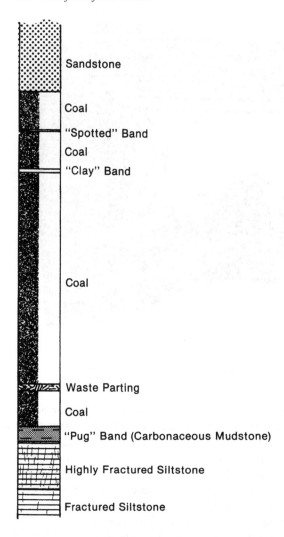

Sandstone

Coal

"Spotted" Band

Coal

"Clay" Band

Coal

Waste Parting

Coal

"Pug" Band (Carbonaceous Mudstone)

Highly Fractured Siltstone

Fractured Siltstone

Fig. 4.4.9. Generalized section of coal seam, with subcoal strata profile (Ellalong Coal Basin, Australia).

1. General considerations of the potential mine floor within subcoal strata showed that lithological units at or near the base of seam in the longwall area may be unstable as a working environment. For this reason several lithological units of subcoal strata have been investigated as an alternate to mine floors, with the following implications in mind:[29]

    a) Likely floor conditions in relation to equipment performance;

    b) Total proportion and nature of waste rock that might be handed;

    c) Possible roof conditions commensurate with mining on certain floors;

    d) The height of the coal face and coal quality.

In that consideration the 'pug' band was dismissed as a potential floor. This was based on observations of near zero bearing strength during wet conditions and its insufficient thickness (Fig. 4.4.9). The investigations of bearing capacity of individual lithological units were carried out in the laboratory and in situ.[29] A matter of further consideration is in situ investigations.

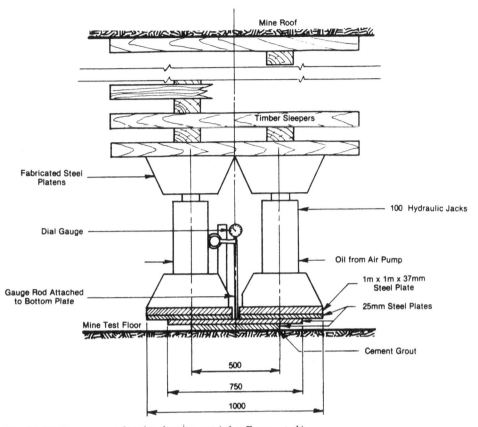

Fig. 4.4.10. Arrangement for plate bearing test (after Enever et al.).

2. Plate bearing tests in situ have been conducted by J. R. Enever et al. on several lithological layers of subcoal strata. The loading conditions were simulated to a load magnitude typical for longwall support (499 tonnes). The two testing arrangements were adopted and are briefly described below:[29]

a) Exposed layers of the mine floor of present mining workings (not damaged substantially) were loaded and tested as shown in Figure 4.4.10.

b) Exposed layers on the coal face required a rearrangement of loading. In this case the continuous miner should be used to cut a slot in a standing face up to 0.30 m from the top of the lithological band.

The test arrangement employing the 0.5 m diameter plate was preferred because it maximised the extent of floor tested for the loading available. The general test procedure consisted of applying increments of load to the test horizon and then recording the incremental deflection indicated by the dial gauge. The load increment was recorded until failure of the mine floor occurred or until total deflection reached a limit where no further load could be applied.

These tests included cyclic loading, where fifteen cycles were selected as representing the greatest number of loading cycles that would be applied by supports at any one location on the longwall face. After fifteen cycles of loading the load was increased and the procedure repeated. The results of the cyclic loading test indicated no signs of appreciable increase in deflection over fifteen cycles of loading for up to 80 per cent of the bearing capacity indicated by a single cycle of loading.

3. Bearing behaviour of lithological units is represented diagrammatically in Figure 4.4.11, and was evaluated by J. R. Enever et al., and is briefly discussed below:

a) Siltstone below coal splits (Number 1 as mined, Number 6 dried out) exhibits a bearing capacity up to $15\,MN/m^2$, which is in agreement with laboratory testing (uniaxial strength $18\,MN/m^2$). Saturated rocks have reduced a strength of $5\,MN/m^2$, which is still sufficient as a mine floor for powered supports with a reasonably sized bearing pad. The load at the powered support could cause a deflection of the mine floor of about 2 mm, and should be a negligible deformation from the practical point of view of a self-advancing support.

b) Lower coal (Number 3) exhibited satisfactory bearing capacity and could form a sound mine floor.

c) Siltstone bands (Number 7; Number 8) were tested at two different levels. Both tests indicated a 'quasiplastic' behaviour, with no distinct peak bearing capacity. For a base pressure load of $2\,MN/m^2$, typical for powered a support, the test results indicated a probable deflection of 20 mm (Number 7). This lithological unit should be considered as marginal for a potential mine floor.

d) Carbonaceous mudstone (Number 4; Number 5) tests indicated somewhat different responses due to the different confinement of the rock. Test Number 5 where the location was relatively well confined, and bearing behaviour was similar to siltstone below the coal split (Number 1), represented a workable mine floor for longwall mining. Test Number 4 where stratum was relatively unconstrained suggested a bearing pressure of $3.7\,MN/m^2$ and then a yielding stage with mechanics similar to that observed for the siltstone band. This lithological unit offerred a net bearing capacity without significant deflection and it may be considered able mine floor.

The bearing capacities expressed by maximum loads at failure and floor deflection

CONTACT PRESSURE (t/m^2)

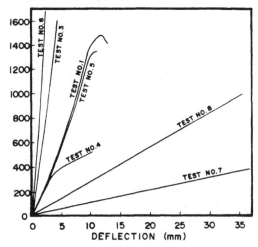

Fig. 4.4.11. Results of single cycle plate bearing tests (after Enever et al.).

obviously are related to the mechanical properties and the behaviour of individual lithological units.

4.3.3. Stability of subcoal strata

Very often in mine stability analyses or design of mining methods consideration is given to the strength and bearing capacity of a mine floor.

The mine stability analyses and the design of mining methods is done with the aspect of interaction between coal pillars or powered support (primarily shield support) and mine floor are based mostly on the strength and bearing capacity of the exposed floor. Mine stability could be better understood by first giving consideration to the profile of subcoal strata and then to the exposed mine floor as discussed below.

1. The thickness of beds of individual lithological units, which might belong to firm and soft rock strata, and their distances below the mine floor are important elements in coal mine stability. An example, with increased thickness of a soft layer overlying a firm one, mine stability decreases, while with increased thickness of a firm layer overlying a soft one, there is increased stability. Thus there are variations in mine stability due to stress distributions beneath the coal pillars. More rigid layers can accumulate and redistribute stress, but softer layers have to flow when the stress is higher than their strength (immediate floor) or else distribute their load to more rigid layers (beneath the immediate floor).

The strengths of firm and soft layers are directly related to the bearing capacity of the mine floor. The calculation of the bearing capacity of a mine floor should be in conjunction with the subcoal strata profile rather than the exposed mine floor. A concept proposed by A. S. Vesic, for example, makes it possible to evaluate the influence of the bearing capacity of a soft soil overlying a firm soil, and is expressed by the following equation:[30]

$$q_{ult} = C_1 N_m + q \qquad (1)$$

where

q_{ult} = bearing capacity;
C_1 = undrained shear strength of upper layer;
N_m = modified bearing capacity factor which depends upon the ratio of shear strengths of two layers, $K = C_2/C_1$;
q = surcharge.

Table 4.4.1. Safety factors for different types of mine floor

Subcoal strata	Safety factor
Clay-shale, thickness up to 5m	0.8 (floor problems)
Silt-mudstone, thickness up to 1.5m	0.9 (floor problems)
Bone-coal and coal, thickness up to 1.5m	1.6 (no floor problems)
Coal and shale layers, of individual thickness up to 0.5m	0.7 (severe floor problems)

Using the equation above have been calculated a safety factor of subcoal strata for four geological settings of the Plains Region in Western Canada, as illustrated in Table 4.4.1.

Investigations showed that for the same lithological unit, safety factor could be very little increased with thickness of the layer, because stability of mine floor is rather governed by position of firm and soft layers within subcoal strata than their individual thickness.

2. Alternate beds for mine floor can be considered if thickness of the coal seam is below or close to workable height of the coal face. Under these circumstances the bearing capacity of mine floor can be increased by eliminating incompetent rock stratum and exposing a competent one. This is possible only if mining technology, physical properties of strata and economics is in favour of implementation of alternate mine floor. In the case when seam thickness exceeds the workable height for conventional full face mining, there is the possibility to optimize coal face to avoid incompetent floor stratum where alternatives are available. The alternative floor design must take in consideration to possible roof stability and the quality of the mineable coal section.

3. Wearing capacity of rock stratum due to movement of continuous machines or rubber-tire equipment could be directly related to stability of subcoal strata. The mine floor wear causes fracturing and dispersing fragmented rock and formation of bumps, which could expose the bed below stratum of mine floor. Exposing incompetent beds in the bumps of mine floor could deteriorate mine stability particularly if the water is collected in them. The adverse effects of exposed beds is difficult to observe, because the mine floor usually is covered by mine dust and coal, and in many cases due to their invisibility this deformations are not getting adequate attention.

The evaluation of wearing capacity of the mine floor is done usually from the swelling test and slake durability test, conducted on rock samples in the laboratory. In many cases the evaluation is not realistic, because it is difficult to relate geotechnical indices obtained in the laboratory to a real mine situation.

CHAPTER 5

Primitive strata pressure

The pressure concentrated within strata in the earth's crust is referred to as 'primitive' when it has not been disturbed by any man-made openings. The determination of such primitive pressures is complicated by our limited knowledge of the stress distribution, the geological history and even the geological structures present in a given sequence of strata. Strata pressure often results from the interaction of several stress fields with different origins. In order to simplify the analysis and emphasize the importance to mining, four types of primitive stress fields are proposed.

1. Geological stresses which are caused by such forces as earth's gravitational field, orogenic movements and glaciation.

2. Tectonic stresses generated by the lateral component of thrust during folding of earth's crust. This is of particular significance to the stability of coal mining operations in fold belts.

3. Stresses in coal which result from changes in physical properties due to different depth of strata.

4. Hydrodynamic stresses due to the pressure of such pore fluids as water or gas.

These stress fields and their relationship to strata mechanics and mine stability are discussed in the following paragraphs.

5.1. GEOLOGICAL STRESSES

The depth of coal mining operations is generally in the order of 300 to 600 m but may be greater than 1000 m. At these depths the strata are frequently deformed, the distribution of stress is complex and a complete analysis of the stress-strain relations is difficult to achieve. Regional stress fields influence the local distribution of stresses which affect mining operations. Consequently, some knowledge of the regional geology is necessary in order to understand the possible causes and orientation of stress in the mine strata.

Recently nearly everyone agreed that the natural stresses in the upper part of the earth's crust is due to integrated stresses. The indices of individual stress fields are relative and in some cases highly variable.

5.1.1. *Geostatic stresses*

Geostatic stresses are formed due to the gravitational force and lateral constraint of the

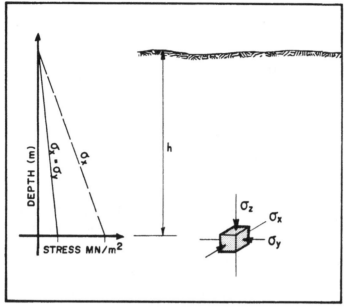

Fig. 5.1.1. Concept of state of geostatic stresses.

rock (Fig. 5.1.1). The magnitude of the geostatic stress is represented by equation:

$$\sigma_V = \gamma h \tag{1}$$

where

σ_V = geostatic stress;
γ = unit weight of the rocks;
h = depth of the point from the surface.

Assuming an average density of the sedimentary rocks containing a coal seam at normal depths is 2.5 g/cm³, the geostatic stress (the vertical compressive stress or the vertical load) would increase 2.5 MN/m² for every 100 m of depth.[1]

If the rock is isotropic, the magnitude of the horizontal component of stress (the confining pressure) would depend on Poisson's ratio. That is:

$$\sigma_H = v\sigma_V \tag{2}$$

where

σ_H = the horizontal stress;
v = Poisson's ratio;
σ_V = the vertical stress.

The magnitude of Poisson's ratio has been found to vary widely in the range from 0.15 to 0.45 and to depend mostly upon the rock type. For underground mine design purposes, Poisson's ratio is generally taken to be 0.33, implying that the horizontal stress is one third of the vertical.

Despite convincing data regarding the existence of residual stresses within strata, many workers persist in thinking that vertical stress corresponds solely to the weight of strata and that horizontal stresses may be determined by using Poisson's ratio.[2] It should be pointed out that in mine design it is equally important to consider the existence and influence of tectonic stresses as discussed below.

5.1.2. *Tectonic stresses*

Tectonic stresses can be considered in two groups each of which have a different influence on strata mechanics and mine stability.

1. Active tectonic stress. This is related to earthquake phenomena which are due mainly to stress release at the boundaries of the lithospheric plates of the earth's crust. The direction of the maximum compressive stress at these boundaries, as deduced from earthquakes and ground deformation, is consistent with the direction of relative motion of the plates. Away from the plate boundaries the stresses are more difficult to analyze because large intraplate earthquakes rarely occur and often do not fit the pattern of those at the rim.[3] Evidence from earthquake focal measurements indicate that the Australian Continent is in a state of horizontal compression. Faulting associated with Australian earthquakes indicates that they are caused by compressive stress acting almost in a horizontal plane. In situ stress measurements in Australian mines, tunnels, quarries and on rock outcrops confirms the presence of horizontal compressive stress in all areas of the continent.[3]

The increase in magnitude and redistribution of stresses in strata both during and after earthquakes are well known, but are of short duration. Whilst these stresses are tectonic in nature their manifestations are varied often resulting in nonuniform and nonsteady horizontal movements and uplifts.

Such active tectonic stresses are difficult to predict and measure, are transient in nature and usually are not taken into account in mine stability analyses.

2. Passive tectonic stresses. These consist of the remainder of the orogenic forces which have not been entirely spent on crustal deformation. Such stresses are significant in the evaluation of strata pressures since they may be of equal or greater magnitude than the geostatic stresses. Residual tectonic stresses in coal bearing strata are mainly due to orogenic movements of the earth's crust manifested by uplift. The main compressive stress may be resolved into two components: one vertical and the other horizontal which serves as a basis for classifying the residual tectonic stresses.

a) The existence of vertical tectonic stress has been recognized only very recently. Such stresses are significant because they act in opposition to the geostatic stress (γh). In situ stress measurements have shown that vertical stresses may be greater, sometimes much greater, than γh. For example, in the Donbass Coal Basin (U.S.S.R.) it was determined that the vertical stress $\sigma_v = 3.5\gamma h$ at the Petrovskaya Mine and $\sigma_v = 4.5\gamma h$ at the Kochegarka Mine. The presence of remnant tectonic stresses is the most likely cause for such a great discrepancy between the gravitational stress and the observed one. Unfortunately, at the present time there is very little data which demonstrates the magnitude of such stress fields, indicating the pattern of their distribution in strata.

b) Horizontal and lateral tectonic stresses have been the subject of investigations in many countries for more than two decades. In situ stress measurements within fold belts have shown the presence of the major principal stress (σ_1) approximately perpendicular to the fold axis and the minor principal stress is (σ_3) parallel with them. In this stress pattern the vertical stress is usually greater than gravitational force. For example, measurements in the Shaorskii coal deposit (U.S.S.R.),[6] indicated the presence of considerable tectonic stresses distributed as follows: $\sigma_v = 21\,\text{MN/m}^2$, $\gamma h = 16\,\text{MN/m}^2$ and principal horizontal stress $\sigma_H = 61\,\text{MN/m}^2$. A large number of stress measurements suggest that the lateral stress might be two to three times greater than the vertical stress.

Determination of the tectonic stress fields in coal bearing strata have a practical importance for the efficiency and safety of mining operations. There is little information of the stress-strain state in rocks due to tectonic stress because the parameters which could separate this stress field from the total stress field are not yet developed.

5.1.3. *Frozen stresses*

Frozen stresses usually are found in coal basins which were not subjected to orogenic forces but were effected only by glaciation and erosion. The general model of frozen stress distribution is illustrated in Figure 5.1.2 and briefly described below.

1. Loading of the coal formation was by slow sedimentary processes with a gradual increase of pressure. When the accumulation in the deep basin was about at completion, the pressure on the coal formation was hydrostatic at a depth of 4,000 m or more. As water squeezed out, particles compacted, cement crystallized and the uncohesive rock mixture slowly petrified into the solid rock. Simultaneously with this process a hydrostatic pressure of possibly $100\,\text{MN/m}^2$ was locked in the coal formation.

2. Unloading of the coal formation was by a slow process of uplift and erosion to the level of outcropping of the formation. Consequently, the original thickness of the strata

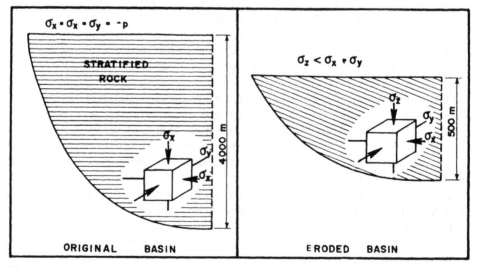

Fig. 5.1.2. Concept of development of frozen stress in rock strata.

which may have been over 4,000 m has only a small part of it left. As the load was being removed from the layers of rock, it allowed expansion upward relieving some part of the vertical stress. Relief of the vertical stress was not in direct proportion to the load removal due to elasto-plastic properties of the coal formation which stored part of the strain energy. Also, the lateral stress which may have been as high as 100 MN/m^2 still stayed locked in the coal formation because the possibility for lateral expansion was minimal. Today geostatic stress is not only composed of the gravitational body force but also of the unrelieved part of the original hydrostatic stress, i.e. the frozen stress.

Frozen stresses have been investigated in the coal strata of the Plains Region of Western Canada, which are not effected by the orogenic movements which produced the Rocky Mountain fold belt. These coal strata are almost in the original attitude of sedimentary deposition and the only changes which they have experienced are erosion of the overlying sediments and glaciation. The coal strata are shallow, gently dipping, and lie immediately below glacial till. The following characteristics of the frozen stress field have been identified:

a) The horizontal stress is greater than the vertical stress. The ratio between horizontal and vertical stresses decreases with strata depth. In general, the ratio between horizontal and vertical stresses at shallow depth (50 to 150 m) is greater than 3 (i.e. $\sigma_H/\sigma_V > 3$).

b) The magnitude of the compressive horizontal stresses in both the X and Y directions are approximately equal.

c) The presence of horizontal stresses has been observed at very shallow depths (about 10 m) in the Wabamun strip coal mine. Here the floor of a box cut buckled up due to the action of the horizontal stresses. A similar phenomena has been observed in the floors of open pit mines in Australia.[3]

Frozen stresses deserve as much attention as tectonic stresses in which group they are often incorporated by many authors. They should be distinguished from tectonic lateral stresses particularly in studies of mine stability.

5.1.4. *Structural stresses*

Structural stresses are very difficult to evaluate because they originate by tectonic deformation of the strata. Many authors emphasize the importance of analyzing tectonic structure as a factor which influences the variation of stress fields and consequently the stability of mine openings.[5] The geological-structural aspect has been investigated by theoretical analysis which showed that if the strata can slip, the tectonic stresses would be partly relieved and partly stored in deformations. Structural stresses can be classified in several groups related to particular types of geological structures.

1. Stresses of shattered strata. In this case we must bear in mind that openings produced by fracture redistribute stress in their vicinity and change its orientation in accordance with the direction of the openings. A fractured zone represents a more or less-destressed zone with some limited bearing capacity similar to that which been observed along the perimeter of coal pillars still capable of supporting a super-imposed load. The behaviour of shattered rock stressed beyond the elastic limit should be investigated as a discrete rather than as a continuous medium. This type of analyses is widely used for evaluating the stability of mine structures which are in the region of plastic deformations.

2. Stresses of faulted strata are usually sources of violent deformations or bursting of the rock. Accumulation of potential energy results from the influence of both external and internal factors (the regional and the local stress fields) which create the overall stress field. The accumulation of potential energy develops only if the rock is competent and this characteristic can be determined as outlined below:[5]

$$K = w \rho \tag{3}$$

$$U_{pot(max)} = K U_c \tag{4}$$

where

K = factor of competence;
w = fracture energy;
ρ = density;
U_c = energy concentration factor;
U_{pot} = potential energy.

A method of evaluating the strength of rock liable to bursting has been developed in Poland. It involves testing the rock profile in the vicinity of the fault by a penetrometer,[8] as illustrated in Figure 5.1.3. Rock bursts have been experienced in the Spring Hill Number 2 Mine and the Sunny Side Coal Basin when development headings driven in virgin strata came close to a fault.[9,10] The most violent rock bursts were experienced when headings were driven perpendicular to the fault zone in competent strata. Headings driven at low angles to such a zone gradually release the accumulated stress and eliminate the violent effects of bursting ground.

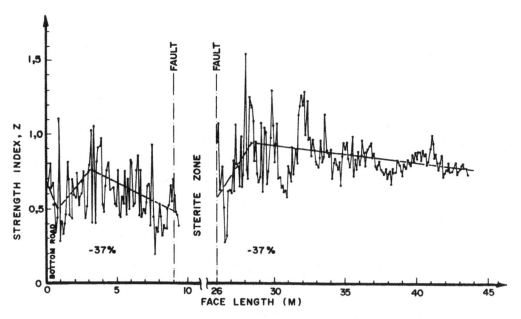

Fig. 5.1.3. Interaction between rock strength and rock burst liability in fault zone (after Kidybinski).

Elevation m

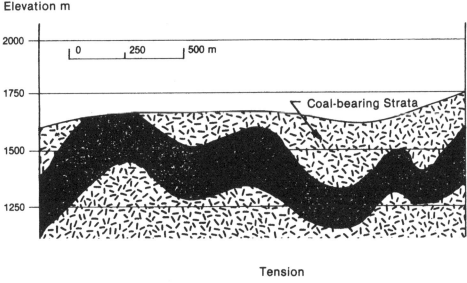

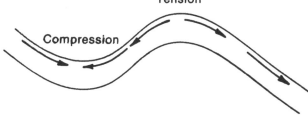

Fig. 5.1.4. Overstress in folded strata (Cardinal River, Western Canada).

3. Stresses of folded strata have been investigated in structural units which are bent and overstressed. When the strata is folded in the shape of a syncline or anticline, there are additional stresses, caused by flexural deformation, added to existing strata pressure. Under these circumstances the outer part of the anticlinal structure of a fold would be in lateral tension and the inner part of the synclinal structure would be in lateral compression which will increase already existing lateral compression (Fig. 5.1.4). Stresses governed by such fold structure have been observed in roadways in the Knurow Coal Mine (Upper Silesian Basin, Poland). Roadways driven in the anticlinal portion of the folds have not experienced any adverse ground pressure, however, the roadways driven in synclinal portions exhibited large-scale floor heave due to high lateral compression.[1] It was also established that the load on interchamber pillars is reduced by 30 to 60 per cent for mine workings which are located in the anticline structure. There are several theoretical analyses of stresses in folded rock. J. J. Golecki proposed a numerical solution for the stress-strain state in a fold structure. He calculated stress values for folded sandstone and came up with a value for the quality and quantity of stress increase in the bent area.[11]

4. Stresses of overthrusted strata. Overthrust strata exhibit two distinct zones of stress redistribution. The upper mobile thrust sheet is relieved of tectonic stresses but lower

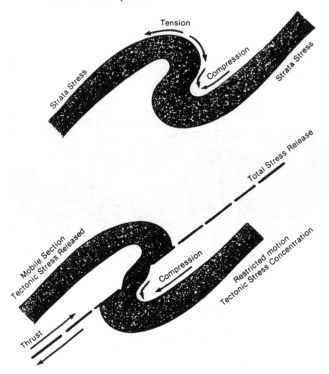

Fig. 5.1.5. Model of overstress and stress release of thrust fold.

stationary sheet exhibits a concentration of stress (Fig. 5.1.5). For example, at the Concordia Mine (Upper Silesian Basin, Poland), where the Number 3 coal seam was folded and overthrusted along a zone 450 m. The seam below the overthrust is subjected to violent bursting while the part above is free. The difference of 32 m in depth is not sufficient to explain a stress anomaly.[1] It is assumed that due to release of stresses the mine workings in the upper region are safe while the lower section is still under pressure and represents a hostile environment for underground mining.

In situ stress measurements indicate that the stresses of deformed strata are generally higher than the stresses calculated by equations of linear elasticity for flexural or bent structure. It should be noted that for the extrapolation and interpretation of stress in rock it is necessary to have a good knowledge of the structural geology of the sedimentary formations.

5.1.5.   *Residual stresses*

Residual stresses have a local character and they remain in rock strata after all external tectonic and gravitational forces and their moments are removed. The magnitude and orientation of these stresses are not well known. From in situ investigations of the virgin strata pressure it could be inferred that residual stress exists only on a mesoscopic scale. Under these conditions, residual stress does not have a noticeable influence on ground stability in a mine. The volume of rock strata which contains this unbalanced system of forces must surely be dependent on the continuity of the strata.

The existence of residual stress has been indicated in core samples from exploration drillholes in Mount Head Basin, Alberta. It has been assumed that the residual stress is related to the direction of tectonic shear stress and will have been relieved to some extent by fracturing. For example, some core samples of sandstone indicate that there are stress planes of the same family on which relief by failure does not occur. These planes of residual shear stress have been physically indicated on the core samples when, after a certain period of time, they broke up (delayed fracturing). Residual shear stress tends to be concentrated in local sets of curvilinear surfaces, with higher strain than average in sandstone strata.

5.1.6. *Thermal stresses*

Thermal stresses depend on the state of strata stress, the properties of the rocks and the thermal effects within the earth. There is a division of opinion on the influence of thermal stresses in underground coal mining at the present depths. Some investigators believe that thermal stresses play a subordinate role while others suggest that with an increase in the geothermal gradient, thermal stresses are increased and are further amplified by self heating processes in the coal. We must bear in mind that the temperature field of coal strata was formed over geological periods of time and temperature anomalies are induced when an excavation is made. There has not yet been any research in which an adequate effort was made to analyze the influence of temperature fields and their variations on strata mechanics. However, there is remarkable evidence of changes of stress-strain in rock strata due to seasonal variations. For example, in underground coal mines of Western Canada, the maximal tensile stresses and related roof falls occur in summer and the maximum compressive stresses and related roof falls occur in winter. Between these two seasons the maximum surface temperature difference could be up to 75° C.

Besides the influence of seasonal and climatic factors there is also an influence of some coal extraction factors which affect the thermal stress-strain state in the strata. There is evidence that the temperature rises in coal pillars when it is deformed by impact force due to instantaneous stress transfer from suddenly collapsed mine structures. Analysis of the causes of spontaneous combustion in coal has indicated that one of the factors is mine depth. Also, with an increase of deformation and compression of coal pillars there is a rise in temperature. These effects are interrelated and could be analyzed using the principles of thermodynamic theory.[12]

5.2. LATERAL TECTONIC STRESS

Underground investigations show, that besides geostatic and mining stresses, there are also a lateral tectonic stresses which influence the stability of mine openings.

Deformation and failure of the mine roof and floor structure in relation to lateral tectonic stress has been studied in underground coal mines using the room-and-pillar system of mining in the Canadian Rocky Mountain fold belt. These studies suggest that there are three limiting cases for the orientation of an opening relative to the direction of principal lateral compressive stress, as listed below:
1. Mine opening perpendicular to lateral stress;
2. Mine opening parallel to lateral stress;
3. Mine opening oblique to lateral stress.

Each case of stability of mine roof and floor structure is individually discussed.

5.2.1. *Consideration of lateral stress*

Comprehensive studies and underground measurements of geological stresses have been described by many authors.[13] The general conclusions of these studies are: firstly, that there is a relationship between stresses and geological structure; secondly, that lateral stress might exceed the vertical stress particularly at relatively shallow depths.

Studies of the influence of lateral stress on structure stability have been carried out by in situ investigations, physical models and digital models.[14] For example, R. A. Yeates, by finite element modeling, showed the development of a different stress field in the immediate roof of mine openings as a function of variations in the ratio between vertical and lateral stresses (K). At a ratio of K = 0.5, the immediate roof is subjected to tensile stresses and ribs suffer vertical loading. However, at a ratio of K = 3.0, intense lateral stress concentration is elongated to the mine roof (Fig. 5.2.1). The latter situation (K ≃ 3.0) might approximate stress conditions encountered in the Western Canadian coal field where lateral tectonic stress is oriented perpendicular to the Rocky Mountain range. It is due to the horizontal-shear thrust component of the active tectonic stress which caused folding and thrust faulting of the coal bearing strata (Fig. 5.2.2).

Underground investigations in coal mines of this region have exhibited the effect of lateral tectonic stress on the interaction of yielding deformation between pillars, roof and floor. The presence of lateral stress is particularly well exhibited by a broad

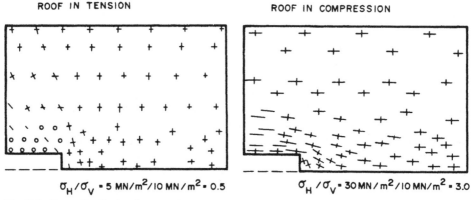

ROOF IN TENSION

ROOF IN COMPRESSION

$\sigma_H / \sigma_V = 5\,MN/m^2/10\,MN/m^2 = 0.5$

$\sigma_H / \sigma_V = 30\,MN/m^2/10\,MN/m^2 = 3.0$

Fig. 5.2.1. Stress distribution due to ratio σ_H/σ_V.

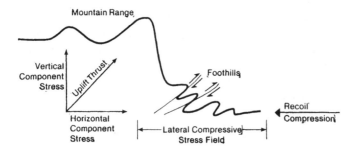

Mountain Range

Vertical Component Stress

Uplift Thrust

Foothills

Recoil Compression

Horizontal Component Stress

Lateral Compressive Stress Field

Fig. 5.2.2. Lateral compressive stress field.

pressure arch which is subjected to high shear stresses. Here low angled fracturing occurs as a function of the strata strength and the magnitude of the shear stresses. Further discussion particularly emphasizes this mechanical phenomena.[15]

5.2.2. *Mine openings perpendicular to the lateral stress*

About eighty per cent of roof falls in underground mines are in openings perpendicular to the main lateral tectonic stress direction.[16, 17]

1. Roof falls. Horizontal load (P_H) on roof beams is made of:
  a) 1/3 of the vertical load (rock beam weight);
  b). Friction and cohesion between coal and roof stratum;
  c) Horizontal tectonic compressive pressure.
Roof failure can be analyzed by the principles of solid mechanics as follows:
  a) The acting horizontal stress of the beam is:

$$\sigma_H = \frac{P_H}{wt} \tag{1}$$

  b) Compression is followed by shear and slip at the edges and can be expressed by Hooke's Law:

$$\sigma_H = \frac{E\Delta l}{l} \tag{2}$$

and

$$\frac{E\Delta l}{l} = \frac{P_H}{wt} \tag{3}$$

so that

$$\Delta l = \frac{P_H l}{wtE} \tag{4}$$

where w, t, l, are the width, thickness and length of beam and Δl is the increment of contraction.

  c) There is a tendency for the rock beam to buckle. The critical stress (σ_{cr}) can be represented by Euler's equation:

$$\sigma_{cr} = \frac{\pi^2 EI}{l^2} \quad \text{and} \quad I = \frac{wt^3}{12} \tag{5}$$

where
  I = the moment of inertia;
  l = length of the roof span.

  d) Deformation is a maximum at this critical stress and is determined by:

$$\sigma_{cr} = \frac{Ewt^2}{12} \frac{\pi^2}{l^2} = \frac{E\Delta l}{l} \tag{6}$$

so

$$\Delta l = \frac{wt^3 \pi^2}{12\ l} \tag{7}$$

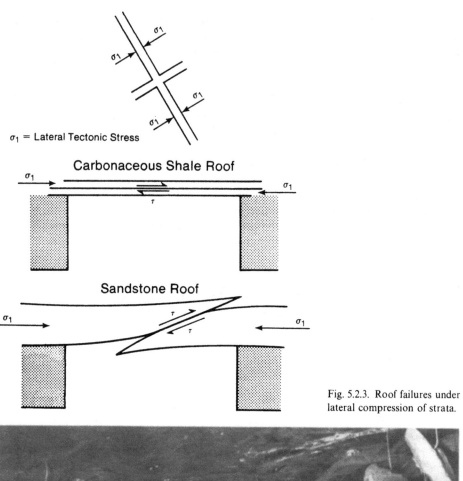

σ_1 = Lateral Tectonic Stress

Carbonaceous Shale Roof

Sandstone Roof

Fig. 5.2.3. Roof failures under lateral compression of strata.

Fig. 5.2.4. Roof falls in compressive stress field (sandstone).

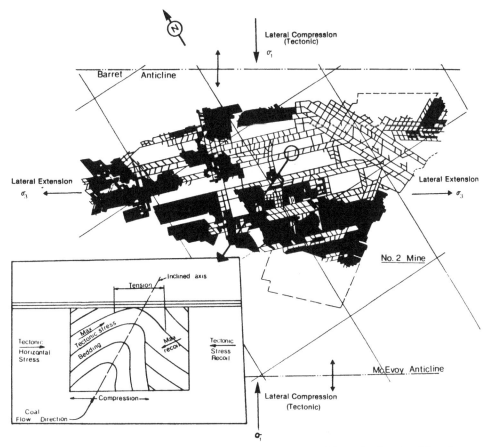

Fig. 5.2.5. Orientation of Number 2 Mine to lateral tectonic stress, with detail of mode of floor heave (McIntyre Porcupine Mines Ltd.).

There are two modes of roof failure depending on rock types: 1) slip along bedding planes in thin bedded carbonaceous shale and siltstone and 2) shearing at low angles or along small faults in thick bedded siltstone and sandstone (Fig. 5.2.3).[16] The massive sandstone roof of the mine openings will cave in large blocks because internal support by rock bolting is insufficient to resist the concentrations of lateral shear stress (Fig. 5.2.4).

2. Floor heave. Floor deformation in entry ways driven perpendicular to the lateral tectonic stress is accelerated, particularly in thick soft coal seams (Fig. 5.2.5). The characteristics of floor heave are as follows:

a) The floor coal is in a visco-plastic state and has a tendency to flow into roadway openings.

b) The heaved coal parallel to roadway axis forms anticline structures with an inclined axis.

c) Heave deformation is accelerated by the presence of high lateral compression, which is partially released by coal expansion into roadway openings.

Floor heave structures and large folds of the coal field must have tectonic forces as a common cause since the results of the geological structures are similar to those of the small scale releases of stress seen in the roadways.

3. Pillar convergence. Pillar convergence by coal extrusion is briefly explained below.[18, 19]

a) **Rib sides** of such structures as pillars which are perpendicular to the lateral tectonic stress exhibit coal extrusion of up to 10 cm into the opening immediately after entry excavation.

b) The rate of coal extrusion into the opening is synchronized with the rate of floor heave and with the rate of roof and rib separation (up to 2.5 cm) and roof displacement (Fig. 5.2.6).

c) Intensity of extrusion depends on the shearing resistance between layers of the coal.

Investigations in underground mines show that the after one year the pillar is totally fractured because of this continuous extrusion.[20]

5.2.3.   *Mine openings parallel to the lateral stresses*

The lateral compression parallel to the pillar-entry structure causes extension perpendicular to this direction which results in the formation of lateral tension cracks parallel to the entry way.

1. Roof falls. Deformation and failure of the roof span is less severe than in openings driven perpendicular to the lateral compressive stress, as discussed below:

a) The roof beam begins to sag immediately due to the excavation, with slip occurring along bedding planes.

b) Roof sag is followed by bed separation as gravity-induced tensile stress acts perpendicular to the bedding.

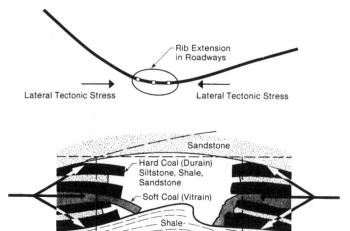

Fig. 5.2.6. Rib deformation in direction of lateral compression.

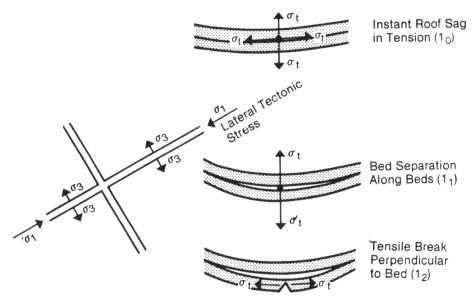

Fig. 5.2.7. Mechanics of roof failure under lateral extension.

Fig. 5.2.8. Stable roof conditions after displacement and limited sagging.

c) Breaking finally results when the tensile strength of the rock parallel to its bedding is less than the tensile stress acting in this direction (Fig. 5.2.7).

In accordance with this phenomena, unsupported false roofs of carbonaceous shale or bedded siltsone will fall down in a matter of hours, but unsupported sandstone roofs will stand for days or even weeks. If a massive sandstone roof is supported by rock bolts and wire mesh immediately after entry excavation, displacement deformation will take place only at the contact between the coal pillar and the roof span. Under these circumstances the roof might experience a limited sag deformation and it may stay stable for the entire life of the mine (Fig. 5.2.8).

2. Floor heave. Some aspects of floor heave are listed below:

a) The intensity of heave depends on the magnitude of the overburden load and on the rock type of the floor.

b) Heave of the coal floor at Number 2 Mine (McIntyre Porcupine Mines Ltd.) has the form of gentle anticlines having concentric internal structure, but heave of the shale floor at Number 10 Mine (McIntyre Porcupine Mines Ltd.) shows displacement along joints (Fig. 5.2.9).

c) The floor heave structures indicate that both vertical and lateral stresses are nearly equal and that both stresses are above the geostatic level.

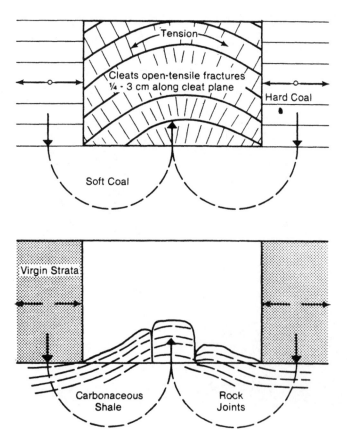

Fig. 5.2.9. Modes of floor heave of soft coal and carbonaceous shale under lateral extension (McIntyre Porcupine Mines Ltd.).

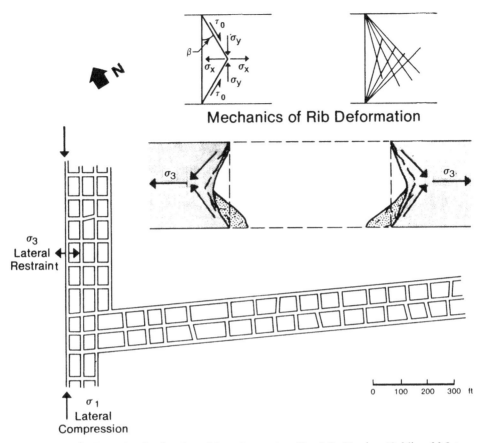

Mechanics of Rib Deformation

Fig. 5.2.10. Rib deformation in direction of lateral extension (Panel E. Number 10 Mine, McIntyre Porcupine Mines Ltd.).

3. Pillar convergence. Openings parallel to the maximum tectonic compressive stress experience deformation caused by lateral stress restraint, which influences pillar failure as described below:

a) The ribs start slabbing, as is typical for pillars which have frictional restraint of the outer zone.

b) Fracturing typical of the elasto-plastic state of coal develops with failure (Fig. 5.2.10).

c) The intensity of pillar slabbing depends on the magnitude of the acting stress and also on the direct shear strength of the coal in the cleat direction.[18, 19]

Deterioration of these pillars in one year is only half that of roadways driven perpendicular to the lateral tectonic stress.

5.2.4. *Mine openings oblique to the lateral stress*

Openings oriented in a direction between that of the maximum lateral compression and the maximum lateral extension avoid the effects of these two extremes to a certain degree.

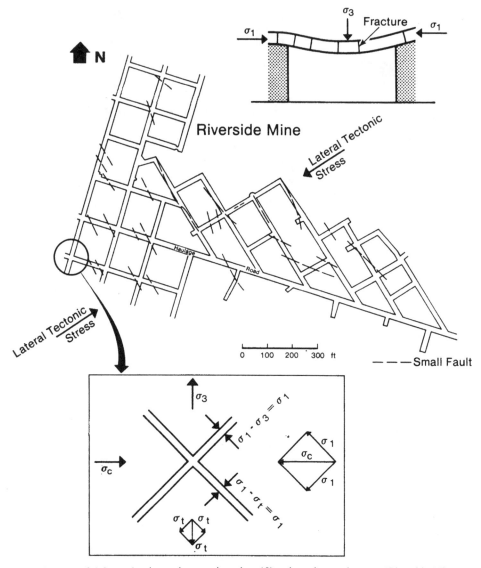

Fig. 5.2.11. Roof deformation in roadways oriented at 45° to lateral tectonic stress (Riverside Mine).

1. Roof failure. Some aspects of roof deformation and failure are as follows:

a) The active stress on the roof is compressive and of moderate intensity.

b) The magnitude of the active horizontal stress is decreased, because it represents the component of compressive tectonic stress.

c) Openings in this direction have a limited exposure to small faults because they are intersected at an angle. The fractures are compressed and the roof behaves as a Voussoir beam, so that the adverse effects of fractures on roof stability are eliminated (Fig. 5.2.11).[22]

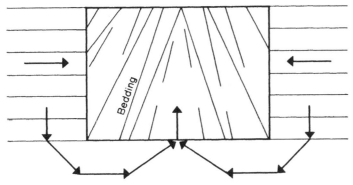

Fig. 5.2.12. Mode of floor heave of soft coal of roadway in virgin ground oblique to lateral tectonic stress (Number 2 Mine, McIntyre Porcupine Mines Ltd.).

d) The roof is stable if the compressive strength of the rock beam is greater than the active compressive stress:

if $\sigma_c > \sigma_1$,

and if the roof span is less than that permitted by the following equation:

$$1 = \left(\frac{5\sigma_c t}{3\gamma} \right)^{1/2} \tag{8}$$

where

  t = the beam thickness;
  c = unit weight;
  σ_c = the rock's compressive strength.

Orienting openings oblique to tectonic compression is an excellent example of the elimination of adverse natural stress conditions to achieve a stable roof.

2. Floor heave. Some elements of floor heave are as listed below:
  a) Coal in-flow from beneath pillars has a tendency to close mine openings.
  b) Coal heave has the form of typical "chevron-type" folds (Fig. 5.2.12).
  c) Coal heave along longitudinal axis (fold axis) produces steep shearing planes and coal fragmentation. This can be observed underground below sequeezed ribs in the brushed roadways.

This type of floor deformation suggests that lateral stress is greater than vertical and that it has a major influence on floor heave.

3. Pillar convergence. Pillars are stable, except at the intersection crosscuts (Fig. 5.2.13).
  a) The corners of pillars in the direction of maximum compressive stress are in an intense state of fragmentation and sloughing, which can progress up to 3 m in one year.
  b) If the corners of pillars in the direction of the main compressive stress are constrained, then they could violently and suddenly burst (Riverside Mine).
  c) Corners of the pillars in the direction of maximum lateral extension are also in a state that causes coal sloughing, but to a limited extent only.

McIntyre Porcupine Mines Ltd.

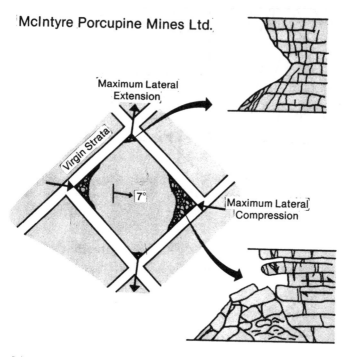

Canmore Mines Ltd.

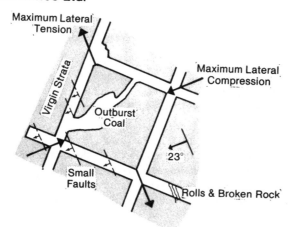

Fig. 5.2.13. Rib deformation by yielding and by outbursting at roadway intersections (oriented 45° to lateral tectonic stress).

To achieve satisfactory stability of pillar-entry roadways it is necessary to decrease the number of intersections.

The orientation of mine openings relative to that of tectonic stresses as well as small faults and cleats is an important element for design of the room-and-pillar system in mines.

5.3. STRESS STATE OF SOFT COAL SEAM

In situ stress measurements in sedimentary strata from a number of regions show differences in the magnitude of stress in hard and in soft stratum. For example, vertical stress measurements at a depth of 225 m in sandstone exhibited a magnitude of $\sigma_v = 9.3\,\mathrm{MN/m^2}$ but in shale the value was $\sigma_v = 3.7\,\mathrm{MN/m^2}$ and the calculated geostatic stress was $5.2\,\mathrm{MN/m^2}$. It is apparent that the soft stratum could not support the superimposed load and it was transferred to the stronger rock unit.[23]

In situ stress measurements in virgin coal seams confirmed the phenomenon described above and indicated that the stress distribution within a coal seam is characterized by an equalization of the vertical and horizontal stresses regardless of the state of stress in the surrounding strata. Under these circumstances a hydrostatic stress state has been induced in the coal seams.

Generally speaking, a soft coal exhibits a decrease of strength and elasticity with increase in mine depth. For example, it has been determined that for sections at 250 m depth there is a 25 per cent decrease of strength and a 50 per cent decrease in the modulus of elasticity relative to sections of surface.

There is a direct relationship between mine depth and the stress state in coal seams and there is a relationship between the state of stress, coal properties and behaviour. Three types of behaviour are proposed for coal seams:

1. Shallow coal deposits are stress relieved and the coal exhibits an elastic-viscous behaviour;

2. At moderate depth the deposit is in a transitional stress state where a plastic-viscous behaviour predominates;

3. In deeper deposits the coal exhibits viscous behaviour indicating a hydrostatic stress state.

This analysis of state of stress in a coal seam is largely concerned with the soft coal seams of the Rocky Mountain fold belt of Western Canada. The philosophical approach of this analysis is based on the mechanics of time dependent deformation and coal viscosity and takes into account the effects of long-term geological loading and the relaxation of upper crustal strata due to erosion.

An analysis of the mechanics of the contemporary stress state in this case follows the same principles that were presented in the publication "The state of geological stresses in the Athabasca oil sand deposits" and are discussed in the following paragraphs.[24]

5.3.1. *Stress relieved seams (shallow deposit)*

The majority of shallow deposits are in a state of relieved stress because a great deal of the overlying load has been removed by erosion. Such coal seams have been investigated under two sets of conditions as follows: firstly, as an intact deposit where the maximum overburden thickness is 250 m (dead load up to $6.5\,\mathrm{MN/m^2}$); secondly, where the overburden has been stripped and the coal is exposed to the surface (dead load is zero). To simulate these states of loading and unloading, laboratory investigations have been carried out by progressive loading of the samples and time dependent loading at constant load. For example, the time dependent loading for constant load conditions suggest the following behaviour:

– When a load is applied, the material is instantaneously strained. This is the elastic component of deformation;

– The strain rate increases with loading time after the instantaneous stress is exhibited. This is the viscous component of deformation.

These inelastic properties of coal are anologous to a Maxwell body which is a mechanical model composed of a spring (elastic component) and a dashpot (viscous component) in series. This model was introduced by Maxwell to describe the behaviour of materials such as pitch which show instantaneous elasticity and flow under stress if the stress is applied for a sufficiently long time period. Our investigations also suggest that this model, to a certain degree, describes the stress state of shallow coal deposits because the overburden load acts simultaneously on the elastic element and on the viscous element, so that the total strain is a sum of the elastic and viscous deformations, which is written:

$$\varepsilon = \frac{\sigma_0}{E} + \frac{\sigma_0 t}{\eta} \tag{1}$$

elastic   viscous
strain   strain

A similar relation can be written for the shear stress and strain since the total shear strain must be the sum of the elastic and viscous shear strains

$$\gamma = \frac{\tau}{G} + \frac{\tau t}{\eta} \tag{2}$$

Both equations suggest that the rate of deformation also depends on the coefficient of viscosity of coal which is not a constant due to variation in petrographic composition and intensity of cleat.[26]

The relation between elastic and viscous strain can be expressed by a strain/time diagram (Fig. 5.3.1). First of all, for different values of instantaneous strain the intensity of viscous deformation varies. This phenomenon suggests that the magnitude of viscous flow of coal increases with depth and that its shear strength decreases.

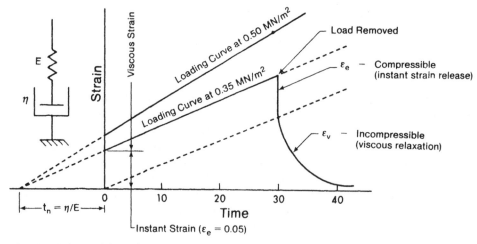

Fig. 5.3.1. Soft coal deformations for various loading values and stress relief conditions.

The straight lines of viscous linear deformation for various values of σ_o approximately emanate from a single point of the time axis given by $t_n = \eta/E$. This suggests that for a constant viscosity the coal has the same time of relaxation for various load magnitudes after the pressure is removed (for example, at different stripping ratios). However, the relaxation time for a shear stress is $t_t = \eta/G$.

The relaxation of the coal due to mine excavation might be an important stability factor in two cases: first, development of extension cracks until relaxation time has passed; and secondly, formation failure (slumping) due to deterioration of shear strength because part of the viscous deformation is permanent. These mechanics might be explained by the fact that the elastic strain (ε_e) relaxes instantaneously, but the viscous strain (ε_v) relaxes with time (Fig. 5.3.1). The stress magnitude can be expressed by the assumption that at time zero the strain applied is $\varepsilon = \varepsilon_o$ and is held constant, then

$$\frac{d\varepsilon}{dt} = 0; \quad E\frac{d\sigma}{dt} = -\frac{\sigma}{\eta} \tag{3}$$

and, integrating

$$\int_{\sigma_0}^{\sigma} \frac{d\sigma}{dt} = -\frac{E}{\eta} \int_0^t dt \tag{4}$$

gives

$$\sigma = E\varepsilon_0{}^{-Et/\eta} \tag{5}$$

From equation (5) it can be concluded that stress falls off exponentially and relaxes with time. Also an exponential equation can be written for the shear stress:

$$\tau = G\gamma_0{}^{-Gt/\eta} \tag{6}$$

Equation (6) states that:

a) The magnitude of the shear stress depends on a constant strain value, the modulus of elasticity $(G = E/3)$ and the coefficient of viscosity (η);

b) And that part of the viscous deformation is permanent, consequently the shear strength of the original material has to be decreased.

It should be noted that relaxation time observed under laboratory conditions is much shorter than under conditions of geological loading and unloading by mine excavation.

5.3.2. *Transitional stress state* (*deposit at moderate depth*)

This stress state may be related to depths between 250 m and 450 m at which the majority of underground mines in the Rocky Mountain fold belt have been located.

In situ investigations[27] and laboratory tests for moderate loading conditions (up to 12 MN/m^2)[28] suggest the following behaviour of the coal:

– The shear strength of the material decreases with depth while the percent strain at failure increases;

– The shear stress equals the yielding stress and further deformation follows a pattern of shear flow resistance.

The increase of shear strain rate with an increase of confining pressure suggests that

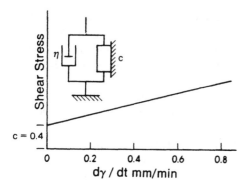

Fig. 5.3.2. Resistance to flow at rupture.

the coal tends to be in a plastic-viscous state. For these stress conditions, it can be assumed that a yield stress exists.

Such behaviour of coal may be explained by a plasto-viscous substance which can be modeled by a friction block (plastic component) and a dashpot (viscous component) in parallel (Fig. 5.3.2). This model was introduced by Bingham[25] to describe the behaviour of plastic material under yielding stress conditions.

The shear stress of soft coal at the yielding point (below the failure stress) has to satisfy the following conditions:

$$\tau > C \quad \tau = C + \eta \dot{\gamma} \quad \text{yield deformation}$$
$$\text{(no shear flow)} \tag{7}$$

because for

$$\tau \leq C \quad \dot{\gamma} = 0 \quad \text{yield deformation}$$
$$\text{has ceased (shear flow)} \tag{8}$$

where

η = viscosity of coal;
C = initial yield stress (below failure stress).

Actually, shearing resistance is a combination of the substances strength and viscosity.

From laboratory testing at the yield stress, it may be assumed that both a shear stress and a normal stress will act on a shearing plane. This suggests that the coal particles are pressed together forming a compact medium which could explain the increase in shearing resistance. This behaviour is in accordance with the principles of elementary mechanics concerning dry friction between rough surfaces. The relation between a shear and normal stress can be represented schematically by Figure 5.3.3 when the total yield stress consists of an initial yield stress and a shearing resistance. This is given by:

$$\tau_{tot} = C + \tau_f \tag{9}$$

From laboratory tests it is known that the initial yield stress 'C', which is approximately equal to the cohesion,[7] has a very small magnitude (something between 0 and 6 KN/m^2) which suggests that the main stress component is a shearing resistance (τ_f). If τ_f is considered as a value proportional to the normal stress, where the coefficient of

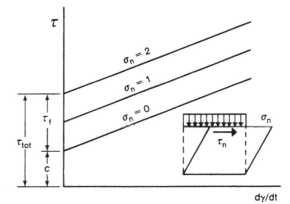

Fig. 5.3.3. Total yield stress deformation.

proportionality is the shearing resistance (friction), equation (8) can be rewritten as

$$\tau_{tot} = C + \tau_f \sigma_N \tag{10}$$

This equation represents a Coulomb failure criterion. If the equation of the Bingham model is generalized by Coulomb's postulation, the behaviour of coal under plasto-viscous conditions may be expressed as:

$$\tau_{tot} > C + \tau_f \sigma; \quad \dot{\gamma} = \frac{\tau_{tot} - C + \tau_f \sigma_N}{\eta} \tag{11}$$

$$\tau_{tot} \leq C + \tau_f \sigma_N; \quad \dot{\gamma} = 0 \tag{12}$$

If the magnitude of shear stress is below the shear strength, the rate of deformation given by the equation will depend on three parameters:
 – Initial yield strength (approximately equal to cohesion, C);
 – the angle of internal friction (shearing resistance $\tan \phi = \tau_f$);
 – Viscosity of the coal (η).
At the point where the shear stress equals or exceeds the magnitudes of the coal's shear strength, the material will begin to deform continuously. This deformation will continue as viscous flow (dashpot) like a fluid without shear resistance.

5.3.3. *Hydrostatic stress state (deeper deposits)*

This state of stress occurs within coal seams in areas where the depth is over 450 m. Such cases have been exhibited in several underground mines, which ceased production because of uncontrollable strata.

Laboratory investigations have shown that some soft samples under triaxial compressive loading cannot support vertical loads greater than 20 MN/m^2 because the horizontal and vertical components of stress became equal. Coal in such a hydrostatic state of stress is without shear strength. This corresponds to fluid behaviour where there is no shear resistance (C = 0) and can be illustrated by the Newtonian model (dashpot). The relationship between the stress and strain rate for a perfectly viscous material is proportional to the coefficient of viscosity (Fig. 5.3.4).[25] According to the Newtonian Law, the soft coal cannot withstand any shear stress without deforming

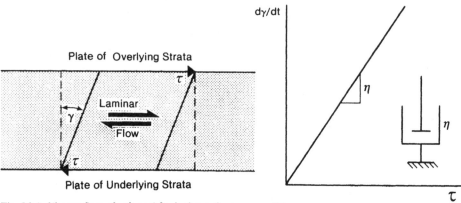

Fig. 5.3.4. Viscous flow of soft coal for hydrostatic stress conditions.

permanently. Such flow deformation depends on the magnitude of shear stress, the coefficient of viscosity and time. It is given by:

$$\tau = \frac{\eta \mathrm{d}\gamma}{\mathrm{d}t} = \eta \dot{\gamma} \tag{13}$$

And when integrated (for $t = 0 - t$ and $\gamma = 0 - \gamma$)

$$\tau = \int_0^t \mathrm{d}t = \eta \int_0^\gamma \mathrm{d}\gamma \tag{14}$$

is given by:

$$\tau t = \eta \dot{\gamma} \tag{15}$$

$$\dot{\gamma} = \frac{\tau t}{\eta}$$

Further analysis of the shear stress deformation can be obtained by applying fluid mechanics theory and assuming that under high pressure the coal behaves like an incompressible fluid ($K = 1$).

When an underground mine is excavated, the natural hydrostatic stress within the coal deposit will be relieved by coal flow (extrusion) at a rate which would depend upon the coal's viscosity.

Laboratory investigations of coal samples subjected to triaxial pressure showed that with an increase of the ratio between σ_1 and σ_3, the shear rate is also increased as illustrated in Figure 5.3.5. The strain rate acceleration is directly proportional to the ratio between σ_1 and σ_3. This phenomenon suggests that equilibrium conditions within virgin soft coal deposits after openings are excavated will be maintained by the viscous flow rate. Subsurface mine openings in the initial excavation will look like a stable structure but, as time progresses, the accelerated shear rate will cause it to collapse within a relatively short period of time. This has been monitored and observed in underground coal mines of the Number 4 seam in the Grande Cache coal basin.

In this analysis, prediction of the nature of strata stresses in soft coal deposits due to

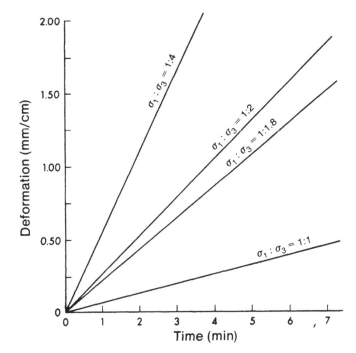

Fig. 5.3.5. The rate of viscous deformation of soft coal for confined stress conditions ($\sigma_1 + \sigma_3$ = 27 MN/m²).

overburden pressure represents an extrapolation of results from laboratory tests on a number of samples as well as field tests and observations. Defining the nature of the shear stress and deformation of virgin coal deposits for mine design should be based on investigations within a pilot subsurface mine opening.

5.4. HYDRODYNAMIC STRESSES

The stress-strain state of coal-bearing strata will be altered by hydrodynamic stress if water and/or gas are present. The hydrodynamic stress due to these fluids in porous and jointed rock strata and coal seams is dependent upon the geological history and depth of burial of the strata. Under these circumstances the effective normal stress is

$$\sigma_{(\text{eff})} = \sigma_v - mp \tag{1}$$

where
  σ_v = vertical virgin stress of strata;
  m = coefficient of the void ratio;
  p = the pore pressure of the fluid.

Stresses due to fluid pressure acting upon the surfaces of pores, joints, or cavities within the rock strata are also influenced by the state of stress of the strata.

 The geotechnical and mining literature contains numerous descriptions of these hydrodynamic stresses, their origin and associated strata mechanics so that only a brief summary is presented here.

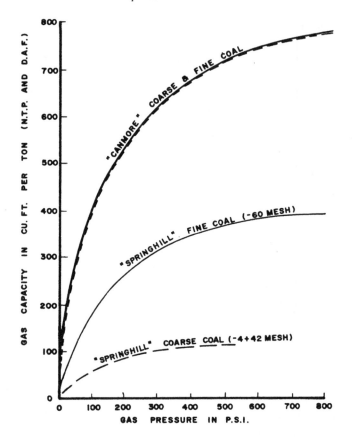

Fig. 5.4.1. Gas capacity isotherms of Canmore and Springhill coals (Canada).

5.4.1. *Gas pressure*

The presence of gas in underground coal mines creates the hazards of fire, explosion, bumps and asphyxiation ever since mines extended to depths below the surface. The development of new mines at greater depths with rapid excavation rates have increased the concern about gas outbursts.[30]

1. Coal gasification. It might be related to free gas within cracks, fissures and pores or to an adsorbed layer on the interior surface of fissures and pores and to gas desorption and its emission. The free gas in coal represents only 10 to 20 per cent of the total gas present and its appearance is not important for coal strata mechanics but, sorbed and desorbed gases are of the greatest importance and they should be briefly represented as follows:

a) The term 'sorbed gas' refers to mechanism of adsorption and absorption where molecules of gas may also be dispersed between the coal molecules. The amount of sorbed gas depends on the pressure, temperature, internal surface and nature of the coal seams. The saturation point of coal by methane will occur at different pressures. Highly porous coal with pores of small radii requires a pressure greater than 500 atmospheres for saturation but coal of lower porosity reaches the saturation point at pressures of 150

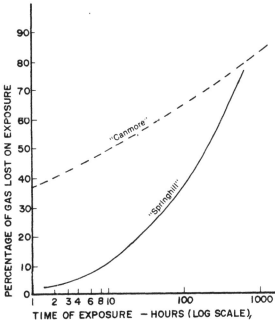

Fig. 5.4.2. Comparison of gas desorption rates of Canmore and Springhill coals (Canada).

to 200 atmospheres. For example, Canadian bursting coking coals reached equilibrium after 24 hours soaking in methane (Canmore), but non-bursting coking coal did not reach equilibrium after 24 hours of soaking time (Spring-hill). As illustrated in Figure 5.4.1 bursting coal has one isotherm for both coarse and fine coal, but non-bursting coal has two isotherms and the coarse coal has a lower isotherm than the fine coal. This phenomena is attributed the lower porosity of non-bursting coal.[31] The sorption capacity of a coal for CO_2 gas is far higher than for CH_4 gas. The amount of CO_2 adsorbed is almost 2 to 3 times that of CH_4 gas.

b) Desorption of gas is a reversible phenomena which takes place in coal gel structures. Any changes in the equilibrium conditions occurring in the coal mass (pressure, temperature, structure of pores due to cracking) will tend to reverse this state. The rate of desorption of gas is proportional to the rate at which conditions are changed and the intensity of change of these conditions. Desorption capacity with rapid gas release is an important factor for coal outburst phenomena. For example, bursting coking coal exhibits a very quick gas release (Canmore) in relation to non-bursting coking coal (Springhill) as shown in Figure 5.4.2. Methods of utilizing desorption of gas based upon determination of the amount of gas present in coal samples has been described as indices of coal outbursts. In this case the measure of desorbed gas could be either the amount of gas liberated over a given period of time under controlled field conditions, or pressure developed in the chamber containing the sample.[30]

The coal gasification conditions and types of gases present primarily depend on the origin of the coal and its coalification path. It should be noted that rocks adjacent to coal seams might also contain gas derived from organic matter in the rocks or gas which has migrated into porous space in the rocks.

2. Flow mechanics. The study of gas emission from unconfined fragmented coal and from coal faces in mines are the most common aspects of gas flow mechanics which have been studies in coal mining. In the first case, the studies are carried out in a laboratory under controlled conditions where mathematical analyses are of primary concern. The second approach is more applicable to practical mining, but control of physical factors in the field is more difficult. Gas flow through solid, unfissured coal is by very slow diffusion of gas molecules through the pores due to the difference in concentration. In cleated coal, gas flow is much faster and due to the pressure difference. Both types of gas flow occur in the coal and it is difficult to distinguish between them.

The mechanics of gas flow through coal is described by Darcy's Law. The gas mechanics around mine excavations can be defined by the following parameters:

a) The flow of gas primarily depends on the permeability of the coal and internal pressure of the gas, as expressed by Poisenille's Law:

$$Q_x = -\frac{k_x}{\eta} + \frac{dp}{dx} \tag{2}$$

where

k_x = permeability of the coal;
p = internal pressure of gas;
x = distance from the face boundary;
η = coefficient of viscosity of the gas.

The gas flow is influenced by coal porosity which depends on the natural porosity composed of capilarity and tectonic cleating as well as induced porosity produced by fracturing due to mining stresses, where strength of coal is an additional factor. According to this the permeability characteristics at a distance could be expressed by the following equation:[32]

$$k_x = k_o(1 + Je^{-\beta})^3 \tag{3}$$

where

k_o = natural coal porosity;
J = coefficient due to tectonic coal fracturing (> 1 fractured coal);
β = coefficient of coal strength.

This coefficient could be given by the following equation:

$$\beta = \frac{2.3\,\sigma_c}{wh(1 + \sin\alpha)} \tag{4}$$

where

σ_c = uniaxial compressive strength of coal;
w = width of the mine opening;
h = depth of opening;
α = angle of seam inclination.

The outflow of gas is maximum at the boundary of the mine opening and exponentially decreases to the virgin gas flow in the solid coal (Fig. 5.4.3/a). The rate of gas flow is directly proportional to coal permeability and inversely proportional to the internal gas pressure.

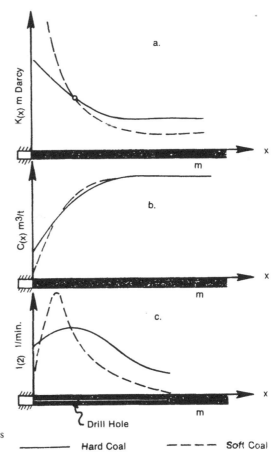

Fig. 5.4.3. The diagrams of gas flow (a), gas and horizontal (σ_x) stresses of the pillars.

——————— Hard Coal — — — — Soft Coal

b) Gas capacity is expressed by volumes of gas per unit volume of coal as shown in Figure 5.4.3/b. It is obvious that broken coal at the free boundary of a mine opening has a low gas capacity due to an increase in gas desorption and gas outflow and is described by the following relationship:[33]

$$C_x = \frac{k_x}{\eta} \tag{3}$$

The depth of zone which is appreciably degasified is in the range between 0.2 and 0.8 m.

c) Gas emission or discharage can be represented by the velocity of gas per unit area at a distance from the free boundary of an opening, as given by the following equation:[32]

$$i_x = \frac{k_0}{2\eta RT} 1 + Je^{-\beta x} \frac{{}^3 dp^2}{dx} \tag{4}$$

where

　　R = gas constant;
　　T = temperature of the seam.

Analysis of Equation (4) suggests that gas discharge would depend on two functions: firstly, on the natural porosity of the coal seam; secondly, on the distribution of gas pressure in an area around the opening. The gas emission also depends on coal strength (Fig. 5.4.3/c) and is going to cease at a shorter distance from the free face in hard coal than in soft coal. This suggests that for coal degasification the area of drainage from a drillhole is greater in soft coal seams than in hard coal seams.

The hydrodynamic stresses of coal strata induced by gas flow are important for mine safety and stability. Coal deformation and gas outburst are the result of an integrated action between abutment stresses and gas pressure. The controlling factor for triggering of a violent deformation is the length of the zone (x) behind a free face which is influenced by the mine excavation. The most common technique for indicating this length by drillholes or by instrumentation using electric resistivity where an increase of resistivity coincides with a stress increase. The disadvantage of this method is that individual coal lithotypes have different resistivity, for example fussain 10^5 ohm/cm and durain 10^{10} ohm/cm. The most accurate method involves monitoring the degree of increase in acoustic emission activity which indicates those seams wherein gas outbursts are imminent. Analyses of the amplitude-frequency distribution of such acoustic emissions might also indicate variations of the stress distribution in these seams.[34]

5.4.2.   *Water pressure*

The water flow and water pressure in underground coal mines will depend on the hydrogeological characteristics of the coal bearing strata as on mining factors. From the point of view of mine stability three problems are important as discussed separately below:

1. *The flow of water in virgin strata* is governed mainly by rock properties and geological structural defects of the coal deposits. Generally speaking, the amount of water present and the rate of flow depend upon the water source and the permeability of rock strata. An example of this problem exists in by the Mount Head coal deposit in the Rocky Mountain fold belt.[35]

The water bearing capacity and related hydrodynamic stresses of coal strata at this location have been analyzed in relation to elevation of the ground surface by two cases (Fig. 5.4.4):

Firstly, coal bearing strata above an erosion valley are affected by several hydrogeological phenomena, some of which are described below:

a) Surface water run-off occurs mostly along the mountain slopes into the valley. Contours of the water table follow approximately a configuration of the ground surface forming an accumulation basin in the valley. However, most of the water accumulated in the valley is drained off by existing creeks.

b) The ground water flows through an inclined aquifer (sandstone) and accumulated below the valley surface. Hydrodynamic pressure in coal bearing strata above the drainage level should have little affect on mine stability because of water outflow from the level entry.

c) The hydrodynamic stress state of water bearing strata can be altered if faults are present. For example, those faults which intersect coal seams could be water feeders to an underground mine.

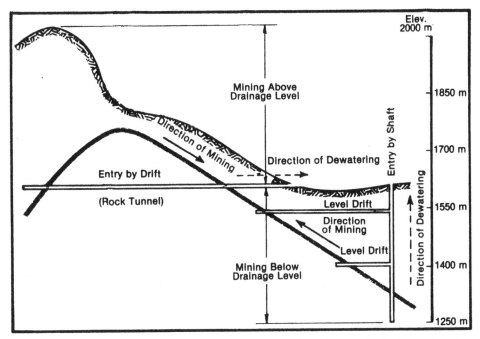

Fig. 5.4.4. Model of dewatering of coal deposit in relation to drainage levels at Rocky Mountain region.

The hydrogeological conditions above drainage at Mount Head coal basin are similar to other underground coal mines above drainage level in this region, which have not experienced any difficulties in relation to hydrostatic stresses of coal bearing strata.

Secondly, the coal bearing strata below the valley floor exhibit a rather complex hydrogeological case, which was observed during drilling B-1 and B-2 holes. This drilling was prematurely stopped due to high hydrodynamic stresses in the coal strata, which progressively increased with depth (Fig. 5.4.5).

Hydrogeological investigations should be carried out as part of the planning and development of underground mines. Mine design should be based on extensive hydrogeological field investigations (pump tests and others) so that the data can be processed by computer modeling to provide a versatile and reliable simulation of hydrological behaviour and the state of hydrodynamic stress of the formations.

2. *The water flow system altered* by mining activities may have a significant effect on underground coal mine stability. In this case the hydrodynamic stress state of rock strata is governed by both water pressure due to the natural pore system of the rock and a seepage force within joints and fractures induced by mining stresses.

Mine openings can be seriously damaged due to loss of stability when water movement is generated through a fractured zone. 'Damming' of the water which is trying to drain into an excavated space generates hydrodynamic stresses which act in the direction of flow i.e. into the mine opening. Under these circumstances, collapse of the mine structure can often occur.

The alteration of flow systems and the development of new ones is particularly

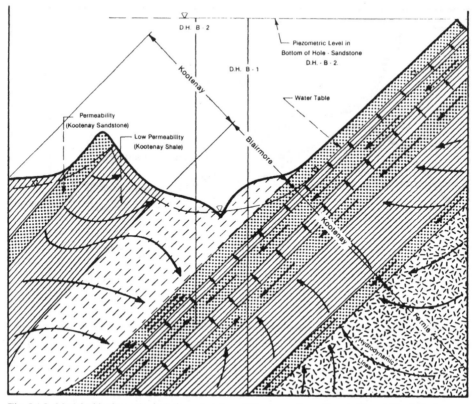

Fig. 5.4.5. Model of hydrogeological characteristics of Kootenay formation at Mount Head Basin.

common in mining systems plagued with roof caving followed by strata fissuring to the ground surface. Such roof strata caving accompanied by subsidence can divert surface run-off water into an underground mine. In this case the large quantities of water in the mines not only cause significant deterioration in the stability of rock strata but also create the nuisance of wet working places and hazard of slippery footing. Often such subsidence is the cause of underground mine flooding and the formation of surface depressions filled with water.

A number of computer models for investigation of the altered water flow patterns due to underground mine excavation have been developed. Of the particular interests are models which could analyze deep mine conditions. Models of this type usually analyze the relationship between the amount of water infiltration from the watershed to the subsurface flow which is further related to the daily mine flow.[36, 37]

3. Water drainage as a tool to decrease hydrodynamic stresses has been applied in underground mining with a variety of results. Those cases of limited success are probably due to incomplete knowledge of hydrogeology and hydrodynamics rather than in the effectiveness of the drainage method. Generally speaking, hydrogeology and the control of water pressure in underground mines was ignored in the past because of lack of expertise in this field. When large water inflow occurred the mine was usually

Fig. 5.4.6. Dewatering of the water saturated sand which is immediate floor of the
thick coal seam (Kreka Coal Mine, Yugoslavia).

abandoned and flooded. Three types of drainage methods should be considered:

a) Drainage by drillholes from the ground surface should be done in advance of underground excavations and should aim to decrease water pressure in the strata before mining is commenced (same as the gas drainage). Under these circumstances the requirement for decrease of the water pressure is at least 3 to 4 atmospheres. The main concern is reducing water flow into underground coal mines. For example, a study in Central Pennsylvanian coal mines, indicated that a sandstone aquifer 0.3 to 11.0 m thick overlying an underground mine, can be slightly dewatered. Twenty-five holes were drilled for aquifer dewatering and the flow system and leakage rate were evaluated in the mine. A negligible reduction in leakage was observed in sandstone with a permeability equal to 0.3 m/day, the leakage was reduced by 2.5 per cent.[38]

b) Drainage by drillholes from underground mine openings has been used in a variety of ways. For example, the miners in Western Collieries (Australia) after drift excavation immediately drill short holes in the roof strata to drain it and make more comfortable conditions for work. Dewatering of strata of steep pitching coal seams has been accomplished by drillholes in fan pattern ahead of the workings (U.S.S.R.). For decades, Kreka Coal Mine (Yugoslavia) has carried out drainage of the water-saturated sand in the mine floor by holes drilled from dewatering drifts (Fig. 5.4.6)[39, 40]. The state of the stress is hydrostatic and increases with mine depth. Under these circumstances there is a potential for dangerous underground conditions which has to be remedied by drainage as well as by other techniques, such as leaving protective layers of coal toward the mine floor or roof.

c) Drainage by dewatering drifts and adits is more common in open pit mining than in underground mining. If this technique is applied in underground mining then the correct location of workings is essential. If a satisfactory reduction of water flow can be achieved it will decrease hydrodynamic stresses around the mine structure and increase its stability.

Analyses of water drainage in underground coal mines should be carried out to obtain satisfactory solutions for mine stability with a full understanding of the mechanics of integration of strata stresses with hydrodynamic stresses.

CHAPTER 6

Deformation and failure of coal structure

The discussion in the previous chapters represents a very clear picture of the relationship between the geological characteristics of coal seams and their mechanical properties, as well as the behaviour of coal structure.

The most complex subject of strata mechanics is to determine the deformation and failure of coal structure as a function of the adequate stress-strain state at the coal face.

Generally, the deformation and failure of the coal structure could be classified in four groups:

– Coal and gas outburst with sudden and violent deformation, which could cause dangerous underground conditions.

– Yielding fracturing of coal structure which is the most convenient type of failure with the aspect of mine stability.

– Shear displacement of coal with a particular consideration in orogenically deformed strata.

– Extrusion deformations, which are the most devastating. In this case mine stability should be controlled by pillarless mining systems.

Each classified type of mechanics of deformation and failure is individually discussed in the following subheadings.

6.1. COAL AND GAS OUTBURSTS

The coal and gas outbursts phenomena is recorded in the number of coal fields all over the world. The coal liable to outbursts belongs primarily to higher rank. This coal has elasto-brittle properties and behaviour, and at critical stress magnitude fails suddenly and violently. This phenomena usually is followed by expulsion of finely fragmented coal and release of large amount of the gas.

Loading conditions, internal stress concentration and related pore pressure are of particular interest in the consideration of coal and gas outbursts.

6.1.1. *Vertical stress concentration*

The build-up of vertical stresses in the vicinity of coal faces follows the principles of solid mechanics of stress concentration around openings. However, with the aspect of excavation in coal seams, certain criteria have been established, which are discussed below.

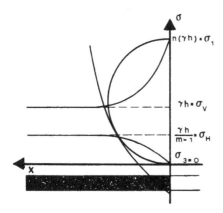

Fig. 6.1.1. Stress concentration at free boundary of mine excavation exhibited by Mohr's circle of stress and Mohr's envelope.

1. Stress concentration at free boundary of excavation, is the most simple case which considers the coal seam as an anisotropic solid body which totally supports external load. The stress concentration is illustrated in Figure 6.1.1, where Mohr's circle of stress represents the stress state below the failure envelope.[1] The build-up of stress at the free boundary, governed by these elements, is of principal importance and is discussed below:

a) Stress redistribution governed by coal excavation and the transition of omnidirectional compression in the coal seam to a maximum uniaxial stress compression at the free boundary is given as,

$$\text{Vertical stress } \sigma_v = n(\gamma h) \tag{1}$$

$$\text{Horizontal stress } \sigma_H = 0 \tag{2}$$

where
γ = unit weight of overburden rocks;
h = coal seam under depth of the consideration;
n = coefficient of stress concentration.

b) The magnitude of stress concentrations is directly proportional to the size of excavation and it increases as the excavation advances. The interaction between the magnitude of stress concentration and the size of openings is diagrammatically represented in Figure 6.1.2.

c) The property, behaviour and matrix of the coal itself is also a key factor in the accumulation and liberation of potential energy, as well as formation of the pore (gas) pressure.

However, the stress concentration at the free boundary of excavation in coal seams is very seldom exhibited, because the coal near the face is fractured and the stress is liberated. This will be discussed in the next subheading.

2. Stress relaxation at the free boundary of an excavation is caused by coal fracturing under an uniaxial compressive stress state, because horizontal thrust at the edge of the seam is not present. Under these circumstances the stress concentration at the free boundary is diminished and external stress is insufficiently supported inward of seams as illustrated by Mohr's circle of stresses in Figure 6.1.3. The fractured zone fails in the

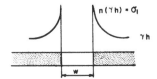

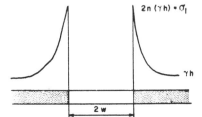

Fig. 6.1.2. Stress concentration increment as function of width of mine excavation.

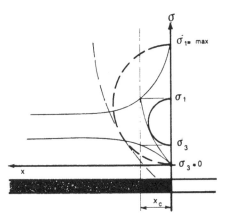

Fig. 6.1.3. Stress redistribution from free boundary inward of seam.

region above the failure envelope of the stress, and represents the stress liberated zone. With the aspect of mechanics of stress relief, three particular elements are of importance.

a) The absence of horizontal thrust at the free boundary is the cause of formation of uniaxial conditions of coal loading. The uniaxial strength of the coal is lower, and loading type of fractures are introduced. Due to coal fracturing at the free face, there is a reduction in the side thrust and coal farther from the free face is fractured.

b) The extension of the inward fracturing of coal seams will depend on how far the reduction of horizontal thrust will be promoted along the planes of stratification. In addition, the length of the fractured zone is also influenced by the height and width of openings, mine depth and relative effect of elasticity and plasticity between the coal seam and the roof and floor strata.

c) The length of the fractured zone is a key factor in facilitating outbursts, and it is called critical length (x_c). If a sufficient extension of the fractured zone occurs in coal, the stress will be dissipated and conditions for sufficient build-up of strain energy to produce outbursts will not exist.

The understanding of stress relaxation is still not well defined. Besides other factors such as properties of coal and the length of the fractured zone, the mechanics of the stress transmitted and the build-up of abutment pressure is one of the key factors. This will be further discussed in next subheading. It is necessary to point out that in a fractured zone, there exists a balance of energy, which has been accumulated in the coal before it reaches its ultimate strength. Part of this energy has been spent for fracturing of the coal and the rest of it is transferred in the abutment zone.

3. Abutment stress concentration is located in the solid coal seam in the vicinity of the fractured zone. The stress distribution is similar to the one at the free boundary of excavation,[2] where the abutment zone is in omnidirectional compression (Fig. 6.1.4). The interaction of stress changes in the abutment area can lead to outbursts if the following conditions are exhibited:

a) All the transferred stresses are concentrated within the strip of coal, which is elastically deformed (contraction). This narrow area is distinguished from the rest of the coal seam, with high accumulation of elastic energy.

b) There is a relationship between the vertical stress accumulation and the stress liberation in the coal seams of coal of sufficient strength to the accumulated transmitted stresses. In such coal seams when overload of abutment stresses is exhibited, their release is sudden and violent. For example, this phenomenon has been experienced in the Eastern Australian coal fields (Fig. 6.1.5).

c) Besides the overload conditions due to the transmission of vertical stresses, it could be a high horizontal stress, facilitated by friction between the coal seam and the hard rock strata. This is discussed in the next heading.

The strata mechanics in relation to outburst could be particularly attributed to imperfect closure of gob, delayed and incomplete roof caving, collapse of pillars and presence of strong and massive immediate roof and floor strata, as well as, coal seams.

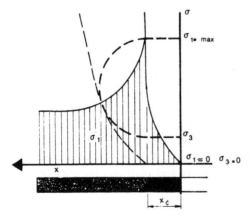

Fig. 6.1.4. Abutment stress of maximum concentration at distance x_c from free boundary of excavation.

Fig. 6.1.5. Cavity left by a coal outburst at Leichhardt Colliery Australia (courtesy of Lama).

6.1.2. *Lateral stress concentration*

The lateral stress concentration in the vicinity of the coal face is a matter of interest for outburst mechanics, and might be related to the formation of high shear stress concentration in the coal seams. The concentration of vertical stresses build-up of abutment pressure was described to a certain degree in previous heading. Build-up of lateral stress requires additional analyses, which are presented below.

1. Closure magnitude has a certain influence on the mechanism of the accumulation of high stresses in the vicinity of the coal faces, which is exhibited as follows:

a) Before outburst, a rapid closure deformation could be exhibited by simultaneous roof sag and floor heave, for example, in Western Canada in an underground mine liable to coal and gas outbursts (Canmore Mine). The convergence is up to 2 cm per metre within 24 hours, which is 20 times greater than in case of mine openings uneffected by outburst.

b) There is insufficient evidence that immediately before outbursts the convergence rate is appreciably decreased as some researchers suggested. For example, Maxnii suggested that before outbursts, sudden changes in convergence might be exhibited even by divergence. A new accumulation of potential energy closure of the roof and floor, continues with normal rate of convergence until, outburst is manifested, and it is repeated (Fig. 6.1.6).

It should be pointed out that closure mechanism with the aspect of outbursts is effective only if the conditions which prevent lateral deformation of the coal arise, as discussed in the next subheading.

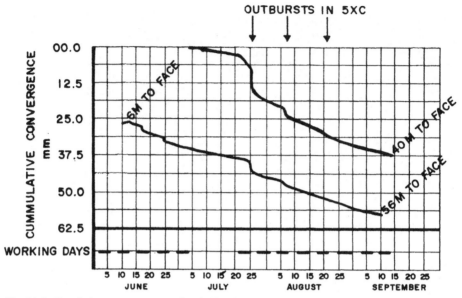

Fig. 6.1.6. Cumulative convergence reading in Number 5 crosscut (Canmore Mine, Alberta).

2 . Motion of the coal is limited, in contrast to large convergence deformations; which actually facilitates 'jamming' of coal by roof and floor rocks,[3]

a) The brown stains in roof rocks observed in Pennsylvanian coal mines indicate relative motion between the roof strata and the coal seam and high stress concentration at interface, which did not permit lateral coal restraint.

b) Limited displacement of coal (Canmore Mine, total displacement up to 6 cm), with prevention of elastic deformations, results in an increase of coal supporting capacity beyond the critical state. This facilitates an accumulation of a large quantity of potential energy in the vicinity of the coal face, and it might actively participate in forming conditions for the coal and gas outbursts.

c) The degree of locking of the coal face is inversely proportional to the rate of face advance, the stiffness of support and the friction between stratification planes of the coal seam and rock strata. In addition, the coal 'jamming' depends on the caving mechanics of the roof strata. For example, the degree of locking is directly proportional to the sequence of the roof caving and position of the cave line to the coal faces. Delayed caving or absence of caving, might have a devastating consequence on an outburst, because the amount of energy stored varies as the fifth power of roof span with clamped edges, and thirty six times greater for roof span as a cantilever beam.

Internal stress concentration is possible only for coal seams which deform with recoverable elastic deformation, rather than definite plastic deformation.

3. Shear resistance between bedding planes of the roof strata, coal seam and floor strata are of great importance. However, outburst mechanics is exhibited only for an adequate geotechnical profile of the coal seam, where, besides the coal rank the lithology of the roof and floor strata is also important. This will be discussed below.

a) If a coal seam is sandwiched between hard rock strata (sandstone or siltstone), then the coal material will be 'jammed' because strong stratum will set up forces to oppose motion. This can happen only if the coefficient of friction between the bedding planes of coal and the rock strata is in the range of 50° or greater.[4]

b) Existence of soft strata layers (shale, mudstone) with a lower coefficient of friction, which is below 35°, will not facilitate coal 'jamming'. The absence of lateral confinement results in the sloughing and squeezing of the coal face, and conditions do not exist for coal and gas outbursts.

c) The frictional forces induced between various members of the coal seam profile resist the restraint of the coal face. Restraint of the coal seam might or might not alleviate outburst phenomena. For example, in the case of a shorter length of fractured zone close to the coal face, the restraint is high because it is concentrated over a smaller area and outburst is possible. On the contrary, when the length of the fractured zone is greater, the restraint is smaller because it is deconcentrated over a larger area, and outburst cannot be expected.

The friction, strength of coal and adjacent strata are the basic elements for the formation of stress concentration, which is a key factor in facilitating outburst phenomena.

6.1.3. *Potential energy*

The build-up of potential energy directly depends on the ability of coal to have greater stress accumulation. For example, coal which can have a small stress concentration cannot store great amounts of potential energy and cannot fail violently. For this reason, in this analysis it is assumed that coal can store greater amounts of potential energy, and that it is perfectly constrained and stressed in compression. The magnitude of potential energy depends on the degree of stress concentration, which is represented by numerical expressions as follows.[4]

1. Major principal stress approximated for flat and gently dipping coal seams, should be perpendicular to planes of stratification. Under these circumstances, it could be assumed that the major principal stress (σ_1) is equal to the overburden pressure (γh). For this set of conditions the stress concentration corresponds to vertical abutment stress (σ_v), as represented by equation below:

$$x_c \sigma_v = \int_0^{x_c} \sigma_1 \, dx \tag{3}$$

It is obvious that maximal magnitude of major principal stress depends not only on the overburden pressure (γh) but also on the length of fractured zone (x_c).

2. Minor principal stress acts in a direction parallel to bedding planes. It is compressive, and it is located in the vicinity of the abutment area, as illustrated in Figure 6.1.7. The magnitude of the minor principal stress (σ_3) can be derived from analysis of total shear resistance of the roof stratum and floor stratum to coal seam, which is 'jammed' between them, and is expressed as:

$$\sigma_3 = \tau_1 + \tau_2 \tag{4}$$

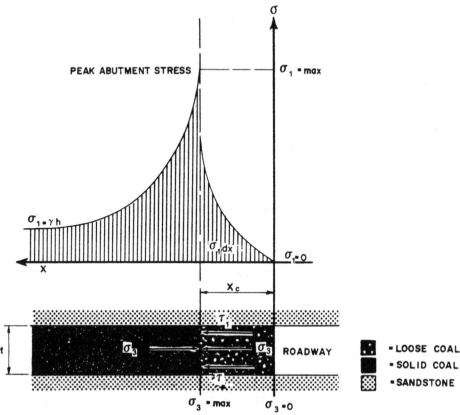

Fig. 6.1.7. Development of lateral confining stress due to friction between coal and rock strata.

$$\tau_1 = \frac{\phi_1 \displaystyle\int_0^{x_c} \sigma_1 \, dx}{t} \tag{5}$$

$$\tau_2 = \frac{\phi_2 \left[\displaystyle\int_0^{x_c} \sigma_1 \, dx + \gamma_c x_c \right]}{t} \tag{6}$$

where

τ_1 = shear stress at the roof-coal interface;
τ_2 = shear stress at the floor-coal interface;
t = workable thickness;
γ_c = unit weight of coal;
x_c = length of fractured zone;
ϕ_1 = coefficient of friction at coal-roof contact;
ϕ_2 = coefficient of friction at coal-floor contact.

From the above equations it is obvious that the coefficients of friction between the roof and floor stratum and the coal seam are the main factors in forming shear resistance. The shear stress for the equal seam conditions is going to be greater for a seam of smaller thickness, than for thicker seams. If someone approximates that the frictions between coal-roof rock and coal-floor rock are equal ($\phi_1 = \phi_2$), then the total shear stress, which corresponds to lateral compressive stress can be written as follows:

$$\tau = \phi \frac{2 \int_0^{x_c} \sigma_1 \, dx + \gamma_c t x_c}{t} = \sigma_3 \qquad (7)$$

The integral equation for the calculation of minor principal stress can be successfully used if the coefficient of friction and the length of the fractured zone are determined.

3. Strain energy can be represented on the basis of the established equation from a mechanics of solid body, for uniaxial loading conditions as given below:

$$U = \frac{\sigma_1^2}{2E} \qquad (8)$$

This is proportional to the square of the stress in the coal. The strain energy accumulated within the elastic boundary per unit of volume for triaxial loading conditions is expressed by the well-known equation:

$$U = \frac{1}{2E} \left[\sigma_1^2 + \sigma_2^2 + \sigma_3^2 - \frac{2}{m}(\sigma_1 \sigma_3 + \sigma_1 \sigma_2 + \sigma_2 \sigma_3) \right] \qquad (9)$$

if $\sigma_2 = \sigma_3$, then Equation (9) can be rewritten for following loading conditions

$$U = \frac{1}{2E} \left[\sigma_1^2 + \sigma_3^2 - \frac{2}{m} \sigma_1 \sigma_3 \right] \qquad (10)$$

Fracture strength of the coal is higher under constrained conditions of loading than under uniaxial loading, and depends on Poisson's number and the magnitude of lateral compressive stress, which is given below:

$$U = \frac{1}{2E} \left[\sigma_1'^2 + \sigma_3^2 - \frac{2}{m} \sigma_1' \sigma_3 \right] \qquad (11)$$

where

$\sigma_1' =$ fracture strength;

$\sigma_3 =$ lateral stress,

$m =$ Poisson's number $= \dfrac{1 - v}{v}$;

$v =$ Poisson's ratio.

For idealized case where the expansion of coal is not permitted due to the constraint, it can be written $\sigma_3 = \sigma_1'/m$, and Equation (11) is:

$$U = \frac{1}{2E} \left[\sigma_1'^2 + \frac{\sigma_1'^2}{m^2} - \frac{2\sigma_1'^2}{m^2} \right] \qquad (12)$$

From experimental work it is known that fracture strength of coal under constrained conditions increases over two times ($\sigma'_1 = 2\sigma_1$), which can be written[5] as:

$$U = \frac{\sigma_1^2}{2E}\left[4 + \frac{4}{m^2} - \frac{8}{m^2}\right] \tag{13}$$

A study of this equation showed, when the magnitude of Poisson's number is 1, then there is no potential energy for outburst stored in stressed coal.

The underground observation in several mines showed that for conditions of uniaxial compression in the abutment zone, the stored strain energy may be released suddenly and violently in the form of decompression waves. The violent and unstable equilibrium could occur either because of impulsive leveling and redistribution of the stresses at the edge of abutment or to the instantaneous decrease of the constraining lateral stress, which kept coal 'jammed' between the roof and floor strata. The underground mining practice showed that a large number of outbursts is caused due to transmission of mining stresses from active mining area on abutment zones of pillars and coal faces. This is exhibited in the Pennsylvania coal fields where 80 per cent of the outbursts occur due to the impact of stress transmission on solid coal from abutment areas.[4]

6.1.4.   *Kinetic energy*

The discussions in previous paragraphs offer sufficient knowledge for understanding mechanics of deformation and failure of the coal faces liable to coal and gas outbursts. The main points of such mechanics are briefly represented below.

1. Phenomena of coal and gas liberation are governed by instant relief of the stresses from the 'jammed' coal faces between strong roof and floor strata. The tremendous stress is relieved due to the integration of strata dynamic stresses and gas hydrodynamic stresses. With this aspect there are several factors of particular interest which are listed below:

a) The instant relief of the elastic energy and pore pressure causes coal breakage within the mechanics of tensile stress in the decompressive wave, giving a rapid restoration of elastic deformation of the coal structure. Also, the simultaneous gas release increases the volume of coal, which is a porous medium, and is followed by the expansion phenomena, which additionally induces a tensile stress in the coal face area. This dynamic phenomenon was indicated underground by coal popping and scaling before gas outbursts. The underground monitoring indicated that impulsive coal fracturing and the gas outburst is controlled by stratification planes, due to changes of the stress within rock strata in the vicinity of the coal face.[8] This phenomena has to be contributed also to the concentration of elastic energy due to the compression of coal and gas.

b) Fracturing of coal is greater in the case of coal and gas outbursts than in the case of only coal outbursts (bump), because there is a release of a greater amount of accumulated energy. The coal fragmentation might cease at the abutment zone, when exposure of the tightly locked section of the seam is exhibited at the free boundary (Fig. 6.1.8). However, the impulsive nature of violent stress release usually promotes continuation of outbursts further in solid coal beyond the abutment zone. Due to

Fig. 6.1.8. Coal and gas outbursts at the rib face, West Cliff Colliery, Australia (courtesy of Lama).

differential anisotropy of the coal seam profile, coal fragmentation and breakaway may start deeper in the coal seam within stronger coal layers (for example, durain lithotype), and is accompanied by ejection of nearby soft or loose coal with poor cohesion (for example, vitrain lithotype), which does not give appreciable resistance to moving mass. This explains why the axes of cavities are close to parallel of the coal faces, with their collaring in softer coal layers. Usually the strong coal is coarsely broken and soft coal is finely fragmented.

c) Laboratory investigations suggest that mass velocity of particles behind the front of the decompressive waves vary most rapidly up to the pressure drop of $(3.44 - 8.40) \times 10^6$ N/m^2. However, beyond this stress magnitude the relative changes are negligible except in the case of anthracite for which the pores are significantly deformed, before the attainment of high stress concentration. In this case it gets the greatest increment in the accumulation of potential energy. If the external pressure is further increased, the relative changes in the accumulation of strain energy becomes less important.

As described earlier, coal and gas liberation exhibits brittle fracture mechanics of a tensile stress field which is caused by the decompressive wave arising in an elastically compressed coal saturated by gas. After sudden and impulsive liberation there is a stress redistribution in the coal seam with leveling out stresses.

2. Mechanics of energy liberation as stated in the previous paragraph, could be analyzed by the assumption that under described stress circumstances the lateral expansion is imposible.[9] The deformation should coincide with the direction of the mass movement, which is opposite to the direction of propagation of the decompressive wave. The deformations which correspond to directions perpendicular to the direction of motion are equal to zero:

$$\varepsilon_2 = \varepsilon_3 = 0 \tag{14}$$

$$\varepsilon_1 \neq 0 \tag{15}$$

Further, generalized from Hooke's Law

$$\sigma_2 = \sigma_3 = \frac{v}{1 - v}\sigma_1$$

$$\varepsilon_1 = \frac{(1 + v)(1 - 2v)}{E(1 - v)}\sigma_1 \tag{16}$$

where

σ_1 = stress in the direction of motion;
σ_2, σ_3 = tangential stresses;
v = Poisson's ratio;
E = modulus of elasticity.

To transfer these mechanics in the dynamic process, the dynamic Poisson's ratio and the modulus of elasticity could be used, and are expressed as,

$$v_d = \frac{0.5 - R^2}{1 - R^2} \tag{17}$$

$$R = \frac{V_s}{V_p} \tag{18}$$

where

v_d = dynamic Poisson's ratio;
V_s = velocities of transverse waves;
V_p = velocities of longitudinal waves.

$$E_d = \frac{\gamma_c}{g}\frac{(1 + v_d)(1 - 2v_d)}{1 - v_d} \tag{19}$$

where

E_d = dynamic modulus of elasticity;
γ_c = unit weight of coal;
g = acceleration due to gravity.

Substituting parameters for deformation in Equation (14) is obtained as below

$$\varepsilon_1 = \frac{\sigma_1}{E_d} \tag{20}$$

and further substituting Poisson's number by the expression as given below

$$K_\tau = \frac{v_d}{1 - v_d} \tag{21}$$

the relationship between the stress and deformation may be written as follows:

$$\sigma_1 = E_d \varepsilon_1 \tag{22}$$

$$\sigma_3 = K_\tau \sigma_1 \tag{23}$$

where

K_τ = coefficient of lateral thrust.

According to the theory of elasticity, the stress arising in deformed media can be calculated by the following equation:

$$\sigma_1 = \frac{\gamma_c V_p u}{g} \tag{24}$$

where

$u =$ is the mass velocity of the particles.

The minimum stresses behind the decompressive wave front are σ_{refl}, at which a violent fracture may occur in the samples near their strength within static loading conditions, which is written below:

$$\sigma_{refl} = \frac{\gamma_c V_p u_{cr}}{g} \tag{25}$$

where

$u_{cr} =$ critical particle velocity.

The final consideration should be given to the energy of the coal and gas outburst, which is actually an integration of the potential and kinetic energies:

$$
\begin{aligned}
U_{tot} = \quad & U_{pot} \quad & + \quad & U_{kin} \\
\text{total energy} = & \text{potential energy of} \quad & + & \text{kinetic energy of the particles} \\
& \text{elastic deformations} \quad & & \text{moving behind the wave front}
\end{aligned} \tag{26}
$$

The total energy equation can be composed by Equation (8) for potential energy and by modified Equation (24) for kinetic energy, as expressed by Khanukaer below:

$$U_{tot} = \left(\frac{\sigma_1^2}{2E} + \frac{(\gamma_c/g)u^2}{2} \right) 10^{-6} J \tag{27}$$

Equation (27) can be used for calculation of total energy of the decompression wave per unit volume of the coal if the mass velocity of particles is known.

The represented mechanics of coal and gas liberation exist only if potential energy stored is not spent for permanent face deformations. The coal and gas outburst is common for coal seams of higher rank, where the mechanics of violent fracture could be observed by overcoming the tensile strength of coal and its stress concentration further from free boundary. Although, near each newly formed solid surface, tensile stress again arises which is manifested by repeated sudden fracture. The coal fragmentation is by solid slab which loses adhesion to solid coal face. The instrumentation showed that the slab strata to move freely at velocity equal to twice the particle velocity in the wave $(V = 2u)$.

6.1.5. *Prevention of outbursts*

It should be stated at the beginning of this paragraph that coal and gas outburst can be decreased but cannot be completely eliminated. This dangerous phenomena in coal mines proves that outbursts should be accepted as an occupational hazard, which has to be controlled by certain number of preventive techniques.

1. Hydraulic fracturing of the hard and brittle cohesive coal mass, where a penetrative system of planar weaknesses such as a cleat is often facilitated. The cleat orientation is initially stress field controlled. Rapid injections of water cause changes of the stress field, fracture direction and consequent physical property of coal. A traditional elastic continuum approach is usually employed to predict fracture density and propagation, which could be utilized for the evaluation of the degree of coal seam de-stressing. Hydraulic fracturing might be an integral part of the mine operation. It is possible to change the property of the coal from elastic-brittle to elasto-plastic, by water infusion. Therefore, instead of sudden and violent fracturing, yielding deformations are induced, which can be successfully controlled. For example, for de-stressing of an anthracite coal seam ('Ataman') in Odessa Mine at a depth of 127 m required water pressure of $12.5 \, \text{MN/m}^2$ and at a depth of 278 m, $30 \, \text{MN/m}^2$. It should be pointed out that the concept of coal 'pre-softening' by hydraulic fracturing can be utilized in an early phase of extraction. After that, 'pre-softened' coal might be mined by monitor jets establishing a regular hydro mine operation.

2. Prolonged wetting of the seam with water in which optimum amounts of surfactant added. By this technique the elastic properties of the coal are changed, and are decreased to the level of non burst failure. Under these circumstances the dynamic phenomenon of failure is exhibited by particles moving with equal accelerations and almost equal velocity, which does not facilitate conditions for the formation of sudden and violent breaks in coal. There is only partial failure of the individual slabs detached from each other, in extension inward of the coal face.

3. Advance drilling was the first method adopted to control outburst. The success of this technique depend on the direction of drilling to the geological-structural features of the coal seam, drillhole length, diameter and others. For example, outburst control in Eastern Australian coal fields is achieved by drillholes of 100 mm diameter at intersecting angles to the shear zone. Usually three holes were drilled at 2 m intervals on the face, where length of individual holes depends on the distance of shear zone (20–30 m)[5]. A similar pattern of drillholes have been used in Canmore Mine (Western Canada), for pressure release. However, success was limited due to the small diameter of drillholes, where gas was liberated only from around the borehole and not from the solid body of surrounding coal.[7] Advance drilling for longwall face degasification and release of hydrodynamic stresses indicated that gas concentration and distribution within the seam might vary in wide range as illustrated in Figure 6.1.9. The changes in the physical properties of coal and related de-stressing of seams sometimes depends on the uniformity of fracture propagation and gas release. For example, existence of large pockets of weak porous coal surrounded by strong homogeneous coal or vice versa, could facilitate coal and gas outbursts after de-stressing of coal seams.

Finally, it should be pointed out that advance drilling either from surface or underground working is also used for coal seam degasification to decrease the hazard of methane gas emission which can lead to a dangerous explosion. At present coal seam degasification by drillholes is tried in order to produce a gas for commercial use.

4. Camouflet blasting is a popular method in some coal mines to control coal and gas outburst. Its effectiveness depends on matching the blast parameters to the local coal seam conditions. There are different approaches to the blasting of coal seams with the aspect of de-stressing. For example, in Western Canadian coal fields the blasting was

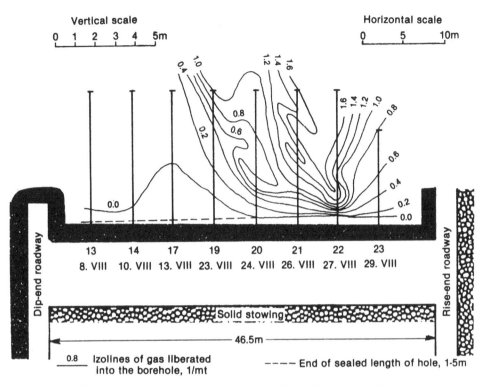

Fig. 6.1.9. Gas liberation by boreholes on the central longwall face, Roman Seam (Poland).

carried out by heavy rounds of shots fired simultaneously during off-shift periods. The release of stored energy per unit volume is greater than or close to the magnitude with respect to the sudden and violent fracture. This technique becomes less important with the introduction of mechanised coal mining and the elimination of cyclic mining of drill blast technique. In some coal mines in the U.S.S.R., firing of small explosive charges has been carried out to create camouflet fracturing similar to the hydraulic fracturing. In the outburst prone seams, the blasting produced radial air gaps as a result of fracturing and decompaction of coal. By this blasting the elastic behaviour of coal is changed to the inelastic one. The success of coal property transformation depends on the hole spacing, hole length, hole diameter and the pattern of hole charges.[10]

5. Controlled mine layout is an important parameter for avoiding adverse effects of the outbursts on mine stability. The mechanics of the relationship between cause (mine layout) and effects (burst or non-burst) has been investigated by many authors. They primarily gave several good practical solutions, because the theoretical solutions have not been particularly successful. These topics are broadly described in the literature,[11] and some very brief comments are presented below:

a) Adoption of mining methods requiring minimum development drivage, with maximum possible regular shape of coal production faces devoid of pointed projections.

b) Working of protective seams, when multiple seam mining is in effect.

c) Mine layout should be adapted to natural conditions of coal seams, which will not cause adverse effects on mine stability. For example, the geological irregularities (folds, rolls, faults), as well as, seam structures (pitch, thickness, coal and adjacent rocks) might in interaction with mine openings cause violent and sudden fracturing.[11]

d) The irregular width of advancing front extraction of large slice in time, rapid advancing without allowing sufficient time for gas liberation and other phenomena are common cause of outbursts in coal mines. It is a well-known rule that nonbursting (protective) seams have to advance ahead of seams liable to outburst in order to de-stress ground before extraction. The complete caving of the gob is essential for the elimination of stress transfer from mined-out area and formation of high stress concentrations in the working area.

e) The size, formation and excavation of coal pillars greatly govern violent and sudden pillar failure (Fig. 6.1.10). The pillar outburst is a most dangerous phenomenon, where all sides and adjacent openings are affected, particularly at cross intersections. Pillar failure is usually accompanied by a collapse of an already weakened roof strata. The pillar design should avoid a critical pillar size for outbursts and implement pillars of uniform sizes. The layout should facilitate a large time lapse between the formation of pillars and their extraction, and the decrease of sudden roof collapse behind pillar to avoid impact loading.

Techniques of the prevention of outbursts and stress concentrations in underground mining have to be an integral part of the mine design, particularly at the present time where new technologies exhibit rapid coal extraction at high production rates with increasing mining depth. Deep mining is becoming a critical factor for outbursts,

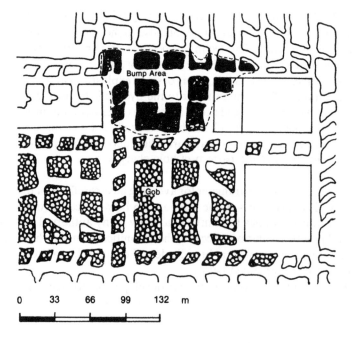

Bump Area

Gob

0   33   66   99   132   m

Fig. 6.1.10. A district burst, which might be triggered by deriving cut in large pillar (after Holland).

particularly in mountainous terrains, which cannot be eliminated and under these circumstances, effective burst prevention should be incorporated in the mine layout.

6.2. YIELDING FRACTURING OF COAL

The most common mode of deformation of mine structure is by non-violent failure, which can be related to the elasto-plastic state of stresses. This type of deformation has been extensively studied in laboratories (coal samples) and underground mines (coal pillars). Particular consideration in these studies has been given to coals of sub-bituminous and high volatile bituminous rank. The property and behaviour of these coal ranks can be appreciably changed due to weathering.

Considerations are given to deformation and failure of pillars, stress distribution during structure deformation, as well as on principles of fracture mechanics.

Finally, the points which are of interest for reinforcement of coal structure, and which are of much lesser extent than in the case of the outbursting failures, are briefly listed. Non-violent and gradual failure of the coal structure deserves investigations in greater extent than carried out here.

6.2.1. *Behavioural phenomena*

Coal with yielding fracturing exhibits non-bursting behaviour, due to its inability to deform by reversible deformations. Generally, this behavioural phenomenon is particularly exhibited in low rank coals, for example, sub-bituminous coal. This coal rank could further decrease elasticity and increase the inelastic deformations by coal weathering which has been studied at the Star-Key Mine located about 20 km north of Edmonton, in Western Canada. The mine workings are in the Cover Bar coal seam, which is a flat lying lense with a thickness of 3.35 m at its centre. The floor of the seam rolls very gently, which could have been caused by glacial thrusting. The weathering front in the coal seam might be parallel to the crest of post-glacial river valley. This weathering may reflect leaching or desiccation of the strata above the water table.[12]

1. Observations of the coal in underground mine have been particularly concentrated on the sample sites taken throughout the mine from the large central rib pillars (Localities 5–10, Fig. 6.2.1), and close to the recent workings (Localities 1–5, Fig. 6.2.1). We believe that the condition of the central rib pillars is only slightly affected by load transfer from the gob. Further, if these loads are due to overburden and are in any case small, less than 1.03 MN/m². The deformation of some coal from these pillars, which will be described later, is attributed to the alteration rather than the loading conditions. Certainly coal from sample localities 1–4 in the butt entries where loads might be expected to be much higher than in the central ribs is unfractured and appears unaltered.

The main entries of the mine are developed perpendicular to the cleat of the coal which have a NE-SW trend. The rib walls are thus parallel and perpendicular to cleat. In the coal exposed at localities 1–5 close to recent workings, the rib walls are smooth and parallel to the cleat (Fig. 6.2.2). Flake off along conchoidal fractures is limited to

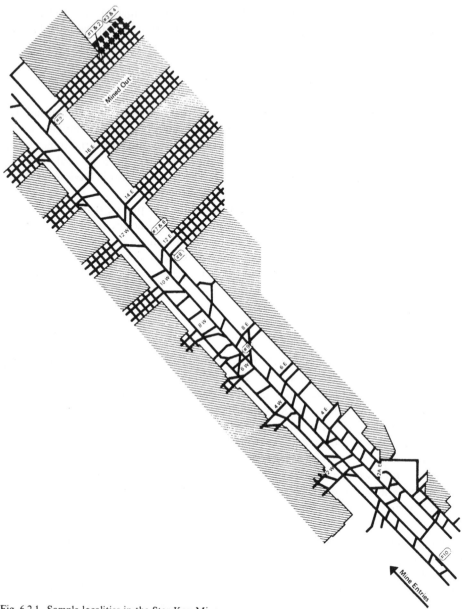

Fig. 6.2.1. Sample localities in the Star-Key Mine.

depths of 3–5 cm. There is some sloughing and fragmentation of coal on the rib walls perpendicular to the cleat but the cleat fractures can be seen to be closed and tight. We think that this coal is unaltered.

Coal exposed at localities 6–10 is more fragmented. Besides the well-expressed cleat in a NE–SW direction, there is an approximately perpendicular set of cracks that are also vertical.

Fig. 6.2.2. Rib wall along cleat in unweathered coal.

Fig. 6.2.3. Rib wall along cleat in weathered coal.

Deformation of rib walls parallel to the cleat is illustrated in Figure 6.2.3. Cracks open normal to the cleat and the slabbing off of coal along them is the usual pattern of rib deformation. Such slabbing does not usually extend more than 10 cm into the rib wall.

When the rib walls are perpendicular to the cleat the volume of coal sloughing may triple that of the walls parallel to the cleat. Again, fractures open parallel and perpendicular to the cleat but failure is by falls of coal cubes rather than slabs (Fig. 6.2.4). The fractures that open along the cleat can be up to 10 cm. wide but the extension of these fractures into the body of the pillar is limited to 15 cm.

Tests were planned on specimens of altered and unaltered coal to determine whether properties of the coal used in mine design showed significant differences between the two types. Tests of the compressive strength of the coal followed procedures outlined by Hustrulid.[13] Brazilian tests on coal have been described by Evans.[14]

Fig. 6.2.4. Rib wall
perpendicular to cleat in
weathered coal.

2. Test methods are described in the following order. Blocks of coal up to 0.5 m in dimension were taken from rib walls in the mine. Specimens were cut from coal blocks in the required sizes with a circular saw.

For uniaxial compressive strength tests, the coal was cut in cubes ranging from 1.25 to 12.70 cm in size. Fifteen unaltered coal specimens and ten altered coal specimens were prepared. The preparation of cubes of altered coal was very difficult, 80 per cent of the samples broke during cutting. Loads were applied perpendicular to bedding.

For tensile strength tests by the Brazilian method cylinders of coal with diameters from 2.15 to 2.25 cm were prepared. In four specimens of weathered coal, disc thickness ranged from 1.49 to 2.99 cm. In twelve specimens of unweathered coal, disc thickness ranged from 1.80 to 2.03 cm. In different specimens, the bedding plane was placed at different angles to the loading direction.

3. Test results – Uniaxial compressive strength. Tests have been carried out on cylindrical and cubical coal samples. It should be noticed that the strength of the 2.45 cm weathered coal cubes is about 13.79 MN/m² lower than the strength of the unweathered coal. Larger cubes show smaller differences. This finding is important in the design of ribs and pillars in underground coal mines. Strength of these structures is usually estimated from the uniaxial compressive strength of the coal material.

It is also important to know how weathered and unweathered coal material behave under load. Different load-displacement curves are obtained from laboratory tests of weathered and unweathered coal cubes (Fig. 6.2.5).

Unweathered coal samples produce a linear loading curve with a high tangential modulus and with an easily distinguished peak strength at which there is a clear failure. This material might be characterized as elasto-plastic.

Weathered coal samples do not give a linear load-displacement curve and the approximate tangential modulus is low. Peak strengths are poorly defined and yielding is observed over a range of loads. The material is more plastic.

Clearly, unweathered coal can support relatively high loads in the mines. Weathered coal will yield slowly under lower loads without clear failure.

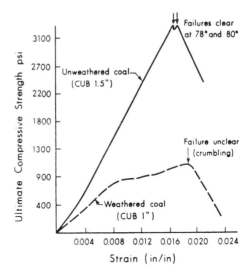

Fig. 6.2.5. Load-displacement curves for weathered and unweathered coal under compression.

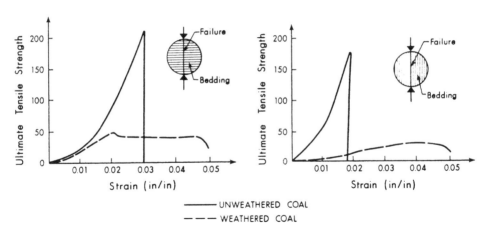

Fig. 6.2.6. Load-displacement curves for weathered and unweathered coal under tension.

4. Test results Brazilian tests (σ_T) of the coal discs were performed on altered and
unaltered coal under direct tension, gave results similar to those under compression
(Fig. 6.2.6). Unaltered coal is brittle. Altered coal shows yielding even under very low
loads.

Both these tests and our rib deformation observations suggest that altered coal
under tension will support only very low loads.

The tensile strength of the coal varies with the angle between the load and the
bedding as shown in Table 6.2.1. This is an important factor for the evaluation of bed
separation and the limit of bed sag.

Bed separation may also occur in the Plains Region in the weak rock strata
overlying the coal. A foot of coal is often left in the mine roof to increase its supporting
capability.

Table 6.2.1. Results of Brazilian tests on altered and unaltered coal

Angle of bedding trace to applied load	Range of strengths in unaltered coal (MN/m^2)	Range of strengths in altered coal (MN/m^2)
Tests on cores with axes parallel to bedding		
90°	1.45–2.13	0.35–0.38
45°	0.86–1.55	–
0°	0.79–1.14	0.15–0.24
Tests on cores with axes perpendicular to bedding		
0°	1.24–1.93	–

It is interesting to note that a similar property and behaviour of the coal has been observed at Collie Basin in Western Australia. The coal in this basin is of the same rank, subbituminous, as coal at Stark-Key Mine in Western Canada.

6.2.2.   *Mode of deformations and failure*

The non-violent deformation and failure of mine structures have been investigated in many underground mines for almost a century. However, a significant understanding and solutions have been achieved recently by implementing elasto-plastic state of stresses and related failure mechanism. With the aspect of practical mining, which is based on the underground observations and phenomenological investigations the deformation and fracturing of coal pillars can be described by two main aspects.

1. Transitional pillar deformation is discussed on the basis of monitoring deformation and failure. In this case, a particular consideration is given to the sub-bituminous and bituminous coal seams which are not liable to sudden and violent fracturing. The

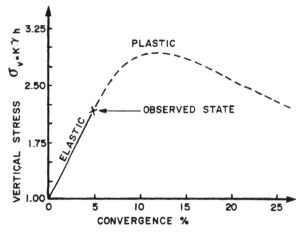

Fig. 6.2.7. Coal pillar convergence observed up to failure (St. Barbara Mine, Yugoslavia).

Fig. 6.2.8. Pillar within plastic state of deformations with fracture development (McIntyre Porcupine Mine, Canada).

quality of coal pillar deformations to total failure is given in the order of events as below:

a) The pillar stability at initial stage of loading have also been studied at Star-Key Mine, where pillars are located at a very shallow depth (25 m). In this case the vertical and horizontal stresses are far below the strength limit of coal, but deformations were indicated by limited convergence (1.0 cm/1.0 m). From this, it can be concluded that the pillar is in the state of elastic deformation (Figs. 6.2.2 and 6.2.4).[12]

b) The monitoring of a pillar convergence due to the progressive loading during its formation at St. Barbara Mine (Yugoslavia) showed that vertical stress reached the proximity of the strength limit with deformations within the elastic region (Fig. 6.2.7). However, all deformations were within the elastic region, because cracking and spalling had not been exhibited.[15] The investigations could not define a form of the state of stresses, except that they are omnidirectional.

c) The pillar observations at Grande Cache Mine, showed failure development due to additional loading (stress transfer from extraction area). Under these circumstances, the outer part of the pillar came to a critical state and plastic deformations are exhibited, by coal spalling and fracturing (Fig. 6.2.8). It should be noted that in this case only the outer pillar zone is in the plastic state, and the core of the pillar is in the elastic state of deformation. The vertical and horizontal stresses are deviated in accordance with the development of different zones of deformations.

d) It has been observed in coal mines, that progressive pillar fracturing decreases the

Fig. 6.2.9. Pillar in collapsing state, Asercleer East Colliery, Australia (courtesy of Lama).

size of the core of the pillars. Under these circumstances, the pillar is fractured and is in plastic state. Before the total collapse the pillar is in a de-stressed state because it supports only the load just superimposed above it.

General underground observations also suggest the existence of shear stress along bedding planes due to compression. The monitoring showed that the increase of elastic deformation and the development of the plastic state is followed by an increase in the intensity of shear stress.

2. *The pattern of failures* has to be related to the above described transitional mechanism of pillar deformations from elastic to plastic state and their stability analyses should reflect this state, where the pattern of pillar fracturing can be as follows:

a) The failure is by longitudinal cracks parallel to the action of a compressive load.

b) The failure is facilitated if the roof and floor strata are composed of soft rocks, which do not offer sufficient resistance for coal pillar constraint.

c) The fractured slabs can be geometrized, as two truncated pyramids joined by a smaller base, which suggest that pillar slabbing is governed by shear stress acting at a certain angle to the vertical free face of the pillar.[16]

d) The shear stresses directed towards the pillar center, do not permit pillar fracturing along its median horizontal axis and it is responsible for the formation of confined stress nucleus (core).

e) After shearing of the slab-prism adjoining the free pillar face, the next slabbing might be underway until the central part of coal pillar is fractured, where individual slabs of coal are split off due to tensile stresses (Fig. 6.2.9).

The intensity of cracking and slabbing of coal pillars by non-violent and sudden fracturing is primary governed by the magnitude of external stresses and pillar sizes. Also, the cleat orientation to the free faces of the pillars in great deal influences the intensity and geometrization of the slabs.[17]

On the basis of the acceptance of the listed parameters of deformation and failure of the coal pillars, further analysis of mechanics of their deformation and failure has been carried out, as discussed in following two headings.

6.2.3. Yielding mechanics

The deformation of pillars without the formation of violent fracturing, can be studied by the elasto-plastic stress analyses. Such an analysis has been done by Fadeev and Abdyldayev[18] who used the finite element method to determine the distribution of stress and strain within underground mine structures. The results of their investigations are briefly presented below:

1. *Theoretical analysis* of elasto-plastic state of stresses are represented by numerous description in present literature, where a majority of them regard the plastic flow as equivoluminal, i.e. proceeding under a constant state without voluminal changes. However, elasto-plastic analysis by 'incremental' and 'initial' stresses suggested by the authors above is most optimal. They found out that the 'initial' stress procedure proposed by Zienkiewicz et al.[19] gives most optimal solution for problems in rock mechanics. Brief excerpts of their theoretical analyses are given below:

a) The conditions of plain strain are examined, where compression is denoted as

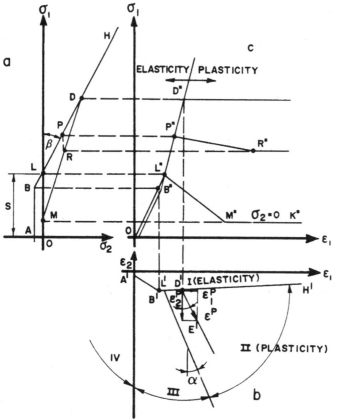

Fig. 6.2.10. The graph of constitutional law – 'holograph of state':
a) initial (ABDH) and residual (OMDH) strengths,
b) relationship between ε_2 and ε_1
c) relationship between σ_1 and ε_1.

positive. The strength properties of media are presented in Figure 6.2.10 in the axes of main stresses. The line ABH represents the initial (or limit) strength. The segment AB has the equation:

$$\sigma_2 = \sigma_1 \tag{1}$$

where

σ_1 = tensile strength

The line BH has the Coulomb equation:

$$\sigma_1 = \sigma_C + \sigma_2 \cot \beta \tag{2}$$

$$\sigma_C = 2c \cot \beta$$

where

σ_C = uniaxial compressive strength;
c = cohesion;
β = angle of failure;

b) Due to work softening during inelastic deformation, the strength of the media decreases to a residual level characterized by the line OMDH with parameters of residual strength $\sigma_T = 0$, for segment OM, S' and for line DH. If the stress state of an element of media changes in such a way that it never goes out of the contour illustrated in Figure 6.2.10, then this element might be analyzed by constitutional law.

c) Further, these authors considered distribution of the stresses and strains in mine structure in elasto-plastic state by known constitutional law expressing relationship between stresses and strains, which is rather complicated for rock masses, because when tensile failure occurred, the mutual dependence of stresses and strains vanishes.

d) The dependence between strains and stresses was demonstrated by using the finite element method (FEM) in which the solution is obtained by the procedure of 'initial' stresses. The results for every element were stored in the memory of the computer and were for each cycle introduced into the vector of nodal forces by adding to it the 'initial nodal forces' corresponding to initial stresses of elements. Then a new elastic solution was obtained with new values of nodal forces. The previously accumulated initial stresses are subtracted from the newly calculated elastic stresses and resulting 'real stresses' are compared with new theoretical stresses corresponding to new strains. The iteration process continues until the real stress in all elements would differ from their theoretical values up to a predetermined precision.

The graphical and mathematical description of the constitutional law of analyzed medium and the digital procedure of 'initial' stresses as a part of general FEM process have been considered satisfactory for stress analyses of strip-like coal pillars which are under load in the plastic zone of deformation, as discussed further.

2. *Solutions of analyses* are examined for barrier pillars in the Kuznetsk Coal Basin (South–West Siberia). They are located between mined out and active mine areas. Due to the flooding of old workings with water, the evaluation of filtrational properties of pillars is necessary to define the form of plastic zones in pillars with increased coefficient of filtration. The pillar evaluation has been carried out in following order:[18]

a) The average pillar stress is assumed to be in the same manner as for abutment pillars (trapezoid shape of dead load). Also the pillar loading is represented by finite element mesh in net form commonly used for a numerical model analysis.

b) Mechanical properties of coal and rocks have been determined by in situ testing, and they are:

$$\text{coal:} \quad E_0 = 2 \times 10^6 \, KN/m^2, \quad V_0 = 0.35, \quad c = 2000 \, KN/m^2, \quad \phi = 30^\circ$$
$$\text{sandstone:} \quad E = 7 \times 10^6 \, KN/m^2, \quad V = 0.35, \quad c = 7000 \, KN/m^2, \quad \phi = 30^\circ$$

The coefficient of expansion is assumed to be equal to 3. The load onto half of the pillar determined before was equally distributed along the upper boundary as nodal forces. The weight of elements were neglected and taken equal to zero.

c) The analysis has been done for pillars at three depths (h = 250, 370 and 470 m) and two different widths (w = 40 and 60 m). The principal stresses in the rock strata and contours of pillar displacements are illustrated in Figure 6.2.11. The analysis showed that lateral displacement of a 40 m wide pillar is approximately 1.7 times greater than that for the width of 60 m, for the same depth (h = 470 m).

d) The distribution of vertical stress σ_y and horizontal stress σ_x along the pillar axis is illustrated in Figure 6.2.12. It is clearly exhibited that the relation of the peak vertical stress in the pillar at the contact of inelastic and elastic zones to the stress in the middle

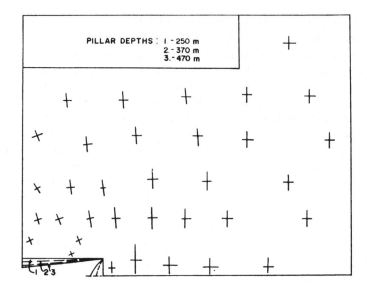

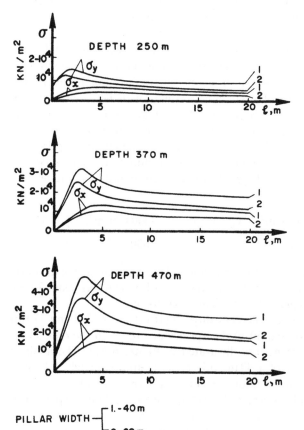

Fig. 6.2.11. The vectors of principal stresses (σ_1 and σ_3) and displacement contours for pillar 60 m width.

Fig. 6.2.12. Distribution of vertical (σ_y) and horizontal (σ_x) stresses of the pillars.

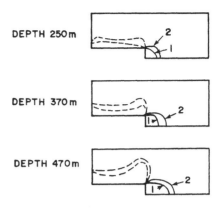

DEPTH 250 m

DEPTH 370 m

DEPTH 470 m

PILLAR WIDTH : I.- 60 m
2.- 40 m

Fig. 6.2.13. The zones of inelastic states.

of the pillar does not actually depend on vertical load (i.e. mine depth) but is mainly defined by the size of the pillar. In the analysis of Fadeev and Abhyldayev, for pillars with the widths of 69 and 49 m, these relations are 2.25 and 1.8, respectively.

e) The inelastic zones formed due to the increase in the load on the pillar are illustrated in Figure 6.2.13, where elements of the inelastic zone are represented by a continuous line. These elements are in a state of biaxial compression and their strains belong to zone II in Figure 6.2.10. These elements are torn at least in one direction and coal from these zones would fall due to gravity.

The lengths of plastic zone (outer zone) of the pillars observed in underground coal mines of Kuznetsk Basin showed an approximate agreement with the calculated values. For example, the calculated width of the plastic zone in the 49 m pillar increased from 2 to 4 m and in the 60 m wide pillar increased from 2 to 3 m, when the mine depth was increased from 250 to 470 m. Both, the mine instrumentation and model calculations showed that expansion of coal in the plastic zones of the pillars exhibits considerable displacements of their faces into the excavated areas (roadways).[18]

6.2.4. *Fracture mechanics*

The failure deformations of the outer pillar zone could be analytically explained under the following boundary conditions. A coal face exhibits shear stress concentration along the slip plane at an angle β to the vertical axis, and horizontal stresses acting towards the free face are equalized. This could be written as follows:[20]

$$\tau_\beta = \sigma_x \varepsilon_x = \varepsilon_z = 0 \tag{1}$$

where τ_β is the projection of the shear stresses on the σ_N axis in the coordinate system $\sigma_N - 0 - \tau$ (Mohr's circle) and σ_x is the projection of horizontal stress on the σ_N axis, arising at a point B and directed towards the free face (Fig. 6.2.14); ε_x and ε_z are the relative deformations along the corresponding axes of an element at point B.

From the principles of the theory of elasticity, the relative deformation along the x axis is:

$$\varepsilon_x = \frac{1}{E}(1 - v)\sigma_x - v\sigma_y \tag{4}$$

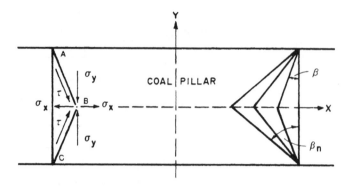

Fig. 6.2.14. Mechanics of coal pillar fracturing.

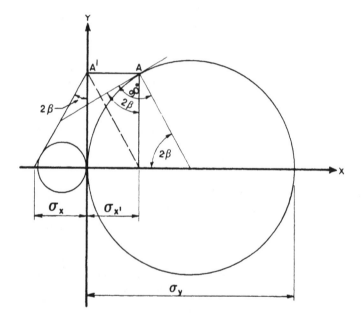

Fig. 6.2.15. Stress state of outer pillar zone given by Mohr's circles.

if $\varepsilon_x = 0$ then

$$\sigma_x = \frac{\nu}{1 - \nu} \sigma_y \qquad (5)$$

Further, from the Mohr's circle of the stress, for slip line at angle β, as illustrated in Figure 6.2.15, the shear stress can be represented by equation below

$$\tau_\beta = \tfrac{1}{2}\sigma_y \sin 2\beta \qquad (6)$$

By projecting shear stress on the τ axis and on the σ_x axis at angle 2β, the stress component so obtained should balance σ_x. Thus it could be written as below:

$$\sigma_x = \tau_\beta \tan 2\beta \qquad (7)$$

According to the condition of equilibrium at point B we now have:

$$\tfrac{1}{2}\sigma_y \sin 2\beta \tan 2\beta = \frac{v}{1-v}\sigma_y \tag{8}$$

hence

$$\sin 2\beta \tan 2\beta = \frac{2v}{1-v} \tag{9}$$

After angular transformations, we get

$$\cos^2 2\beta + \frac{2v}{1-v}\cos 2\beta - 1 = 0 \tag{10}$$

Solving this equation for $\cos 2\beta$ and discarding the negative roots, we can determine the angle β

$$\cos 2\beta = \frac{-v + 2v^2 - 2v + 1}{1-v} \tag{11}$$

$$\beta = \tfrac{1}{2}\text{arc}\cos\frac{-v + 2v^2 - 2v + 1}{1-v} \tag{12}$$

Knowing β, the angle of friction may be determined as,

$$\phi = \frac{\pi}{4} - \beta \tag{13}$$

where β is the angle of the slip line and $\pi/4$ is the angle for an ideally plastic material. For the given state of stress, the angle of friction ϕ is constant.

After determining angles β and ϕ from the boundary conditions, the parameter K for coal material may be found in an elasto-plastic state (Fig. 6.2.16). This parameter could be used to establish the relationships between compressive and tensile stresses. They are as follows:

$$\sigma_C = 2K(2\beta' + \sin 2\beta') - \sigma_T \tag{14}$$

$$\sigma_C = K\frac{2\cos\phi}{1-\sin\phi} \tag{15}$$

$$\sigma_T = K\frac{\cos\phi}{1-\sin\phi}\sin 2\beta' \tan\beta' \tag{16}$$

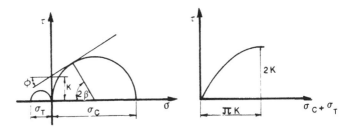

Fig. 6.2.16. Determination of parameter K, by sinusoid describing the elasto-plastic state of coal material.

The angle β between the slip line and the direction of principal stress at any point of the stressed coal pillar is:

$$\beta = \frac{\pi}{4} - \frac{\phi}{2} \tag{17}$$

The value K could be taken as the radius of the generating circle of cycloid. The arc of the cycloid can be given in parametric form using the function as below:

$$\sigma_C + \sigma_T = K(\pi - 2\phi - \sin 2\phi) \tag{18}$$

$$\tau = K(1 + \cos 2\phi) \tag{19}$$

or by the arc of the sine curve given by equation below:

$$\tau = 2K \sin\left(\frac{\sigma_C + \sigma_T}{2K}\right) \tag{20}$$

with

$$0 \le \sigma_C + \sigma_T = \pi K$$

The main element of structural stability is controlled by shear stress concentration and the shear strength of the coal in a direction parallel to the vertical axis of the free face e.g. planes along which all slabbing deformations take place.[17]

Finally, it should be concluded that pillar slabbing is a continuous process, which starts before the load reaches a critical value, probably at its conversion from the elastic to the elasto-plastic state. The continuous process of coal pillar slabbing results in the pillar weakening, decreasing pillar dimensions, and related decreased bearing capacity of the structure.

6.2.5. *Reinforcement of coal structure*

The main characteristic of non-violent fracturing is the gradual deterioration of the strength of the coal structure. This stability phenomenon is in contrast to the coal outburst with violent fracturing which instantly decreases the strength of mine structure. It is obvious that under the circumstance of non-violent fracturing different elements are of importance in mine stability, here the main task is to reinforce rather than de-stress the mine structure. A few methods are briefly listed below.

1. External support by leaving the slabs in place, because removal of the slabs may accelerate the formation of subsequent slabs as discussed in previous heading. The underground observations showed that continuous removal of slabs could result in the destruction of the pillars. The accelerated and progressive slabbing due to stress redistribution and concentrations is a process which cannot be eliminated by external support, but could be, to a certain degree, remedied. For example, mining experience suggests that reinforcement should be done by piling all broken and fallen coal and placing it next to pillar along entries and roadways to provide lateral support to the pillars and to catch any material that might fall. It has been observed that the wrapping of the pillars by old cables offers a constraining effect and decreases the slabbing deformations.

Fig. 6.2.17. Rock bolting of ribs and spans within coal seam (Collie Basin, W. Australia).

2. Internal support is primarily achieved by rock bolting. However, rock bolting is effective only if the thickness of the slabs is less than 1.0 m, and the bolt anchors are at least 1.0 m into solid coal. Rock bolting is not effective if a second slab has formed behind the first and the bolt anchors are in the second slab.[21] The rock bolting for rib reinforcement becomes important when piles of coal are to be removed to obtain an access to the face, as in the case of the monitoring of marginal slabs by visual and audible warning devices. Rock bolting is equally important for entries driven in thick coal seams, which are surrounded by coal material. The effectiveness of such a support is illustrated in Figure 6.2.17, where the mine opening stands up stable for almost two decades of excavation. This is a good example of the effectiveness of rock bolting within coal material, and the degree of reinforcement which can be achieved by this type of support.

3. The pillar design based on the increase of width-to-height ratio will improve the overall strength of the coal structure. The ability of a pillar to resist slabbing depends to some degree on the coal strength.

6.3 SHEAR DISPLACEMENT OF COAL

The preferential planes of shear displacement in the coal seams are stratification continuities. This displacement is particularly facilitated if the bedding planes of the coal seams have been orogenically sheared. The deformation and failure of the coal structure is in the domain of shear mechanics controlled by stratification and orogenic features of the seams.[22] Points of particular interest in the analyses of the shear displacement is the mechanics of movement of coal blocks along the stratification planes and thrust planes as well as sliding deformations due to gravity.

The influence of shear displacement of the coal on mine stability in the first place depends on the competency of the coal and shear strength of geological continuities as well as the angle of inclination of coal seams. For example, Western Asutralian coal seams are little affected by this type of deformation and failure due to their satisfactory shear strength and mechanical stability, which is in contrast to the mechanically unstable and mostly pitching coal seams of Western Canada.

6.3.1. *Displacement along stratification planes*

This mechanical phenomenon has been studied in great deal in the Rocky Mountain fold belt which contains coal of high rank. The main structural feature of these deposits is their severe orogenic deformations which had been followed by movement along stratification planes, which at present have mainly residual shear strength. However, the intensity of shearing due to tectonic processes depends on the type of coal adjacent to the bedding planes. For example, along durain bedding planes, displacement varies up to 0.3 m, but along fibrous coal planes displacement is greater, reaching as much as 1.5 m. Brittle, bright coal bands generally disintegrate due to shearing. The shear displacement along these planes is mostly less than 1.0 m. Finally, the amount of shearing along discontinuities depends on the strength of individual petrographic layers of coal, as well as on the intensity and direction of tectonic stresses which caused the folding and shearing.

The roughness of interlocking shear surfaces has a significant influence on the shear strengths both between coal bands and between parting bands. Four geometries of asperities or roughness can be considered:

a) High angle of asperity ($\alpha_1 = 30° - 50°$; $\alpha_2 = 70° - 90°$ to bedding planes) and limited width of teeth (up to 2.0 cm). This sharp, tooth-like geometry of asperities is

Fig. 6.3.1. Interlocking surface irregularities of sheared hard coal layers (Mount Head Area).

Fig. 6.3.2. Interlocking surface irregularities of sheared soft coal (Grande Cache Mine).

found more in hard durain coal and less in semi-hard durain and vitrain banding in coal and coaly shale. It influences mining induced shearing, which is of limited displacement (Fig. 6.3.1.).

b) Low angle of asperity (approximately $\alpha_1 = \alpha_2 < 20°$ to bedding planes) and extended width of the teeth (up to 10 cm). This geometry of interlocking surface is found in soft coal (vitrain, clarain) and relates to appreciable shearing displacement during mining (Fig. 6.3.2.).

c) Asperities with almost no angle to bedding planes ($\alpha_1 = \alpha_2 < 5°$) where shearing occurs along friable, very soft fusain bands (Fig. 6.3.3). The joint surfaces are compacted with filing of powder-like disintegrated coal.

d) Low angles of asperity with shearing along very smooth interlocking surfaces between either mudstone partings or partings of adjoining coal layers. Joint planes are not filled. Shearing continuity along the joints may be most important regarding mine stability.

The shear displacement deformations of mine structures are greatly affected by gross geological structural defects within coal seams, particularly along planes previously sheared by orogenic movement. A point of particular interest for strata mechanics is to quantify shearing displacement along stratification continuities, in addition to direct shearing strength between them. For this reason, primary consideration has been given to coal seams with heterogenous internal structures, from which coal samples were taken for laboratory testing. Sampling of the coal ribs has been carried out in development roadways of several underground mines of the Rocky Mountain fold belt. The testing procedure and testing results the discussed individually in the following order:

Fig. 6.3.3. Disintegration of coal seam due to intensive shear movement along bedding planes of soft friable coal bands (Cardinal River).

1. Shear tests were carried out on the samples cut from coal blocks by a circular saw in blocks of sufficient size so that the area to be subjected to shear ranged from 15 to 25 cm². The rate of shearing was held constant at 1.27 mm/min for all tests.[24] For each test the peak value of shear resistance was chosen as the highest stress applied during shearing. The peak shear stress is plotted against the corresponding normal stress. All peak values are plotted in a normal stress-shear stress diagram.

2. Shearing resistance of coal bedding planes produced the data which clearly exhibits the peak strength-ultimate strength behaviour in the relationship between shear stress and displacement of the stratification plane. From the graphs of normal stress-maximum shear stress (Fig. 6.3.4) for all three tested samples, the frictional angle is nearly constant (34°) whereas cohesion varies appreciably (0.35 to 1.20 MN/m²). This data can be represented by the Coulomb criterion of failure:

$$\tau = c + \sigma_N \tan\phi \tag{1}$$

where

$\qquad \tan\phi = 0.65 =$ coefficient of friction

The cohesion of the sheared coal samples could be called an intercept cohesion (c') whose magnitude depends on the pervasiveness of the coal bridges.[23] Samples with a higher number of coal bridges have a higher cohesion than samples with a lower number of bridges. However, it seems very likely that differences in intercept cohesion of the tested samples also depends on the nature of individual sheared petrographic units where hard coal (c' = 1.2 MN/m²) and brittle coal (c' = 0.8 MN/m²) have a greater cohesion than fibrous coal (c' = 0.35 MN/m²). As the great majority of sheared bedding planes are vitrinite (brittle) bands, cohesion is of lesser importance in determining mine stability because it resists only initial shearing stress. The main strength component is frictional resistance. In evaluating soft layers and their slip

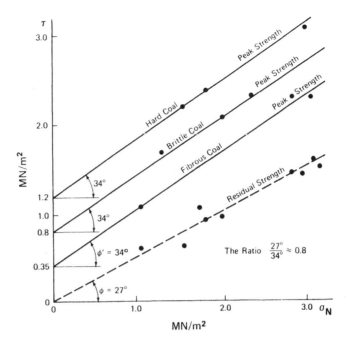

Fig. 6.3.4. Evaluation of shear strength of coal stratification planes.

deformation, the residual strength is probably most important. The residual friction angle is approximately 27°. Shearing resistance can be written as:

$$\tau = 0.5\,\sigma_N \tag{2}$$

Generally, both the peak and the residual friction angles in the Rocky Mountain coal belt have lower values than most other North American coals.[24] This correlates to the lower mine stability in this region.

3. Shearing resistance between parting bands or between coal and parting bands also influences the stability and flow of mine structures, particularly for partings in the roof or floor of the coal seams. The test data did not reveal a peak strength in the relationship between shear stress and displacement. The graph of normal stress-maximum shear stress shows no cohesion between layers and that shear resistance comes only from friction (Fig. 6.3.5). A certain number of the stratification contacts have only residual shear strength. Between parting layers or partings and coal bands the orogenically sheared planes exhibit smooth and shiny surfaces. The angle of friction ranges from 27° to 29°, angles similar to the residual shear angle along coal planes. This number of tests is limited so this finding cannot be considered definitive.

'Dirty coal' seams or coal seams with mudstone close to the roof or the floor or both are most liable to extrusion deformations, particularly if the majority of the coal layers are already sheared by orogenic events. The peak shear strength, governed only by frictional resistance (cohesion = 0) of moderate magnitude, causes instability of mine structures particularly when movement is propagated by squeezing deformations in soft coal layers, as discussed in the next subheading.

It is a point of interest that shear strength between stratification planes deteriorates with time.[25]

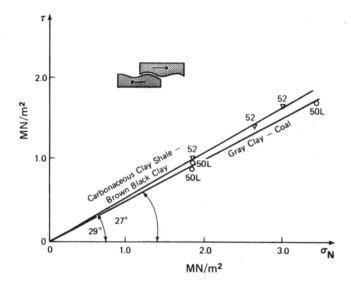

Fig. 6.3.5. Shear strength between parting layers and between partings and coal layers (Number 2 Mine, Grande Cache).

Mine stability is greatly controlled by shearing displacement between individual bands with differing petrographic characteristics within coal seams. Shear motion along stratification planes is mobilized by excessive overburden pressure redistributed in the mine structure as well as by frictional effects between individual layers of different mechanical properties.

The classical approach for designing structures in coal seams using the uniaxial compressive strength of coal or coaly material is of lesser significance in the Rocky Mountain Belt. This is because the shear strength of coal seams in this area is lower than in other coal regions of the world and shearing stresses along stratification planes govern the mechanics of deformation and failure in mine structures.

6.3.2. *Displacement along thrust plane*

The interaction between instability of coal structure and thrust fault planes have been analyzed for the Rocky Mountain fold belt. The main structural feature of the fold belt is expressed by a multitude of thrust faults (low angle reverse faults or contraction faults).[26] Large thrust and fold structures are easily detected and delineated. However, the thrust faults of smaller displacement which influence coal pillar stability are usually discovered during mining operations and adversely affect the planned mine layout and coal pillar stability. Underground observations suggest several modes of pillar deformations and failure due to folding and thrusting which are briefly discussed below:

1. Low angle thrust which persists throughout the mine structure is a well exhibited case of their deformation at Rocky Mountains and Foothills regions. The thrust fault represents a potential glide plane of the pillar. The pillar stability depends on the angle of fault inclination towards pillar horizontal axis and the shear strength of continuities (Fig. 6.3.6). The laboratory testing to evaluate a shear strength of the thrust planes

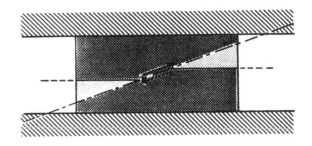

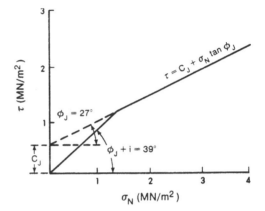

Fig. 6.3.6. Diagram of shear strength
of thrust fault and related coal pillar
deformation.

(Number 2 Mine, McIntyre Mine) have shown the following results:

Firstly, under low normal load (2 tonnes shear box, Department of Mineral Engineering) the angle of friction was 39° and cohesion close to zero.

Secondly, under high normal load (shear box, Department of Civil Engineering), the angle of friction was 27° and cohesion 0.61 MN/m².

Under these circumstances the pillar bearing capacity is not controlled by the compressive strength of coal, but by the shear strength of thrust planes as given by the following equation

$$\tau = c_J + \sigma_N \tan \phi_J \tag{3}$$

$$\tau = 0.61 + \sigma_N \tan 27° \tag{4}$$

$$\tau = 0.61 + 0.50 \sigma_N \tag{5}$$

In the final instance a mine structure stability depends on the time factor, e.g. time required that asperities along a thrust plane are going to be destroyed, and only residual shear resistance to be present.

2. The complex thrusting is represented beside low angle faults also by displacement of upper coal block by shearing, which happens after the main thrusting period.[27, 28] This is the worst type of coal structure weakening because the coal is highly sheared with sliding wedges exposed on the rib walls (Fig. 6.3.7). Additional fragmentation of such a structure has been observed along failures perpendicular to thrust planes, which are propagated from the upper fault. This type of thrusting is mainly implemented in

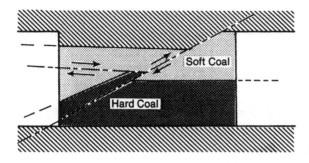

Fig. 6.3.7. The coal pillar sheared and fractured in wedges.

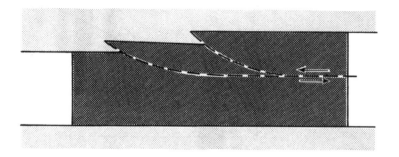

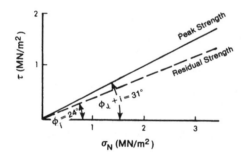

Fig. 6.3.8. Diagram of peak and residual shear strength of the orogenically sheared bedding planes and related coal pillar geometry.

thicker coal seams, where individual coal layers have differential properties. To avoid collapsing deformation in such a tectonically disturbed ground, the only solution is to leave large pillars of the unmined coal.

3. Thrusting parallel to the bedding of coal seams was by faults, which intersect coal beds and then tend to die out along their stratification planes (Fig. 6.3.8). The faults parallel to bedding planes are easily detectable in coal core samples obtained during exploration by diamond drilling. The fault plane preferably aligns with soft coal layers, which are completely fragmented and pulverised during shearing. The laboratory testing of shear strength of this fault planes have shown an absence of cohesion, where the shear strength depends only on the angle of friction (ϕ_J) and the angle of inclination of asperities (i). The peak shear strength including the geometrical component is $\phi_J + i = 31°$, but the residual shear strength has an angle of friction $\phi_i = 24°$ (Fig. 6.3.8).

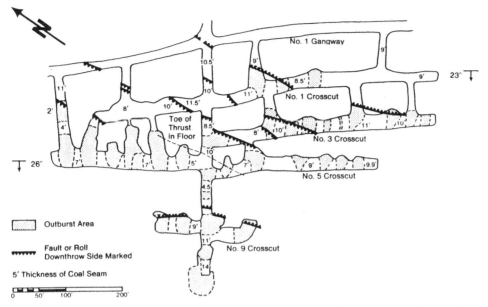

Fig. 6.3.9. Location of coal and gas outbursts at Upper March seam, Canmore Mine, Alberta (after Patching).

The low residual friction angle might be due to the lubricant role of pulverised coal along the sheared bedding planes. This type of orogenic shearing is very common in coal seams which particularly influence the pillar stability.

Under such structural conditions the coal seam thickness is abruptly changed in steps, which also influences unstable pillar conditions.

4. Thrust faults which die out laterally into folds are mapped in the semi-anthracite coal seams, which have a rolling geometry. Patching studied in great detail the relationships between these structural irregularities and the failure at Canmore Coal Mine, Alberta,[29] as illustrated in Figure 6.3.9. Due to the structural irregularities and outburst phenomena the room-and-pillar mining method in this mine was not successful and finally after several decades of the active mining, it has been shutdown.

6.3.3. *Sliding displacement*

The sliding displacement of coal structures occur in inclined coal seams of different thicknesses. The mechanics of sliding deformation have a particular consideration in engineering geology, and are known as wedge stability analyses. These analyses consider the wedges with multiple planar surfaces, but sliding-slip surface occurs always along only one side of the wedge. In underground coal mining the sliding of coal blocks is along stratification planes or fault planes. The sliding along stratification planes is facilitated if they are orogenically sheared. The coal blocks have multiple planar surfaces within the coal structure, where movement along the slip surface is in the direction of gravity. Generally, there are two groups of sliding deformation as further discussed.

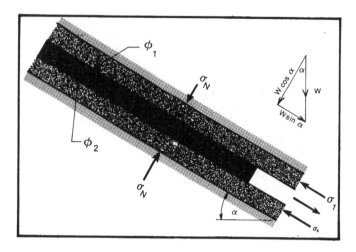

Fig. 6.3.10. Sliding mechanics of thin and inclined coal panel (longwall mining).

1. Single block sliding is common for thin and inclined coal seams of higher strength. The slide movement of the coal blocks in dip direction corresponds to the pillar or panel structure, which is sandwiched between rock strata. The slip surface and sliding usually takes place along the stratification contacts between the coal seam and roof and floor strata. The stability of the coal structure primarily depends on the magnitude of the vertical stress, e.g. its component perpendicular to stratification planes and frictional resistance between coal seam and roof and floor strata. The instability of coal pillars or coal panels is induced when mobilized shear stresses between coal seam and roof and floor strata overcome the shear resistance between their beds, as illustrated in Figure 6.3.10. The block sliding deformation is going to be for the following shear stress relationships:

$$W \sin \alpha > \tau_1 + \tau_2 \qquad (6)$$

$$W \sin \alpha > \sigma_N \phi_1 + \sigma_N \phi_2 \qquad (7)$$

$$W \sin \alpha > \sigma_N (\phi_1 + \phi_2) \qquad (8)$$

These equations suggests that stability of coal pillars or panels of inclined thin coal seams would deteriorate with an increase of mine depth, e.g. increase of the overburden pressure.

2. Multiple block sliding is common for thick and pitching coal seams of lower strength. In general, coal structure consists of numerous mating blocks delineated by stratifications planes and perpendicular fracture induced or propagated by mining stresses. The sliding movement of blocks in the dip direction is along larger and smooth slip surfaces. Besides smoothness of slip surface, the block sliding is facilitated by the angle of inclination of coal seams. For example, the underground studies in coal mines in Rocky Mountain Belt, suggests that block sliding occur in seams with angle of inclination greater than $35°$ ($\alpha > 35°$), whereas blocks crushing combined with sliding occurs at lower inclinations ($\alpha < 35°$), where the critical angle is arbitrarily determined.[9]

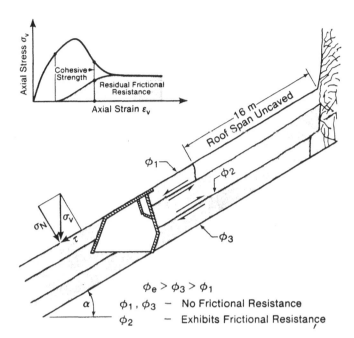

Fig. 6.3.11. Sliding deformation of inclined coal pillar (Coleman Colliery Limited).

The investigations of multiple block sliding within coal pillars at Vicary Creek Mine, Coleman Colliery suggests that in addition to the factors discussed above, other parameters are also important as represented in Figure 6.3.11 and discussed below:

a) Lower frictional resistance adjacent to mine roof (orogenically sheared planes) contributed to rapid sliding of the coal block.

b) Residual frictional resistance in the middle of the seam exhibited initial sliding of the coal block which subsequently ceased.

c) Higher resistance adjacent to the mine floor occurred i.e. a slow sliding of the coal block.

The differential frictional forces along the bedding planes creates freedom for individual coal blocks to slide independently.

The point of interest regardless of seam thickness is to evaluate slip deformations of the coal seams which were already orogenically sheared along stratification planes, as for example coking coal deposits of Western Canada, where a majority of coal reserves are locked at inclinations between 25° and 65°. The laboratory investigations of coal samples from this area indicate the following slip failure mechanics.[30]

a) At inception of loading the stress-strain curve has a very gentle slope due to the closing of cracks, but as the sample becomes compacted the curve steepens and the coal exhibits elastic deformations (Fig. 6.3.12).

b) The hysteresis of repeated loading and unloading of the sample suggests some loss of strain energy, probably due to the internal friction which is developed between inclined soft layers.

c) Rupture takes place along a soft band when surface irregularities on the bedding planes are destroyed and slip occurs.

d) The stress-strain curves of coal samples tested at different angles of bedding

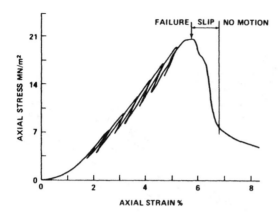

Fig. 6.3.12. Stress-strain loading and un-loading curves (coking coal sample, bedding at 45°).

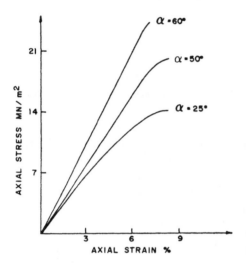

Fig. 6.3.13. Stress-strain curves for coal at different bedding inclinations (α).

inclination suggest that coal at lower angle ($\alpha = 25°$) exhibits elasto-plastic behaviour whereas coal at higher angle of bedding ($\alpha = 65°$) exhibits elasto-brittle behaviour (Fig. 6.3.13).

e) Also, the stress-strain curve suggests that with progressive shearing, the shears eventually lock, which resulted in development of residual strength as a product of frictional resistance.[31]

The laboratory data suggests that coal samples under loads inclined to bedding exhibit strength which depends on cohesion and frictional resistance along bedding planes. Orogenic shearing along bedding planes has reduced the strength of the coal. Such phenomena have been observed in underground mines.

Although, the laboratory testing of the coal samples taken from three different coal mines have been carried out to evaluate direct shear strength along stratification planes. The investigations indicated that the angle of friction, ϕ, along bedding planes is constant at 34°, that the intercept cohesion, c', varies from 0.35 to 1.25 MN/m² and that the residual angle of friction, ϕ_i for all tests averaged 27°. A safety factor in mine design should be involved in estimating the possibility of sliding of coal blocks[32] with the sliding related to angle of inclination by

$$\frac{\tan \phi}{\tan \alpha} = k = \text{coefficient of safety} \tag{9}$$

When evaluating mine stability for pitching coal seams in the Rocky Mountain region one must consider that the coal has low shear strength along stratification planes already sheared by orogenic movement, and that layers of hard coal have a tendency to slide as rigid blocks. Mine structures during yielding can support loads for a certain time due to frictional resistance.[33]

6.3.4. *Prevention of sliding deformations*

To prevent displacement deformations is difficult and to eliminate them is impossible. Under these circumstances the displacement of structure has to be anticipated, but with shearing magnitude below collapsing deformations. At the present time in the domain of civil engineering, there are several techniques of rock structure stabilization but some of them are not applicable to underground coal mining because of the prohibitive cost or differences in construction techniques. Prevention of structure displacement in mining is primarily based on the philosophy of layout and technology of coal extraction. The short duration of coal structure favours the gifted ideas rather than expensive engineering solutions to avoid adverse effects of displacement deformations.

Some possibilities of sliding displacement prevention of coal structure are briefly listed as follows:

1. Artificial reinforcement by rock bolting or cable bolting primarily should be considered for gliding planes of thrust faults. The effectiveness of support of the coal structure depends on geological rather than mining conditions. For example, mine structure in thick coal seams fractured during orogenic shearing are difficult to be reinforced either by rock bolting or cable bolting.

The 'pneuma-balloon' or pneumatic bag was developed for roof support for coal faces in steep seams, and was first manufactured in 1974 in the U.S.S.R.[12] This type of support could replace the coal pillars which are in a state of sliding displacement and without the ability to support the superimposed load. The pneumatic bags are made of rubber-cord cloth and joined by rubber belts, in the pattern required. These pillars operate in a simple way. When they have been filled with compressed air they are set between roof and floor of the face. In the process of convergence of roof and floor strata the height of the pillar gets less, the pressure of compressed air inside increases, the area of contact with roof and floor increases and the load on the pneumatic pillar is increased appreciably. The 'pneuma-balloon' support can be used in thin coal seams, less than 0.9 m, and also very steep seams. Their application improves safety and ensures considerable economic advantages compared with maintaining sliding coal pillars.

2. Mine layout should anticipate a design philosophy of minimal coal seam development with large blocks, which should contain structural irregularities. Under these circumstances the orogenically sheared planes will have a minimal exposure on the

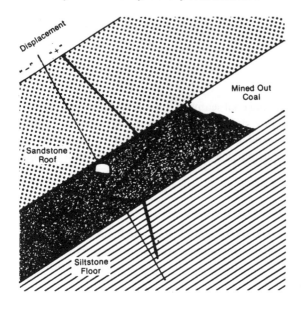

Fig. 6.3.14. Diagram of differential displacement of coal seam (full face mining).

walls of mine structure. To achieve this goal it is necessary to know the local structural features in great detail, which is often not possible.

3. Mining technology could be instrumental in dealing with mine stability, when displacement deformations are strongly pronounced, particularly in the case of soft, thick and pitching coal seams. In such coal seams large displacement along stratification planes or thrust planes are most common. Under these circumstances the stability of coal structures could be jeopardized. The instability of coal structure along bedding planes particularly affects mine roadways and extraction stopes (Fig. 6.3.14). Hard mining experience suggests, that the stability of mine in slice (bench) mining should be maintained by the application of stowing technology, where the excavated coal is replaced by hydraulic fill. If full face mining is applied, then the monitor jet technology should be applied.[34] Rapid coal extraction by monitor jets prevents development of accelerated displacement and collapse of stope structure before coal has been recovered.

6.4. EXTRUSION DEFORMATIONS

Generally speaking, the strata associated with coal mining are relatively soft compared to the strong rocks encountered in hard rock mining. In coal mining, the coal seam could be the weakest member of the seam profile, which results in coal extrusion. Underground mining in weaker coal seams, will have the tendency to close the excavation, particularly in the case when the immediate floor is equally as soft as the coal. Hence, the problems of outbursts in the stronger coal and progressive fracturing in coal of moderate strength is replaced by that of total closure in the soft coal.

The considerations in this heading are given to several aspects, for example determination of closure deformations of the openings where extension deformations of the ribs are in effect, and determination of length of the zone within the rib, which is involved in extrusion deformations. In addition, consideration is given to the mechanics of coal flow, with an attempt to produce equations for calculating the rate of its flow and velocity. The possibilities of strata stabilization will be discussed at the end.

Finally, it should be pointed out that at the present time there are several approaches in representing and solving the problems of extension deformations. However, here preference is given to visco-plastic mechanics rather than the theory of perfectly plastic substance which is commonly used for design in rock salt and potash mining.

6.4.1. *Closure of entries*

The weak and soft coal seams facilitate closure deformations, but their intensity depends also on the strength and lithology of immediate roof and floor strata of the mine opening,[35] pitch of the coal seams, and duration of the mine opening.

1. Character of the immediate roof and floor strata, is defined mainly by the rock lithology and strength, influences the two types of closures.

a) Uniform closures exist when the roof and floor rock lithology and strength are identical. Under these circumstances the intensity of convergence of the mine roof will be approximately the same as floor heave and control of entry stability is predictable.

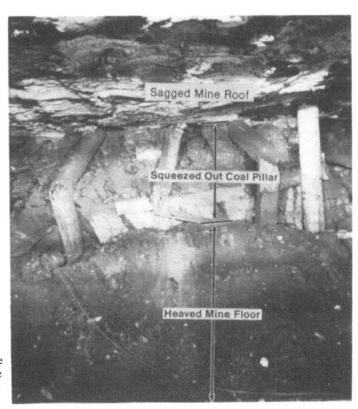

Fig. 6.4.1. Differential closure deformations due to existence of sandstone roof and coal-shale floor stratum.

b) Non-uniform closure exist when the rock lithology and strength of the mine roof and mine floor differ to a certain degree. The intensity of the convergence will depend on the rock type exhibited in the mine roof or floor, as illustrated in Figure 6.4.1. From this illustration it can be seen that the roof strata exhibits a yielding deformation (rock fracturing), which obviously must have an influence on the formation of differential closure.

It should be pointed out that the extrusion deformation of the coal faces and ribs are accelerated, if either roof and floor rocks are soft or if one member (roof or floor stratum) is represented by soft rocks.

2. *Closure of the pitching coal seams* also depends on the lithology and strength of roof and floor rocks. The most common case of closure is differential because of the differences between the mine roof and the mine floor. Under these circumstances of pitching coal seams the evaluation of entry closure requires a recalculation of the measured convergence data, obtained by simple height differences as illustrated in Figure 6.4.2.

$$\Delta h = h_1 - h_2 \tag{1}$$

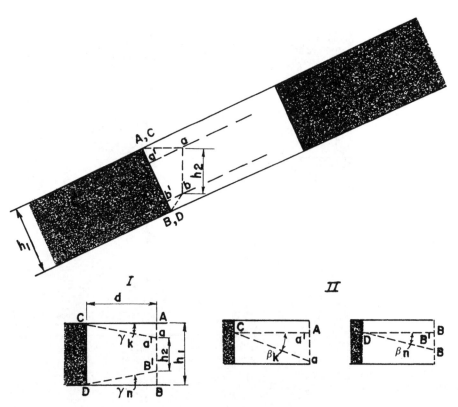

Fig. 6.4.2. Diagrams of closure deformations of pitching coal seams.

where

 Δh = total closure;

 h_1 = original height of entry;

 h_2 = height of entry at the time of instrumentation.

The closure in Figure 6.4.2 is represented also by the relation:

$$\Delta h = Aa' + Bb' \tag{2}$$

where, the parameters Aa' and Bb' depend upon the angles of convergence along strike (γ_k and γ_n) and along dip (β_k and β_n). On the basis of these data (distances and angles) which can be easily measured underground, the calculation of closure deformation is as follows:

- Along strike: roof strata Aa' = $d \tan \gamma_k$
    floor strata Bb' = $d \tan \gamma_n$
- Along dip: roof strata aa' = $d \tan \beta_k$
    floor strata bb' = $d \tan \beta_n$
- Total convergence: roof Aa = $d \sqrt{\tan^2 \gamma_k + \tan^2 \beta_k}$
    floor Bb = $d \sqrt{\tan^2 \gamma_n + \tan^2 \beta_n}$

The closure measurements and their recalculation to the values of integral convergence in some pitching coal seams in Zenica Colliery (Yugoslavia) showed differential displacement of the roof and floor strata of mine openings.[36] In a majority of the cases the greater convergence was along the dip rather than along the strike. The mine roof and mine floor composed of similar lithological rock units and approximately of the same strength, exhibited greater convergence (55 to 75 per cent) of the roof than floor. This could be attributed to geological factors as well as mining factors.

3. Acceleration of closures in the mine roof and mine floor showed a hyperbola or exponential function (Fig. 6.4.3). For example, in some Western Canadian coal mines the measurement of convergence closure in the mine stope area (room-and-pillar mining) exhibited appreciable strata sagging before the roof caving (width of stope span over 15 m). The total closure deformations can be well investigated in the mine entries, which is discussed separately in the subheading of coal flow deformations.

——— CENTER OF STOPE

– – – – RIB AREA

Fig. 6.4.3. Acceleration of roof convergence and floor heave.

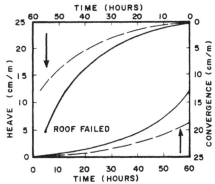

6.4.2.   *Length of flow zone*

The extension of the coal into the excavation is also a function of the width of the yield zone, and there is a direct relationship between the length of this zone and the amount of closure. For example, fragmented and loosely piled coal can occupy about 1.5 times its original volume, which explains the fact that during extrusion deformation entries can be totally closed.

The coal seam which is weak and soft and as such liable to extrusion deformations, could be due to geological conditions of loading (creep over million years), considered that is in a state of lithostatic stress. For these stress conditions, the width of the flow zone around circular openings is given by Wilson[37]

$$r_b = r_0 \left[\frac{2q - \sigma_c^{1/(K-1)}}{p(K+1)} - 1 \right] \tag{3}$$

where

r_b = radius of yield zone;
r_0 = radius of opening;
p = restrain on opening boundary;
q = unit load;
σ_c = unconfined strength;
$K = \sigma_1/\sigma_3$.

This author concluded, that in case of weak rock and weak support, the flow zone may be appreciably extended from the entries edge at a greater mine depth. For example, in Western Canadian coal mines due to the very weak nature of the coal, the extension of flow zone has been observed in lengths up to 30 m at a mine depth of 500 m. Also, an underground mine observation showed that the benefit to be achieved by the further increase of support load is small (maximum decrease of flow zone was observed below 20 per cent of the total length of this zone). The intensity of extrusion deformation would in great deal depend on the length the flow zone, here two extreme cases could be suggested:

a) Restricted extrusion, relates to the fact that the mine roof and the mine floor are composed of hard rock strata, particularly in case of strong and elasto-brittle sandstone. These conditions are described by Wilson, who suggested the following equation for the calculation of the length of flow zone[37]

$$\sigma_y = K\gamma h = pK \exp\left[\frac{x_b}{t} \quad FK\right] \tag{4}$$

$$x_b = \frac{H}{FK} \ln\frac{\gamma h}{p} \tag{5}$$

$$FK = \frac{K-1}{\sqrt{K}} + \frac{(K-1)^2}{K} \tan^{-1}\sqrt{K}$$

where

H = height of working place.

From Figure 6.4.4 it can be seen that the length of the flow zone is limited (a little over 5 m for maximum depth) and also, that it does not appreciably increase after 200 m depth.

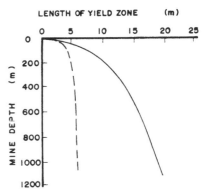

Fig. 6.4.4. Length of flow zone.

b) Unrestricted extrusion is related to the existence of the soft and weak roof strata, coal seam and floor strata. The rib deformations under these conditions are also described by Wilson, which is represented by the equations:[37]

$$\sigma_y = K\gamma h = pK \frac{x_b + H/2^{K-1}}{H/2} \tag{6}$$

$$x_b = \frac{H}{2}\left[\left(\frac{\gamma h}{p}\right)^{1/K-1} - 1\right] \tag{7}$$

Also, from Figure 6.4.4 it can be seen that the length of the flow zone is much greater than in the case of soft coal seams sandwiched between rigid rock strata. At the same time the tendency of the increase in the length of the flow zone with an increase of the mine depth, is clearly exhibited.

However, if hard strata is in the mine roof and soft strata in the mine floor or vice versa then differential extrusion of coal layers should be expected.

Wilson, pointed out that recognition should be given to the intermediate state between soil mechanics and rock mechanics, as it is exhibited by existence of weak and soft coal seams. It is necessary to put more effort into defining the principles of the so-called soft rock mechanics. Particularly, for explanations of individual mechanical phenomena observed underground, as for example coal flow from ribs and working faces, where an attempt has been made to explain the mechanics of this deformations.

6.4.3. *Mechanics of extrusion deformations*

The extrusion deformations of the coal from pillars and working places have been observed particularly in coal fields of younger geological age (Jurasic/Cretaceous) where the coal is of the higher rank (coking coal of the Rocky Mountains and Kusnetsk Coal Basin).

The mechanics of extrusion deformations are based on the assumption that the coal seam is a body situated between two plates (which represent mine roof and mine floor stratum) with a tendency of continuous extrusion in the mine opening. This phenomena could be postulated as a steady flow of Bingham substance between two rigid plates, which exhibit a process of continuous creep or visco-plastic flow, as illustrated in Figure 6.4.5.

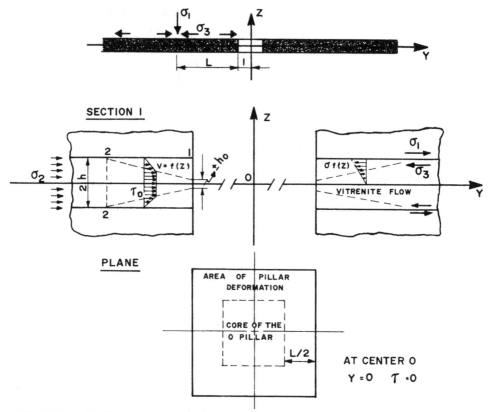

Fig. 6.4.5. Model of mechanism of extrusion deformations.

From the same figure it can be seen that the z-axis is the center line between two coal pillars, and the y-axis passing through the middle of the soft layer in the virgin coal seam conditions is denoted by h, and during extrusion in mining conditions, it becomes $\pm h_0$.

If someone assumes that extrusion of the soft layer from pillar or working face is due to mining pressure, which is induced by pillar formation and coal excavation, then the differential pressure can be written as,

$$\Delta\sigma = P_2 - P_1 \tag{8}$$

where

P_2 = mining stress;
P_1 = virgin stress;

which are uniformly distributed on the pillar surface along the y-axis of length L/2. This corresponds to the assumption that 25 per cent of the mine structure is influenced by extrusion deformations (Fig. 6.4.5). Let us suppose that a substance flows mostly laminar and constant, and that the rigid layers converge in the direction of the line $z = 0$, then the principles of fluid mechanics are acceptable. For example, the flow described above has been observed in McIntyre Mine (No. 4 coal seam) which suggested that it is a very slow process. Under these conditions the Reynolds number should be less than one (R < 1). For these flow parameters the mass forces and inertia could be assumed to be equal to zero. With these assumptions, the general differential equation of the continuous flow of visco-plastic layer can be written[38] as,

$$\frac{d\sigma}{dy} + \eta\left(\frac{d^2\omega}{dy^2} + \frac{d^2\omega}{dz^2} + \frac{1}{y}\frac{d\omega}{dy} - \frac{\omega}{y^2}\right) = 0 \tag{9}$$

$$-\frac{d\sigma}{dz} = 0 \tag{10}$$

$$\tau_{yz} = \sigma_0 + \eta\frac{d\omega}{dz} \tag{11}$$

$$\frac{d\omega}{dy} + \frac{\omega}{y} = 0 \tag{12}$$

The boundary conditions might be written as below:

$$\omega_y(y = \pm h) = 0 \quad \frac{d\omega_y}{dz}(y = \pm h_0) = 0 \tag{13}$$

If it is assumed that each element of the soft material (in this case coal) for the section of height h, is representing rigid flow laminae, then the shear stress would be limited, as given below:

$$\tau_{yz}(y \neq h_0) = \tau_0 \tag{14}$$

Also, each element is under the same state of stress, so that for conditions of stress equilibrium, it could be written:

$$h_0 = \tau_0\left(\frac{d\sigma}{dy}\right)^{-1} \tag{15}$$

Using the system of solution of equations for laminar visco-plastic flow (10), the above equations are solved in the following order:

$$\sigma = f(y) = \frac{d^2\omega}{dy^2} + \frac{1}{y}\frac{d\omega}{dy} - \frac{\omega}{y^2} = 0 \tag{16}$$

For these conditions equation (9) is given:

$$\frac{d^2\omega}{dz^2} = -\frac{1}{\eta}\frac{d\sigma}{dy} \tag{17}$$

and by integrating this equation, the velocity equation is obtained as,

$$V = -\frac{1}{\eta}\frac{d\sigma}{dy}\frac{z^2}{2} + C \tag{18}$$

If the velocity profile is examined for the region z > 0, then this equation is:

$$V = -\frac{1}{\eta}\frac{d\sigma}{dy}(h - z)\left(\frac{h + z}{2} - h_0\right) \tag{19}$$

The equation for the maximum velocity will be

$$V_{max} = -\frac{1}{\eta}\frac{d\sigma}{dy}(h - h_0)\left(\frac{h + h_0}{2} - h_0\right) \quad (20)$$

Using the approximation that the parameter of coal pillar area where extrusion deformations are in effect has the following dimension.

$$L^2 x(1/2L \times 1/2L)$$

then the rate of flow of soft mass (coal) between two rigid layers (sandstone) can be calculated by the following equation:

$$q = L^2 h_0 V_{max} + L^2 \int_{h_0}^{h} V_{dz} = -\frac{d\sigma}{dy}\frac{L^2 h^3}{3\eta}\left(1 - \frac{3 h_0}{2 h} + \frac{h_0^3}{2h}\right) \quad (21)$$

Substituting in equations 19 and 20, the expression for $d\sigma/dy$ by ultimate shear stress conditions for the sector of the moving layer h_0, the equations for velocity and maximum velocity are given by:

$$V = \frac{1}{\eta}\frac{\tau_0}{h_0}(h - z)\left(\frac{h + z}{z} - h_0\right) \quad (22)$$

$$V_{max} = \frac{1}{2\eta}\frac{\tau_0}{h_0}(h - h_0)^2 \quad (23)$$

where

  z = depth of the pillar;
  h = half of height of the opening;
  η = coefficient of viscosity of soft layer.

If the final element of equation (21) is disregarded and the expression $d\sigma/dy$ is substituted by lateral stress of the pillar (σ_H) by length of the area of pillar extrusion ($L/2$), this equation can be rewritten as follows:

$$q = \frac{\sigma_H}{L/2}\frac{L^2 h^3}{3\eta}\left(1 - \frac{3 h_0}{2 h}\right) = -2\sigma_H\frac{L^2 h^3}{3\eta}\left(1 - \frac{3 h_0}{2 h}\right) \quad (24)$$

The velocity and the rate of flow mechanics of soft layers of the coal pillars is governed by the depth of the coal seam, lateral stress in the coal pillar, thickness and viscosity of soft layer.

Finally, it should be concluded that velocity and rate of flow of the coal material which is under extrusion deformations obviously depend on the intensity of the horizontal stress (σ_H), which in flat coal seams coincides with the principal stress.

Besides this parameter, the coal viscosity (η), as well as, the height of the mine opening (2h) are equally important in evaluation of mine workings particularly regarding the entries and roadways.

6.4.4. *Shear flow of coal structure*

Extrusion deformation of coal structures in response to the removal of lateral confinement is facilitated by reduced strength of the coal or coaly material. This reduced strength results from both the effects of the previous orogenic movement and the inelastic behaviour of most coal bands during and after mining. Three main types of deformations may be defined in terms of the frequency of orogenic shear planes, the thickness of the coal seams, and the proportion and number of soft and hard coal bands.[39]

1. Deformation of type 1 occurs in the extrusion of individual coal layers due to their differing elastic properties. The mechanics of extrusion of individual coal layers can be explained by frictional effects between hard and soft coal bands:

a) The integral expansion of pillars is resisted by hard coal layers which have a higher Young's modulus and a lower Poisson's ratio and thereby set up forces opposing the expansion of soft layers.

b) The initiation of shearing motion in soft layers depends on the magnitude of cohesion and frictional resistance between bands which in turn are influenced by interlocking surface irregularities.

c) When shearing motion starts the amount of horizontal movement will be controlled by the elastic properties of hard layers and by the magnitude of the shearing force generated along the stratification planes.

Extrusion of individual layers in seams is unimportant in extracting coal in short-term structures such as longwall faces. The short duration of a free coal face permits 'instantaneous' extrusion of only a few small layers (Fig. 6.4.6). However, for semi-permanent and permanent structures such as pillars, extrusion of individual layers is of

Fig. 6.4.6. Small extension of soft layers (darker in color) immediately after coal excavation (Number 4 coal seam, Grande Cache).

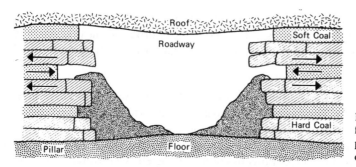

Fig. 6.4.7. Profile along rib faces showing a stage of partial extrusion deformations.

paramount importance because surfaces are exposed for significant periods of time. Deformation by the shear flow of individual layers is characteristic of medium-thick seams of 'dirty' coal, with a heterogenous internal structure, or of seams with an appropriate number of dull, hard coal layers. Due to the progressive extrusion of soft layers, unsupported fragments of hard coal layers in the outer zone of the pillars have a tendency to break and fall down into the roadways (Fig. 6.4.7).

Individual layer extrusion has been specifically observed in Number 10 Mine at Grande Cache, where coal seams contain several benches of high ash content:

a) An estimated 50 to 60 per cent of the total pillar volume fractures immediately during mining. This is the result of shear flow of up to 5 cm in soft layers. Six months after a pillar is formed, individual layers may have extruded up to 50 cm. About 90 per cent of the total pillar surface is in a fractured state, and further deformation may progress by the extrusion of the whole pillar.

b) The magnitude of shearing depends on the location of the mid plane relative to the horizontal axis of the pillar, on the shear resistance between layers, and on the number and extent of orogenically sheared planes (residual shear strength). CANMET personnel[40] have made measurements of the total lateral displacement of a 24 m × 30 m × 2.5 m pillar over one year. The incremental displacement from pillar core to outer zone was as follows:

9–12 m – pillar core – no motion; high stress concentration; 10 per cent constrained area (non-fractured) ⎫
6–9 m – semi-fractured zone – 2 cm displacement ⎬ 90 per cent restrained area (fractured)
3–6 m – fractured pillar zone – 10 cm displacement
0–3 m – extruding pillar zone – 20 cm displacement ⎭

c) Underground observations suggest that the shearing flow of individual soft coal layers also depends on the size of pillars and on the orientation of pillar faces to the dominant cleat direction. For example, pillars with a width-to-height ratio of less than 5 under equal stress conditions have accelerated shearing flow of the individual layers, and in less than one year, it will be in an integral extrusion and a collapsing state.

2. Deformation of type 2 involves integral extrusion of all coal layers. Thus all faces of a mine structure may be considered to be composed of soft coal which represents a continuously fragmented substance. The mechanics of integral shearing flow deformation may be explained as follows:

a) The expansion of pillars is influenced by the inelastic properties of all coal bands

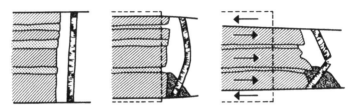

Fig. 6.4.8. Model of sequences of coal pillar deformation.

present. Individual coal layers are not laterally constrained; the forces along bedding planes opposing extension are therefore negligible.

b) The integral shearing flow of ribs may be facilitated by two phenomena: first, because roof and floor strata are equally as weak as the coal or coaly material (equal elastic constraints), they tend to flow into the roadway together with the pillars; secondly, interlayer displacement took place during orogeny along planes of weakness near roof and floor contacts.

Integral shearing flow of pillars from underground observations in three coal mines (Canmore, Coleman, Kaiser) in the southern part of the Rocky Mountains has been observed. Results can be summarized as follows:

a) During the formation of pillars, the magnitude of redistributed overburden load progressively increases within the pillars. This phenomenon is followed by limited extensions of pillars. Timber posts set close to pillar faces take some of the load as shown by their compression (Fig. 6.4.8).

b) After their formation the extrusion of pillars with soft roofs and floors is accelerated, followed by cracking and opening of cleats and further coal fragmentation. During this sequence of shearing flow roofs of roadways sag, beds separate (2.5–5.0 cm), and timber posts buckle (Fig. 6.4.8).

c) Due to progressive fracturing, the extrusion of coal and coaly material is accelerated, with pillars expanding laterally up to 2.5 m, and contracting vertically up to 0.5 m. In this state of pillar deformation, timber posts are broken and pushed into the roadway (Fig. 6.4.8).

d) In the final stage of deformation the completely crushed pillar remains standing probably due to frictional resistance between coal blocks with the resulting forces opposing total disintegration of the pillar. The crushed state of the pillar is indicated by roof falls in surrounding roadways resulting from increased spans. In the Canmore Mine roof falls in roadways have been observed as far as 75 m from a crushed pillar.

3. Deformation of type 3 involves differential shearing which is particularly well exhibited in thick coal seams. The mechanics of this type of deformation is also governed by frictional effects between stratification planes:

a) The differential deformation results from differential shear resistance at the top and bottom of the seam.

b) The differential movement of coal slices is governed by the distribution of slices of hard and soft coal in the seam profile.

c) The differential shearing flow of the mine structure is facilitated by a shear plane of orogenic origin between harder and softer units of rock and coal.

This type of deformation is well exhibited in the Number 4 Mine at Grande Cache

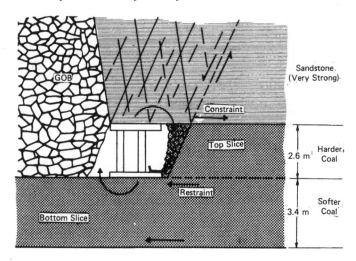

Fig. 6.4.9. Differential shearing of longwall face during top slice extraction (Number 4 coal seam, Grande Cache).

where extraction of a thick seam by slicing was halted approximately six months after mining commenced. The differential flow of coal slices during longwall mining of the upper slice (Fig. 6.4.9) caused the failure of this method. The sequence of events was as follows:

a) Roof caving started and progressed in advance of the longwall face. The lack of a cantilever rock beam above the supports caused the sandstone to cave in the working area.

b) The powered support system rotated to some extent due to floor uplift and limited resistance of the caved roof. The rotated support tended to fall down.

c) Longwall mining had to be discontinued because differential extrusion of the coal seam resulted in an inclined coal face, the extracted coal was diluted by caved rock, and the powered supports were left with very little supporting capability.

Differential shearing also affects the stability of ribs in semi-permanent or permanent mine structures. Two typical examples of deformation in mine ribs caused by the thickness of the coal seam were observed where soft coal sections displayed continuous yielding (Fig. 6.4.10). In coal seams of medium thickness differential shearing generally will be within the coal seam along sections of soft and hard coal, but in thick coal seams it generally occurs throughout the entire coal face, not in individual coal sections. In both examples shearing in the seam deforms the roadway into a trapezoidal-shaped profile instead of a rectangular one. During extrusion flow the roadways maintain their trapezoidal shape although the height continually decreases and the width increases.

The shear strength between stratification discontinuities is the determining factor of mine stability and can be quantified by laboratory testing. Afanasea et al.[41] suggests that in existing mines, long-term loading reduces the shear strength between coal contacts up to 10 per cent and between shale contacts up to 30 per cent. Thus shear strengths determined in the laboratory should be appropriately adjusted for designing mines in addition to the regular adjustment from lab testing to field conditions.

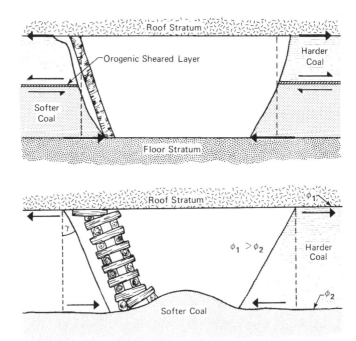

Fig. 6.4.10. Differential
shearing of coal pillars.

6.4.5. *Stabilization of mine structure*

The soft and particularly thick coal seams could be defined as mechanically unstable. They mainly belong to geologically younger coking coal deposits, in which the coalification part has been rather different to that of carboniferous coal. With this type of coal deposit of insufficient strength and bearing capacity with a tendency to fail and deform by visco-plastic flow, there is not a satisfactory method of artificial reinforcement, as well as, particular stable mining system or mining technology. The options available at present are briefly discussed below:

1. Artificial reinforcement is mainly by rock bolting or by fire proof plastic sealant by which coal pillars are constrained. It should be pointed out that soft coal pillar coated by plastic sealant in Kuznetsk Basin (Siberia) were satisfactory with aspects of dust depression and ventilation, but with limited success with the aspect of mine stability. It should be pointed out that pillars which were constrained immediately after their formation by injecting adequate plastic sealant experienced very limited degradation because the deformation and large displacement have been eliminated. The stability ceased when plastic sealant has been broken and the coal starts to squeeze out from pillars.

2. A mining system which could offer maximum stability for mechanically unstable structure is not in existence. The pillarless mining, which is briefly discussed in other headings, should be most convenient, because the number of coal pillars would be decreased. Also stability can be increased if the coal extraction is carried out with limited development workings, which in very unstable conditions might be excavated in stable rock strata. The application of hydraulic mining technology, might be essential in achieving mine stability.[42]

Coal extraction by monitor jets results in the formation of large cavities and a lack of artificial support of stopes. This results in total load transfer to entries, which are usually in differential stress fields. The entry design in relation to the hanging wall or footwall of the seam is an important factor for their stability.

Due to unmanned mining by remote control, the roof sags and falls should not affect coal production, providing the hydromonitor has not been buried under rocks.

The monitor jets will flush out coal mixed with rock, until a hollow sound is heard suggesting that coal is mined out and that a large cavity has been formed. The key to the success of hydraulic mining is that the large cave should collapse as soon as possible so that adjacent entries and extraction blocks will be de-stressed.

The coal seams with exceptionally soft and weak coal, which is deformed and crushed due to orogenic movement until now are considered unmineable. However, hydraulic mining by drilling-and-reaming from underground and surface might avoid such mechanical instability of coal material and be applicable for extraction of such seams.

CHAPTER 7

Room and pillar mining

This is the oldest method of mining, where pillars are utilized to support ground as excavation takes place. With general advancement of engineering technology the room-and-pillar method has been developed to a state of high sophistication, as for example, coal excavation using programmed continuous miners operated by remote control. A major barrier to furthering the technological development of this mining method is the difficulty of controlling strata during rapid coal extraction. Roof control comparable to self-advancing shield support in longwall mining is required. This chapter deals with the following topics:
  – Current room-and-pillar systems represented in four groups including large pillar and shortwall mining.
  – Stability of stope pillar structures analyzed for various conditions of pillar loading.
  – Strata mechanics analyzed for various stress distribution within pillars, with the use of physical and digital models.
  – Caving mechanics discussed for shallow and deep coal deposits with controlled and uncontrolled ground surface subsidence.
  The understanding of the above listed topics are important for controlling the mine stability and mine safety.

7.1. PRINCIPAL MINING SYSTEMS

Room-and-pillar mining, one of the oldest methods in coal mining, has been extensively described in mining literature. In this book only the most fundamental aspects are considered with the sole objective of allowing a better understanding of strata mechanics. Generally speaking there are numerous variations of this method, under different classifications and with different titles. It is the intention at this time to consider only thin and moderately thick coal seams which are flat or gently dipping. The mining approaches to such coal seams are considered as follows:[1]
  – Board-and-pillar system along the lines of the South African approach and using a technology of cyclic coal excavation.
  – Room-and-pillar system along the lines of the North American approach and using a technology of continuous coal excavation.
  – Chamber system, along the lines of the European approach and using a technology of cyclic coal excavation.

233

– Large pillar or shortwall mining as a transitional system between room-and-pillar mining and longwall mining, along the lines of the Australian approach and using a technology of continuous coal excavation.

Each of these systems is discussed individually with practical examples.

7.1.1.   *Board-and-pillar system*

This mining method has been used primarily for shallow coal mining situations with limited overburden pressure. It is convenient for shallow mining where it is essential that overlying strata do not cave and surface subsidence is not damaging.

For example, at Collie Basin in Western Australia this mining method is chosen to prevent caving of overlying strata, which consists of water bearing horizons capable of flooding the mines. The basic principles of board-and-pillar mining in Western Australia are summarized as below. Many of these could be considered applicable to other mines in the world where this system is practiced.

a) Development working divides the coal seams into blocks, from which a coal is excavated by two series of boards which are parallel to each other. The width of boards is 6 to 7 m with pillars 7 × 18 m. The coal excavation consists of a drill, blast and load

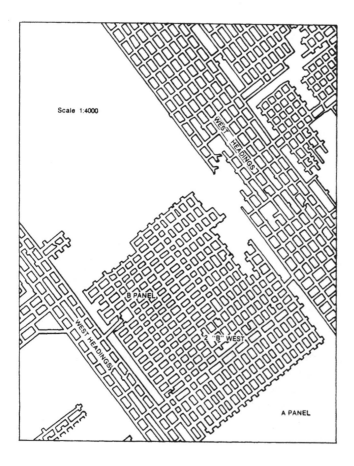

Fig. 7.1.1. Layout of board-and-pillar system (Western No. 2 Mine, Collie Basin, Western Australia).

Fig. 7.1.2. Board-and-pillar configuration with protective bed of coal toward incompetent roof strata (Hebe Mine, Western Australia).

cycle. Coal is drilled with electric auger drills and blasted using half second delay detonators and semi-gel non-permitted explosives. The broken coal is loaded by small 'Bobcat' front end loaders onto a chain conveyor, which delivers coal to a belt cenveyor system for transportation to the surface.[1]

b) Extraction is on a retreat basis from the board by splitting pillars into blocks 6 × 7 m which are left for permanent roof support. The layout of unsplit and split pillars is illustrated in Figure 7.1.1.

c) Ventilation has to accommodate the crosscutting boards between the adjacent boards along strike. Most crosscutting boards have to be blocked off when they have served their purpose because they are potential leakage points between intake and return air currents. The ventilation system uses brattice walls to divide headings longitudinally in order to provide an inlet to, and an outlet from the face.

d) Support is by timber square sets. Local jarrah timber has been used for roof support within the boards over the life of mine. The necessity for permanent support is due to weak rock strata, of low or no tensile strength, which is totally saturated with water. Wet, incompetent strata under hydrostatic pressure due to water saturation cannot be supported by timber sets alone. For this reason a one metre thick coal layer is left in the roof to act as a support (Fig. 7.1.2), as discussed under previous headings in some chapters. For those faces where continuous miners have recently been implemented, hydraulic props and beams are used to allow continuous coal excavation (Fig. 7.1.3).

e) The productivity of this system is low, because the coal extraction is obtained from entry headings. Productivity ranges up to OMS 8 tonnes. The production of individual underground mines is up to 250.000 tonnes/year, with coal recovery up to 50 per cent.

The board-and-pillar mining in South Africa is the main mining method, because of the following advantages:

– It is a convenient method for very shallow coal deposits with aspect to mine stability.

– It is capable of recovering a high percentage of reserves.

– It is a simple and flexible system.

– It requires a low capital commitment.

– The ground surface is protected from subsidence.

Fig. 7.1.3. Coal excavation by continuous miner (Western No. 2 Mine, Collie Basin, Western Australia).

The disadvantage is that coal reserve recovery percentage decreases rapidly with depth, because the size of pillars has to be increased. The overall extraction could be as low as 15 per cent in deep mines. To offset such low coal recovery it has been used mining with stowing, where ash filling follows the extraction process.[3]

7.1.2. *Room-and-pillar system*

Room-and-pillar mining usually refers, in a general sense, to situations where coal is extracted around pillars which may or may not be removed at a later date. In some instances this mining system could be called pillarless as can be seen from an example of this method as briefly discussed at Grande Cache coal mine in Western Canada:

a) Development is by three entries which are driven in parallel and connected at intervals by crosscuts driven at right, or acute, angles to provide flexibility of excavation for continuous miners. The entries are 5.5 to 6.0 m wide and driven parallel at 24 and 39 m spacing. Crosscut connections are made also at 24, and 39 m spacings to permit access for mining machinery and for ventilation purposes.[5] The equipment used for entry excavation is basically continuous miners (Fig. 7.1.4) followed by rock bolters. The continuous miner cuts the coal from the working face (heading) and loads the broken coal into shuttle car, which haul and discharge coal into feeder breaker, from which it is transferred onto a belt conveyor and transported to the preparation plant.

b) Extraction represents a depillaring phase, where pillars are extracted by lift slice method or split and fender method (Fig. 7.1.5). In underground mines of the Grande

Fig. 7.1.4. Continuous
miner at Cliff Terrace Mine
of McIntyre Mine (Western
Canada).

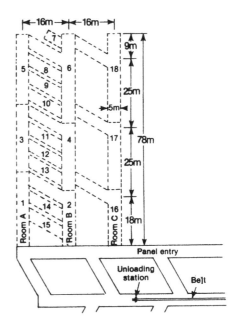

Fig. 7.1.5. Entry excavation and depillaring layout.

Cache Basin, however, coal pillars are partially extracted, as can be seen from mine
layout (Fig. 7.1.6).

c) Support of the rooms-entries, is by rock bolting and wire mesh. Due to
mechanical instability of coal pillars supplementary timber support is also required,
particularly during the depillaring phase.

d) Ventilation of mine workings is by auxiliary fans and vent tubing. The major
problem is that contaminated air from one face is used to ventilate another face. To
solve this problem auxiliary blowers and exhaust fans are used, if the mine is gaseous,

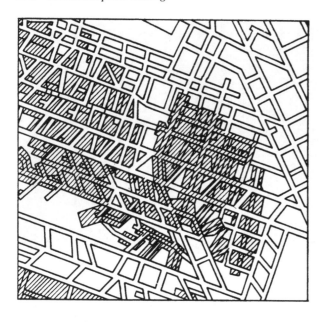

Fig. 7.1.6. Layout of room-and-pillar system at Grande Cache Basin, Western Canada (Number 2 Mine).

and the rate of gas emission throughout the mine, ranging between 1.8 and 18 cubic metre per minute.

e) The productivity of this system is very satisfactory if the coal is of the adequate strength. Productivities of up to OMS 20 tonnes and 200 to 350 tonnes/shift for a continuous miner are realistic.

The technology of room-and-pillar mining is particularly well developed in Eastern U.S.A. in the Appalachian region, where geological conditions favor a such a mining system. The coal seams are flat and of limited thickness with hard coal sandwiched between competent rock strata (Pittsburg bed). The advance is usually with a three entry system and the retreat is with depillaring of the pillars formed in the advance phase.

The room-and-pillar mining method is very flexible and permits considerable variation in the layout from one mining area to another. The following points may be observed:

a) Variations in gradient and thickness of the coal seam can be accommodated, for example, by driving entries obliquely to seam strike.

b) The success of the mining operation depends upon the properties of the roof strata. A weak roof may create difficulties in maintaining an economic roof span of the room and roof bolting. A very strong roof may inhibit the caving of the roof on retreat creating higher pressure peaks in the neighboring coal pillars which are to be recovered.

c) The capital cost per unit face is low compared with other methods but operating costs tend to be rather high due to roof support in the rooms and during the pillar extraction. Also, ventilation, power and material handling costs are higher.

d) In shallow conditions, an uneven subsidence profile may lead to sharp depressions or 'sink' holes at the surface. When mining under surface structures, lakes, rivers, railways or the sea, the room-and-pillar system should be replaced by board-and-pillar system keeping coal pillars in place as permanent support.

It should be pointed out that the system of room-and-pillar mining is applicable to extraction of thin and moderately thick coal seams which are also mineable by longwall mining. To mate the advantages of room-and-pillar mining and longwall mining a hybrid system has been developed known as a shortwall mining, which is briefly discussed in subheading 7.1.4.

7.1.3. *Chamber system*

This system is known as a room mining, komora mining or chamber mining.[6] A brief description of this method is given for the Star-Key Mine, near Edmonton, for which various aspects of strata mechanics have been discussed in this and other chapters.

The three-entry system has been used for development of extraction panels. These entries allow coal haulage, man-travelling, electric cable installation and ventilation. The application of the chamber system at Stark-Key Mine is as follows:

a) Development of extraction openings is by the main and butt entries. There is one of each entry type. The butt entries have centre line approximately 151 m further inbye, so as to develop a coal block 114 m or so wide (Fig. 7.1.7). The length of this block depends on coal seam boundary and varies from 125 to 625 m. Locomotive haulage is

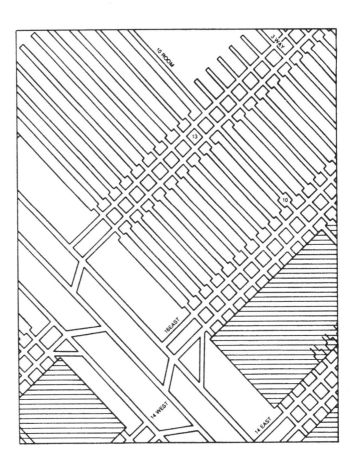

Fig. 7.1.7. Layout of chamber system at Star-Key Mine (Alberta, Canada).

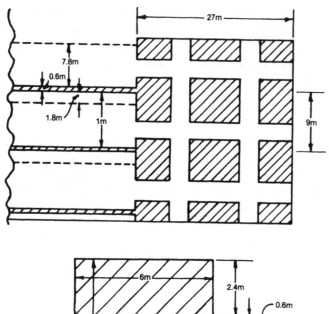

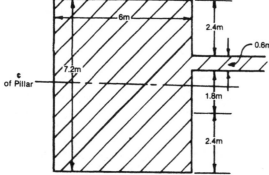

Fig. 7.1.8. Extraction of chambers
by leaving interchamber pillars.

provided along the butt entries. The coal excavation in this particular mine is by drilling and blasting. In some European mines, the development for a similar layout of the coal excavation is carried out by continuous miners.[6]

b) Extraction is carried out in chambers driven at right angles to the butt entries, and commences by advancing 6 m with a width of the standard roadway (3 m), then widening out to 14 m. The chamber is then driven for 114 m to meet the adjacent butt entry. The rib of coal left between adjacent chambers is 2.5 m wide. After the chamber has been driven the 114 m length, 1.8 m is cut of one side of the pillar (Fig. 7.1.8).

c) Support of development working is by timber square sets. However, in strong roof strata support by rock bolting is adequate. The support of the chambers during extraction is by timber props and beam (Fig. 7.1.9) or by hydraulic props and I beams. On retreat during partial recovery of interchamber pillars temporary support by wooden props is used.

d) Ventilation is simple. One butt entry is for intake and other for exhaust air. Ventilating currents of air pass through chamber during extraction.

e) Productivity in manual operations using cyclical advance is up to OMS 12 tonnes. Mechanized operation with continuous advance in the development phase could increase productivity to OMS 15 tonnes. Chamber mining with narrow

Fig. 7.1.9. Timber support
of the chamber at Star-Key
Mine (Western Canada).

interchambers pillars, as in case of shallow mining in Star-Key Mine, can result in a coal recovery as much as 85 per cent. With increase of depth coal recovery decreases rapidly.

The chamber system is not as widely used at the present time as the board-and-pillar and, particularly, room-and-pillar systems. The reason is that the continuous mining technology is not fully utilized in extraction of chamber configurations.

7.1.4. *Large pillar-shortwall system*

The development of large pillar mining resulted from integrating individual pillars into one large pillar. When coal face with close end was replaced with open ends the system was called shortwall mining. The transition from room-and-pillar mining to large pillar mining and from this method to shortwall mining has been carried out in Eastern Australia in the Wongawilli seam where the system was named. Due to the variations in layout and coal extraction of large pillar and shortwall systems they will be discussed in separate paragraphs.

1. Pillar system discussion is based on the Eastern Australian mining experience. The original room-and-pillar method was based on entries of 5.5 m width and pillars of 6 to 24 m least width. The width of the pillars depends on mine depth, which is between 60 and 399 m. With the concept of large pillars, the pillar length is 100 m and the width of 50 to 60 m. The principal aspects of this method are as follows:

a) Development is carried out by entries driven in-seam to block out large pillars (Fig. 7.1.10). Multi-entry panel development of the main entries is used where pillars are partially or completely extracted. The coal excavation is by continuous miner. Coal haulage away from the miner is by shuttle car, which transports to a belt conveyor. All continuous mining machines are equipped with canopies for the protection of the operators from roof falls, hydraulically powered jacks for raising and holding cross bars to the roof before props are placed under them, and water sprays for both dust reduction and ignition prevention. Remote control of continuous miners is in the experimental stage in many mines. This trend is the result of trying to improve operator safety in the area of the immediate face.

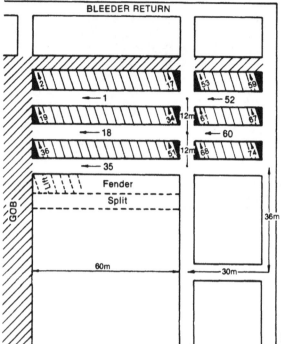

Fig. 7.1.10. Layout of long pillar system (Wongawilli method).

b) Extraction relates to blocks 100 m long and 50 to 60 m wide, which are formed by multiple entries (Fig. 7.1.10). The width of the block largely depends on the reach of the cable of shuttle car and ventilation considerations. While driving the split, a breach the width of the continuous miner head is sometimes driven through the fender to shorten the length of brattice required. The method of extraction is by splitting the block leaving a fender which is subsequently removed by open ended lifting or by lifting to leave a small fender on the gob side (Fig. 7.1.10). This small fender may or may not be pocketed. The viability of the system is determined to a large extent by the ability of the split and fender to remain stable under the mining and strata stresses.[7] The mining practice suggested that a narrow fender has the advantage that the continuous miner driver stays under the supports in the split during the extraction of fender.[7] Usually support rules suggest width of splits of 5.5 m and lifts also up to 5.5 m.

c) Support of roadways is by roof bolts. Past practice has seen the continuous miner cut 5 to 6 m before putting up any bolts, but, because roof conditions in Australia do not allow areas of the roof to remain stable long without support, practice has evolved to the point where there is usually no more than 1 m of unsupported roof in front of the face at any given time. Therefore, roof bolting and coal cutting are much more interconnected in Australia than, for example, in the United States, where continuous miners advance up to 6 m before bolts are installed.

d) Ventilation is based on safety regulations in the New South Wales mandate that there must be 3 cubic metres of air per second over a continuous miner operator. This requirement is easily met with the use of line brattice at the face. In gaseous and dusty mines, ventilation regulations insist on 10 cubic metres of air per second over the operator. In this circumstance auxiliary fans in the face area are used. Such a panel has a separate split of air for its ventilation.

e) Production and productivity of this system which is common for coal mines should be considered as satisfactory for Australian conditions. The productivity is OMS 10.5 tonnes but production is large. For example, Statistical Report for 1978/79, NSW underground mines produced 38,205.000 tonnes out of a total state coal production of 50,557.000 tonnes, a whopping 76 per cent.

In order to avoid inevitable losses of coal in stooks, an open ending method came into existence, employing continuous miners and shuttle cars operating alongside a gob in a passageway up to 50 m long protected from the gob by rows of breaking off timber. The success of this method in NSW coal field (Australia) at a depth of 200 m resulted from the design of self-advancing support as is discussed in the next paragraph.

2. *Shortwall system* discussion is based on experience in the United States, which has been implemented in this country on the basis of Australian layout and experience. The brief outlines of the system are as follows:

a) Development is by driving three entries 6 m wide on 15 m centers with 6 m wide crosscuts on 22.5 m centers to create a 45 m wide and 500–1000 m long shortwall panel.[8] The development entries serve first as head gate entries for the next panel. At the end of the development entries, 45 m of additional three-entry bleeders are driven perpendicular to establish a shortwall face. The entry excavation is by continuous miner as described for long-pillar system. The similar development is carried out in Moscow Basin for shortwall mining of shallow coal deposits. Due to soft rock strata the development is done by double entries (Fig. 7.1.11).[6]

b) Extraction of the shortwall face is by continuous miner where coal was originally transported from the face by a flexible conveyor and more recently by shuttle cars to the conveyor belt (Fig. 7.1.12). The coal winning is in lifts of the width up to 3.5 m, which depends on type of continuous miner used for coal extraction housed under self-advancing support. It should be pointed out that the equipment used in shortwall mining is identical to that used for room-and-pillar operation with one significant addition, the powered roof support.

c) Support of entries is identical to other room-and-pillar systems, where use of the rock bolting is predominant. The face support, by self-advancing powered support, is

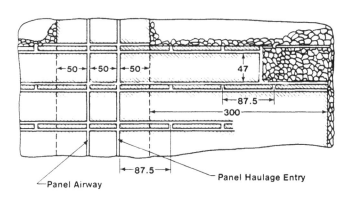

Fig. 7.1.11. Layout of development workings for shortwall (large pillar) mining in Moskow Basin.

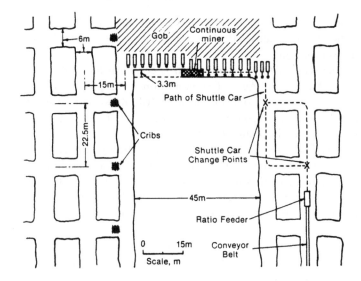

Fig. 7.1.12. Layout of shortwall face in extraction.

similar to longwall mining, but with the particular problem of roof control of wide cut of the web because it exposes a greater area of roof between the hydraulic chock supports and the coal face. The powered support designed in Australia with width of passageway 'prop free front' to as much as 4.5 m, experienced difficulties in roof control for increased mine depth, which was 450 m.

d) The ventilation system is similar to one described for large pillar mining, and it has to fulfil the same requirements.

e) Productivity and rate of extraction of this system is somewhere between room-and-pillar mining and longwall mining. For example, producing mines in the United States recorded that average daily production from the shortwall ranged from 700 to 1200 tonnes and productivity OMS from 20 to 47 tonnes. The highest daily production before 1977 was 2.027 tonnes.[9]

It is estimated that the produced coal by shortwall mining in Australia and the United States is less than 5 per cent of the total coal production. Prospects for this method are for a further decrease, particularly with development of short faces for longwall mining with a shield support. For these reasons consideration of shortwall mining is not given a separate chapter.

7.2. STABILITY OF STOPE PILLAR STRUCTURES

The stability of stope pillar structures has been considered only in flat and gently dipping coal seams, because in such seams room-and-pillar mining is the predominant system. Generally speaking, the key factor in underground structural stability is the coal pillar whose bearing capacity should equal the load superimposed upon it. The superimposed load relates directly to average pillar stress which is discussed with reference to the following aspects:

– Average pillar stress which is considered for single pillars (Tributary theory) and multiple pillars (Beam concept and Salomon's concept).

– Abutment pillar stress which is considered in relationship to extraction panel width and extraction depth.

– Internal stress which is considered in photoelastic models where, instead of average stresses, vertical and horizontal stresses are given at various pillar sections.

The point of interest is that stability of stope pillar structures depends, amongst other factors, on extraction depth, pillar size and length of the room span, as will be now discussed.

7.2.1. *Average pillar stress*

The average pillar stress due to overburden load is an important parameter in determining final mining depth for room-and-pillar systems.[10] For example, low strength cooking coal deposits in Western Canada may be successfully mined by room-and-pillar methods only to a depth of 600 m.

The pillars have an open face on all four sides and thus are uniaxially loaded. The magnitude of super incumbent load increases with mine depth and extraction ratio.

When pillar size is adequate in relation to the depth of overlying rock, the average pillar stress for flat deposits can be readily calculated by a simple method based on a three-dimensional concept of the Tributary theory. This concept is also proposed for pitching deposits, but with a correction related to the dip angle, as briefly discussed in Chapter 11. A more advanced method to calculate pillar loads is suggested by Salomon who bases calculations on an analytical analyses.[11, 12] Sheorey and Singh estimated a pillar stress by a complex approach in terms of differential equations.[13] These equations provide for overburden deflection by treating the rock strata as a thick beam composed of layers having different elastic properties.

1. Tributary theory. Average stress is calculated by assuming that pillars uniformly support the entire load of the overlying both the pillars and the mined-out areas. The effects of deformation and failure in the roof strata resulting from the mining operation are disregarded.[14] The tributary area in relation to pillar area is illustrated in Figure 7.2.1. From this relationship a set of equations have been established for calculating the average pillar stresses for a horizontally bedded deposit.

$$\sigma_v = \frac{\gamma h}{WL}(W + w)(L + 1) \tag{1}$$

$$\sigma_v = \gamma h \left(1 + \frac{w}{W}\right)\left(1 + \frac{1}{L}\right) \tag{2}$$

$$\sigma_v = \gamma h \frac{A_t}{Ah} \quad \text{or} \quad \sigma_v = \gamma h \frac{A_t}{A_t - A_c} \tag{3}$$

$$\sigma_v = \gamma h \left(\frac{1}{1 - R_e}\right) \tag{4}$$

where

γ = unit weight of overburden strata;
h = height of overburden;
W = least width of pillar;
L = length of pillar;
w = least width of excavated area;
l = length of excavated area;
A_t = tributary area;
A_c = extracted area;
A_h = pillar area;
R_e = recovery factor.

An example of the application of Tributary theory calculation with aspect to pillar design is given for the Kipp Mine near Lethbridge, Alberta in Canada. The Kipp Mine is owned by the crown corporation Petro-Canada. The coal seam is flat and of moderate thickness (1.5 to 1.8 m).

Room-and-pillar mining is one of three mining methods considered for future coal extraction. It has the advantage of considerable flexibility for mining around or through faults, in changes or seam height, under varying angles of dip and for leaving unmined any areas where hazardous or uneconomic conditions occur. Disadvantages of the method are related to the high cost of roof support required in the rooms, the time

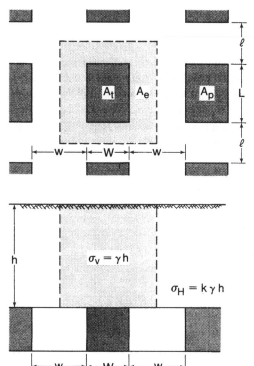

Fig. 7.2.1. Pillar loading by the tributary area concept.

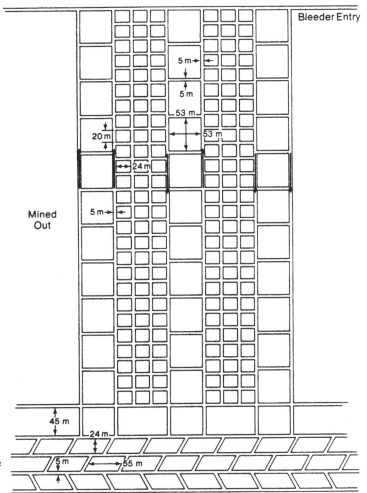

Fig. 7.2.2. Layout of the room-and-pillar with bleeder entries.

required to install roof support and the lack of continuous coal haulage from the face to the main mine haulage system.

The concept of the room-and-pillar layout has been based on three entries for each extraction panel (7.2.2). One main gate entry from the previous panel will be left open and double entries developed. The entries compose one intake and two returns.

Rooms are turned off at right angles to the panel entry. Rooms are driven, starting 45 m from the main entry, and advanced to the limit of the entry. When the limit is reached, bleeder entries are driven. The bleeder entries are turned to the right and left and they are driven approximately 45 m in either direction. Thus, the bleeder openings connect with the bleeder entry driven in the preceding panel.

Two sizes of pillars are used depending on the area being mined. When rooms have reached their limit, the pillars are removed. The method used to recover pillars is to enter the center of the pillar and drive through it. Coal is extracted on the left and right until the pillar has been virtually removed. Minimum dimensions were used for the

Table 7.2.1. Average pillar stress at various mine depths

Pillar width (m)	$\sigma_p(KN/m^2)$					
	245 m	230 m	215 m	200 m	185 m	170 m
10	11,165	10,481	9,798	9,114	8,431	7,747
12	10,545	9,899	9,253	8,608	7,962	7,317
14	10,101	9,483	8,865	8,246	7,628	7,009
16	9,769	9,171	8,573	7,975	7,377	6,779
18	9,511	8,928	8,346	7,764	7,182	6,599
20	9,304	8,734	8,165	7,595	7,025	6,456
22	9,135	8,576	8,016	7,457	6,898	6,338

pillars so that the half pillars can be mined through in one advance of the continuous miner. It is essential that blocks be layed out in such a manner that a pillar is maintained in a straight line over several blocks. This permits failure of the roof behind the pillar line. If this were not done, then very heavy roof pressures would result which could cause roof failure in the working area.

The average pillar stress has been calculated by the Tributary theory for various depths. Rooms are 5 m in width, which is defined by the capacity of the continuous miner, and pillar length is 24 m, as given in Table 7.2.1.

On the basis of the strength of coal pillars the final layout was established at 5 m wide rooms with 12 or 20 m by 24 m pillars.

The evaluation of average pillar stresses of flat or gently dipping by Tributary theory is oversimplified. This concept does not take into account abutment stress distributions and deformation or failure of the pillar. Also, if there is displacement interaction between the surrounding strata and the pillar itself, stress may be redistributed within the system, resulting in a stress state considerably different to the theoretical state.[5]

For practical design purposes, the suggested equations for average stress calculation are acceptable if the designer appreciates the limitations.

2. Analytical method. This method established by Salomon[11, 12] for horizontal coal bearing strata where virgin stress is up to $25\,\gamma h\,(KN/m^2)$. The total induced tensile force is distributed uniformly over the roof and floor of the mined-out area, and their magnitude depends on extraction ratio. This force must be supported by pillar and by the solid coal around the mining area. The estimation of the force acting on a single pillar is a complex problem, because it depends also on the deformation characteristics of both the surrounding strata and the pillar themselves. To simulate these conditions Salomon and Oravecz[12] used a mechanical model of an elastic beam with three supports, ends held rigid, and an elastic spring acting at the centre (Fig. 7.2.3/a). The centre support provides resistance only if the spring is compressed (Fig. 7.2.3/b). This compression occurs only as a result of the deflection of the beam, as shown in Figure 7.2.3/c where the spring has been replaced by an upward acting reaction force P. The deflection (η_b) at the centre of the beam is increased if the pressure (p) is increased or the reaction (P) is decreased. Hence, deflection can be written as:

$$\eta_b = \alpha p - \beta P \tag{5}$$

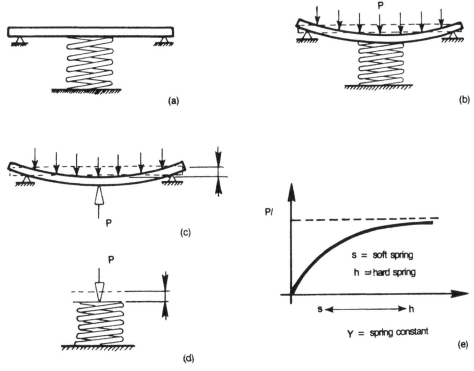

Fig. 7.2.3. Load transfer by deformation (after Salamon & Oravecz).

where
  α and β = constants (deformation characteristics of the beam).

It is obvious that the compression of the spring (η_s) is proportional to force P (Fig. 7.2.3/d), that is

$$\eta_s = \frac{1}{\gamma} p \tag{6}$$

where
  γ = spring constant (deformation characteristics of the spring).

Assuming that the beam and spring interact with each other, than $\eta_b = \eta_s$, hence

$$\alpha_p - \beta P = \frac{1}{\gamma} p \tag{7}$$

Solving this equation for P gives

$$P = \frac{\gamma \alpha p}{1 + \beta \gamma} \tag{8}$$

The equation shows that force P depends on the characteristics of both the beam and spring. Figure 7.2.3/e shows the variation of the ratio P/p as a function of the spring characteristics, where a soft spring supports smaller loads than a stiffer spring.

By using several springs the mechanical model could be used to simulate board-and-pillar configuration, where the beam represents rock strata and the springs correspond to the pillars. In order to estimate pillar load, it is necessary to determine the deformation characteristics of both the roof strata and the pillar. The development of electric analogue model permits the determination of pillar load in the general situation, by means of electric resistance.

Salomon and Oravecz[12] described that in the analogue model, the strata is represented by an orthogonal network of resistors of unit resistance R, forming a prism of $12 \times 12 \times 51$ unit nodes. On the face of the prism formed by the 12×12 nodes, additional resistors (r) are connected in series with every node and the other ends of these resistors are earthed. R and r represent the deformation characteristics of the strata and of the coal seam, respectively (Fig. 7.2.4/a). A fixed electrical voltage, V_1, is connected between the earthed resistors and the other face of the prism with 12×12

a. Pre-mining Stresses (virgin stress)

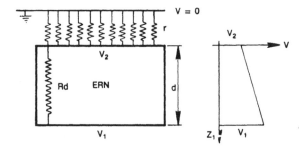

b. Mining Stresses (induced by excavation)

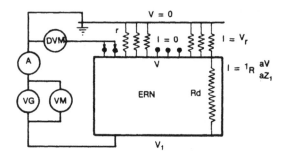

ERN = electric resistance network	R = unit resistance of network
d = depth	r = resistance representing the seam
V_1 = applied voltage	I = current
V and V_2 = voltage on undisturbed network	Z_1 = vertical coordinate
VG = voltage generation	VM = voltmeter
A = ammeter	DVM = digital voltmeter

Fig. 7.2.4. Modeling mining on the analogue (after Salamon & Oravecz).

nodes. The uniform voltage distribution, V_2, obtained on resistors r correspond to the pre-mining or virgin stress of the coal seam (g). 'Mining' on the analogue is modelled by the removal of the series resistors r from the nodes that represent the mined-out board areas of the working section (Fig. 7.2.4/b). From the new voltage distribution on the nodes modelling the pillars, V, the average resultant stress, σ, over a pillar area represented by a series resistor r is obtained simply from the following expression;

$$\sigma = g\left(1 + \frac{V - V_2}{V}\right) = g\frac{V}{V_2} \qquad (9)$$

A large number of mining configurations could be investigated by the analogue model at various ratioes of r/R. This ratio depends on the scaling factor of modelling and the ratio of elastic modulii of the strata (E) to that of the seam (E_s) and is given by following equation:

$$\frac{r}{R} = \frac{E}{4(1 - v^2)E_s} \cdot \frac{NH}{l} \qquad (10)$$

where

    v = Poisson's ratio of the strata;
    N = number of the nodes representing half board-centre distance;
    l = length of working;
    H = height of the working.

The analogue model has been used to produce diagrams of maximum pillar loads in a panel of pillars, as a function of extraction ratio and overburden load. The diagram gives a reliable estimate of pillar load only if the panel width does not greatly exceed the mining depth.

3. Beam deflection. This concept for the estimation of average pillar stress was suggested by Sheorey and Singh.[13] It is primarily for narrow workings, where loads are transferred onto pillars due to deflections of the overburden strata. Under these circumstances the rock beam deflections are a function of the flexural rigidity of the strata and the compressibility of pillars.

The approach for calculation is based on the theory of thick beams on elastic support. A thick beam is considered to be one where thickness is greater than five times the distance between the supports (roof span).[15, 16] The evaluation of average pillar stress is convenient for narrow workings or mining configurations where the tributary area concept does not apply (Fig. 7.2.5).

Two kinds of deflections are produced in any beam, thick or thin, those due to bending moments and those due to shear forces.

In the pillar deflection equations derived in the analytical method shear forces have been neglected. But in addition, the beam deflection concept considers shear forces through the use of differential equations for beam deflections.

The derivation of an equation for beams on elastic supports, that takes into account both bending moments and shear forces is given by Tincelin and Sinou.[17] The equation is:

$$D\frac{d^4v}{dx^4} - \frac{3(1 + m)}{Eh}Dk\frac{d^2v}{dx^2} = \gamma h - kv \qquad (11)$$

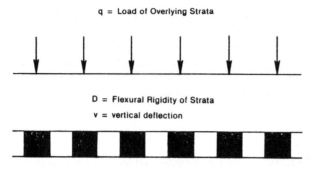

Fig. 7.2.5. Layout of interoom pillars, with model of loading and other parameters.

where

  D = flexural rigidity of the overburden rock;
  v = vertical deflection;
  k = foundation modulus of the pillar;
  E = modulus of elasticity of the overburden rock;
  m = Poisson's number of overburden rock;
  h = height of overburden;
  γh = overburden pressure.

The importance of differential equations in determining the deflection of beams on elastic supports was realized by Hofer and Menzel,[16] Stephasson[15] and Sheory and Singh,[13] who produced deflection solutions for superincumbent stratified rock. The mathematical treatment produced the differential equation for thick beams on elastic supports, as given below:

$$D\frac{d^4v}{dx^4} + \frac{(2+m)h^2}{10(1-m^2)} \cdot \frac{d^2}{dx^2}(\gamma h - kv) = \gamma h - kv \tag{12}$$

This is the complete equation for deflections of thick beams on elastic supports. Equations derived for solutions of bending moments and shear stresses separately are as follows:

$$M = D\frac{d^2v}{dx^2} - \frac{(2+m)h^2}{10(1-m^2)} \cdot (\gamma h - kv) \tag{13}$$

$$Q = -D\frac{d^3v}{dx^3} - \frac{(2+m)h^2}{10(1-m^2)} \cdot \frac{d}{dx}(\gamma h - kv) \tag{14}$$

It should be pointed out that for horizontal coal deposits, the load due to strata weight γh is independent of x and E_q. Under this condition a given differential equation is sufficient for deflection calculations (v).

In this case the average pillar load can be simply calculated by multiplying the deflection v by the modulii of foundation k.

Foundation modulus k_p of the pillars could be determined by replacing the pillars by an imaginary continuous bed such that compressibility remains the same.[17] Thus:

$$k_p = C_c\frac{E_c}{t} \tag{15}$$

where

    C_c = per cent of coal left in pillars $(1 - R)$;

    t = working thickness of coal seam;

    E_c = modul of elasticity of coal pillars, considering their W/H ratio.

Finally, the replacement of pillars by a continuous bed is justified only for pillars within room-and-pillar configuration, but not for abutment, barrier and other pillars.

7.2.2. *Abutment pillar stress*

The concept of abutment pillar design was introduced a long time ago to improve ground conditions in mine extraction areas. Stress was diverted from the mining extraction panel onto the abutment pillars by using yielding extraction pillars which support only a limited load.

The magnitude of the load resting on the abutment pillars depends on the geometry of the caved roof and its position relative to the ground surface. This is a factor of the ratio of mine extraction panel width to mine depth.

Abutment pillars are discussed primarily from the aspect of stability of pillar-stope structure, which depends on the size of the abutment and extraction pillars as well as the overall pillar layout.

Abutment pillars serve two ground control purposes: firstly, to absorb stress from the extraction panel; and secondly to control the vertical extension of the caved area. The latter is important as the intersection between the cave and an aquifer or ground surface could be detrimental to mine production or to surface structures.

Leaving coal in the abutment pillars lowers the overall mine recovery. As much as 25 per cent of the coal may be lost. Yet in some cases where pillarless mining has been attempted, losses are higher due to severe ground control problems with as much as 50 per cent of the coal being lost.

1. Stress diversion mechanics is largely a result of the basic mine layout which consists of paired abutment pillars which share the load from the extraction panel located between them. The extraction panel may be mined-out by any system (sublevel, shortwall, room-and-pillar mining, etc.) which uses interstope extraction pillars (Fig. 7.2.6).

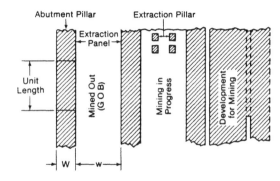

Fig. 7.2.6. Principal layout of abutment pillar concept.

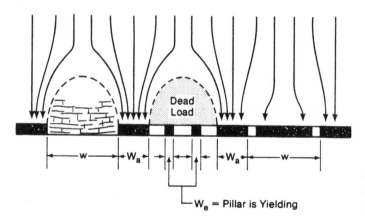

Fig. 7.2.7. Stress diversion on abutment pillar due to arching action above extraction panel.

Stress is diverted from an extraction area onto the abutment pillars in the following manner:

a) Abutment pillars are overdesigned so that they can support the overburden load without failure (stiff pillars).

b) Interstope extraction pillars are underdesigned so that they cannot support the overburden load without failure (yielding pillars).

c) As the yielding pillars are mined-out, the load above the panel is slowly transferred onto the abutment pillars and as the remaining yielding pillars fail the total load is transferred.

The stress diversion is described by the arching action of the roof strata above a mine panel (Fig. 7.2.7). The yielding pillars carry only the load of the rock arch above them because the rest of the overburden load is shifted onto the abutment pillars. The dead load on the yielding pillars is equally redistributed if the roof remains intact i.e. the roof is slowly lowered to the floor without failure.

The mechanics of stress diversion are not well understood and contradict the theory of elasticity which postulates the formation of rock mass equilibrium above an arch where failure does not occur.[18]

2. Average abutment pillar stress could be calculated on the basis of pillar layout and caving geometry. Roof caving is controlled by the compressive stress at the excavation corners (cave failure) and by tensile stress at the centre of the excavation (caving rock blocks). The angle of inclination and propagation of caving laterally and vertically due to tension depends on rock lithology and strength. In relation to these elements, three types of caving geometry are suggested, each defining average stress on the abutment pillars.

a) In sedimentary strata consisting of relatively homogeneous low strength material, caving tends to form a triangular gob area above the extracted panel (Fig. 7.2.8). The average stress per unit pillar length can be calculated using the following equations:

$$\sigma_p = \sigma_{p1} - \sigma_{p2} \tag{1}$$

$$\sigma_p = \gamma h (w_0 + W) - \frac{w_0 \times w_0 \tan \beta}{4} \tag{2}$$

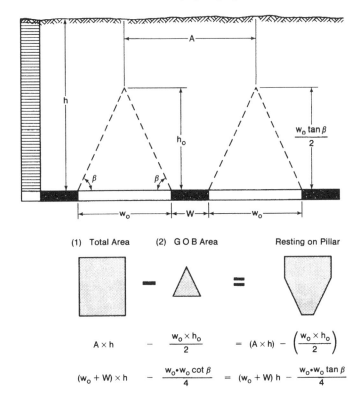

Fig. 7.2.8. Pillar loading
for triangular shape of
caved roof.

$$A \times h \quad - \quad \frac{w_0 \times h_0}{2} \quad = \quad (A \times h) - \left(\frac{w_0 \times h_0}{2} \right)$$

$$(w_0 + W) \times h \quad - \quad \frac{w_0 \cdot w_0 \cot \beta}{4} \quad = \quad (w_0 + W) h - \frac{w_0 \cdot w_0 \tan \beta}{4}$$

$$\sigma_p = \gamma h(w_0 + W) - \frac{w_0^2 \tan \beta}{4} \tag{3}$$

where
  $\gamma = $ unit weight of overburden strata;
  $h = $ mine depth;
  $w_0 = $ panel width;
  $W = $ abutment pillar width;
  $\beta = $ angle of break.

b) A trapezoidal gob area forms when the sedimentary strata is heterogenous and relatively strong (Fig. 7.2.9). Equation 6 is used to calculate the average stress in this case, which is derived as follows:

$$\sigma_p = \sigma_{p1} - \sigma_{p2} \tag{4}$$

$$\sigma_p = \gamma h(w_0 + W) - \gamma h_0 (w_0 + h_0 \cot \beta) \tag{5}$$

$$\sigma_p = \gamma h(w_0 + W) - h_0 (w_0 + h_0 \cot \beta) \tag{6}$$

where
  $h_0 = $ height of caved gob cavity.

c) The caved area in hard rock mines forms an ellipsoid which can be approximated by a trapezoid,[10] therefore, equation 6 can also be used for this solution.

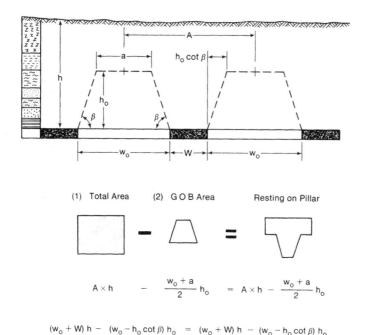

Fig. 7.2.9. Pillar loading for trapezoidal shape of caved roof.

These equations are sufficient for mine design if the extraction panels are reasonably sized and if the roof strata has satisfactory caving properties.

3. Relation between abutment and panel widths is an important parameter because it relates to interaction with ground surface. This parameter controls water penetration into the mine from an aquifer and/or critical subsidence.[19] The width between pillars (panel width) controls the vertical extension of caving and must be chosen so that the gob and related fissures do not intersect the water bearing horizons of the ground surface ($w_0/h_0 = 1.0 - 1.4$).

The panel width is a flexible design parameter which is selected so that both the mine workings and the surface are protected. For example, a preliminary layout for the Kipp Coal Mine (Petro Canada) contains two basic panel and abutment pillar widths: 110 m panels with 50 m pillars were selected for the northern portion of the mine where the total overburden is greater while under the shallower southern portion, 90 m panels were combined with 40 m pillars. Panel length, however, is not a direct function of depth but also depends on other factors such as incubation period of the coal self-ignition, speed of development, coal production rate, and the length of time the main and tail gate roadways remain stable.

The average abutment pillar stress due to panel width and mine depth is calculated on the basis of a classic definition of mine subsidence.[19] This is given below:

a) In deep mines the pillar load is dependent on the panel width (Fig. 7.2.10), and the gradual deformation and failure of the ground without collapse or caving.

b) In shallower mines the pillar load can be dependent on critical panel width when the gob must not intersect an aquifer or the ground surface (Fig. 7.2.10/1). In this case

Critical Subsidence

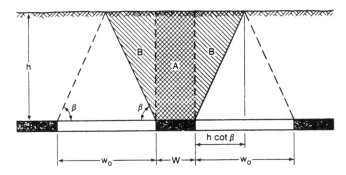

Trough Subsidence

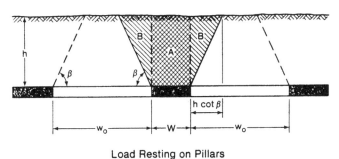

Fig. 7.2.10. Pillar loading
for critical and trough
subsidences.

Load Resting on Pillars
A + 2B

the load resting on the abutment pillars per unit length and the critical subsidence ratio
are given by equations 7 and 8, respectively.

$$\sigma_p = \gamma h(W + h \cot \beta) \tag{7}$$

$$w_0/h = 2 \cot \beta \tag{8}$$

The angle of break (β) is very important in these equations and is a function of rock
strength, joint orientation, depth, rate of extraction etc.

c) When the intersection of the gob and an aquifer or ground surfaces is not critical
the pillar load is independent of panel width. The load resting on the abutment pillars
per unit length is given by equations 9 and 10.

$$\sigma_p = \gamma \left(Wh + \frac{h\,2h \cot \beta}{2} \right) \tag{9}$$

$$\sigma_p = \gamma h(W + h \cot \beta) \tag{10}$$

However, the subsidence ratio is given as below (Fig. 7.2.10/B):

$$w_0/h > 2 \cot \beta \tag{11}$$

This type of pillar loading and subsidence exist in shallow mines where large width to depth ratios over a panel can cause full subsidence often creating flat-bottomed troughs at the surface or pot holes.

Finally, it should be pointed out that success in controlling pillar stope structure by abutment pillars largely depends on the mechanical properties and behaviour of the coal. For example, stiff and brittle pillars are liable to sudden and violent failures (coal outbursts), while weak and fractured pillar yield slowly which, in time, could cause the failure of the whole mine system.[20]

7.2.3. *Internal stress distributions of pillars*

The internal stress distribution of a single pillar in the mining of relatively flat and shallow coal seams has been a matter of consideration for many European in-vestigators for almost half a century. Several decades ago internal stress distribution analyses were carried out by photoelastic modeling. The room-and-pillar modelling for various mining configurations were carried out using optically sensitive material which allowed two-dimensional and three-dimensional stress analysis.[21, 22] With the advent of digital modeling the photoelastic stress analyses become less attrac-tive, due to the superiority of mathematical analysis. It should be noted that photo-elastic stress analyses can still be of value for modeling of brittle fractures in mine pillars.[23] In these cases, the optical polarization method of stress investigation allows determination of the mechanism of brittle fracture and the variation of the stress field during the crack initiation and propagation. Such modeling is best done using the new brittle optically sensitive materials of high strengths ($\sigma_c \simeq 500\,\mathrm{kgf/cm^2}$) and high modulii of elasticity ($E > 37.000\,\mathrm{kgf/cm^2}$) and brittleness index $B = (\sigma_c - \sigma_t)/(\sigma_c + \sigma_t)$ $= 0.7 - 0.8$. This corresponds to the brittleness indices of such rocks as argillite and siltstone. By this method it is possible to investigate the whole process of brittle failure of pillars close to rooms with varying stress fields, and to determine the magnitudes and directions of the principal normal and tangential stresses.[24]

A major concern in photoelastic analyses of room-and-pillar mechanics is the correct choice of the model scale which will satisfy the general stress state of the pillars. Two techniques for photoelastic stress analyses are used to study internal stress distributions of pillars. These are as follow:

1. Two-dimensional photoelastic analysis. This is carried out on optically sensitive plates usually of the size of $25 \times 25 \times 2\,\mathrm{cm}$ and tested for loading conditions which correspond to the mining conditions. The machining of the slots in the celluloid plates to simulate various mine configurations required smoothness of cuts as much as possible in order to avoid distortion of the photoelastic stress patterns. Various aspects of interaction between room-and-pillar structure have been investigated, some of which are briefly listed below:

a) The distribution of the vertical pillar stresses and horizontal stresses, as illustrated in Figure 7.2.11, where the pillar width is twice its height, the vertical stress is more than double the horizontal stress (gravity loading conditions). There is a difference in distribution of stress concentration. For example, the horizontal, stress concentration has a tendency towards maximum at the contact between pillar and roof and floor strata. Maximum vertical stress concentration tends to be at the corners of the pillar. The evaluation of pillar internal stress distribution (vertical and horizontal

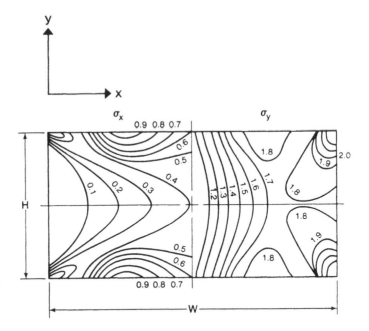

Fig. 7.2.11. Vertical (σ_v) and horizontal (σ_H) stress distribution of the mine pillar model (W/H = 2).

stresses) has been utilized in both soft and hard rock mining to design safe pillar-stope structure since the early 1930's (U.S.S.R.).[21, 22]

b) Stress concentration at the vertical mid point of the pillar varies with width to height ratio as follows:[24]

W/H < 1 high average pillar stresses ($\simeq 3\gamma h$)
W/H = 1 – 2 rapid decrease of average pillar stresses ($\simeq 2\gamma h$)
W/H > 2 the average pillar stresses becomes constant ($\simeq 1.2\gamma h$)
where
γh = overburden pressure.

The vertical and horizontal stresses have a tendency to became more equal and approach a hydrostatic state.

c) Stress concentration at pillar base also varies with pillar width to height ratio as given below:[24]

W/H < 2 gradual increase of average stress ($> 1.0\gamma h$)
W/H > 2 average pillar stress became constant ($0.5\gamma h$)

d) Of particular interest from the research were studies of stress concentration in pillar corners, for different loading conditions. The high stress concentration in the corners of the pillars is a function of radii of curvature. For example, where it is zero, stress concentration is unlimited. It is a well-known fact that the corners of the rooms are the most critical part of a pillar-stope structure, with instability increasing with increasing room width and pillar height. Due to the low strength of the coal, the dimensions of the rooms in coal mining are smaller than in hard rock mining. Where rooms have arch shaped backs larger widths can be employed because the critical stresses at corners are eliminated.

Two-dimensional photoelastic analyses of various room-and-pillar layouts assisted, to certain degree, in learning of interaction mechanics between room and pillar configurations, and has been essential in the design of stable mine structures.[13] It should be a fair statement to say that two-dimensional photoelastic stress analyses have had greater success in the control of mine stability in metal mining than in coal mining, because the properties and behaviour of the hard rock structure are closer to model test conditions of linear elasticity.

2. Three-dimensional photoelastic analyses are mainly based on the principle of the 'stress-freezing' technique. For a three-dimensional scale model, the structure for investigation is machined from a suitable material, usually an epoxy resin, and this model is then subjected to a thermal cycle while under load.[25] The model is gradually heated to a critical temperature, usually in the region of 160°C, and then allowed to cool to room temperature, and the load is removed. A photoelastic pattern will have been 'frozen' into the model. The model is cut into slices, and these are examined in polarized light and the three-dimensional stress distribution in the model can be determined. A high degree of accuracy can be achieved by this modeling technique. However, a major deficiency of three-dimensional modeling is that Poisson's ratio for photoelastic materials ranges from 0.4 to 0.5, while that of rock is between 0.15 and 0.30. Due to this factor very large errors can be introduced, particularly in the case of models in which body forces due to gravitational stress are important. Also for those composite models which simulate a coal pillar structure where different strata are represented by different photoelastic materials. Three-dimensional photoelastic studies for analysis of the stress distribution in mine structures should take into account the fact that errors can be introduced, because all aspects of simulation cannot be perfectly satisfied. Practical results within acceptable limits can often been obtained by this technique.

Great difficulty can be experienced machining a three-dimensional photoelastic model and its loading because of the complexity of the work. For example, modeling a mine structure of a particular layout may take a month of careful casting, curing, annealing, machining, stress-freezing and finally slicing and analysis. The necessity for producing models of high quality cannot be over-emphasized since the results obtained from an unsatisfactory quality of model may mislead mine stability analysis.

Studies of stress distribution around mine excavations by means of the three-dimensional photoelastic stress-freezing technique were carried out by E. Hoek. To simulate the gravitational body forces within the rock mass the Council for Scientific and Industrial Research (South Africa) constructed a special centrifuge, which differed from other centrifuges in use elsewhere.[25] Accurate temperature control is essential for satisfactory stress-freezing results. In the CSIR centrifuge it is achieved by heating the model in an oven which is attached to the end of a rotor. Current for the heating element is supplied through two heavy precision current slip-rings which transmit the signals from thermocouples in the oven to an automatic temperature control, which regulates the heating current.[25]

The epoxy resin (Araldite B or Bakelit ERL 27774) used for frozen stress models requires a maximum temperature of approximately 160°C. In the case of CSIR centrifuge, this requires a heat input of about 2.99 Watts per oven. The photoelastic models become stress frozen at a centrifugal acceleration of 100 times that of gravity.

Photoelastic model investigations, both two-dimensional and three-dimensional, should be considered a useful tool for investigation of internal stress distribution in pillar-stope structures, particularly in the case of the room-and-pillar systems for mining flat and gently dipping deposits. For this reason a brief comment on stress analyses by photoelastic modeling has been made.

7.3. STRATA MECHANICS

Strata mechanics investigations for room and pillar configurations have been carried out all over the world particularly from the aspect of underground instrumentation and monitoring as discussed in various individual publications and books (S. Peng, 'Coal Mining Ground Control'). To avoid duplication and repetition, discussion of strata mechanics is based on data obtained by the author and his graduate students (R. Wright and Z. Felbinger), during their investigations of room and pillar structure. Discussion will be based on the following aspects:
– Abutment stress mechanics of coal pillars on the basis of data obtained from underground mine operations in Western Canada.
– Deformation and failure of mine structures on the basis of data obtained during three-dimensional physical model investigations.
– Stress analyses of mining configurations based on digital model analyses with the aim to developing stability criterion.
Strata mechanics is based on certain mine layouts but for different mine configurations the ground could react differently. The certain differences in strata mechanics is caused also by the coal getting techniques.

7.3.1. *Abutment stress mechanics of coal pillars*

There is a mutual relationship between the yielding deformation of pillars and the abutment stress concentration.[26] This phenomena can be visualized from the coal pillar construction, which has already been discussed in previous chapters, and is summarized as follows:
– A zone of local yielding along the outer zone of pillar rib which is a stress relieved area;
– A zone of transition from a yielding to a solid state in the pillar, which is a stress concentrated area;
– A zone of the pillar core, which has an uniform distribution of confined stress.
The overall coal pillar stability depends on geometrical distribution of these three zones. For example, a decrease in the area of pillar core results in increase of abutment stress and decrease of its stability. As a function of the relationship between constrained pillar area (effective area) and yielding pillar area (outer zone) the pattern of abutment stress distribution might be classified into five categories.

1. Total effective area of the pillar excludes the outer yielding zone, which is typical for a civil engineering structure, but not for mine pillars. Under these circumstances there is an optimal structural capability, and the abutment stress due to axial loading has its maximum at the boundary of structure (Fig. 7.3.1). However, in mining engineering is

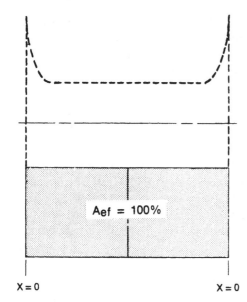

X = 0   X = 0

Fig. 7.3.1. Abutment stress of the pillar with total effective area (A_{ef} = 100 per cent).

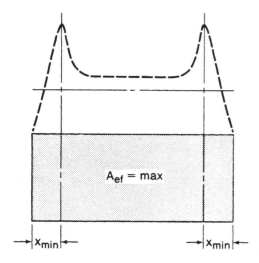

Fig. 7.3.2. Abutment stress of the pillar with maximum effective area (A_{ef} > 80 per cent).

arbitrarily proposed that the length of the de-stressed pillar zone is up to 20 per cent of the least pillar width (Fig. 7.3.2). The pillar area of effective support is large and under constraint.[27] Coal pillar deformation occurs by linear closure at very low rates with no changes in rate before pillar failure. Pillar failure mechanics often occur as a sudden release of elastic energy of bulk and shear deformation accompanied by total coal fragmentation and pulverization. No obvious change in the pillar can be observed before violent collapse. The pillar popping and bumping suggests high stress concentrations within the body.[28]

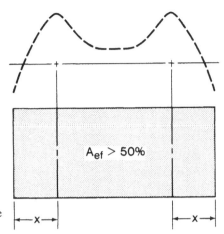

Fig. 7.3.3. Abutment stress of the pillar with moderate effective area ($A_{ef} \geqq 50$ per cent).

2. Moderate effective area of the pillar is indicated with a structure which is clearly fractured to a certain degree. In this case the zones of high stress concentration (abutment stress) lie inward of the pillar edges (Fig. 7.3.3) with vertical and horizontal stresses equal toward the central part of the pillar.[29] Length of the de-stressed pillar zone is less than half of the least pillar width which with the pillar area of effective support is decreased. Pillar deformation in the beginning is by exponential closure and after that continuously as a linear function. Pillar failure mechanics relate to shear and tensile stresses at the pillar edges which are accompanied by progressive pillar slabbing. Pillar deformation can be observed by coal sloughing in the direction of cleat or perpendicular to cleat patterns.[30]

3. Minimum effective area of the pillar is indicated when a large portion of the pillar body is fractured. It has been assumed arbitrarily that effective cross-sectional pillar area is less than half of the total area. Under these circumstances the zones of high stress concentration are located in central regions of pillar (Fig. 7.3.4). Horizontal stresses are equal to, or greater than, vertical stresses in the central region. Pillar deformation starts with accelerated closure which continues at a high rate (pillar squeeze out) accompanied by progressive pillar fracturing. Pillar failure mechanisms are shear displacement along bedding planes due to mobilization of the main shear stress in this direction.[26]

The pillar deformation can be observed by displaced coal layers at different rates, coal fragments falling from individual layers and by bed separation and displacement along the mine roof.

The effective area of the pillar stays intact due to the internal frictional effect of the cleated coal material, and results in a pillar core which can withstand tremendous vertical load. Regardless of the size of the effective pillar area the existence of the pillar core has a certain influence on the yielding pillar stability and its residual strength.

4. Effective area not in existence occurs when the pillar is loaded to such a degree that the fractured zone actually extends throughout the whole cross-section of the pillar and produces overall pillar instability. Under these circumstances zones of high stress

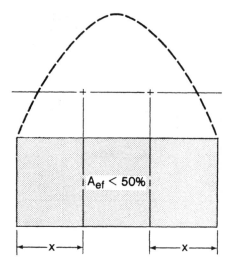

Fig. 7.3.4. Abutment stress of the pillar with minimum effective area ($A_{ef} < 50$ per cent).

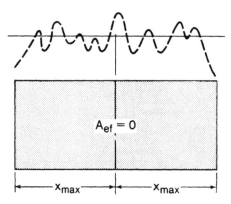

Fig. 7.3.5. Stress of the pillar with total yielding area ($A_{ef} = 0$).

concentration do not exist. The stress for the most part is relieved by fracturing and remaining stress has erratic distribution (Fig. 7.3.5).[29] Pillar deformation is by rapid closure at a magnitude of 5 to 50 cm per day. Pillar failure follows a path of orogenically sheared planes where shear movement is followed by fragmentation of coal layers.[39] It can be observed that integral pillar extrusion with significant floor heave will result in roadway collapse with piles of fragmented coal, broken timber supports and appreciable roof sag. This category of pillar stress distribution is similar to the first category of the total effective area, but with the significant difference of a totally ineffective area, because a pillar core is not present.

From the aspect of strata mechanics the primary consideration is the stress dynamics of the coal pillars, where due to high abutment stresses the coal fractures have been formed in the stress relieved zone. This is followed by an increase of the outer yielding zone and a decrease of the effective area of the pillar. Under these circumstances the magnitude of the abutment stress is increased in accordance with an increase of the stress relieved zone and it is pushed further onto the pillar core

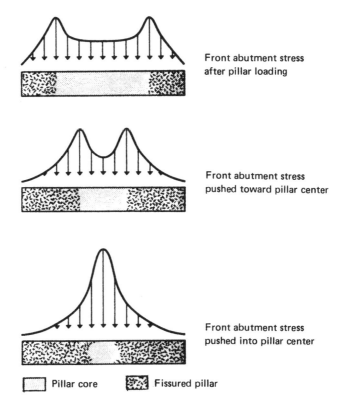

Front abutment stress
after pillar loading

Front abutment stress
pushed toward pillar center

Front abutment stress
pushed into pillar center

Fig. 7.3.6. Stress dynamics
as function of progressive
pillar fracturing.

☐ Pillar core   ▨ Fissured pillar

(Fig. 7.3.6). Similar pillar mechanics have been observed in sublevel coal caving methods, which is discussed in the previous chapter.

The dynamic progressive deformation and failure of coal pillars has been studied in the underground coal mines of the McIntyre Company (Number 4 coal seam). The following observations have been made:

a) The pillars, after development, were stable and able to support the load transferred onto them from the extraction areas (Fig. 7.3.7). With time they exhibited fracturing in the area of front abutment stress so that this part of the pillar could not support its share of the load which had to be transferred forward onto the pillar centre.

b) The progressive fracturing of pillars (width 30 m) was studied at Number 5 Mine (Fig. 7.3.8). The investigations suggested pillar presplitting and abutment front transfer in increments of up to 8 m.

c) Pillar fracturing was followed by coal extrusion until total failure of the whole system occurred (Fig. 7.3.9). When this happened the stress diversion from the mining area ceased and the stress was rebounded back onto the mining area because the yielding pillar could support only the load superimposed directly above it.

It is obvious that pillar mining is successful only if the coal has adequate strength, and could support load. The coal in Number 4 seam, for example, was not strong enough to support the diverted load, therefore, the load was shifted back onto the mining area which caused sudden roof falls and dangerous mining conditions.

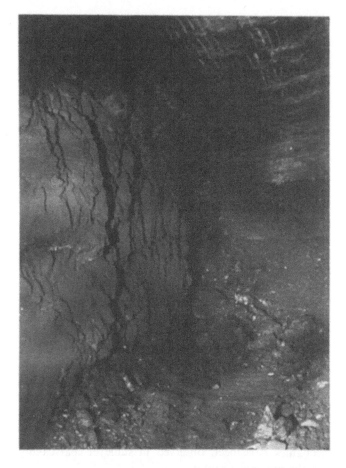

Fig. 7.3.7. Solid pillar stability.

Fig. 7.3.8. Fractured pillar stability.

Fig. 7.3.9. Yielding pillar stability.

7.3.2. *Deformation and failure of mine structure*

The problems associated with deformation and failure of structures in lateral mining methods such as longwall, shortwall and room-and-pillar systems are difficult to simulate by digital modeling techniques. Because of this a three-dimensional modeling technique of equivalent material was developed, which could simulate the failure mechanism of the pillars or panels due to stresses associated with a particular mining layout.[31, 32, 33] The basic problems of three-dimensional modeling as well as the general case of modeling of equivalent material, have been the simulation of the initial stress state and the rock mass behaviour.[34]

For the model analysis of the room-and-pillar mining layout a technique has been used which was developed by Szwilsky and Whittaker.[31] The model was constructed to simulate deformation and failure characteristics of the mine pillars and to observe the interaction between multiple coal seam mining. The model consists of a steel base of approximately, 0.555 metre square area. The coal pillars were modelled in plaster-of-paris and the rock strata by a series of rubber strips. The rubber strips were of two different thicknesses, where the thicker one (1 cm) represented sandstone strata and the thinner one (0.5 cm) represented siltstone strata. The rubber strips offer a lateral strain effect and a high rubber/pillar (strata/pillar) contact friction factor.[35] The loading frame has a 200 tonnes loading capacity and the hydraulic ram is connected to a braced steel plate 0.91 m and 0.61 m as illustrated in Figure 7.3.10.

The example investigated with this model was the case of simultaneous mining of two coal seams by a room-and-pillar system, where coal recovery in the lower seam was approximately 35 per cent. The geometrical scale of the model in the horizontal direction was 1/120 and in the vertical direction 1/90, simulating mining in a 3.8 m thick coal seam. Strength scaling of the pillar could not be matched to a geometrical scale, because the uniaxial compressive strength of model material was 5.24 MN/m². An attempt was made to produce weaker pillars of plaster-sand-water mixture, but it was abandoned after encountering problems with pouring and handling of the model

7.3.10. Set up of three-dimensional model of multiple seam mining by room-and-pillar system.

pillars.[36] Scaling of loading conditions was achieved so that the experiment could be successfully carried out.

The subject of the model investigations was the underground mining by room-and-pillar methods in coal seams of the Grande Cache Coal Basin of Western Canada. The layout consisted of development of extraction blocks with a three-entry system driven along the strike on the downside of the seam. The rooms (entries) are mined in the advancing phase and the pillars are then extracted on retreat down the dip. The rooms are 5.5 m by 3.0 m, the height varying with the thickness of coal seams. The average pillar width is 24.5 m (30.0 m entry centres).

The model of the room-and-pillar structure was loaded for different mining depths, a four sequence of loading has been examined as illustrated in Figure 7.3.11 and briefly described below.[36]

a) Ram pressure 2.0 MN/m², simulated a mining depth approximately 85 m. The average model stress was 0.96 MN/m². Under these loading conditions no pillar deformation nor failure was observed. It is obvious that for shallow mining conditions pillar stability is adequate.

b) Ram pressure 12.0 MN/m², simulated a mining depth approximately 510 m. The average model vertical stress was 5.76 MN/m², which is in the range of the ultimate stress of model material. These loading conditions showed critical stress conditions for this particular layout, which was evident from the entry deformation and pillar failures. Over the pillar the predominant stress field due to the frame loading was in the form of a direct vertical stress. In the region of the entries, the shear stress and bending stress were predominant, as exhibited in the model by roof sag and floor heave. The main pillar slabbing failures were induced and propagated in the upper seam because of additional

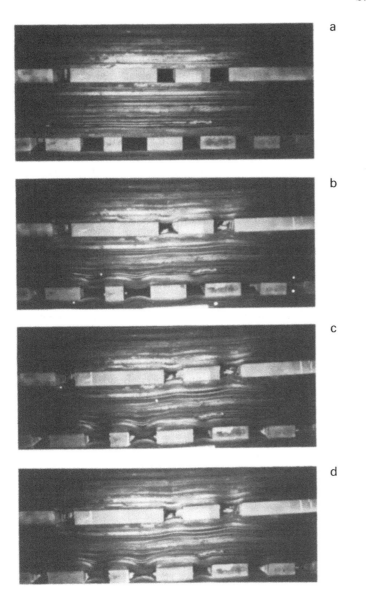

Fig. 7.3.11. Sequences of model loading and related deformation and failure of mining configurations (2, 12, 18, 20 MN/m² Ram pressure).

stress interaction from pillars below of lower seam. Some limited pillar slabbing was exhibited also in lower seam. The observed failures were mainly formed in a direction perpendicular to the model face.

c) Ram pressure 18.0 MN/m², simulated a mining depth approximately 765 m and was critical for model pillars. The pillar vertical average stress was 8.64 MN/m² and about 40 per cent greater than the uniaxial compressive strength of model material. The entry closure deformation occurred, particularly in lower seam. It can be concluded that the model showed an excessive roof convergence which would cause difficulties in ground control. The pillar fracturing at these stress levels extended the original failure patterns and formed another set, which was parallel to the model face. Under these circumstances, two sets of failures running parallel to pillar free edges showed relaxation of vertical stresses of the outer pillar zone and their transfer onto the solid part of the pillar. This phenomena of model pillar failure and stress distribution is in agreement of the fundamental concept of pillar abutment stress.

d) Ram pressure 20 MN/m^2, simulated a mining depth approximately 850 m. The average model vertical stress was 9.6 MN/m^2, which could be considered ultimate for the investigation of the mine configuration, and under such conditions the mine structure could be in a state of collapse. The closure deformations and pillar slabbing were further increased. At this stage of the investigation the model was dismantled and the pillar failure observed in detail. A plan view of the model showed either a minimized size of pillar core from progressive slabbing or fracturing with failure more or less oblique to the pillar rectangular shape (Fig. 7.3.12). The fractured material from the pillar sides largely filled the entries. The investigations of both failure sets which can parallel with the free edges of the pillars showed that they were almost vertical or at very steep angles to the ribs; and some of them were intersecting each other. It was noticed there was a greater intensity of fracturing in the direction perpendicular to the model face.

This technique of model investigation for various mining layouts of lateral mining methods could be a valuable tool for the study of failure mechanisms of longwall face-ends,[31] rib pillars, room-and-pillar structure[35, 36] and others. Besides failure mechanisms it is possible to study strata movement both above and below panel or pillar structures. Particularly, it could be of interest to evaluate the inter-strata displacement above the entries and roadways. For example, in the described model investigation, it was indicated that roof strata displacement extended to about two coal seam thicknesses. However, it must be noted that the models are a simplified representation of the room-and-pillar mechanics at various depths of mining but they do offer the potential to gain an insight into the possible behaviour of pillars and the related roof and floor strata.

Fig. 7.3.12. Pillar failures and geometry of pillar core.

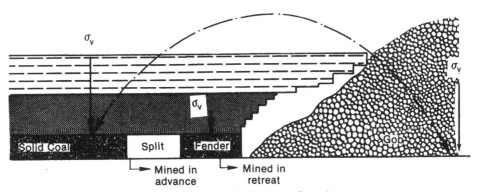

Fig. 7.3.13. Stress redistribution for split and yielding fender configuration.

7.3.3. *Stress analyses of mining configurations*

The stress analysis of room-and-pillar mining system commenced almost a half a century ago, with the application of the arching theory.[37] As a matter of fact, this concept is still valid and it is used by some researchers in analysis of stress redistribution in roof strata, as for example in Eastern Australian long-pillar mining (Wongawili system) as represented by Hargraves and Kininmonth.[38] In Figure 7.3.13 a typical extraction layout is given from rock mechanics aspects. A driving split of the long-pillar is in a high stressed zone, but the newly formed fender could be de-stressed if it is narrow and has deteriorated strength, so that stress has to be transferred onto the solid coal of the long-pillar. Under these circumstances the fender experiences limited stress concentration, which results from the rock bridging between solid coal and gob. Under these stress conditions it is possible to extract the fender without great difficulties (retreat) and achieve a greater output than by excavation of split (advance). It should be pointed out that a key factor which controls the stress redistribution is the width of the fender and the deterioration of strength due to fracturing under high stress concentration (abutment stress), when it was still part of the solid coal block.

At the present time stress analysis in the majority of cases is carried out by digital models based on the principles of continium mechanics. For example, Crouch's program of the Displacement Discontinuity Method, which was implemented and adopted by F. E. Eves, as shown in Table 7.3.1 for stability analyses of multiple pillar-stope structure. The code name of the program is DDSEAM applicable for analysis of up to five seams with up to eighty elements per seam and it is convenient for both interactive and batch operating modes. A particular feature of the program is that it can produce graphical output of stresses and displacements within the area of con-sideration whilst in the interactive mode. The computer results may be obtained in the form of printed tables and plotted graphs.

The DDSEAM program was used by undergraduate and graduate students for room-and-pillar stability analyses of coal seams of various rank in Western Canada (Department of Mineral Engineering, The University of Alberta). Three practical examples of the application of this program are briefly discussed below:

1. Boundary element – normal and shear displacement – investigations were carried out for a four entries mine configuration and corresponding layout of chain pillars in

Table 7.3.1. Adopted Crouch's program of displacement discontinuity analysis.

Title:	'Analysis of Stresses and Displacements Around Multiple. Parallel Seam-type Deposits by the Displacement Discontinuity Method'.
Code name:	DDSEAM
Author:	Steven L. Crouch
Implemented/	
Adapted by:	Fred E. Eves
Version:	0 Date: February, 1978

Program DDSEAM computes the displacements and stresses induced by the mining of multiple, parallel seam-type deposits using Crouch's 'Displacement Discontinuity Method'. The program can be run either interactively or in batch mode. The computed results may be obtained in the form of printed tables and/or plotted graphs. The IG and PLOTLIB library routines are used to implement the graphics option.

Aside from the changes made to incorporate the plotting capability, parts of the program have been rewritten in order to increase efficiency.

Subroutines called by DDSEAM:

ONSEAM	– computes the stresses & displacements along a seam
OFSEAM	– computes the stresses & displacements along any parallel line in the half-space y < 0 not on a seam
GFUNC	– computes the G-function values as indicated on pages 232–234 of Crouch's report
GRAPH	– plots up to 5 curves on a single labeled graph
GETTY	– moves the curve data into columns of a single matrix for the curve plotting routine, GRAPH
STATS	– plots the specifications describing the multiple seam mine
MINEX	– plots a cross-section of the multiple seam mine

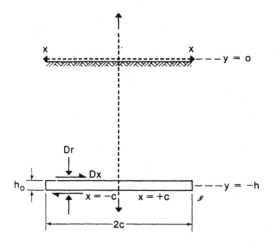

Fig. 7.3.14. A finite element segment parallel to the surface of semi-infinite homogenous elastic body.

the Kipp Mine (Lethbridge). The model represents a constant displacement discontinuity over a finite element segment parallel to the surface of a linear elastic body (Fig. 7.3.14). The distribution of displacements within the structure was calculated as a function of the closure (roof to floor convergence of the entries) and ride (shear displacement between roof and floor strata). In reality, the closure and ride components of an excavation within a seam are not constant over the whole span, but are functions

Table 7.3.2. Average values of elastic constants (Lethbridge Coal Basin)

Elastic constants		Hard coal	Soft rock
Modulus of elasticity		2.4×10^3 MN/m²	1.9×10^3 MN/m²
Poisson's ratio	(xy)	.25	.35
" "	(yz)	.35	.40
" "	(zx)	.30	.38
Bulk shear modulus		$.86 \times 10^3$ MN/m²	0.55×10^3 MN/m²

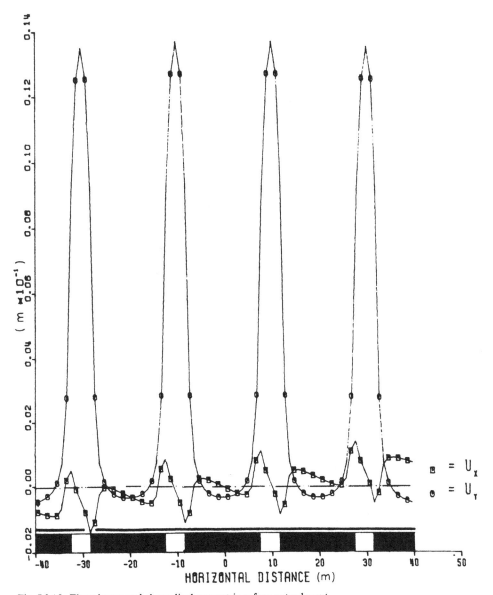

Fig. 7.3.15. Floor heave and shear displacement in a four entry layout.

of the position along the opening. In order that the digital model analysis could be carried out, the elastic constants of the high volatile bituminous coal and rocks were assumed as a result of tested data of this coal field (Table 7.3.2.)[39] The investigation in this particular case was limited to only displacement analysis, in order to compare the floor heave predicted by the digital model to the actual heave measured in the mine.

Figure 7.3.15 shows the results of the run using the assumed data which computed a floor heave of 0.013 m. This number appeared reasonable considering the nature of the material and the preliminary obsevations of floor heave in the Kipp Mine (Lethbridge). Many combinations of elastic parameters would probably have achieved the same results given the fixed dimensions of depth and opening size. For example, limited analytical analysis suggests that functional relationship might exist between mine floor deformations and coal pillar strength degradation.[40]

2. Boundary element – normal and shear stresses investigations were carried out for room-and-pillar mining configurations at the Number 2 Mine (Grande Cache). The digital model study considered normal and shear displacements,[40] and also distribution of the normal and shear stresses as function of their positioning relative to the pillar-stope structure.[41] These stresses were calculated from the solution of the constant displacement discontinuity over a finite element segment. In this case, an image solution was superimposed to create a shear stress free surface (ground surface), and a supplemental solution to reduce the normal stresses to zero along the same surface. The boundary conditions for a given mining configuration are assumed that the displacements were continuous everywhere in the body except over pillar-stope structures. Also, that the normal and shear stresses, as well as all stresses and displacements at infinity, were zero. The elastic constants of the low-volatile bituminous coal seam and rock strata were defined during laboratory testing of the samples from this coal field (Table 7.3.3). The digital model investigations of normal stress distribution in-seam pillar (Fig. 7.3.16) and shear stress distribution on-seam strata (Fig. 7.3.17) showed appreciable differences from data of underground observations of this particular mining configuration. For example, smaller coal pillars were fractured due to appreciable mine depth ($\simeq 500$m) and low strength of coal ($\simeq 0.5 \text{MN/m}^2$) and they were actually stress relieved structures. Under this condition,

Table 7.3.3. Elastic constants of tested samples (Grande Cache Coal Basin)

Elastic constants		Soft coal	Hard rock
Modulus of elasticity		1.95×10^3	7.00×10^4
Poisson's ratio	xy — yz — zx	0.40	0.25
Bulk shear modulus		0.90×10^3	3.90×10^4

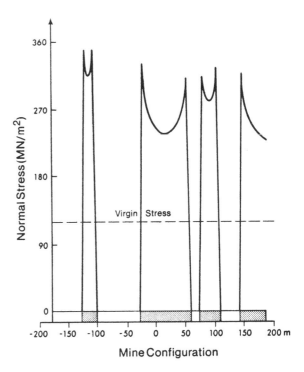

Fig. 7.3.16. The normal stress distribution with coal pillars.

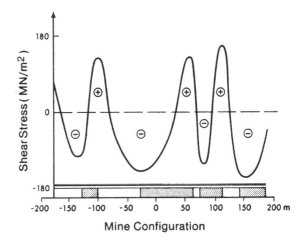

Fig. 7.3.17. Shear stress distribution along roof stratum.

the vertical stresses were transferred onto large pillars where average pillar stress is greater than results from numerical modeling. Underground observations clearly showed that maximum stress concentration (abutment stress) is inside the pillar body and not close to the edges as the digital model suggested (Fig. 7.3.16). The modeled shear stress distribution showed a maximum concentration in the roof strata in the vicinity of the pillar edges (Fig. 7.3.17). This might correspond to underground situations where a number of roof displacement were observed at pillar ends. Validity of the digital model shear stress distribution is decreased a great deal due to orogenic shearing along bedding planes and presence of residual shear stresses. Our present experience does not favour digital modeling for normal and shear stress distribution for pillar-stope structure of coal strata, particularly in orogenically deformed regions. This numerical method with certain limitations might be satisfactory for stability analysis of room-and-pillar configuration in hard rock mining, under conditions that the deterioration of pillar elastic constants due to excavations are taken into account. In such cases, the deformational response of pillars can be simulated by an elastic material with modulus of deterioration elasticity:[42]

$$E_p = \frac{A_p}{A_p + A_e} E_m \qquad (1)$$

where

E_p = reduced modulus of elasticity of coal in pillar area;
E_m = modulus of elasticity of coal;
A_p = pillar area;
A_e = mined out area.

Further mine pillar investigations in this direction are necessary before this digital technique can be utilized for mine design.

3. Finite element – vertical and horizontal principal stress investigation was carried out for multiple seam mining configurations of the foothills Region (medium-volatile bituminous coal). In this case, a two-dimensional finite element program was used for nonlinear elastic analysis, where the variation of displacement discontinuities can be mathematically modelled by considering the excavation as a series of finite line segments, jointed end to end as illustrated in Figure 7.3.18. On the basis of principles of superposition, the total mining stresses and displacements at any point can be found by adding to the effect of each displacement discontinuity element (seam element). The

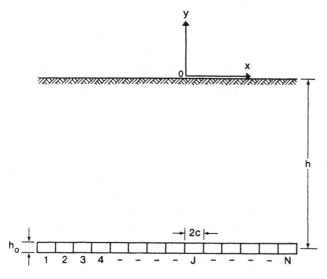

Fig. 7.3.18. Finite line segments jointed end to end.

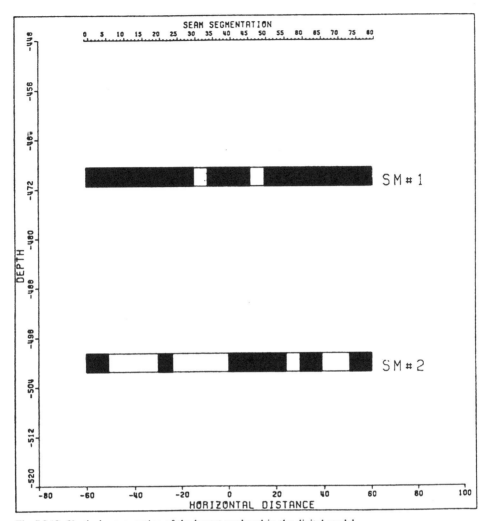

Fig. 7.3.19. Vertical cross-section of the layout analyzed in the digital model.

finite stress analysis was carried out by R. Wright to establish interaction between pillar-opening structures located in two coal seams (Fig. 7.3.19). The influence of undermining of upper Number 1 coal seam by extraction of lower Number 2 coal seam was carried out by the numerical model analysis, where displacements and stresses were diagrammatically represented (Fig. 7.3.20). The in-seam stresses are represented by contours of vertical stress (Fig. 7.3.21) and horizontal stress (Fig. 7.3.22) distribution where concentration is controlled by coal pillar width. Also, the contour plots of major (Fig. 7.3.23) minor (Fig. 7.3.24) principal stress distribution showed that there is very limited interaction between these two seams, because of insufficient length of the excavations in the lower seam, in relation to the vertical distance between seams ($h \simeq 3w$).[43] Further studies of the digital model results from the aspect of mine stability, were carried out by analyses of maximum shear stress distribution within mine

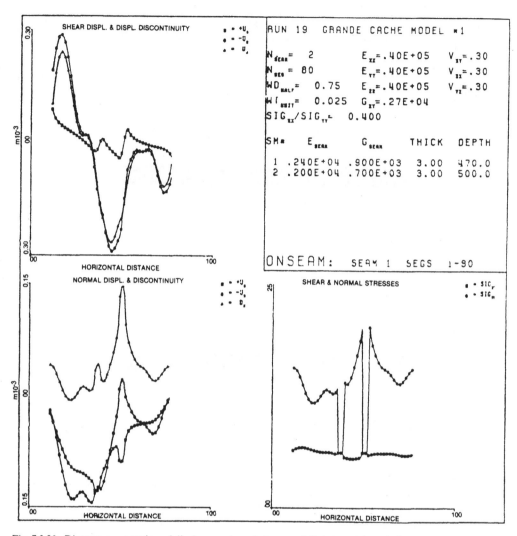

Fig. 7.3.20. Diagram presentation of displacements and stresses of digital model analysis.

configuration (Fig. 7.3.25) of stability criterion by delineation of areas prone to failure (Fig. 7.3.26).[36] Existing criticism of the present capabilities of this modeling technique might be justified. For example, the most recent stability analysis of room-and-pillar coal mining using this modeling technique showed that stress is uniformly divided on the pillars regardless of the pillar width and extraction width,[36] which suggests that this program might be inappropriate for stress redistribution due to extraction ratio, which is a fundamental parameter in any stability analysis.

This conclusion is further reinforced by the fact that the estimation of average pillar stress for underground mining conditions is not simple because individual pillar loads depend on the variation of deformation characteristics of both the surrounding strata and the coal pillars themselves.

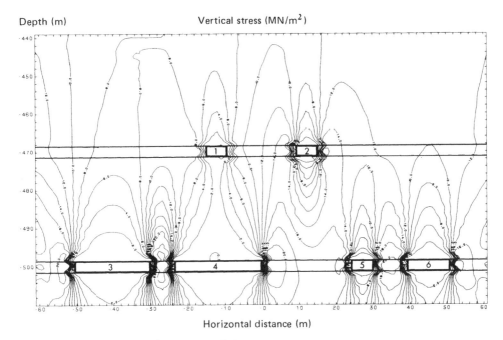

Fig. 7.3.21. Vertical stress distribution around the excavations.

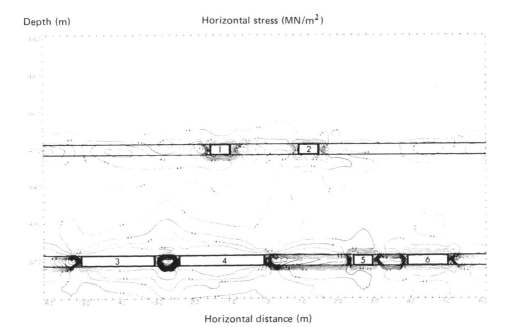

Fig. 7.3.22. Horizontal stress distribution around the excavation.

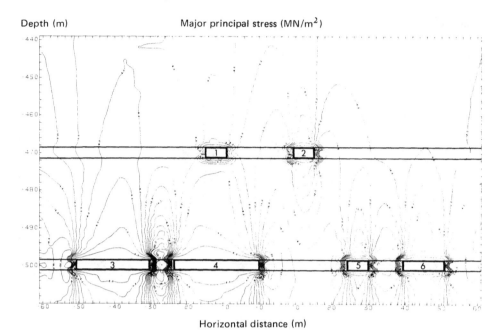

Fig. 7.3.23. Major principal stress distribution around the excavation.

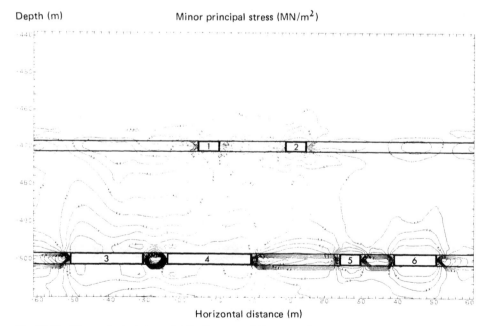

Fig. 7.3.24. Minor principal stress distribution around the excavation.

Depth (m)   Maximum shear stress (MN/m^2)

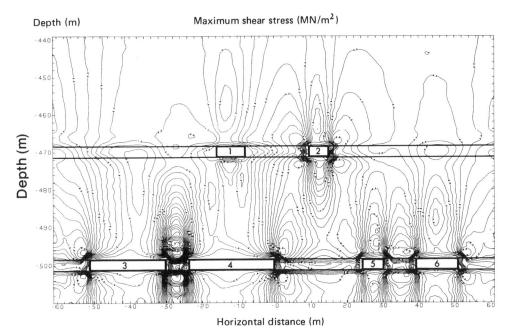

Horizontal distance (m)

Fig. 7.3.25. Maximum shear stress distribution around the excavation.

Depth (m)   Stability criterion (Unstable areas shaded)

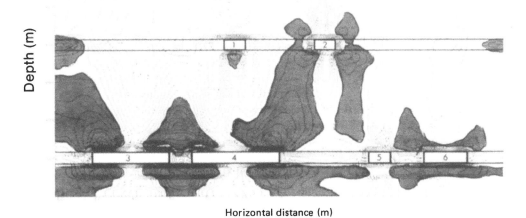

Horizontal distance (m)

Fig. 7.3.26. Unstable areas of surrounding strata defined by stability criterion.

7.4.   CAVING MECHANICS

From the conceptual point of view there is a conflict between room-and-pillar mining and roof strata caving, because this method of mining was designed to leave coal pillars in place with the function of support the overlying strata.

With the implementation of continuous mining and support of roof strata, pillarless mining has become more common and roof caving is an integral part of the system and of strata control.

Discussion in this section is limited to three main areas:

– No caving of roof strata, where coal pillars are used to support overburden and protect ground surface from harmful subsidence.

– Caving to surface in very shallow mining, where coal pillars are extracted on retreat.

– Subsurface caving in deeper coal mines, where coal pillars are extracted on retreat.

7.4.1.   *No caving of roof strata*

Room-and-pillar mining was primarily adopted to avoid caving of overlying strata and subsidence through to surface. The main factors in achieving this control are mining width and the self-supporting characteristics of roof span assuming that coal pillars are of a satisfactory load bearing capacity. The interaction between coal pillars and roof spans represents a complex problem, which is discussed elsewhere in this book.

Investigations into the roof caving characteristics in a very shallow coal mine were carried out at the Star-Key Mine in Western Canada. A summary of these results is given first in this section. The character of surface subsidence over room-and-pillar workings has also been studied in Lethbridge. A summary of these results is given second in this section. Finally, the general case is presented for reinforcement of pillar areas by waste stowing.

1. Stability of roof span as a function of depth is a factor in room-and-pillar mining, particularly because this method was devised for shallower coal seams. Investigations were carried out at the very shallow Star-Key Coal Mine (Depth 25 m), where consideration was given only to openings under the mine roof of the 'bony' coal strata, which is the strongest member of seam profile. On the basis of underground observations of the lengths of roof spans and their state of stability or failure, three parameters of mine depth opening width (h/w) were established. It is interesting that heights of the openings were the same in all the cases investigated (H = 2.3 m).

a) For a ratio h/w > 6, roof spans exhibited equilibrium without any external support. Numerous examples of conditions satisfying this ratio have existed for years without developing fissures or failure (Fig. 7.4.1). The roof is in a compressive state of stress, probably due to frozen lateral strata pressure, which enhances roof stability and limits its displacement. In one example, it was estimated that an unsupported roadway span in existence for 29 years had a maximum displacement v = 3.2 cm/m (vertical) and u = 1.0 cm/m (horizontal). With these magnitudes of the mine roof displacement fissures have not been developed. The established ratio h/w for the roof span ('bony'

Fig. 7.4.1. Self-supporting roof
span.

coal layers) could be considered sufficient for controlling stability of roof and
eliminating strata caving, hence damaging ground subsidence.

b) For ratios of h/w = 3–6, roof spans exhibited limited equilibrium and external
support by timber was required (Fig. 7.4.2). In one instance, the displacement of the roof
strata for a span width of 7.5 m reached a maximum of v = 6.0 cm/m (vertical) and
u = 2.5cm/m (horizontal). The displacements were gradual, and the maximum had been
developed over a duration of one year. The mine roof exhibited intensive fissuring and
it would have failed if timber support had not been installed. It should be pointed out
that for this h/w ratio the caving of roof strata is eliminated only if timber support is
installed. Under these conditions surface subsidence can be controlled and its adverse
effects eliminated.

c) For ratios of h/w < 3, roof spans did not exhibit equilibrium regardless of the
timber support provided, because roof fall was inevitable (Fig. 7.4.3). Intensive roof
strata displacement, as much as v = 18.8 cm/m and u = 5.6 cm/m resulted in breaking
of timber followed by immediate caving of overlying strata. A low ratio of h/w is an
indicator of the impossibility of controlling roof strata from caving because, regardless
of coal pillar size and external timber support, the roof span is too wide to sustain the
superimposed load. Moreover roof instability is increased when compressive stress is
replaced by tensile stress due to enlarging of the span width.

A theoretical analysis has been carried out to analyze stress redistribution in the roof
and floor of a rectangular excavation in shallow coal seams. The results should be
accepted with certain caution, because they are obtained by applying principles of

Fig. 7.4.2. Roof span reinforced by external support (timber).

Fig. 7.4.3. Caved roof span without self-supporting abilities regardless on external support (timber).

linear elasticity, which is not the case in underground mining conditions. Markov and Savchenko reported that when the span of extraction opening (w) becomes comparable to depth below surface (h) then the stress distribution around the excavation takes on certain characteristics, which show the influence of depth of excavation on caving span.[44] Their investigations showed that regardless of lateral geological stresses, the horizontal stresses induced by excavation can significantly exceed the vertical stresses (Fig. 7.4.4).

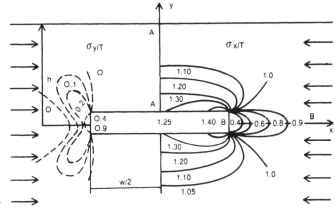

Fig. 7.4.4. Stress distribution of the shallow excavation (after Markov & Savchenko).

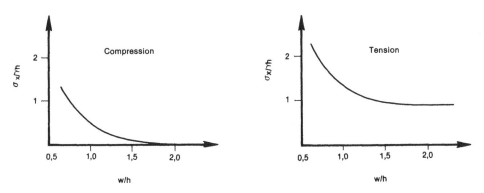

Fig. 7.4.5. Configuration of ground subsidence due to deformation of room-and-pillar structure.

Of primary concern in dealing with span cavibility is the characteristic of horizontal stress (compression or tension). It is proposed that a ratio h/w can be used as; criterion of horizontal stresses quality as illustrated diagrammatically in Figure 7.4.5. The diagrams suggest that if h/w > 2, then the proximity of the surface has no effect on roof cavibility (tensile stress of roof span is $\sigma_x/\gamma h = 1$ and at surface $\sigma_x/\gamma = 0$). However, as the ratio h/w decreases, the tensile stresses in roof increase starting from h/w < 2, and at a value h/w > 2 the increase becomes very significant.[44] It should be pointed out that these theoretical findings are basically in agreement with results of the practical investigations in Star-Key underground coal mine.

2. Surface subsidence of room-and-pillar workings was investigated in the shallow (depth up to 130 m) Galt coal seam in which mining had taken place around the turn of the century. The following conditions of ground surface subsidence due to mining (coal pillars were left in place) were determined:

a) Strata displacement during and immediately after coal extraction was limited, most likely due to the optimal ratio of width of entry ($\simeq$ 3 m) and width of pillar ($\simeq$ 6 m)

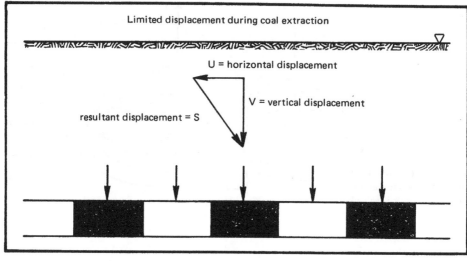

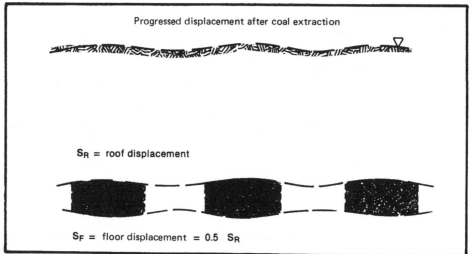

Fig. 7.4.6. Strata displacement due to coal pillar creep deformation.

as well as to timber square set installation after coal removal. In the initial period after mining ground subsidence was evident.

b) Measurable surface subsidence did take place within a year or two of mining, and time of initial surface movement depended on the mine depth.

c) The surface subsidence slowly progressed with time due to creep deformation of the pillars. There is a direct relation between coal pillar contraction, due to time deformation under constant load, and ground surface subsidence, as illustrated in Figure 7.4.6. Under these circumstances the ground surface exhibited uneven subsidence of concentric zonal rolling morphology, with ridges above pillars and depressions above rooms. It should be pointed out that this type of subsidence shows no fissuring. For example, Galt Mine Number 3 mined 50 years ago, showed uneven subsidence. Regardless of subsidence morphology, the part of the mine extracted under a river bottom was not flooded. This suggested continuity of strata above room-and-pillar working as well as absence of fissuring.

d) The measured subsidence at Galt Mine Number 8 was a maximum of 1/4 of the workable seam thickness. Measurements indicated that differences between maximum and minimum of uneven subsidence could be about 2/3 of this magnitude. Such differences in subsidence should be attributed to soft overlying strata with a tendency to plastic flow which facilitates intensive sagging of the roof strata above entries. Results could also occur owing to partial caving of roof strata above entries.

e) The disruption of continuity of the surface subsidence is due to two phenomena; shallow depth of mine workings (depth 30 m) and collapsing of coal pillars due to creep deformation. Evidence is that the earliest coal pillar collapsing appears 45 to 50 years after mining. Under these circumstances, damage of ground surface are unavoidable.

The practicability of room-and-pillar mining, from the aspect of production of ground surface by minimum subsidence, is controlled by mine configuration. In some circumstances to achieve this aim and avoid caving of overlying strata it is necessary to reinforce room-and-pillar structure as briefly discussed in next paragraph.

3. Reinforcing stability by stowing is a common method, particularly in cases where there is a demand for protection of ground surface against the adverse effects of subsidence. Some coal pillars over short or long periods of time undergo failure and experience a considerable magnitude of deformation. Improved stability of the room-and-pillar structure is required.

Salomon and Oravecz discussed as a reinforcement measure the introduction of relatively stiff concrete columns, but they concluded that these could induce stress and cause punching of the roof and floor strata with the resulting loss of all support capacity. Deformation characteristics of reinforcing pillars could be designed in accordance with the rock mechanics of failing pillars, for example by introducing compressible material like wood between the support and roof. In order to achieve satisfactory strata control it would be necessary to install a large number of these pillars at considerable cost.[45]

These authors suggested that a more convenient method to increase stability of room-and-pillar structures would be filling of the void by waste material, e.g. stowing. This method could be advantageous particularly where filling material is a waste which can be obtained without disturbing the surface environment.

From present mining experience it is known that stowing can arrest the gradual deterioration of unstable structures. Salomon and Oravecz examined the mechanism by which stowing assists the pillar but where filling does not extend to the roof (Fig. 7.4.7). Because the stowed material is not in contact with the roof, it will not transmit any pressure between the roof and floor. Under these circumstances, the load on the pillars is not reduced. In any event, the filling, even if it is in contact with the roof, will transmit load only after the roof has sagged, since it is not an active support. A support

Fig. 7.4.7. Room-and-pillar mining with stowing.

is active if it is prestressed so that its supporting action does not depend on the deformation of the workings, e.g. hydraulic props, roof bolts. Stowing, therefore, does not relieve pillar load. This means that its supporting effect is due to a different mechanism.

The filling exerts a confining pressure on pillar walls. This pressure is only a fraction of the vertical pressure due to the weight of the filling material. From this it could be concluded that stowing has a limited ability to enhance the strength of coal pillars. The most significant influence of stowing is at state of pillar failure, because filled material resists lateral strain of the pillars. This lateral strain resistance is several times greater than the pressure exerted by the fill on stationary pillar walls. The effect of stowing is to alter the deformation characteristics of the pillar and prolong its yielding state.[45]

It is a point of interest that effectiveness of stowing depends on the compaction of fill and height of stowing in relation to the pillar height. For example, the maximum later extension of a pillar is approximately at mid-height, and the height of stowing has to be more than half of its height.

Finally, it should be mentioned that in some coal fields the solid pillars are replaced by stowing and pillarless mining is practiced. The stowing reduces the excavated void and if roof strata collapse, the propagation of caving is limited as is surface subsidence. Since the upper strata displacement is proportional to the extracted height, stowing in fact reduces this effective height. For example, in the case of hydraulic stowing where maximum compaction is achieved, the surface subsidence can be 0.15 of the worked height of the coal seam, even if pillarless mining is in effect. The most effective ground protection, particularly in shallow mining, is total coal extraction and replacement by fill, because under these circumstances the surface will subside evenly.

7.4.2.   *Through surface caving*

Through surface caving has been studied at the Star-Key Mine. The roof strata and coal seam of this mine was described in previous chapters. It is discussed because the mechanics of through surface caving is of interest, and mechanical phenomena are easy observable. Results are summarized as follow:

1. The roof strata caving is initiated when the roof span is enlarged due to coal pillar removal during the retreat extraction phase. Several stages of the caving mechanics have been observed:

a) Shear failure initiation in the roof and its propagation is followed immediately by roof falls (Fig. 7.4.8). The caving is more extensive in the direction of the cleat in coal than in one parallel to coal. The main parameter of the caving is the angle of break β, which for the Star-Key Mine, from underground observations is an average 73°.

b) The rate and increment of caving depends on the size of unsupported roof area and the behaviour of the immediate roof (coal) and main roof (rock). It has been established that after major caving there are several successions of minor caving (sloughs), which suggest that shear failure propagation upwards is in steps due to stress redistribution around caved area. It could also be assumed that roof deformation is by flow rather than by brittle fragmentation because roof strata has an appreciable index plasticity (Attaberg's index 11 plasticity).

c) The rock fragmentation is very intensive due to weak rock strata. The void filled

Fig. 7.4.8. Shear displacement
and shear failure propagated
upwards.

Fig. 7.4.9. Pothole on the
surface above a mined-out
room.

with caved rock is very dense and compaction of the void could be sufficient to support overlying strata. For this reason, the extension of the caving above the room is appreciable and it has been estimated up to eight times of the seam thickness. The compaction factor is high and it has been estimated about 0.8 of the seam thickness.

d) Due to plastic properties of the rock, the caved material draws from the rooms in butt entries, which are still supported. It creates an additional void into which additional rock can slough from subsided overlying strata. Thus caving progresses toward surface at a rate determined to a large degree by the rate the caved rock is thrown from the room in the butt entries.

e) The caved zone approaches surface and breaks through to the surface forming a 'pothole' (Fig. 7.4.9).[46] As surface water reaches the pothole, a mudslide can result. In the mine a strong wooden barricade across the entry is installed to prevent the mudslide from spreading. The pothole on the surface has the geometry of an elongated, inverted, cone-shaped, crater. It is usual to fill in these craters.

The caving mechanics at Star-Key Mine for the present mining system indicates: firstly, forming of a broken zone, which has a caving rate upwards in range 3.3 to 9.9 m per hour; and secondly, forming progressive caving to the surface at a rate approximately 3.0 to 9.0 m per month.

2. The roof strata fracturing to surface has been studied for a mined-out room of 91 m long and 18 m wide (h/2 = 1.5). The underground and surface observations indicated the formation of trapezoidal rock block, extended from stope face to ground surface.

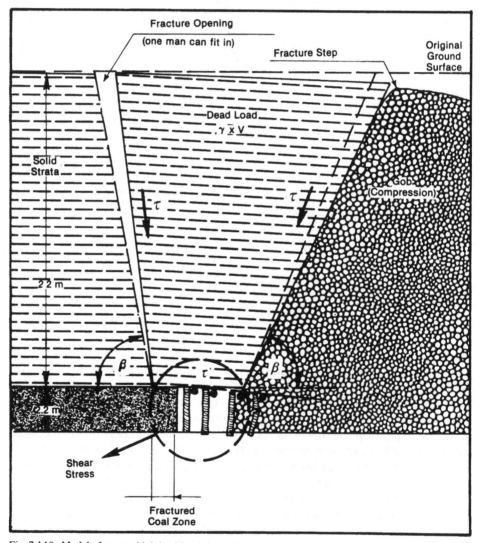

Fig. 7.4.10. Model of trapezoidal dead load delineated by angle of draw (δ) and break (β).

The angle of inclination of fractures along the trapezoidal section is estimated from underground observations and calculations as follows:
– angle of draw (cave line) is observed

$$\delta = 73°$$

– angle of break is calculated by empirical formula

$$\tan \beta = \sqrt{2\pi} \quad \tan \delta = 2.5 \times \tan 73° = 6.875 \qquad (1)$$
$$\beta = 81.5°$$

Both fractures are visible on ground surface at approximately the projected location (deviation of projections probably due to some changes of their angle through glacial till). On the basis of their initiation underground and their projection on ground surface, a model of a trapezoidal section of dead load has been assumed (Fig. 7.4.10).

From practical observations the following have been determined:

a) One day after caving small tensile fractures can be visualized.

b) One week after caving both fractures were clearly seen on the ground surface, along their projected direction.

c) Three weeks after caving, the fracture break has been widened up so that one man can completely fit into it. The fracture of the cave shows a step break which very clearly occurs at the boundary between the compression zone – gob (located down) and the tensile zone – trapezoidal block (located up) as illustrated in Figure 7.4.11.

d) Five weeks after caving the ground subsidence appears complete because visual observations have not given any evidence of further ground displacement.

Both fractures obviously result from the caved unsupported roof and the existence of the supported roof to the gob. It appears obvious that fracture propagation, and its widening, is a function of time. It is logical that propagation and further extension could be controlled by rapid coal extraction such as longwall mining. Due to the shallow depth (h = 10 times seam thickness) and the weak roof strata, avoiding fracture

Fig. 7.4.11. Fracture controlled by angle of draw, which delineates tensile zone and compressive zone.

propagation on the surface is impossible. It is more realistic that it will take place at the rate of face advance.

The rate of fracture propagation and likely rate of caving are also influenced by climatic seasons. For example, during the summer period when most of the overlying strata is soft and partially saturated by water, deformations will be more rapid. During the winter period when most of the overlying strata is in a frozen state and is hard, deformations will be at a slower rate.

3. Ground surface subsidence has been investigated in relation to the roof strata caving and its fracture to the ground surface, which resulted in formation through subsidence. The boundary of the subsidence, the cave fractures (controlled by draw angle) and projections of the gob area on the surface has been estimated from the aerophotograph (Fig. 7.4.12).

The boundary of subsidence are approximately one third of the depth of the seam thickness what corresponds to the following formula:

$$r = \frac{h}{\tan \delta} = \frac{21.35}{3.27} = 6.50\,\text{m} \tag{2}$$

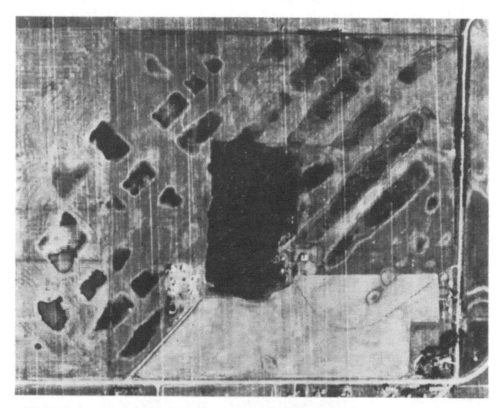

Fig. 7.4.12. Aerophoto of through subsidence above rooms, divided by interoom pillars.

where
>   r = limit subsidence;
>   h = depth of coal seam;
> δ = angle of draw = 73°.

Maximum subsided ground of 1.75 m corresponds quite well to an estimated compaction factor of caved rock, as follows:

$$v_{max} = \kappa \times t = 0.8 \times 2.2 = 1.76 \qquad (3)$$

where
> κ = compaction factor;
>   t = seam thickness.

On the basis of this factor, it could be estimated that a maximum displacement on the ground surface will be as follows:

Compressive displacement (ε')

$$\varepsilon' = \frac{\kappa \cdot t}{h} = \frac{0.8 \cdot 2.2}{21.25} = 0.082 \, m/m \qquad (4)$$

Tensile displacement (ε'')

$$\varepsilon'' = 0.4 \frac{\kappa \cdot t}{h} = 0.4 \times 0.082 = 0.033 \, m/m \qquad (5)$$

Both displacements are significant and they are beyond the limit for any existing surface structures. This means that it is necessary for the protection of pipelines and roads which traverse the property to leave underground protective coal pillars. The design of the protection pillar should take into account besides the draw angle, peculiar ground behaviour due to soft overlying strata and very shallow depth of the underground mine.

7.4.3. *Subsurface caving*

Room-and-pillar mining in some deeper coal mines, where coal seams are sandwiched between strong strata (sandstone and siltstone) primarily employ pillarless mining, with coal pillars mined-out on retreat. The caving mechanics of such roof strata is discussed in the next chapter, and at this place consideration is given only to the subsurface caving aspects of multiple seam mining.

An analysis of strata mechanics was carried out at Grande Cache Mine for the room-and-pillar layout of multiple seams: Number 4 seam (lower seam) and Number 10 seam (upper seam). Mining was first in the Number 4 seam, which undermined Number 10 seam. The interaction mechanics of undermining an upper coal seam due to the mining of lower seam has been discussed in relation to physical model analyses and underground mine studies.

1. The unconfined base friction model technique was originally applied to kinematic studies of the deformation and failure of rock structures. The same technique has been used to study subsurface caving mechanics and its influence on structure stability in upper seams. In the laboratory for Rock Mechanics of the Department of Mineral

Fig. 7.4.13. Front view of base friction model.

Engineering, The University of Alberta, two types of model studies were carried out and each of them is separately discussed.

The base friction modeling technique has been well described in literature and some of the simulated conditions have been defined.[47, 48, 49] The principles of this technique are based on the development of a frictional shear force at the contact of the model material and the frame base to simulate gravity loading. The shear forces are exerted by pushing the model over a roughened surface underneath the model.[47] The principle has been used to build a conventional base friction frame to produce a two-dimensional plane stress model (Fig. 7.4.13). The size of the tilting table of the friction frame is 1.22 m in height by 2.44 m in width, with two 10 cm rollers extending along its entire length. The belt covering the table is composed of a 200 grit sand paper tube, which is 2.66 m long and has a circumference of 3.33 m. To ensure alignment of the belt, there is an adjustable tensioner on each roller. The tilt of the table is done mechanically by twin, 2.5 cm diameter threaded screw jacks with detachable handles. The belt is driven by a 1/2 horsepower reversible motor with a two speed gearing system.

The study carried out by Bray[48] has shown that, providing inertia forces are neglected, the displacement at time in the prototype, it is related to time in the model test (t_b), which is given by equation:

$$t_b = gt^2/2V_c \qquad (6)$$

where

    g = acceleration due to gravity;
    V_c = constant velocity of belt.

The stress conditions for base friction modeling follows scale rule as given below[49]:

$$L_p = \sigma^* \frac{(\cos \Theta \tan \phi' \pm \sin \Theta)L_m}{\rho} \qquad (6)$$

where

    L_p = characteristic length in prototype;
    L_m = characteristic length in model;
    σ^* = ratio of uniaxial compressive strengths of prototype rock and model material;

Θ = angle of inclination of tilting table;
ϕ' = effective angle of friction between belt and model material;
ρ = scaling ratio.

The material used for model investigations was a combination of methyl alcohol and flour, which is optimal for this modeling technique because[50]:
- It is linear elastic till near failure;
- It is sufficiently weak to be utilized in the friction frame (0.2–2.0 MN/m²);
- It fails up to 2 per cent strain;
- It is weak in tension ($\sigma_c/\sigma_t = 8–12$).

The mining models were built with approximately 60 kilograms of flour placed in a ball mill with 35 kilograms of steel balls and 25 litres of methanol. The ball mill was run for 2.5–3.9 hours before model material was separated from the balls and placed on the friction base frame. The model material was well compacted and leveled on the table, before the material was cut using a lab spatula or putty knife in order to simulate geological structural features and mining configurations. Noticeable shrinkage of the prepared model occurs as the methanol evaporates. The model material is consolidated by very small movement to belt forward and adjusting the lateral restraint. The model is allowed to dry at least 36 hours before testing.

The investigations of the block model of the mine strata were mainly qualitative comparisons of deformation and failure produced by the caving processes after stope excavation. An example is discussed of the model which was built to simulate the interaction mechanics of flat and multiple seam mining.

2. Investigation of interaction mechanics resulting from undermining and caving of an upper coal seam were made by physical model analysis and underground phenomenological studies. It was obvious that the mechanics of deformation and failure of the coal pillar structures in the upper seam were governed by subsurface strata displacement and the subsidence caused by room-and-pillar mining, where pillars are recovered on retreat.

Consideration was given to an actual mining configuration which is represented in Figure 7.4.14. The description of the mechanics of deformation and failure of the structures of this particular layout is based on data from both the underground mine[51] and the physical model and is as follows:

a) The pillar-entry structure of the undermined upper coal seam shows clear flexural deformation due to subsurface subsidence effect. The maximum deflection is approximately above the centre of the extracted area of the lower seam. The vertical displacement of the subsided pillar-entry structure was about 70 per cent of the mineable thickness of the lower seam. Underground monitoring of vertical displacement from Number 11 seam to about 30 m above extracted area in Number 4 seam, showed maximum subsurface subsidence of 75 per cent of the workable seam thickness.

b) The plain-strain model showed a bed sagging and separation above flexured pillar entry structure. It can been seen that bed separation is up to four seam thicknesses, which corresponds to data of underground monitoring in Donbas Coal Basin (U.S.S.R.).[52] A similar phenomenon was observed in the mining of Number 11 seam, where at a bed separated plane the dome shaped cave was formed, after a roof fall. The vertical extension of the caved areas was up to three seam thicknesses of the

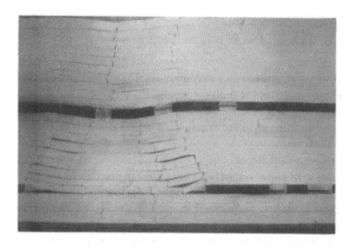

Fig. 7.4.14. Deformation and failure of the pillars in upper seam due to undermining lower seam.

entries with maximum convergence. The strength of the coal pillars in the vicinity of the caved roof was so deteriorated, that they supported only the load directly super-imposed on them.

c) The model showed that in the areas of maximum deflection of upper seam and corresponding strata vertical tensile failure was induced. This failure is a result of existence of extension stresses in opposite directions. This type of failure has been observed in the mine of Number 11 seam, which had maximum openings up to 5 cm.[53]

d) The extracted are of the lower seam is delineated by the planes of angle of draw, which are approximately at 75°. These planes intersect mine pillars in the upper seam and they are further propagated into strata above. In vicinity of the cave planes the pillars are splitted and fractured by tensile failures. This type of strata deformation is most damaging for coal pillars, which are in an advanced state of the extrusion deformation.

e) Underground monitoring showed that in contrast to the pillars above extracted area, which are in extension, the pillars above solid coal below are in compression. This phenomena also was exhibited on the plain-strain model, by absence of tensile failure and compaction of model structure. A small floor heave is noticable in the entry above an unmined pillar in the lower seam. The same phenomena was observed in the entries above solid coal, where floor heave had been up to 25 per cent of entry height. The pillar deformation in these areas was evidenced by coal sloughing.

This modeling technique, in conjunction with underground observation, could be a satisfactory tool for examining caving mechanics and surface subsidence, which agrees with investigations of other mining systems.

CHAPTER 8

Longwall mining

This is the most common method for underground coal mining where engineering has advanced technological capabilities. The extension of the application of longwall mining to pitching and steep pitching coal seams with moderate thicknesses opens up a new era of mining. It would be fair statement to say that to completely cover the topic of longwall mining a separate volume would be required. This section has been limited to the principles of longwall mining and strata mechanics under the four following chapter headings:

1. Principal mining systems with equal consideration to advancing and retreat sequences, layouts and productivity.

2. Stability of panel stope structure with a summarizing concept of longwall support load, types of powered face support and an approximation of hydraulic support as a rheological system.

3. Strata mechanics in relation to stress redistribution in longwall mining and the resulting stress patterns in the abutments, face-ends and gobs. The effect of stowing on stress redistributions and the interaction of stresses between longwall faces.

4. Caving mechanics in longwall mining and physical modeling of strata caving, mining subsidence and controlled caving and subsidence.

8.1. LONGWALL MINING SYSTEMS

Longwall mining has a long history of successful applications, even in thin and pitching coal seams. This type of mining is more mechanized than any other method, and necessitates careful attention to the selection of the expensive equipment required. Longwall mining is a unique method with one principal variation, the direction of coal extraction:

– Longwall advance method. Roadways are driven a short distance ahead of the advancing face and they have to be maintained behind the face in the gob.

– Longwall retreat method. Roadways block-out the extraction panel which is then mined in retreat. Roadways are abandoned in the gob area.

– Coal face descriptions and productivity figures are given to describe basic longwall concepts, but cannot be taken as state-of-the-art because longwall mining is in a state of constant new development.

Longwall mining has been the preferred coal mining method in many countries, especially in Europe. Longwall mining is favoured for such adverse mining conditions as occur in deep seams or where roof and floor conditions are poor. Room-and-pillar mining prevails in the United States which enjoys sound roof and good geological conditions. However, because of the inherent productivity potential and continuity of production, longwall mining is gaining favour in the United States, particularly in deep areas where recoveries are low for room-and-pillar systems.

8.1.1.   *Longwall advance mining*

Longwall advance mining has been primarily used in the deeper underground mines where strata pressures do not permit maintaining roadways for long periods of time.[1]

The majority of coalfields in Europe use the longwall advance system of mining. The coal seam is divided into panels, generally 100 to 230 m wide by up to 1800 m long. Production may commence following a minimal capital outlay for pre-production development. Yet the geological conditions ahead of the advancing coal face may be uncertain, thus introducing an element of risk. Any sudden worsening of geological conditions may cause the production face to halt and an equipment capital outlay of 5 to 6 million dollars (1979) can be temporarily at a stand still. Shallow mining depths are not favoured for longwall advance mining; however, weak strata may require its use even though it may not suit North American requirements for high productivity. There are several approaches to longwall advance mining, two of which are briefly discussed below.[2,3]

1. Advance system with single entry. The single entry is driven only a short distance ahead of the advancing face (up to 3 m) to avoid excessive frontal abutment pressures (Fig. 8.1.1). The advance of roadways has been greatly improved through the use of the longwall shearer for roadway excavation (Great Britain).

The main disadvantage of longwall, advancing single entry mining is maintaining the roadway behind face in the gob for the life of the panel. Roadway support is

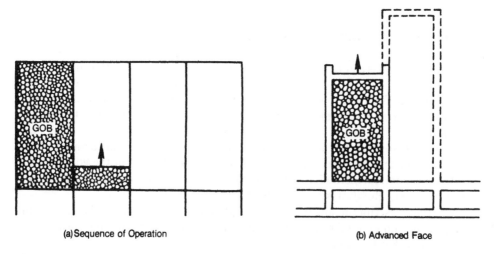

(a) Sequence of Operation   (b) Advanced Face

Fig. 8.1.1. Advance system with a single entry.

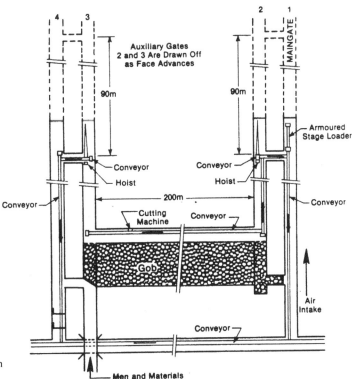

Fig. 8.1.2. Advance system
with double entries.

provided by arches set in the roadways and packs built at least along the gob edge. The
application of the Pump Pack for pack building has reduced the difficulties relating to
roadway maintenance e.g. placing pack material.

There is no opportunity to utilize bleeder return ventilation which is an important
advantage of the longwall retreat system. There is generally a need to install methane
drainage facilities.[4]

2. Advance system with double entries. These have rib pillars with a least width equal to
or greater than one tenth (1/10) of the panel depth separating panels. The ribs provide
roadway protection against strata pressure deformation effects.[1] The driving of double
entries in advance is integrated with the transport of the coal from the longwall face
(Fig. 8.1.2). The main advantage of this system is that there is no need for roadway
maintenance because one collapse is with the gob and the other in the rib is not affected
by gob closure.

This mining system requires more development work, but this is more than offset by
the savings in roadway maintenance.

8.1.2. *Longwall retreat mining*

Longwall retreat mining is basically the same as longwall advancing extraction, except
that the coal seam is blocked-out and then retreated in panels between development

roadways. Its advantages over advance mining are low risk and consistently high output. However, there are factors which limit the application of retreat mining. The most important of which is the development of high stress levels due to the influence of nearby workings which affect the stability of development roadways in soft strata, especially at depth. The life of the coal face depends upon the life of the roadway gate support. Reinforcement techniques are available to assist in stabilizing the mine roadways.[5]

1. A retreat system with a single entry is similar to the advance system with one entry, except that the panel is fully developed before extraction starts. There is a problem of roadway maintenance near the gob.

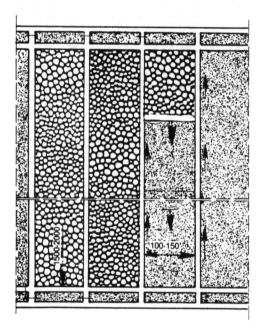

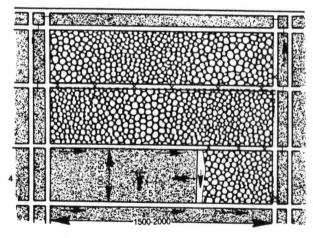

Fig. 8.1.3. Retreat system with a single entry along dip.

This method is prevalent in the U.S.S.R. and has the advantages of economical use of roadways and the efficient recovery of coal reserves. The mining direction is either down-dip or along strike (Fig. 8.1.3). The disadvantages of the system are that the developed roadways in solid coal are liable to interaction from neighbouring workings in the same seam: and the panel in extraction must be mined-out before the next one can start to avoid short circuiting ventilation.[6]

This system as well as other single entry systems offer total coal extraction in the panel area because rib pillars are eliminated.

2. Integrated advance and retreat systems are used mostly in deeper and gaseous coal mines. Single entries are used resulting in limited development and easier face-end operations. Alternate faces advance in opposite directions (Fig. 8.1.4). This method, as in other single entry longwall methods, re-uses the roadway of the mined-out panel for extraction of the adjacent panel. In some countries, of which Poland is one, integrated single entry systems have been used to control surface subsidence strains. Towns have

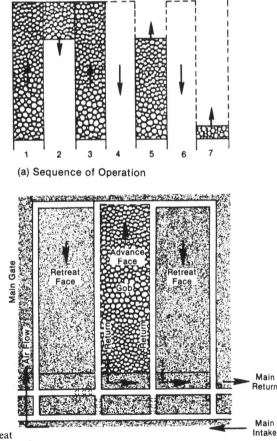

(a) Sequence of Operation

Fig. 8.1.4. An integrated advance and retreat system.

(b) Ventilation System

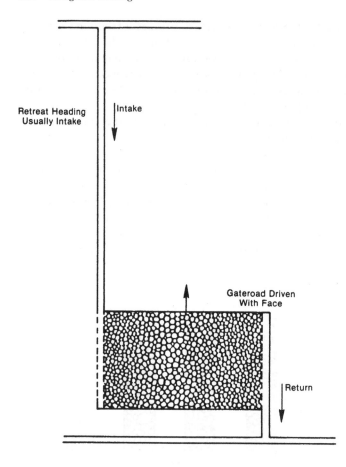

Fig. 8.1.5. Retreat 'Z' system.

been subsided without structural damage. The width of panel faces is between 100 and 150 m and lengths can be up to 2.500 m.

3. The retreat 'Z' system has often been used in coal mines to retreat the starting panel. One roadway is developed either the maingate or tailgate before the coal face commences production. In Figure 8.1.5, the maingate is driven and the tailgate is formed at the face-line.

This system is applicable in weak friable strata because one roadway is formed behind the face-line and development work is reduced by 50 per cent. The retreating maingate may use a bleeder roadway where high methane emissions in the face and gob are expected. The advancing main roadway may be used again for the next face. A disadvantage is that advancing of the roadway is a limiting factor in rapid face advance.[2]

4. The retreat system with double entries is the most commonly used longwall mining method in European countries where strata conditions are favourable. It is similar to the longwall advance system with double entries (Fig. 10.1.6).

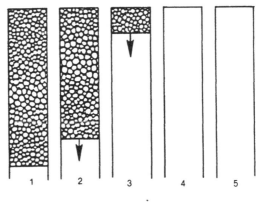

(a) Sequence of Operation

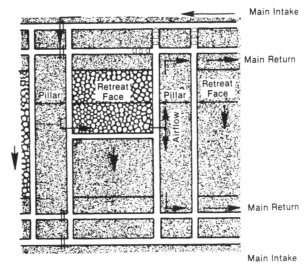

Fig. 8.1.6. Retreat system with double entries.

(b) Ventilation System

Ideally, work in the first retreat face should be finished before the headings for the next adjacent retreat face are driven to avoid roof control difficulties in the headings when they advance within 70 m of the retreat face. This is not generally possible because of the need to have a replacement face ready to take over immediately each retreat face finishes.

Where double entry development headings are driven by crosscuts, the outer headings can be used to service the retreat face on both sides of the intervening pillars, due to account should be taken of the depth of cover to plan adequate pillars. It is important to prevent the first retreat face from influencing the stability of the crosscut junctions.[2] Such junctions should be kept in minimum to avoid support problems at face-ends when the second face retreats. This can be a major problem at depths greater than 300 m.

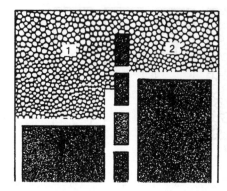

Fig. 8.1.7. Re-use of the gate roadway system.

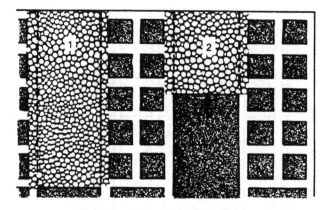

Fig. 8.1.8. Retreat systems with multiple entries.

5. A retreat system with the re-use of entries is applicable to shallow depths and good strata conditions. Some form of pack area support is required to protect the roadway against closure while the gob caves and consolidates. Materials for pack support usually takes the form of wood chocks or concrete blocks. A double row of wood props supporting parallel lines of half-round wood splits set almost touching each other, has been a low-cost and effective alternative method used extensively in West Germany as well as in the UK. Re-use of a retreat roadway in this form of layout requires competent roof conditions preferably sandstone or a strong slatey siltstone. Most layouts require a rib of coal and newly formed retreat headings to avoid delays arising from roof control problems at the face-end of the second retreat face (Fig. 8.1.7).[1]

6. A retreat system with multiple entries is the typical layout for longwall mining in the U.S.A. This system separates entry development from coal face operations and represents an integration of both pillar and longwall mining concepts (Fig. 8.1.8). The development entries usually consist of triple roadways and continuous miners with room-and-pillar mining technology are used to develop the entries. The longwall is mined in retreat with a shearer type of machine and the roadway towards the gob area is allowed to collapse with the gob.

The ventilation system can become complex due to the existence of many entries and the need to provide positive ventilation to individual workings.

The least width of the chain pillars has to be determined to protect the roadways of the retreating face against excessive closure effects from the abutment strata pressure zone. This method also requires good roof conditions to avoid roof control problems at crosscut intersections during longwall retreating.[5]

The large variety of longwall retreat systems exist as a result of unique coal seam conditions and the philosophy to minimize coal extraction. Any chosen longwall system for a new coalfield is usually modified to suit local geological and mining conditions.

8.1.3. *Longwall face and production*

Generally, the coal face is the key factor in mine productivity which depends a great deal on equipment reliability, and the training and experience of the operators. The original design of the coal face must be flexible enough to cater for potential improvements to minimize downtime.

The presentation of the technology of the coal face and its productivity should be viewed with the understanding that information on productivity lacks behind improvements in mining equipment and that productivity is also influenced by local geological features.

1. Coal face cutting machines. In the 60's and 70's single drum shearers were used, some of which are still in use. However, by the late 70's the most commonly used coal cutting machines for longwall mining were the Shearer Loader and in thick seams, the Double Ended Ranging Drum Shearer (DERDS).[3]

Ideally, cutting machines should be designed to cut at low drum speeds with no breakdowns and a minimum of maintenance. Attention should be given to visibility to enable the operator to control and manipulate the machine with ease. Radio control has been used.

There should be an emphasis on robustness and simplicity in the design and an abosolute minimum usage of hydraulics. There should be provisions for dust suppression.

Automatic sensing devices, such as nucleonic steering, have been incorporated on a fixed drum shearer to cut to the required level at the roof. Good results can also be achieved by an experienced operator. The use of shield supports may offer the extra room for the operator to maneuver so that he can concentrate on cutting coal. This extra room will be a decided disadvantage if a haulage chain is used. The violent vibration of 'whipping' of the haulage chain is always a potential hazard as is the breakdown of the chain under tension. Hence, horizontal rack-a-track system, eliminating the haulage chain, be considered when using 'caliper' type shield supports.

In the near future it is anticipated that coal face cutting machines will require up to 1200 HP, 4160 volts motors and weigh up to 50 tonnes. The power of the machines increase with the increased demand for slower shearer drum speeds to reduce dust production. Drum speeds of 33 revs per min. have been used successfully but the optimum is considered to be 21 revs per min. The optimum is a compromise between the speed of the drum required to maintain a cutting advance of up to 0.11 m/s and the need to minimize dust.[5]

2. The armoured flexible conveyor (AFC) is as important as the cutting machine.

The AFC should be designed to transport coal continuously and be strong enough to withstand the forces of the cutting machine and the push-over rams of the face powered supports. A large shearer weighs as much as 25 tonnes and can warp the pan deck plates, especially wide 760 mm pans. Closed bottom pans with the bottom plate welded across the width of the pan to from a strong box section helps prevent warping.

Generally, the AFC is the weakest link in the production capacity of the longwall coal face. There are three factors to consider:

– The cutting capacity of the shearer. For example, a machine cutting rate of 9.95 m/s, a width of cut of 9.7 m and a height of extraction of 3 m, gives a production rate of about 780 tonnes per hour. Table 8.1.1 gives the details of typical capacities for two standard sizes of conveyor.

– For a DERDS, cutting in both directions along the coal face, the coal has to pass beneath the underframe of the shearer (Fig. 8.1.9). The coal conveyance capacity depends upon the relative velocity of the DERDS and the AFC. The capacity of the AFC is about $A(Vc - Vm)$ cutting towards the maingate and $A(Vc + Vm)$ cutting towards the tailgate. Assuming speeds of 9.95 m/s and 0.95 m/s for the DERDS and AFC respectively the above relations would give: $A(Vc - Vm) = 0.9A$ and $A(Vc + Vm) = 1.0A$.

As an example, a single longwall coal face might produce up to 24.000 tonnes of coal in a day. To achieve this figure AFC would have to run for 24 hours continuously at its capacity of 1000 tonnes per hour. Any stoppages would result in lost production as the AFC cannot run at overcapacity.

The current trend for AFC design is towards heavier duty pans. It is anticipated that the AFC will have a width of 1.0 m, open-section pans, a chain section of 26 to 34 mm, a conveyence speed of 1.37 m/s and a capacity of up to 1.500 tonnes per hour. The pans will be strengthened with a centre plate 2.5 cm thick.

3. Total mine and face productivity are the best measures of longwall mining performance. Here are several examples of underground coal mining performances in the U.S.A. and Australia. The extraction face geometries are illustrated in Figure 10.1.10 and the production rates are listed below:

a) Robinson Run Mine (Consol, No. 95). The coal seam, the Pittsburg, has a weak

Table 8.1.1. Typical values of AFC Capacities

AFC speed m/s	Capacity AFC size (760 mm)	Tonnes per hour AFC size (610 mm)
0.86	455	300
0.95	500	330
1.08	570	375
1.20	640	420

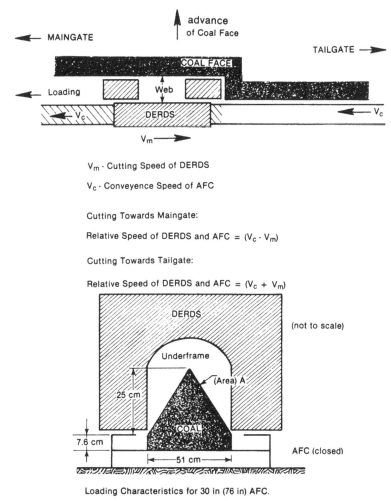

Vm · Cutting Speed of DERDS

Vc · Conveyence Speed of AFC

Cutting Towards Maingate:

Relative Speed of DERDS and AFC = $(V_c - V_m)$

Cutting Towards Tailgate:

Relative Speed of DERDS and AFC = $(V_c + V_m)$

Loading Characteristics for 30 in (76 in) AFC.

Loading Rate Cutting Towards Maingate   A. $(V_c - V_m)$

Fig. 8.1.9. Cutting concept of a coal face.

Loading Rate Cutting Towards Tailgate   A. $(V_c + V_m)$

shale roof and originally a roof coal layer was left. The introduction of a shield support system, Eickhoff EDN 300L DERDS, permitted full coal face extraction. The following production peaks have been achieved:

Advance – 24 m in day;

Production – 12.395 tonnes in 24 hours

(January 1976);

Productivity – 23 miners and 4 supervisors for 3 shifts of production.

b) Shoemakes Mine (Consol, Wheding, W. Va). The coal seam was mined together with a weak slate false roof. With shield support only the coal seam face is mined and rock rejects have been reduced from 30 per cent to 10 per cent of the face production.

Produced – April 1975 to April 1976 – 750.000 tonnes;

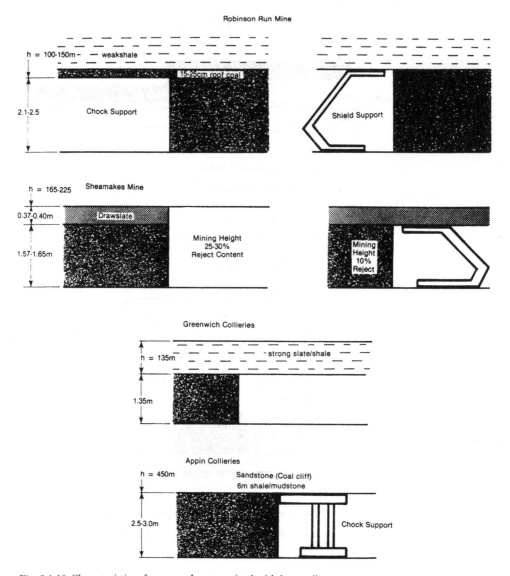

Fig. 8.1.10. Characteristics of some coal seams mined with longwall systems.

Daily production – 7.999 tonnes in 24 hours.

c) Greenwich Colliery, Pennsylvania. The coal seam with a height of 1.35 m has a strong slate/shale roof. The longwall has the following characteristics:

– Face support: G. D. 6 leg 500 tonnes chocks with a support density setting of 12 tonnes/m and a yield of 36 tonnes/m the setting pressure is 20.68 MN/m².

– Machine: 700 mm Eickhoff EW 170L SERDS taking a full cut in both directions in a 1.2 m thick seam of coal. The tailgate end needs hand loading.

– Chain Conveyor: AFC 760 mm wide, 2 × 125 HP drives Static ramp plates, single stand 26 mm chain, and a speed of 65 m/min.

– Gates: 1.3 m. Panel retreats average 1520 m and the average face length is 140 m.

It took 4 months to extract one panel, working 15 shifts per week. The face manpower per shift consisted of: 2 shearer operators, 3 chock operators, 2 maingate men, 1 tailgate man, 1 mechanic, 2 supervisors.

The performance statistics are listed below:

Average number of shears per day = 30;

Average production per day = 5000 ton (US);

Record production, October 1977 in one day = 7800 ton (US) on 3 shifts (47 shears).

Record production, 17/21 October 1977 in one week – 25.800 ton (US) on 3 shifts/day, 154 shears.

d) Appin Colliery, Australia. The coal seam is 3 m high with a 6 m shale/mudstone falce roof and is mined by the following longwall system:

Face Support-Chock Type 4 leg 720 tonnes C. D. type. Setting support density 15 tonnes/m at 22 MN/m² yields 45 tonnes/m.

Machine: Anderson Struthclyde DERDS-200 HP Drums = 1.95 m dia

Chain Conveyor: AFC 9.75 m wide 2 × 150 HP drives. 63 m/min Westfalia MM-600 twin centre strand. 26 mm chain. Open bottom with 20 mm deck plate-Static ramp plates.

The performance of this longwall face is listed below:

18 shifts per 5 day week on production;

2 shifts per 5 day week on maintenance;

Face manpower per 24 hours: 25 miners, 4 supervisors, 4 electricians, 7 mechanics, (40 total).

Total manpower 350–363;

Colliery productivity OMS 9.9 tonnes;

Face productivity OMS 54.2. tonnes;

Best continuous performance over 180 days was 3030 t/day;

Best single day's performance – 20.5000 tonnes;

Overall cutting length – 129 m;

Face retreat distance – 1515 m;

Total face extraction – 780.600 tonnes.

This list of face size, equipment and productivity should be consider only for instructive purposes. There was no attent made to update the list to current performance levels. Coal production and productivity should be estimated for high speed retreat faces with a large production potential at the optimum capacity of the coal handling system.

8.2. STABILITY OF PANEL-STOPE STRUCTURES

Panel stability became the subject of investigation for many researchers, mostly in Great Britain and the Germany, after the introduction of hydraulically operated powered supports in the late 40's. Practical mining experience showed that the effectiveness of powered supports was dependent on their interaction with the panel-face structure. This is discussed under the following headings.

– The concept of longwall support load. The key factor for the design of an effective support system for the given geological conditions.

– Descriptions of longwall face supports. The correct choice is of paramount importance. The capital investment for setting up a production longwall face is very high and once the support has been installed there is a very limited scope for change.

– The approximation of hydraulic support as a rheological system. The interaction between self-advancing supports and geological setting is becoming more and more appreciated as a design criterion.

8.2.1. *Concept of longwall support load*

The fundamentals of support load have been studied by Ashwin et al.,[7] and Whittaker.[8] Generally speaking, there are two approaches used to estimate support requirements:

a) The size of the largest block liable to need support is estimated, with the strata load being the dead weight of the rock. The size of the block is influenced by the workable seam thickness, the strength characteristics of the roof strata, the support residence time, rate advance of coal face, and the spread of props along the face (Fig. 8.2.1).

b) Measure the loads on supports which have been found to be satisfactory in practice. This concept is based limiting roof strata convergence to experienced safe values (Fig. 8.2.2).

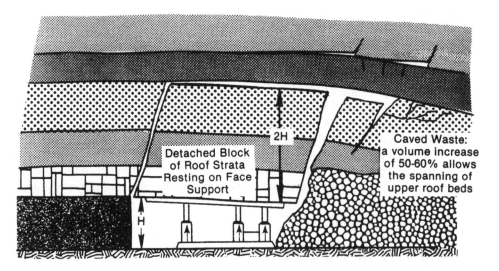

Support Loading = 2H (9.8 x 2.3 x 10⁻³) MN/m²

= 0.04H MN/m²

Min. Setting Load = 0.08H MN/m²   For S = 2

Fig. 8.2.1. Roof caving and strata loading on supports (after Ashwin et al.).

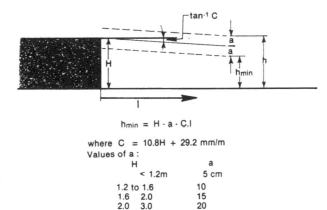

hmin = H - a - C.l

where C = 10.8H + 29.2 mm/m
Values of a :

H		a
	< 1.2m	5 cm
1.2 to	1.6	10
1.6	2.0	15
2.0	3.0	20

Gig. 8.2.2. Roof convergence and strata loading on supports (after Ashwin et al.).

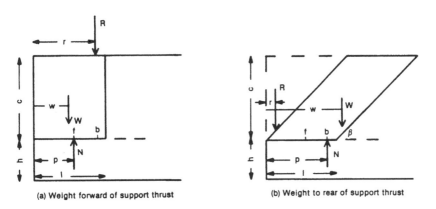

(a) Weight forward of support thrust   (b) Weight to rear of support thrust

Fig. 8.2.3. Basic roof block assumptions and forces involved for flat coal seams.

The following section on estimating the support loads in flat and inclined seams is based on the concepts discussed by Whittaker[8, 9, 10] and Wilson.[11, 12]

1. The support load in flat seams is estimated using a simplified version of the equations derived by Wilson.[12] To analyze the forces acting on the roof block, it is necessary to examine the equilibrium of the basic system as shown in Figure 8.2.3. In the calculations of the dead load, it is assumed there is a face break and a parting exists between the lower and upper roofs. The angle of the face break is assumed to be equal to the break angle β as this is indicative of the geology of the roof. The caving height c is taken as twice the seam thickness, unless information to the contrary exists. The width l of the block is taken as the width of the working area.

The support must keep the freed block in contact with the upper roof. It is important to note that the block itself does not need to be solid; it can be made up of many smaller blocks.

The forces acting on the block are:

a) The body weight which is taken as a single force acting at the centre of gravity, a distance w from the face line.

b) The support resistance which is taken as the resultant force N acting at a distance p from the face line and in a direction normal to the roof.

c) A reaction R between the freed block and the strata above, acting at a point a distance r from the face line, and assumed to be vertical.

The support resistance N is then divided into the individual forces F and B supplied by the front and back legs at distances f and b respectively from the face line.

From static equilibrium considerations:

$$N = W + R \tag{1}$$

$$N \cdot p = R \cdot r + W \cdot w \tag{2}$$

$$F + B = N \tag{3}$$

$$F \cdot f + B \cdot b = N \cdot p \tag{4}$$

Which give:

$$F + B = W + R \quad \text{and} \quad F \cdot f + B \cdot b = R \cdot r + W \cdot w \tag{5}$$

This means that $F + B$ will be a minimum when R is a minimum. If the weight acts between the front and back legs ($f < w < b$) then R can be reduced to zero by:

$$F = \frac{W(b - w)}{(b - f)} \tag{6}$$

$$F = \frac{W(w - f)}{(b - f)} \tag{7}$$

If the weight acts ahead of the front legs ($w < f$) then R will be a minimum when:

$$F = \frac{W(r - w)}{B(r - f)} \tag{8}$$

where

r is as large as possible.

If the weight acts behind the back legs ($w > b$) then R will be a minimum when:

$$F = 0 \qquad B = \frac{W(w - r)}{(b - r)} \tag{9}$$

where

r is as small as possible.

If the angle of break is equal to the caving angle then:

$$W = \gamma \cdot S \cdot \ell \cdot c \quad \text{and} \quad w = \frac{\ell}{2} + \frac{c(\cot \beta)}{2} \tag{10}$$

where

γ = unit weight of the roof material;
S = support spacing along the face.

To prevent excess convergence the support yield loads must be at least the calculated loads. If the support unit load capacity is known then its adequacy in a given situation can be tested from the equation (5) as follows:

$$Rr = Ff + Bb - Ww \tag{11}$$

$$r = \frac{Ff + Bb - Ww}{R} \tag{12}$$

$$r = \frac{Ff + Bb - Ww}{F + B - W} \tag{13}$$

If r lies outside the postulated free roof block then the support may be inadequate. It must be sufficiently far over the block to prevent corner crushing.

If the face break is assumed to be vertical high caving angles have little influence on the value of r. Low caving angles, on the other hand strongly influence the system. In thicker seams it may be necessary to increase the back leg load by up to 80 per cent for cave angles in the order of 30°.

2. The support load in step seams is dependent on whether the extraction direction is up-dip, down-dip or along strike. As illustrated in Figure 8.2.4, the forces acting on the

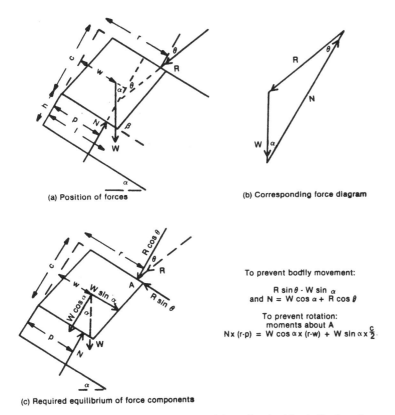

(a) Position of forces

(b) Corresponding force diagram

(c) Required equilibrium of force components

To prevent bodily movement:

$R \sin \theta \cdot W \sin \alpha$
and $N = W \cos \alpha + R \cos \theta$

To prevent rotation:
moments about A
$N x (r-p) = W \cos \alpha x (r-w) + W \sin \alpha x \frac{c}{2}$.

Fig. 8.2.4. Basic roof block assumptions and forces involved for inclined coal seams.

block are similar to those in a flat seam except that the seam inclination is introduced as a new variable. The forces acting on the block are as before, except that the reaction R is now inclined to the normal by the angle θ, (Fig. 8.2.4). A sliding mechanism of failure is possible.

For equilibrium, the force triangle in diagram (Fig. 8.2.4/b) must form a closed figure and the projected lines of the forces in diagram (Fig. 8.2.4/a) must meet at a point or equilibrium:

$$N = W \frac{\sin \alpha}{\tan \theta + \cos \alpha} \tag{14}$$

where

$\alpha =$ the seam inclination;
$N =$ the support reaction.

The reaction R can be found from:–

$$R = \sqrt{W^2 + N^2 - 2W \cdot N \cos \alpha} \tag{15}$$

Eliminating θ for the condition that the three forces should intersect at one point:

$$N = \frac{W(c \sin \alpha)}{2} + \frac{(r - w)\cos \alpha}{(r - p)} \tag{16}$$

Using the component diagram (Fig. 8.2.4/C) the following failure mechanisms are defined:

To prevent the block sliding down the slope

$$R \sin \theta \geq W \sin \alpha \tag{17}$$

To prevent movement normal to the roof

$$N \geq W \cos \alpha + R \cos \theta \tag{18}$$

For the limiting case when both these movements are just about to occur:

$$N = \frac{W(\sin \alpha)}{\tan \theta + \cos \alpha} \tag{19}$$

The limiting case for safety is when the angle θ is equal to the angle of friction ϕ:

$$N = \frac{W(\sin \alpha)}{\tan \phi + \cos \alpha} \tag{20}$$

where

N is the minimum setting load required.

This equation is independent of p, r, w and β, it depends only upon the magnitudes and directions of the forces.

If ϕ is assumed to be 22°, then the Mean Setting Load Density S can be found from the equation:

$$S = \frac{h}{20} \left(\frac{\sin \alpha}{0.4 + \cos \alpha} \right) \tag{21}$$

(taking $\gamma = 2.5 t/m^3$, $\tan \phi = 0.4$ and c = 2h).

The value for the seam inclination used should be the true dip of the seam, and not the apparent dip of the face.

For the calculation of minimum support yield loads a graphical solution is preferable. A scaled cross-section of the face with assumed roof block is used.

For individual front and back leg loads the following relationship is used:

$$F = \frac{N(b-p)}{b-t} \quad \text{and} \quad B = \frac{N(p-f)}{b-f} \tag{22}$$

N will be a minimum when θ is made as large as possible subject to it not exceeding 68°.

The concept of support load has been developed for the powered supports which revolutionized longwall mining. It should be pointed out that similar considerations can be given for estimating the dead load on the gateside packs by assuming that the roof block is supported entirely by the pack system. Pack loading in flat seams and pitching seams have been described by Whittaker and Woodrow.[10]

8.2.2. *Face powered roof support*

The object of the face powered support is to separate the caved area from the coal face, providing a safe working area. For the travel of mine personal and the cutting machine the cross-sectional area of the supported opening must also be adequate for mine ventilation.

Powered self-advancing supports can be classified into two groups:

1. Frame and chock supports are advanced forms of the hydraulic prop. They became a major milestone in mining technology by introducing preloading and the capacity to yield without loss of stability. They also provided the basis for the development of hydraulically operated self-advancing roof supports.[11, 12]

A frame type support is a set of hydraulic jacks sharing a common roof canopy and floor base. The support advances when the front set of jacks are released and pushed forward by the shifting cylinder. The front jacks are then set. The rear jacks are released and then pulled forward by the shifting cylinder and reset. The loading conditions of the frame complex are illustrated in Figure 8.2.5. The advance of the frame support is independent of the face conveyor or spill plates. At the present time they are used mainly in shortwall faces, which are restricted to the frame and chock type of support supplemented with a retractable extension canopy.

Chock type supports also have hydraulic jacks connected by a common roof canopy and floor base, but self-advancement depends on an attachment to the face conveyor spill plates. They are advanced by horizontal double-acting hydraulic cylinder into the base and coupled to the conveyor panline. Chock advance begins when the cylinder is extended to its full length to advance the conveyor panline or spill plates. The chock legs are lowered until the roof canopy is free and then pulled forward for jacking against the roof in the new position. The loading conditions of chocks are illustrated in Figure 8.2.6.

A range of hydraulic jacks or power-set legs for chock support are available from 30 to 200 tonnes. They may be single or double telescopic, and both single or double acting.

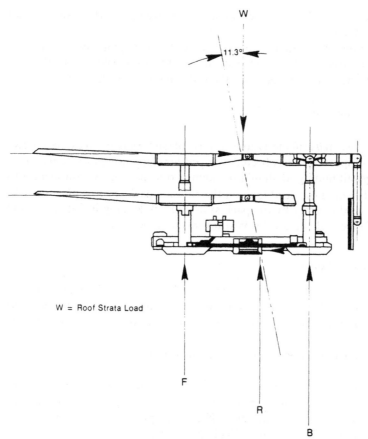

W = Roof Strata Load

Fig. 8.2.5. Frame type support.

There are two factors in loading, the setting load and the yielding load. High setting loads may damage a weak roof by either the direct load or the cyclic setting and releasing operations. On the other hand, too low a setting load may allow the strata to generate high loads on the powered supports. The yield load determines the overall structural strength of the support unit.[13]

Powered supports must be designed to accept sudden (dynamic) loadings. A safety factor of 2 is applied in the design of support units. For times when the rate of loading exceeds the yield rate of the unit.

Table 8.2.1 gives a summary of the recommended setting and yielding load densities, based on the relation given by Whittaker.[8]

Although there are many variables to consider, it is generally accepted that chock supports should be used where the roof is a strong sandstone and shield supports where the roof is weak and friable and a high flushing factor is prevalent.

In the U.S.S.R. chock supports are preferred for longwall mining in flat and thin coal seams. For example, the Donbass coal mines generally use chock supports characterized by a high ratio of square canopy to the area of supported roof. This produces increased side stability and low specific pressures exerted on both the roof and floor for high setting loads. The characteristics of this support are listed as below:[14]

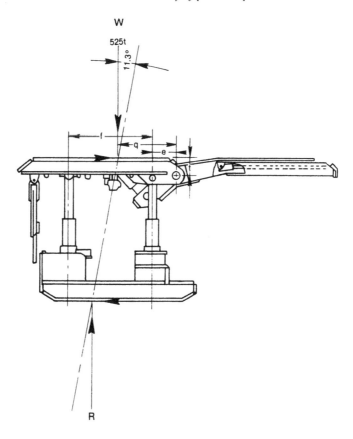

Fig. 8.2.6. Chock type
support.

Table 8.2.1. Recommended setting and yield load densities

Seam thickness	Min. Setting Load Density (SLD) MN/m	Nominal SLD MN/m	Nominal Yield Load Density MN/m
2 m	0.16	0.21	0.26
3 m	0.24	0.32	0.34
4 m	0.32	0.42	0.52

Minimum settingload density (SLD) = Nominal (SLD) + 33%*; *Losses in hydraulic system in setting load; Nominal yield load density = Nominal SLD × 1.25

a) The chock self-advancing support is equipped with a hydraulically extended canopy to provide immediate support of newly exposed roof without releasing the legs. The canopy is extended immediately after the power-loader passes and supports the exposed roof in the area of the conveyor snake. Twelve to fifteen metres behind the power-loader, the armored face conveyor is pushed to the face and the chocks pulled to it.

b) It is equipped with double acting double-telescopic legs with a minimum height of 500 mm. Yield is constant over the full travel range.

c) There are mechanical devices for advancing the system which automatically align and position the support along the conveyor to prevent creep and slip.

Self-advancing hydraulic supports integrated with an armoured flexible conveyor having a single chain and ramp plates have the potential for remote control.

It is of interest, that some of the Donbass mines with geological irregularities show a preference for a frame type of support, i.e. no conveyor link. Pairs of double-leg units connected together with the advancing hydraulic jack fitted in the canopy between the units. The upper position of the hydraulic jack ensures that the support can follow the changing lines of the floor and roof. It allows for easy travel along the face, and also reduces the weight of the assembly base. This type of support can cope with seams of variable thicknesses with undulating floors and small fault displacement.

2. Shield supports have an inclined plate whose lower end is hinged to a horizontal base plate setting on the floor. The upper end is hinged to a horizontal roof canopy. The self-advancing mechanism is similar to that of the chock support with the horizontal double acting hydraulic cylinder located between the two hydraulic legs. There have been three main types of shield supports recently developed:

a) The chock-shield support combines the advantages of the two-leg shield and four-leg chocks. Figure 8.2.7 shows a Dowty type. There are several variations

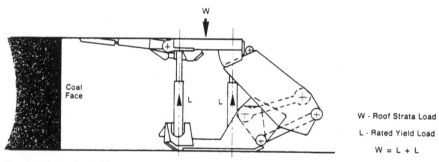

W - Roof Strata Load

L - Rated Yield Load

$$W = L + L$$

Fig. 8.2.7. Chock-shield support.

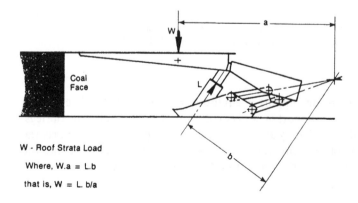

W - Roof Strata Load

Where, $W.a = L.b$

that is, $W = L. b/a$

Fig. 8.2.8. Caliper shield support.

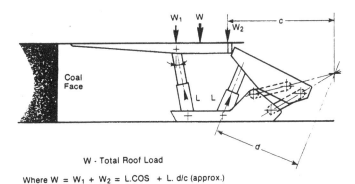

Fig. 8.2.9. Four legs shield support.

W - Total Roof Load

Where $W = W_1 + W_2 = L.\cos \alpha + L . d/c$ (approx.)

Table 8.2.2. Operating data for shield support (after Dowty)

Height Range	Chock shield			2-Leg caliper shield			4-leg shield		
(metres):	I	II	III	I	II	III	I	II	III
Closed	0.94	1.69	1.75	0.89	1.18	1.60	0.70	0.86	1.83
Open	1.93	2.79	3.40	1.80	2.68	8.61	1.60	2.16	3.66
Total yield load (tonnes)									
Minimum	280	300	300	189	226	310	183	250	290
Maximum	280	300	300	224	280	316	207	284	290

depending on whether all the legs react between the top canopy and the base or some of the legs react between the gob shield and the base.

b) Caliper shield supports connect the base to the inclined plate (gob shield) with two power-set legs, Figure 8.2.8. The linkage between the base and the gob shield maintains a constant distance between the face and the canopy tip when the hydraulic cylinders move either up or down.

c) In four-leg shield supports, the two rear legs react between the base and the shield while the two forward legs react between the base and roof canopy, Figure 8.2.9. Placing the top end of the rear legs on the gob shield allows the canopy to remain short while providing extra space between the rear and front legs.

All shield support types have common elements, a canopy, base, power set legs, gob shield and a control system. Table 8.2.2 gives a summary of the characteristics of the each shield support type.

The advantages and disadvantages of shield supports are listed below:

a) Advantages:

– Maintenance costs are low compared with the chock supports.

– Their low closed height is an advantage for transport action underground. This is important in the United States with height restrictions of about 1.3 m on locomotive roads in coal mines.

– Face installations are easier than for chocks.

– They accommodate variations in seam thickness better than double telescopic chock supports.

– There is a short face to waste distance, advantageous for weak friable roofs.

– Construction is simpler.

– They provide complete separation of the caved roof from the working place protecting the face area against dust and heat flow from the caved gob from roof falls at the face.

– The delay in setting supports is low.

– There is easier travelling along the face because there are fewer hydraulic hoses than for chock supports.

b) Disadvantages:

– High initial costs. A shield support may cost in the order of $30.000 to $40.000 per meter (1978, FOB).

– Shields give a low cross-sectional area at the face. This often results in high air velocities and dusty conditions when the face is producing.

– Shields are often employed in poor ground condition faces. Withdrawal (salvaging) may be a problem. Standard practice is to set timber on the face for 3 cuts of advance, leaving the shields stationary to produce a track wide enough to remove the supports.

– The shields are generally close together (skin to skin) and it is often necessary to grade the floor of the seam. This can be difficult for hard floors.

– Gradients of up to 30° have been worked with shield powered supports, but difficulties may be encountered with steeper seams.

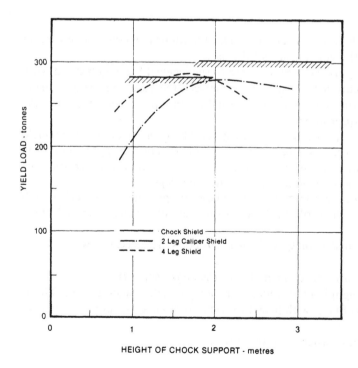

Fig. 8.2.10. Load to height relationships for several types of shield supports.

Shields have a high resistance to lateral movement, and good anti-flushing arrangements. They should be used where ground conditions make roof control difficult. Shields may also reduce the need to leave valuable coal tops.

Typical, yield load/height of shields, values are given for three types of shield support (Fig. 8.2.10).

Shield supports, due to their excellent stability have been extensively developed in the U.S.S.R. for longwall mining steep and thin coal seams which previously were mined by conventional open stope mining systems. As an example, the coal seams of Danbass presented difficult mining conditions because of faulting (frequency 1.45 faults per 1000 m along strike with fault lengths of up to 200 m), weak roof and floor strata (60 per cent of the seams), variable seam thicknesses (0.4–2.2 m), an average mine depth of 665 m, and high gas contents of up to 100 m^3 per tonne of coal. These were the main reasons for delaying the introducing of mechanized mining. However, developments in power-loaders and self-advancing shield supports permitted mechanized longwall coal extraction along both strike and dip under these conditions. Longwall mining with shield supports in the steep and thin coal seams improved working conditions and safety at the face.[14]

It should be pointed out that shield supports for longwall mining have also been extensively developed in West Germany for faces advancing along strike in seams dipping up to 50°.

8.2.3. *The approximation of hydraulic support as rheological system*

Hard rock mining practice suggests that there is an interaction between self-advancing longwall supports and the adjacent rock strata which can vary with the geological pecularities of individual coal seams. For example, where the roof strata does not cave readily, the closure rate may suddenly increase beyond the yield rate of the support. Peak loads will exceed the yield setting and may smash the support. In South Africa, very high yield rate props based on aircraft landing strut designs have been used successfully under rockburst closure conditions.

The interactions of supports and the adjacent rock strata could be studied with rheological mathematical models which could simulate the various types of support and rock strata mechanics investigations suggest there are several rheological models which could be used to represent different rock types. The self-advancing support could be represented by rheological models and under these circumstances the individual elements of the model are, for example: elastic-bearings, roads; plastic-joints, valves, wedges...; viscous-hydraulics.[15] Figure 8.2.11 illustrates the various elements used in rheological models. The operational characteristics of hydraulic supports requires the use of elements characterized by constant stress and variable strain with the maximum strain equal to the sum of deformations of the individual elements. The stress and strain relationships provided below relate to model type VI in Figure 8.2.11.

$$\sigma = \sigma_E = \sigma_{v1} = \sigma_{v2} + \sigma_{p1} \tag{1}$$

$$\sigma_E = E\dot{\varepsilon}_E \; ; \sigma_{v1} = \eta_1 \dot{\varepsilon}_{v1} \; ; \sigma_{v2} = \eta_2 \dot{\varepsilon}_{v2} \; ; \sigma_{p1} = \text{const}$$

$$\dot{\varepsilon} = \dot{\varepsilon}_E + \dot{\varepsilon}_{v1} + \dot{\varepsilon}_{v2} \tag{2}$$

$$\dot{\varepsilon}_{v2} = \dot{\varepsilon}_{p1}$$

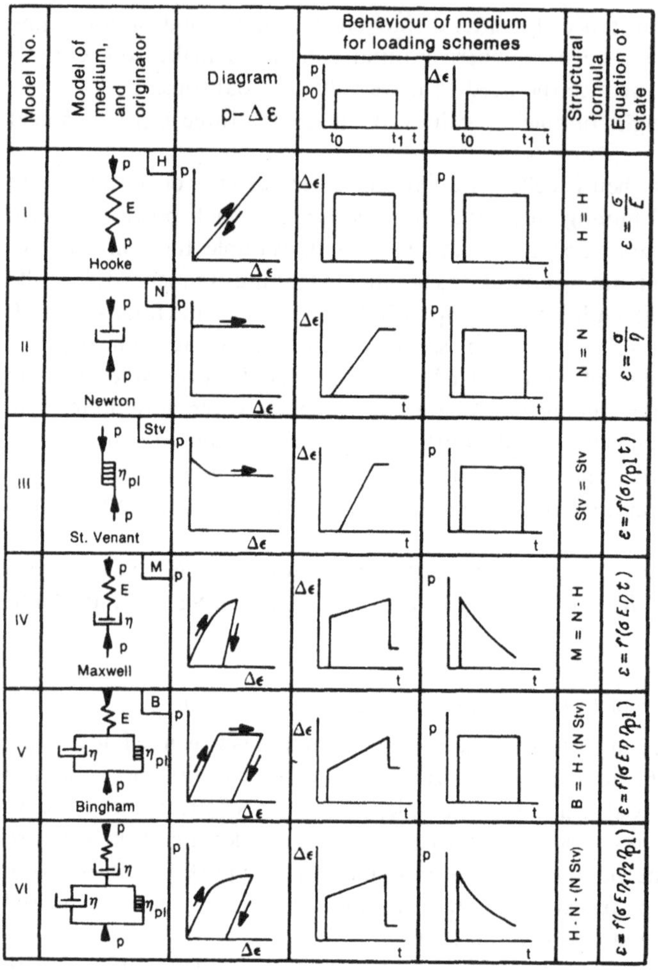

Model No.	Model of medium, and originator	Diagram p–Δε	Behaviour of medium for loading schemes		Structural formula	Equation of state
I	Hooke (H, E)		Δε	p	H = H	$\varepsilon = \dfrac{\sigma}{E}$
II	Newton (N)		Δε	p	N = N	$\varepsilon = \dfrac{\sigma}{\eta}$
III	St. Venant (η_{pl}, Stv)		Δε	p	Stv = Stv	$\varepsilon = f(\sigma/pl \cdot t)$
IV	Maxwell (M, E, η)		Δε	p	M = N · H	$\varepsilon = f(\sigma E \eta t)$
V	Bingham (B, E, η, η_{pl})		Δε	p	B = H · (N Stv)	$\varepsilon = f(\sigma E \eta pl)$
VI	(E, η, η_{pl})		Δε	p	H · N · (N Stv)	$\varepsilon = f(\sigma E_1 \eta_1 \eta_2 pl)$

Fig. 8.2.11. Rheological elements used to simulate operational characteristics of hydraulic supports (after Gritsko et al.).

where

E = Young's modulus;

η_1 and η_2 = coefficients of viscosity;

where the subscripts E, v_1, v_2 and pl stand for elastic, viscous and plastic elements.

Gritsko et al. considered the behaviour of hydraulic support as rheological model in the periods of increasing and constant resistance as illustrated in Figure 8.2.12. They considered the first period of operation of the hydraulic support could be approximated with a Maxwell model. If the deformation is kept constant, $\Delta\varepsilon = 0$ and $\sigma_0 = $ the setting stress, the relaxation curve can be described by the well-known expression:

$$\sigma = E\varepsilon_0 e^{(-t(E/\eta_1))} \quad \text{or} \quad \sigma = \sigma_0 e^{(-t(E/\eta_1))} \tag{3}$$

This shows that under zero closure conditions, the elastic stress could gradually decrease to zero due to leakage in the hydraulic props. Equation (4), relating overall strain to stress, is obtained from the sum of elastic, plastic and viscous deformations.

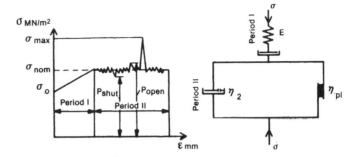

Fig. 8.2.12. Operational
characteristics of a
hydraulic support as
exhibited by a rheological
model.

$$\dot{\varepsilon} = \frac{\sigma_0}{E} + \dot{\varepsilon}_{p1} + \frac{\sigma}{\eta_1} \quad \text{or} \quad \dot{\varepsilon} = \varepsilon_0 + \delta + \frac{\sigma}{\eta_1} \tag{4}$$

In hydraulic prop, at the yield load and with less than critical strain rates, $\dot{\varepsilon}$ is governed not by the prop but by the stress strain relationship, for strain rates exceeding critical yield values. The support will rupture if stresses exceed the ultimate support strength.

It should be noted that constant resistance cannot equate to an increase in stress. The stress is increased because the hydraulic valving is too restrictive to allow the prop to yield fast enough.

If, at this stage, a uniform increase in stress is assumed such that:

$$\sigma = vt \tag{5}$$

and including this in equation (4) one obtains the expression:

$$\dot{\varepsilon} = \frac{v}{\eta_1}t + \left(\frac{v}{E} + \delta\right)t + \varepsilon_0 \tag{6}$$

$$\frac{\text{Total strain}}{\text{rate}} = \frac{\text{viscous}}{\text{strain}} + \frac{\text{plastic}}{\text{strain}} + \frac{\text{elastic}}{\text{strain}}$$

Equation (6) shows that deformation increases by a parabolic function and is determined by the deformation coefficients and the rate of the plastic element. For a constant deformation rate, equation (6) could be written:

$$\dot{\varepsilon} = wt \tag{7}$$

and by putting equation (7) into equation (4) and solving, one obtains:

$$\sigma = \eta_1(w - \delta) + (\sigma_0 - \eta_1)(w - \delta)_e - \left(\frac{t}{T_r}\right) \tag{8}$$

where
T_r = relaxation time.

The above equation shows that with a constant deformation rate the stress in a rheological model of a hydraulic support is largely determined by the difference between rate of deformation of the rock and the rate of deformation of the plastic element.[15]

8.3. STRATA MECHANICS

The strata mechanics of longwall mining have been extensively investigated around the world, particularly in Great Britain (The National Coal Board). There are numerous publications focused on strata mechanics in longwall mining and how better use can be made of existing knowledge to meet the changing strata control requirements of the mining industry. This section will be limited to providing an understanding of the basic mechanics of longwall mining operations under the following topics:

– Abutment stress at the coal face, the fundamentals of its formation and location.

– Strata stress around the longwall structure and the concepts of stress redistribution and concentration.

– Strata stress with stowing support for steep pitching and thin to thick coal seams. Abutment stress concentrations are affected by stowing.

– Stress concentrations around face-ends. This is an important aspect of longwall retreat mining and several layouts are considered.

– Interactive stress mechanics relates primarily to multiple seam mining which is a common case for coal bearing strata with thin to moderately thick coal seams.

8.3.1. *Abutment stresses at the coal face*

The distribution of abutment stresses at the longwall face depends on the structural characteristics of the coal seam and adjacent strata, mine depth, size of extraction panels, type of support and other lesser factors. The magnitude of abutment stress is usually related to overburden pressure which is generally high because the majority of longwall mining operations are at depths ranging between 300 and 1300 m.

1. Abutment stress at a face. In the simple case for coal seams which do not interact with other faces or mine structures, the development of abutment stresses can be explained by equilibrium where the load supported by coal prior to its excavation has to be transferred to solid coal around the opening as the coal is removed. Abutment stress distributions in the unmined coal are characterized by peak stresses near the rib edge which decrease in magnitude with distance into the solid coal.[16]

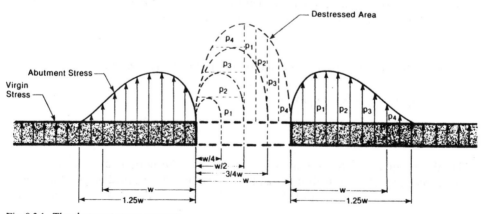

Fig. 8.3.1. The abutment stress concept.

There is a relationship between the initial size of the excavation and the magnitude and extent of the abutment stresses. The governing factor is usually the width of the excavation (w) as illustrated in Figure 8.3.1. The simplified analyses shows that for each lift of excavation, stress is released from the arch above the opening and transferred to both sides of the excavation. With increasing numbers of lifts, the destressing zone enlarges and transferred loads correspondingly increase. As an example for an extracted lift of width 'w' the abutment stresses are spreaded over lengths of 1.25 w on each side of the opening. In underground stress measurements around an initial excavation width of 10 m, the peak abutment stress was 1.3 times the field stress (σ_v). At a depth of 800 m, this amounts to 26 MN/m². Peak abutment stress values increased as the longwall face advanced.[17]

2. Traveling abutment stresses. During coal production the abutment stresses shift in the direction of the face advance. Figure 8.3.2 is a representative model of traveling abutment stresses. The magnitude of the peak stress and its location within the rib conform to Table 8.3.1.

The abutment stress increase on longwall faces is dependent on the geometry of the gob area. Peak abutment stresses increase as the shape of the mined area changes from the slim rectangular opening at the start of mining to a square shape with the advance equal to the longwall face length. Subsequent abutment stress increases are minor. The

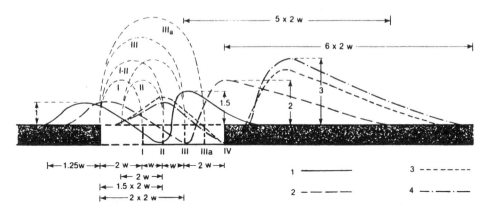

Fig. 8.3.2. The travelling stress concept.

Table 8.3.1. Abutment stress distributions

No. of lifts	Abutment stress concentration	Stress pushed from longwall face
2 w	2.0 σ_v	1 w
3 w	2.5 σ_v	1.5 w
4 w	3.0 σ_v	2 w
6 w	4.0 σ_v	3 w

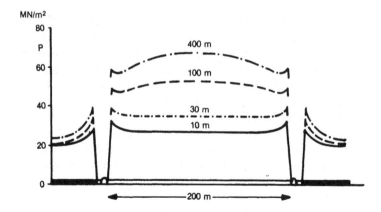

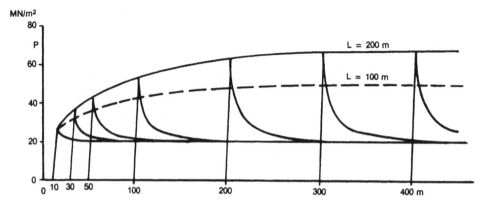

Fig. 8.3.3. The relationship between abutment stress and the length of the extracted panel (after Everling & Jacobi).

magnitude of the stress increase is a function of the face length. The longer the face, the higher the peak stress as illustrated in Figure 8.3.3. The stress diagrams indicate for a 200 m face length that when the coal face advance more than 100 m the stress becomes appreciable higher at the face than in the ribs at the face line.[17]

3. Relation between vertical and horizontal stresses. Vertical stress is approximated as a constant value along the coal face, and is a function of the elements described in the previous paragraph. However, horizontal stress is considered to be a restraint phenomenon and a function of the magnitude of the vertical stress and the degree of excavation (Fig. 8.3.4). Generally speaking lateral stresses at longwall boundaries are not investigated because it is commonly believed they have little influence on strata control and longwall stability. However, the formation of maximum lateral tensile stresses at longwall face ends plays an important role in overall structure mechanics and should not be omitted from stability analyses.

8.3.2.   *Strata stress of longwall structure*

Strata mechanics uses simplified models to analyze complex problems. It incorporates

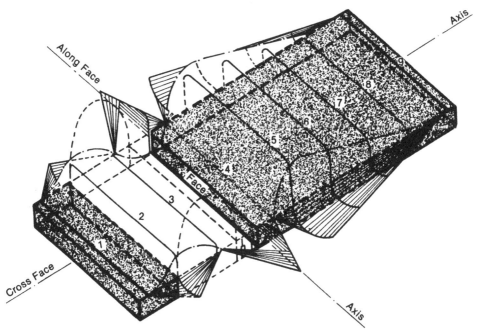

Fig. 8.3.4. Block diagram of vertical and horizontal stresses at the longwall face.

stress dynamics and displacement mechanics as a function of gob de-stressing and mining advance.

1. Strata stress redistribution results from mining. The load that was supported by the removed coal is transferred to the parameter of the mined area, creating abutment pressure. There are two concepts for the interpretation of the redistribution of strata pressure around longwall faces:

a) The old British concept is presented in Figure 8.3.5. Transferred stresses form a corner peak due to the interaction of abutments.[18] The front strata pressure abutment is of appreciably lower magnitude than the side strata pressure abutment. The unsupported roof in the waste area caves and in time reconsolidates and slowly accepts load from the overburden up to the state of virgin strata pressure. Low pressure zones contribute to relaxation at the edges of the extracted region.

b) The German concept does not agree with previous assumptions because it considers that there is no corner peak pressure at the end of a longwall face.[19, 20] The interaction of front and flank strata pressure abutments will induce an extremely high pressure on a projecting angle of a coal seam, where no rib exists (Fig. 8.3.3).

The differences between the two concepts are due to the empirical means of describing the state of the rock stress around longwall faces and the variations in geological conditions between coal fields in these two countries.

2. Stress concentrations above the face can be analyzed using Berry's transverse isotropic theory, where strata is related to induced stresses within the solid adjacent to a longwall face.[21]

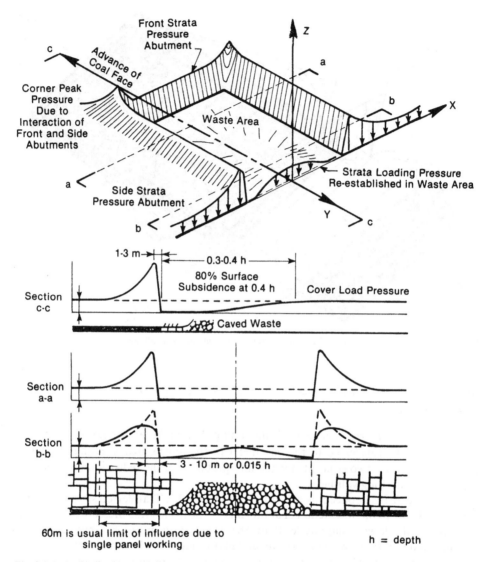

Fig. 8.3.5. Redistribution of strata pressure in longwall mining (after Whittaker).

Vertical induced stress contours are shown in Figure 8.3.6. It can be seen that the line of peak stress gradually departs from vertical over the solid as the height increases above the extraction horizon. There is an area of tensile stress above the gob and compressive stress above the solid coal.[21]

The stress magnitude is proportional to the mine depth, the deeper the coal seam the higher the strata pressure. In shallow mining, abutment pressures of 8.4 to 10 MN/m^2 have been recorded at a depth of 100 m.

The stress contour diagram shows the magnitude of induced stresses at different heights above the seam and can be used to ascertain the significance of superimposed

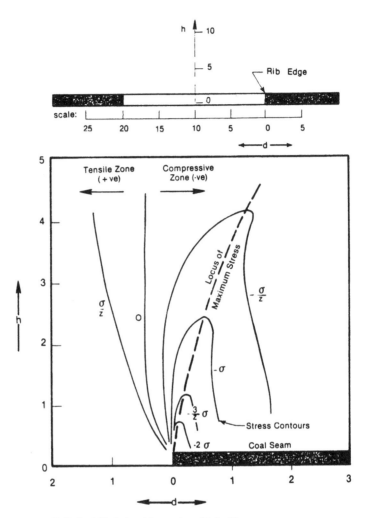

FIg. 8.3.6. Stress
concentrations above the
working face.

σ - Vertical Strata Load (increasing with depth)
h - Height Above Coal Seam
d - Lateral Distance

induced stresses as can occur by making the rib pillar too narrow between adjacent longwall faces.[21] A properly designed rib pillar between successive longwall faces gives protection to the stope structure.

8.3.3. *Strata stress for a stowing support*

Longwall mining with stowing is applied in some areas of Europe. With stowing, the forces and displacement in the rock strata and coal seam differ greatly from the normal cases where full closure is permitted.

The U.S.S.R. carried out experiments to quantify the abutment pressure zone for longwall faces in a steep coal seam, 60° to 65°, with strata control by chocks and stowing. The coal seam thickness was 1.55–1.75 m. The coal was hard and the immediate roof a sandy-clay shale of medium stability. The length of the longwall face was 90 m with advancing coal extraction. The run-of-mine stone was placed in packs 70 m in length parallel the face. The width of the immediate face area was 5.4 m and the stowing was at 2.7 m intervals. The maximum lag of stowage behind the working face was 10.8 m. Roof control in the staggered (bench-out) sector of the longwall, 30 m in length, was by chocks support. The distance between chocks was 4.5 m along strike, and 4 to 6 m along the dip.[22]

Underground instrumentation consisted of four component transverse defor-mometers and hydraulic transducers installed in the roof and floor strata of the coal seam. Zhizlov et al. presented the instrumented data (Fig. 8.3.7):[22]

a) The effect of coal extraction at the longwall face has been measured at up to 50 m from the face, the outer limit of the abutment stress zone. The most intense increase in stress was measured in the section of roof supported on chocks. The stress maximum was recorded 2.2 m ahead of the breast with a vertical field stress of 12.2 MN/m² the maximum stress concentration coefficient is 3.8.

Compression of the seam was also observed along the bedding planes. In this direction the maximum stress concentration factor was 1.9. In the immediate vicinity of the working face a weakened zone was traced with considerable compressive stresses retained at the edge of this face.

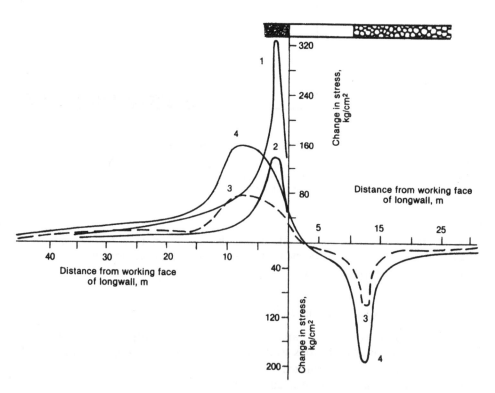

Fig. 8.3.7. Abutment stress distribution (1 and 2: Chock support; 3: Stress at floor; 4: Stowing).

b) The stress distribution was somewhat different in the worked-out area. Near the working face the rock mass was close to virgin stress levels; 2.5 m back of the face stress values dropped below virgin levels. Relief from load reaches a limit 13.5 m from the breast in the stowage area. As a result of the direct interaction of the stowage with the roof, stresses are rapidly restored to virgin values.

The investigators concluded the strata mechanics of stowing differ from those of chock support because the fill undergoes compaction. They also state the qualitative stress diagram fits with existing assumptions for the behaviour of fill material under load. The presence of chock supports close to the working face of the longwall have an appreciable effect on the abutment stress distribution in that the stress peak is not pushed farther from the face as is case for stowing. This is due to redistribution of part of the dead load on the artificial support and to a decrease of deflection of the immediate and main roof strata above the edge of face.

The mechanics of formation of the abutment stresses and stress transfers favours strata control by stowing in thin and medium thick steep pitching coal seams.

8.3.4. *Stress concentration around face-ends*

An integral part of stability in retreat mining is roof control around face-ends. The majority of retreat longwall operations experience stress concentration effects on roadways within 30 m of the face line. Wide rectangular roadways may occasionally require centre props for increased security of the roof girder. Abutment stress concentrations usually cause floor heaves, which greatly increase the thrust between props and girders, creating the potential for the roof girders to deform locally or skew or roll. The roof is also highly likely to sag and deteriorate and is best supported by hydraulic props within the moving abutment zone.

Figure 8.3.8 illustrates common methods of roadway formation in longwall advancing mining in the coal fields of Great Britain.[23]

a) Full face advance headings are located in the front abutment stress zone and experience high levels of deformation. Convergence can be as high as 60 cm at the face line. The major portion of the deformation occurs within 29 m of the face line, with the rate of convergence increasing progressively to a maximum at the face line. Convergence rates then gradually decrease to near zero at a point about 100 m outbye of the face line.

Advance headings with lengths under 20 m are not effected by deformations to the same extent as headings with lengths between 20 and 29 m. Szwilski and Whittaker model investigations show that in advance headings there is an increase in the extent of fracture interaction in the immediate area of the face and where fractures develop over and around the heading. There is a higher degree of roof fracturing at the T-junctions. Fractures in the solid parallel to the face line extend into the advance heading area.[23]

b) The half-heading method of roadway excavation has advantages with respect to strata stability. It is used to limit convergence of the roadway to about 25 per cent of the extracted seam height at the outbye of the face line. Figure 8.3.9 illustrates the layout of a half-heading system. The pillar should be at least half the width of the fully formed roadway and the length of the half-heading should be four to six times the width of the pillar.[18] For mining conditions in Great Brittain, the pillar is approximately 2 m wide, and the length about 15 m. The model studies show that major roof fractures parallel to

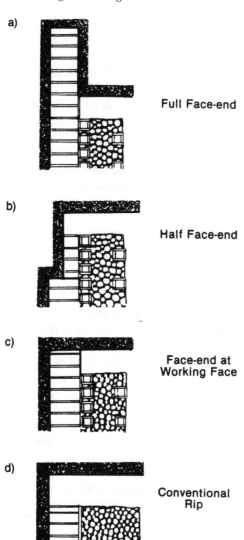

a) Full Face-end

b) Half Face-end

c) Face-end at Working Face

d) Conventional Rip

Fig. 8.3.8. Different layouts for roadway face-ends.

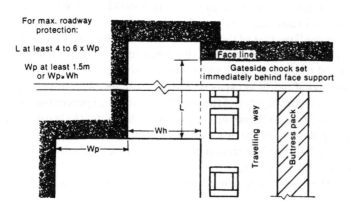

For max. roadway protection:

L at least 4 to 6 x Wp

Wp at least 1.5m or Wp ⩾ Wh

Face line

Gateside chock set immediately behind face support

Travelling way

Buttress pack

Fig. 8.3.9. Layout of a half-heading face-end.

the rib-side are already established at the point where the roadway is formed indicating this region to be significantly de-stressed.[23]

c) The face end at the working face experiences maximum strata movement dates. The main heading stability problem is the provision of an effective support system for this zone of active strata movement. The model investigations indicate two distinct sets of fractures, parallel with the two free edges. The corner of the mine structure is relatively free of fractures which suggests there is no relaxation of vertical stresses in this area.[23] This is also a function of the coal strength and the stiffness of the strata. The lack of fracturing can also be due to a lack of stress – linear elastic 3 D models show low vertical stresses at corners.

d) Ripped debris from widening to the full roadway width can be disposed of in the adjacent packing area, and it lends itself to mechanized packing operations. Its disadvantage is that a powered support system for rapid face advance is not easily incorporated in mining operations.

The choice of the roadway face-end layout influences strata stability and convergence both in front and behind the face line. Roadways should be designed on the basis of their ability to accept the predicted level of convergence and closure. Convergence around face-ends can in most cases be controlled by powered roof supports. The patterns of convergence contours are complex but they are useful for planing the face-end layout.[23]

8.3.5. *Interaction stress mechanics*

Interaction is a factor of longwall mining because every coal face encounters interaction with another face at least once during its working life. Interaction is a word used in many senses, but it largely represents the stress and displacement effects produced by one working face upon another. The interaction of longwall faces is more strata displacement than stress. The main interaction phenomena are closure deformations in the roadways and the deterioration of the strength of roof strata in the work places.

The case of multiple seam mining is given as an example of interaction in Figure 8.3.10. The roadway in the first seam is affected by the new stress state induced by

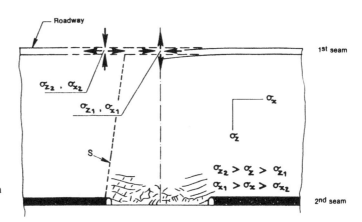

Fig. 8.3.10. Stress and displacement interaction in multiple seam mining.

undermining in the second seam. Also, the roadway is undergoing differential displacements due to subsurface subsidence. The extension and magnitude of the displacements depends upon the size and depth of the undermining panel in the second seam. The roadway is exposed to differential displacements both along and perpendicular to strata that will cause stability problems even though part of the roadway above the solid coal is in a state of high compression due to the surcharge effect of the overburden. The characteristic deformation features of mine structures due to seam interaction are a set of vertical fractures running parallel to the main axis of the pillars and a second set of more widely spaced shear fractures with the same strike but dipping at 45°.[21,24]

Whittaker discussed the implications of a longwall layout in a high pressure zone within fractured ground that can be produced by the interaction of remnant pillars in mined-out seams.[25] The most successful form of layout locates the present face over or under the old gob. The location of the face in this de-stressed situation results in satisfactory stability conditions on both the face and in the roadways. Such a composite layout involves increasing successively the sizes of the rib pillars in lower seams, Figure 8.3.11. An alternative layout which provides roadway protection and the conservation of reserves locates the remnant rib pillars down the central axis of the present face. Whittaker stated that undermining a remnant rib pillar in this way produces the following results:

a) Fractures are kept intact much better since the face is moving 'on end' relative to the orientation of the main vertical interaction fractures.

b) Only a small length of the face is affected at any one time.

c) Convergence on the face is not affected provided the interaction fractures are kept constrained.

d) A more even subsidence profile and a more balanced strain profile are produced at the surface.

e) The successive build-up of localized high pressure regions is avoided.

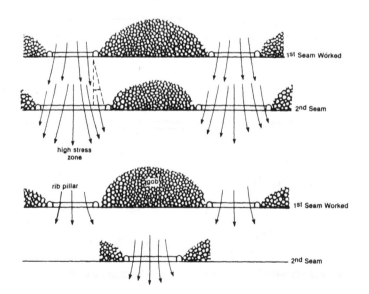

Fig. 8.3.11. Composite longwall layouts with special references to interaction (after Whittaker).

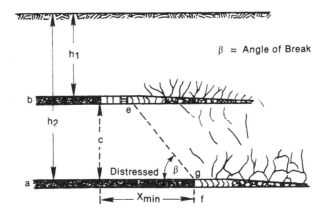

β = Angle of Break

Fig. 8.3.12. Simultaneous longwall mining.

This concept of placing remnants over or under working faces has been successfully carried out where the interval between seams has been as little as 19 m. However, this experience has been confined to British conditions where moderately strong mudstones intervened between the seams.[24]

The most suitable layout with respect to the interaction of multiple seams is simultaneously seam mining.[26] The concept for this layout is illustrated in Figure 8.3.12 which shows the simultaneous extraction of two seams. Point f in seam A and point d in seam B represent the position of the faces. The shortest critical distance between the faces X_{min}, measured along the direction of face advance is selected so that caving of the overlying layers behind the working face in the lower seam does not effect mining operations in the upper seam. The key to the correct location of the upper seam face is the determination of the caving angle which for hard rock is up to 75°, for weak shales 55° and clays and sands 35°. Since the face positions may not be accurately coordinated and the rate of advance of the faces in both seams may not be quite the same, it is necessary to leave a safety margin of 15 to 20 m between point e and the special support in seam b.

Where the production schedule follows the mine layout, multiple seam mining can be superior to mining the second seam only after the first seam is mined out.

Finally, the retreat method, the most economical system of longwall mining, has a limitation imposed by interaction. Interaction problems can be reduced simply by the application of total extraction. Solid drives for retreat panels suffer most from interaction effects. However, a retreat layout can drawn up which eliminates the need for solid drives.[21]

8.4. CAVING MECHANICS

The caving mechanics of longwall mining have been extensively investigated because of the large number of coal faces operating with the roof caving as the support is advanced. This chapter deals with caving as it is related to longwall mining and specifically with:

– Basic caving processes.

– Physical modeling to demonstrate visually the failure and deformation of the overlying strata as the longwall face advances.

– Mining subsidence to show the influence of longwalling on the deformation of the ground surface.

– Controlled caving and subsidence to prevent the surface structure damages.

A prediction of the caving characteristics and ground subsidence should be based on the combined geological-structural data and mining configurations in regions where longwall mining is new and there is no past experience on which to base forecasts.

8.4.1. *Caving processes*

Longwall mining is normally done under sedimentary strata with beds of variable thickness and composition. The roof strata beds are in most cases well defined and can usually be approximated as separate units, so that strata stability and cavibility may be analyzed as a series of planar rock 'plates' or 'beams'.

In the dynamics of longwall caving, support is advanced, as the coal is extracted. This allows the roof to begin to sag behind the support with sag increasing as the span increases. At the start of mining, the span is small and only localized falls may occur (Fig. 8.4.1). As the span increases a critical point is reached where the roof strata will no longer be able to support its own weight and will fail, breaking into blocks and filling up the gob area with fragmented rock.[27] The length of the critical span depends on the mechanical properties and behaviours of the individual lithological units over the mined area.

The caved and fragmented rocks differentially rotate creating void spaces so that the overall volume per unit weight increases and gradually fills up the cavity formed until the upper beds are supported by unconsolidate cave material. As the face advances further, the unconsolidated rocks are gradually compacted by further roof sag until the original vertical stress state develops through the caved material.

The critical span is usually much larger than the working cantilever span after caving starts. The extra loading that the critical span imposes upon the supports is called the first weight. After caving starts, conditions are uniform and any irregularities in loading are caused by geological structural anomalies. The correct nature of the caving process and the method of operating the supports controls the loading that is imposed on the supports and the coal face. Two possible loading conditions are shown in Figure 8.4.2.

Fig. 8.4.1. Model of immediate roof strata bridging.

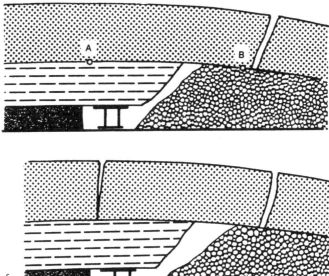

Fig. 8.4.2. Models of main roof deformations.

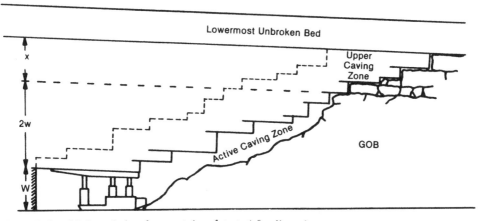

Fig. 8.4.3. Possible boundaries of supported roof strata (after Kenny).

The key factor in longwall mining is the formation of the active caving zone defined by Kenny.[28] The active zone is the cavity or boundary between the caved material and the intact roof strata where failure and caving are occurring. The limits of this zone are delimited by the dotted line in Figure 8.4.3. There is a point where the rock strata fails but the relative positions of the blocks are retained by the caved material from the active caving zone below. The upper limit of caving can be calculated from the volumetric relationship:

$$h = \frac{t}{K - 1} \qquad (1)$$

where

t = seam thickness;

K = bulking ratio or swell factor and is equal to the ratio of the caved volume/intact volume for a unit weight of material;

h = height of active caving limits above the seam.

For example, if the bulking ratio is 1.5, the active caving zone extends above the seam for a distance of twice the seam thickness (Fig. 8.4.3). Local variations in the angle of repose of the caved rock may allow debris to roll down the slope, causing flushing to occur. This may create local variations in the height of the active caving zone to as much as four times the seam thickness.

The relationship between active caving limits and the bulking factor is presented graphically in Figure 8.4.4. It is suggested that free falling wastes will rarely exceed the bulking factor of 1.3.

The upper caving zone is above the limits of the active caving area, but below the position where fracturing of the strata is absent. The strata in this zone is fractured but in contact with the beds below its stratigraphical integrity is maintained. The upper limit of the upper caving zone will generally be controlled by the rigidity and strength of the strata beds within this zone.

The normal case in mining is to have thin or weak layers over the seam which will not stand unsupported (false roof) which fails around the canopy of the powered supports. Stronger beds above the seam in the active caving area, can cause periodic weightings on the face and the supports due to the formation of roof cantilevers with larger than normal spans before failure.

Satisfactory caving of the roof strata, after powered support advance alleviates high strata abutment pressures. The longwall face should be long enough to provide good caving and satisfy the constraints of speed of advance and horizontal control.

Caving processes and the caving mechanics of longwall mining can be studied by numerical models using approximations of the characteristics of the caved rock mass. The fractured strata around longwall faces can also be approximated by a system of

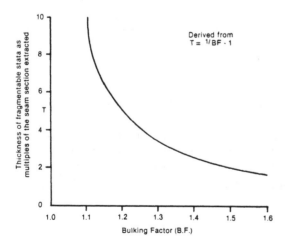

Fig. 8.4.4. Relationship between the bulking factor and the thickness of the fragmentable strata (after Gorrie).

blocks, particularly where there are closely spaced geological structural defects. The study of block models can be either by computational techniques which are still in the early developmental stage or physical models.[29]

8.4.2. *Physical modeling of strata caving*

Base friction modeling is the most commonly used technique for mining methods involving strata caving.

1. A base friction model that proved to be most suitable for studying caving behaviour during longwall mining was built from 45 kg of baking flour and 25 litres of methanol.[30] The model material was spread on the frame, compacted, and cut to simulate beds 0.33 m thick above the seam. The seams and strata were left intact. The mixture was allowed to dry for 36 hours before the Galt and Rider Seams were painted in black, as illustrated in Figure 8.45 (Lethbridge Coalfield, Western Canada).

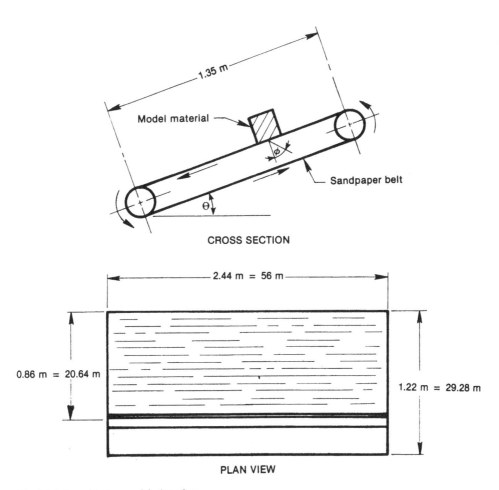

Fig. 8.4.5. Base friction model plan view.

A block of the Galt Seam was cut out to simulate a pass of a double shearer drum at a longwall mining face and the belt set in motion towards the bottom of the model to simulate strata loading. The belt was stopped after caving had occurred. The sequence of block cutting, 190 mm, lift extraction 2.4 m, and caving were repeated 15 times until the gob area reached a length of 1.5 m (36 m).

The compressive strength of the rock strata (weak rock) was calculated using the scaling equation:

$$\sigma_r = \frac{\sigma_m \gamma_r h_r \cos \phi}{\gamma_m h_m \sin (\theta + \phi)} \tag{2}$$

where

ϕ = friction angle between model material and belt;
θ = angle of belt inclination;
γ = specific weight;
σ = compressive strength;
r and m = refer to the rock and model respectively.

The first cut was caved using a belt angle (θ) of 17.8° to correspond to a rock compressive strength of 6.6 MN/m^2. At this angle the model material at the bottom of the frame crushed, causing model distortion. To alleviate this problem, a new model was built with the belt angle lowered to increase the effective strength of the model material to 10 MN/m^2.

The cracks in the centre of the model are tension cracks which opened as the material dried. The entire model was affected but the cracks near the sides were closed by pushing the sides of the frame in towards the centre before the model was run. The tension cracks remaining in the centre of the model are good representations of the planes of weakness found in fault zones.

The scaling parameters between the model material and the real mining situation are given in Table 8.4.1.

Table 8.4.1. Base friction model scaling parameters

Elements	Model	Rock
Specific wt. (γ)	0.95 kg/cm³	2.55 kg/cm³
Seam depth (h)	0.86 m	20.64 m
Model depth	1.22 m	24.28 m
Compressive strength (σ)	0.1 MN/m^2	10 MN/m^2
Seam thickness	7.2 mm	1.75 m
Cut width	127 mm	2.40 m
Roof bed thickness	10 mm	330 mm

Friction angle between
   model material and belt (ϕ)...................... 22
Belt angle from horizontal (θ)...................... 3.2
Model material .... Baking flour and
                                                      methanol

 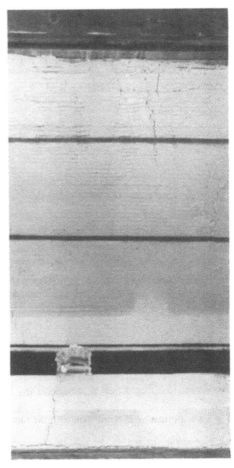

Uncaved span of excavation   Caved span of excavation

Fig. 8.4.6. Model of caving for the initial excavation.

2. Results. The following interpretation of the model study is based on the separate models using the same mining configuration and model scaling parameters.

a) The initial cut illustrated a longwall opening up sequence in a shallow flat-laying, coal seam with a soft-rock roof similar to the roof at the Lethbridge coal field. The first plate in Figure 8.4.6 shows the extraction cut and the second shows the resultant caving. The model indicates the roof span cannot sustain load and caves when the first lift is excavated if support is not installed.

b) When the excavation had advanced three lifts, the roof strata when had been bending, began to cave behind the face and formed a roughly trapezoidal shaped cavity filled with gob. The caved model material exhibited good fragmentation, as is the common case for the coal mines of the Plains Region in soft rock strata. The strata above the caved material is subjected to tensile forces and tends to separate along the stratification planes and to break perpendicular to these planes in a step-wise manner (Fig. 8.4.7).

Fig. 8.4.7. Formation of tensile failure in the roof strata caving area.

c) Beginning at six lifts, the roof strata caved in segments for each advancing lift to form an approximately arch shaped cavity filled with gob. The caved rock exhibits good caving and flushing characteristics. It should be noted that the vertical displacement, bed separation, is of much greater intensity than the horizontal displacement. The cave plane is not uniform because of changes in the angle of inclination, but it can be approximated as close to 70° (Fig. 8.4.8).

d) At nine lifts, the face intersected a small fault. A common structural feature of the Lethbridge coal field. The model suggested that due to the existence of a vertical plane of weakness, the fault, the stresses which are normally transferred onto the coal face were directed onto the roof strata immediately above the extraction area, i.e. onto the chock supports. Stresses cannot be totally transferred through planes of low shear and normal strength. This is indicated by intact blocks of material sliding down along the fault plane when the fault is undercut (Fig. 8.4.9).

e) Excavation beyond the fault zone with the slided block, thirteen lifts, is illustrated in Figure 8.4.10. The roof is caving in the same manner as it did when excavation started (Fig. 8.4.6). Two gob areas have been formed which are separated by the slided block along the fault zone. It should be noted that caving after the fault was intersected is more irregular, and this is mostly due to the development of high shear stress over the solid coal at the longwall face. In the final extraction cycle, fifteen lifts, the shear stresses

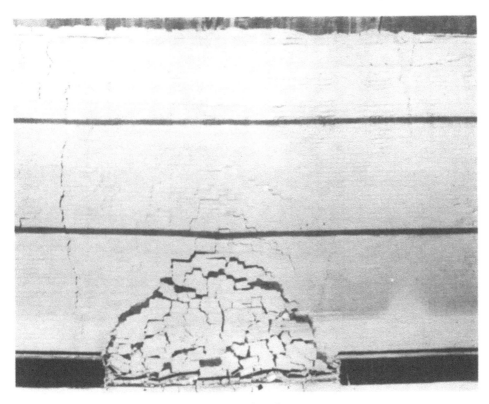

Fig. 8.4.8. Formation of an arching structure above the gob area.

above and ahead of the coal face induced such intensive fracturing that further testing was impossible.

At the present time there is an interest in introducing longwall mining in coal fields with shallow seams and weak rock strata in Western Canada, Western Australia and others. Successful mining under these circumstances, requires planning based on the mechanics of strata caving which takes into account the effects of faults with small displacements. The model investigations showed that due to the formation of a free body, the load placed on chock supports at a fault parallel with the face is a function of the fault height. It should be assumed there is no shearing resistance along the fault plane, no support from the gob and no tensile strength along the top boundary, Figure 8.4.11. The resulting dead weight chock load can be expressed by the following equation.

$$\sigma_v = h_c + 1/2(h_c^2 \tan \beta) W \gamma \tag{3}$$

where

σ_v = load on the chock;
h_c = height of the fault influence;
l = the width of the working area;
β = angle of break;
W = centre width of a typical chock support;
γ = unit weight of rock strata.

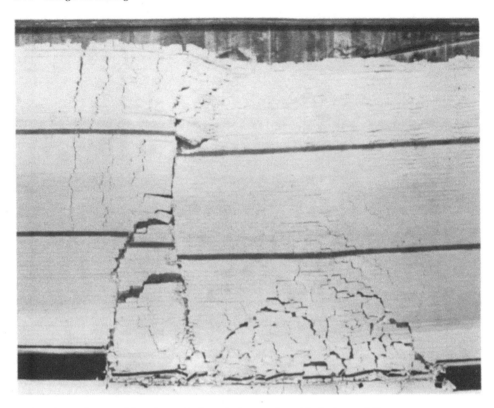

Fig. 8.4.9. Roof strata sliding along a fault at the intersection with the longwall face.

Fig. 8.4.10. Formation of two gob areas separated by the sliding block of the fault zone.

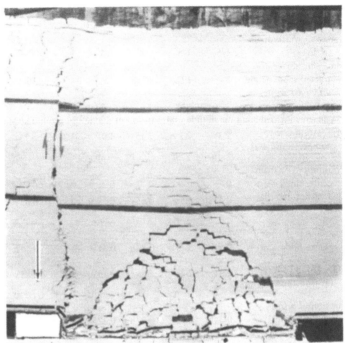

Fig. 8.4.11. Dead load of the
faulted block.

The assumptions stated above are not strictly true, but serve as a model for the calculation of possible chock loads in the worst case situation. Similar loads can also be anticipated in the case where the roof strata stops caving.

Within the scope of strata cavibility, it might be of interest to analyze the stresses induced by caving and by the fault using digital models. Also, one could evaluate chock loads in the fault zone as a function of the mining depth.

8.4.3. *Mining subsidence*

Surface subsidence due to longwall mining in flat coal seams has been well documented and studied in the United Kingdom by the National Coal Board. They have published Subsidence Engineers Handbooks, which are based on principles derived from about 299 case studies involving a wide range of mining and surface geological conditions.[31, 32] Some of the terms and representative values used in the handbooks are discussed below:

1. Maximum subsidence (v) can be evaluated as a function of the ratio of extraction width (w) to the depth of longwall face (h) and the workable thickness of the coal seam (t).[32, 33]

$$\frac{s}{t} = f\left(\frac{w}{h}\right)$$

Table 8.4.2. Maximum subsidence, surface ground strain and tilt due to longwall mining

Longwall width/depth (w/h)	1/6	1/5	1/4	1/3	1/2	3/4	1.0	1.2	1.4
Max. subsidence/extracted seam height v/t caved wastes	8%	12%	15%	25%	45%	70%	84%	90%	90% (max.)

Coefficients for deducing magnitude and position of maximum ground strains and tilt

Max. strain due to compression ($+\varepsilon$)	$\dfrac{2.3\,v}{h}$	$\dfrac{2.2\,v}{h}$	$\dfrac{2.15\,v}{h}$	$\dfrac{1.9\,v}{h}$	$\dfrac{1.35\,v}{h}$	$\dfrac{0.75\,v}{h}$	$\dfrac{0.55\,v}{h}$	$\dfrac{0.5\,v}{h}$	$\dfrac{0.5\,v}{h}$
Position of $+\varepsilon$ from centre line ($+\varepsilon_x$)	0	0	0	0	0.02h	0.10h	0.20h	0.29h	0.39h
Max. strain due to extension ($-\varepsilon$)	$\dfrac{0.45\,v}{h}$	$\dfrac{0.5\,v}{h}$	$\dfrac{0.65\,v}{h}$	$\dfrac{0.75\,v}{h}$	$\dfrac{0.8\,v}{h}$	$\dfrac{0.65\,v}{h}$	$\dfrac{0.65\,v}{h}$	$\dfrac{0.65\,v}{h}$	$\dfrac{0.65\,v}{h}$
Position of $-\varepsilon$ from centre line ($-\varepsilon x$)	0.50h	0.49h	0.42h	0.34h	0.32h	0.40h	0.51h	0.61h	0.70h
Max. ground tilt (g) (at transition pt.)	$\dfrac{1.9\,v}{h}$	$\dfrac{2.2\,v}{h}$	$\dfrac{2.6\,v}{h}$	$\dfrac{3.15\,v}{h}$	$\dfrac{3.35\,v}{h}$	$\dfrac{2.85\,v}{h}$	$\dfrac{2.75\,v}{h}$	$\dfrac{2.75\,v}{h}$	$\dfrac{2.75\,v}{h}$
Position of g from centre line (g_x)	0.34h	0.32h	0.27h	0.22h	0.21h	0.26h	0.37h	0.46h	0.56h

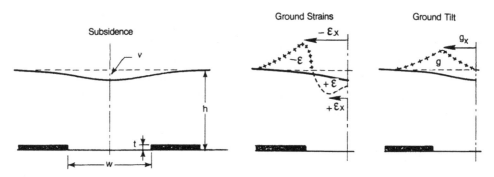

Fig. 8.4.12. Factors of subsidence.

Whittaker and Breeds composed a table, Table 8.4.2, based on selected data taken from the subsidence Engineers Handbook to demonstrate how v/t is related to w/h. The point of maximum subsidence is located directly above the centre of the excavation, Figure 8.4.12 and its magnitude can be simply calculated as follows:

    Longwall width   w = 200 m
    Longwall depth   h = 400 m
    Extracted seam height   t =   2 m

$$\text{since } w/h = \frac{200}{400} = \frac{1}{2}$$

Maximum subsidence $v = 45$ per cent $(2m) = 0.90\,m$. The maximum subsidence magnitude at the centre line is of primary importance for the construction of the subsidence profile.

2. Surface ground strain and tilt are also illustrated in Figure 8.4.12, where subsidence produces regions of ground extension and compression by virtue of relative displacements towards the centre of the trough. The generalized relationship between the principle variables which represents the magnitude of the ground strains is expressed by the equation:

$$\varepsilon = f\left(\frac{v}{h}\right) \tag{4}$$

Maximum tensile strain $(-\varepsilon)$ is located directly above or near but inside the edge of the rib and maximum compressive strain $(+\varepsilon)$, is either at the centre or near, but inside the extraction area. Table 8.4.2 gives a range of practical data for magnitude of $-\varepsilon$ and $+\varepsilon$ together with their respective positions in relation to the centre line of the longwall extraction area.[32] The strain can be simply calculated. For the same longwall case in the previous paragraph, the calculated strain values are:
  – Maximum strain due to compression

$$+\varepsilon = 1.35\frac{v}{h} = 1.35\frac{0.9}{400} = 0.0030 \text{ or } 3\,mm/m$$

at 8 m from the centre line.
  – Maximum strain due to extension

$$-\varepsilon = 0.8\frac{v}{h} = 0.8\frac{0.9}{400} = 0.0018 \text{ or } 1.8\,mm/m$$

at 128 m from the centre line.

The maximum ground tilt as a consequence of subsidence occurs at the transition point between $-\varepsilon$ and $+\varepsilon$ and its magnitude is also a function of v/h. Whittaker and Breeds included values in Table 8.4.2 to calculate maximum tilt (g) for different relationships between w/h. The calculated magnitude of maximum tilt for the example longwall case is given below:

$$g = 3.35\frac{v}{h} = 3.35\frac{0.9}{400} = 0.0075 \text{ or } 7.5\,mm/m$$

at 84 m from the centre line.

From an engineering point of view, the essential data for mining subsidence evaluation are the maximum subsidence (v), the maximum compressive $(+\varepsilon)$ and tensile $(-\varepsilon)$ strains and the maximum tilt (g).[39]

3. The angle of draw represents the limiting plane of subsidence and is drawn from the edge of the extracted area in the seam to the point where subsidence is just measurable on surface. The angle between vertical and the line of draw is known as the 'angle of draw' and it varies from coal field to coal field and from one lithological unit to another. As an example, the angle of draw is approximatly $35°$ in Great Britain. It is of interest

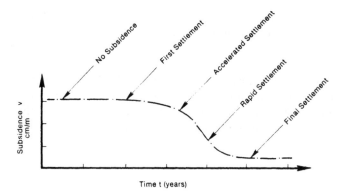

Fig. 8.4.13. Time dependent subsidence.

that in Continental European and North American terminology the angle between the line of draw and horizontal is known as the 'angle of caving'.

4. Subsidence as a function of time is also an important phenomena in stability analyses. The main factor in subsidence duration is the depth of the mine working.[35] Subsidence as a function of time is summarized below (Fig. 8.4.13).[36]

a) In the case of deep coal seams, first phase subsidence is not evident immediately after excavation. It can only be measured after several years. However, in the case of shallow mining it can be noticeable after a short period of time.

b) The second phase of subsidence is characterized by higher settlement rates and it can be considered as accelerated settlement.

c) The third phase of subsidence is abrupt and most of the total displacement occurs at this stage.

d) Post subsidence settlement is the final activity and is associated with very small displacements. It may take several years for the ground to come to rest.

Figure 8.4.13, the curve of a subsidence profile shows many similarities to a generalized creep curve. There may be inferred three stages of creep deformation, with rapid settlement corresponding to the failure deformation stage. Time dependent subsidence can be expressed by the following equation:

$$\frac{v_t}{v} = 1 - e^{-ct} \tag{5}$$

where

v_t = subsidence at time of observation;
v = maximum subsidence;
t = time of observation of subsidence in progress;
c = coefficient of velocity of strata displacement (dependent on the lithological units and geological structural conditions).

Subsidence as a function of time is of a primary concern in heavily populated areas.

5. The classification of subsidence is defined by the method of mining and mine depth. At the present time longwall mining is carried out in two geological settings. The first is shallow depth and weak rock strata, average working depth 50–150 m, length of the longwall face up to 100 m and the face advance about 500 m during its life. The second, is appreciable depth and strong rock strata, average working depth 350–750 m, length of the longwall normally 200 m and the face advances up to 1000 m during its life.

The ratio between the extraction width to the depth of the longwall face and the angle of draw determine the subsidence category as follows:[32]

a) Super-critical subsidence is characteristic of very shallow-mining and affects the ground surface substantially. The large width/depth ratios of the excavated area develop full subsidence commonly in the shape of flat-bottomed troughs at the surface. An important aspect arising from this is the high probability of appreciable surface fissuring occurring mainly along the longitudinal flanks of the subsidence trough. Such fissuring is usually time dependent where pseudoplastic rocks and glacial till occur at the surface. The advancing front of the subsidence trough rarely allows appreciably wide fissuring to develop parallel with the face. If the surface rocks are strong and have a joint or other natural fissure patterns, when the development of surface fissuring with subsidence may be governed more by geological characteristics than coincidence with maximum surface tensile strains. Consequently the geology at the surface needs carefull study.

Figure 8.4.14 shows the general characteristics of surface subsidence to be expected in shallow workings. The relatively narrow surface strain zones are important areas, especially the location of maximum tensile strain which is a risk zone for the occurrence of significant surface fissuring and/or marked stepping in the subsidence profile.

b) Critical subsidence (Fig. 8.4.15) is also related to shallow coal mining, and it has

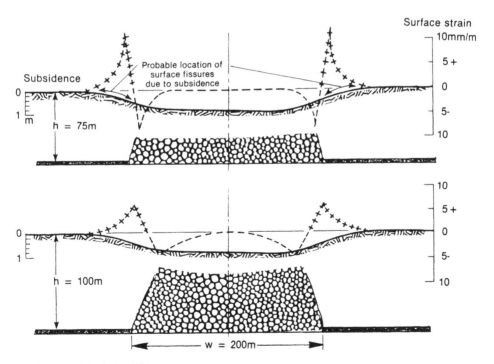

Fig. 8.4.14. Sub-critical subsidence.

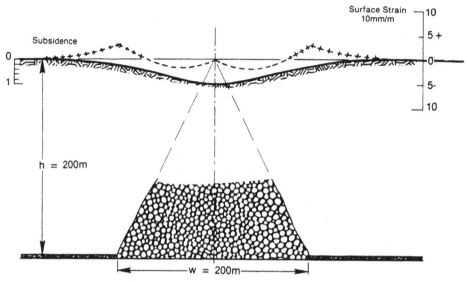

Fig. 8.4.15. Critical subsidence.

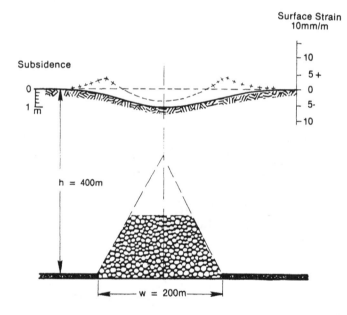

Fig. 8.4.16. Super-critical
subsidence.

similar characteristics to sub-critical subsidence. The geological structural features of
the rock strata are also important with respect to surface ground deformation as
discussed in the previous category. The rock lithologies do not appear to have a
marked influence on subsidence in shallow mining.

c) Sub-critical subsidence is typical of deeper coal mining, where the gob area is
located below the ground surface because the draw planes do not intersect the ground
surface (Fig. 8.4.16). This type of subsidence does not cause damaging deformations and

fissuring of the ground surface because the ground surface gradually subsides without collapse or caving. The subsidence of deep mines continues after mining ceases and in the case of very deep coal mines for example over 1.500 m, it can continue up to 10 years after completion of mining.

The briefly described types of subsidence consider only flat coal deposits which subside with a symmetrical profiles about a vertical plane passing through the centre of the opening. If the coal seam is near flat and of a consistent thickness, the complete subsidence profile can be predicted by the factor developed by the National Coal Board.

8.4.4. *Controlled caving and subsidence*

The main role of controlled caving and subsidence in longwall mining is to reduce strata movement and caving and by this means minimize ground surface effects and damage to surface structures. Several methods are employed in coal fields over the world, as discussed below:

1. Stowing of the excavated void can be by pneumatic or hydraulic means. The type of stowing governs the magnitude of subsidence as given by the equation below:

$$v_{max} = a\,t \tag{6}$$

where

v_{max} = maximum subsidence;

t = workable thickness of the seam;

a = subsidence factor dependent on the type of mining and stowing.

Values for factor 'a' have been developed from surveying data of surface subsidence as follows:[37]

strata caving system	$a = 0.75$
solid stowing (pack)	$a = 0.50$
pneumatic and mechanical stowing	$a = 0.30$
partial extraction (50 per cent)	
and hydraulic stowing	$a = 0.15$

Successful stowing is achieved where the roof does not sag before filling has been introduced into the void.

However, at the present time over 95 per cent of longwall production units have a gob created behind the face by total caving of the roof strata. For example, in Great Britain pneumatic stowing was used to eliminate caving and reduce surface subsidence. However, it is now no longer considered practical since more effective techniques can be used to give greater protection.[32]

In this country, as a general rule, the maximum subsidence had been found to be 45 per cent of the workable seam thickness for w/h ratios greater than 1.2 and with pneumatic stowing. With no stowing, the maximum subsidence is 90 per cent for the same conditions (Fig. 8.4.17).[32]

2. Harmonic extraction usually requires wide longwall faces. Simultaneous advancement of individual extraction faces should be in harmony to each other and two principle cases are considered:

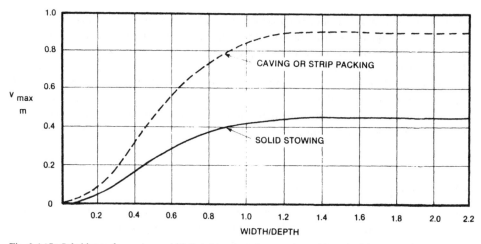

Fig. 8.4.17. Subsidence for various width/height ratios of extraction with and without stowing.

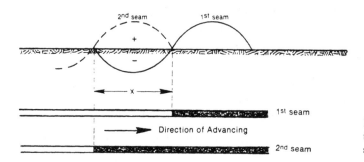

Fig. 8.4.18. Cancelation of tensile and compressive strains.

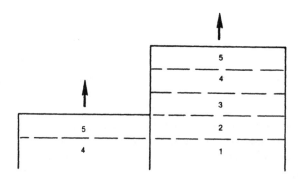

Fig. 8.4.19. Longwall layout with staggered faces.

a) Multiple seam mining should have the spacing of the longwall faces shown in Figure 8.4.18. Mining has to be simultaneous of two or more seams to permit the compressive lateral strains, $+\varepsilon$, from one mining horizon to offset the tensile strains, $-\varepsilon$, from the lower mining horizon. Prevention is by summing strains with opposite signs (Fig. 8.4.18). It is necessary to emphasize that advancing of both faces should be continuous and harmoneous.

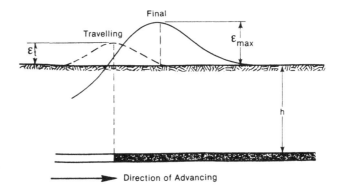

Fig. 8.4.20. Traveling and final strain magnitudes.

Direction of Advancing

b) Single seam mining faces should not be at an angle to each other ('two-winged' faces) because of the danger of summing stresses with the same signs. Mining should be with staggered faces, advancing in a stepped sequence, Figure 8.4.19. The tensile strain traveling ahead of the face for the right side panel is relatively small while the compressive strain behind the face will to some extent be compensated by the traveling tensile strain from the left side panel. With this sequence, the strain induced at a surface structure is maintained at a minimum.[38, 39]

3. Rapid extraction results in a fast moving active subsidence zone similar to a wave. The maximum tensile strain is less than the final obtainable strain from slower extraction and it may be at a permissible level.[38, 39]

The mechanics of controlled caving and subsidence is important due to the requirement for large, continuous and rapid coal extraction. This trend, in the near future, will be further increased with the need for the development of larger underground mining operations.

CHAPTER 9

Slice mining

With the development of new mining methods and new technologies of coal extraction the slice mining system has become less important. However, the existence of a number of coal mines in Europe and Asia with slicing dictates inclusion of this method of mining in this volume.

Representation of the slice mining system is given under four headings, as in the other chapters.

– Mining methods are described by definition of slice structures, longwall slicing with gob caving and slicing with a stowing. Particulars of each systems are given with practical examples.

– Stability of slice-face structures is analyzed for the cases of ascending, descending and simultaneous slice mining.

– Strata mechanics have been directed toward certain specifics of slice mining and are discussed by displacement mechanism of the coal roof stratum; abutment stresses of the gob roof stratum; interaction mechanics between coal structure and fill, and finally, by the influence of geological anomalies on strata mechanics.

– Caving mechanics are described by particulars of roof stratum formation in slice mining, as for example, caving of intact rock strata; caving of the gob rocks stratum; caving of coal slice, and finally, by subsurface and surface subsidence phenomena.

It should be pointed out that a knowledge of strata mechanics for slice mining is of fundamental significance for a general understanding of thick coal seam mining.

9.1. PRINCIPAL MINING SYSTEMS

Thick coal seams could fundamentally be mined out by working either full face (workable seam thickness) or individual slices. The slice, as an extraction structure, divides a thick coal seam into sections which, as a unit, make coal mining easy and convenient. The thickness of slice corresponds to the thickness of the medium thick seam (2 to 3 m). However, implementing thicker coal slices can induce difficulties for coal breaking in upper portions of the extraction face, requiring constant watch over roof stability. From the aspect of mining technology, slicing is represented according to the specific approach, as listed below:

– Elements of slicing structure: firstly, the orientation of the slice structure to the stratification planes of the seam; secondly, the position of the extraction coal faces with respect to the strike and dip of seam; thirdly, the order of extraction of coal slices with respect to the direction of gravity.

– Slicing by longwalling with gob caving is the most common case for extraction of flat and gently dipping thick coal seams, and is briefly discussed as follows:

– Slicing of thick coal seams with stowing mainly relates to two principal methods: longwall mining along bedding planes and horizontal slicing, known as a double winged rising system.

The extraction of thick coal seams by slicing methods, particularly longwalling, could be considered as an engineering technology, which, to a great extent, depends on the geological and structural conditions of coal seams. Under these circumstances, slice mining could be more complex and this usually results in lower coal recovery, less productivity and an increase in production cost.

9.1.1. *Slicing structures*

The slicing element, as a key factor of thick coal seam extraction technology, is also of paramount importance for mine stability analysis in addition to strata mechanics. The success of mining is largely governed by the geometry, site, orientation and order of excavation of individual slices, as further discussed.

1. Intersection of the slice structure with stratification, directly relates to actual slice formation within the coal seam structure. Many ways could be proposed in which a thick coal seam can be split into separate slices by the method of cutting in various directions with parallel planes and spaced at distances equal to the actual thickness of the slice.[1] From the engineering point of view, only the following four methods of coal seam slicing are applicable:

a) When slicing is parallel to planes of stratification, the slice geometry and orientation is governed by geological structural features of the coal strata. The angle of inclination of the slices coincide with the dip of the coal deposit, because they are running parallel to the plane of the stratification (Fig. 9.1.1/a). This is the most common method of mining by slicing, where longwall technology could be successfully applied.

b) Horizontal slicing, excluding flat and thick coal seams, is a most common method for mining pitching and very thick coal seams. In this case, slices are running horizontally from hanging wall to footwall or vice versa, by intersecting stratification planes. The angle of intersection depends on the pitch of the coal seam. It is obvious that, in this case, the slice structure does not coincide with the stratification structure (Fig. 9.1.1/b).

c) Slicing perpendicular to planes of stratification, also known as diagonal slicing or rill cut mining, is represented by short slices. The length of a slice is limited to the true thickness of the coal seam (Fig. 9.1.1/c) and a very small width along the strike of the seam. At present, rill cut mining is becoming rare and, in many mines, is no longer practiced because of a low coal recovery factor and high production costs.

d) Cross inclined or transversely inclined slices show angles of inclination and extension which differ from the thickness, dip and strike of the seam. The inclination of the slices to the horizontal is up to 30°, regardless of the pitch of the seam (Fig. 9.1.1/d). This method has been applied for some variations of room-and-pillar mining in Poland but has been changed to slicing parallel to the bedding plane with longwalling.

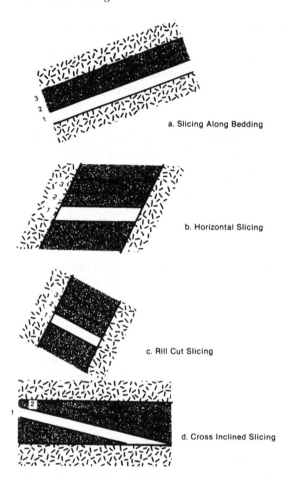

a. Slicing Along Bedding

b. Horizontal Slicing

c. Rill Cut Slicing

d. Cross Inclined Slicing

Fig. 9.1.1. Division of thick coal seam in slices.

It should be mentioned that the method of vertical slicing which runs across the strike and dip of a coal seam, had been in effect for two decades or so, but is no longer practiced. Generally speaking, modern mining technology has a tendency for gradual elimination of slicing methods and their replacement by full face mining methods, in which individual slices are a function of the undercut and are integral parts of the combined mining methods.

2. The orientation of coal face with respect to strike, dip, and thickness of the seam is an important feature of slice mining, particularly in the application of longwall mining technology.[2] Three principal cases are considered, where face orientations also correspond to layouts of longwall mining of thin and moderately thick coal seams:

a) Working along the dip considers two layouts, each with problems which are briefly discussed. Firstly, in working up-dip (Fig. 9.1.2/a), the support loading, by resultant thrust between front leg and face, results in digging into the soft floor of the next coal slice. There are machine instabilities, for example, shearers tend to rotate into the face, where increased strain traps them, and armored face conveyors experience jamming difficulties. The water drains from the gob into the working area. Secondly,

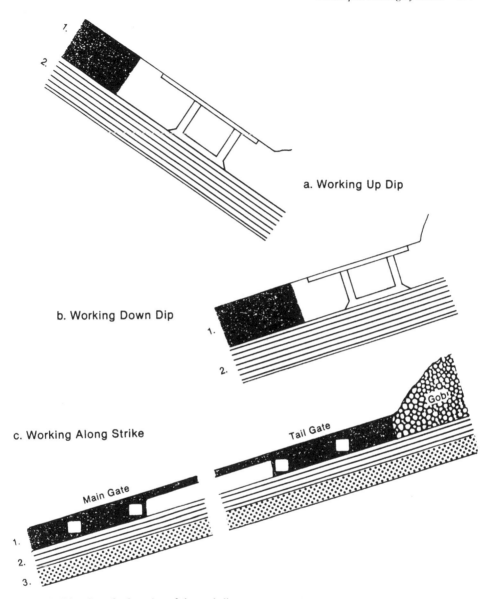

a. Working Up Dip

b. Working Down Dip

c. Working Along Strike

Fig. 9.1.2. Direction of advancing of the coal slices.

working down-dip (Fig. 9.1.2/b) involves coal slabbing and falling towards machines and miners and is thus a more hazardous operation. Although large spans frequently occur between support tip and face, it is difficult to maintain an uniform web depth. The armoured face conveyor runs at decreased capacity, but water is drained well into the gob.

b) Working along the strike or oblique to the strike (Fig. 9.1.2/c) offers a more stable environment for equipment and machines, where the armoured face conveyor grading problems are eliminated. Water drains to a sump in the main gate and coal slabbing

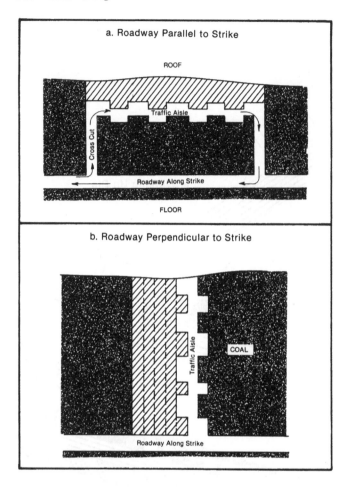

Fig. 9.1.3. Methods of slice face advancing to horizontal section of coal seam.

problems are reduced. However, an increased span between the face and roof support canopy results in cavity formation risks, and large coal blocks falling off the tail gate side of shearers causes coal to pile-up with resulting blockage risk and roof coal instability.

c) There are two possibilities when working along seam thickness: the coal face could be parallel to the strike, with the direction of excavation from hanging wall to footwall (Fig. 9.1.3/a), or it could be perpendicular to the strike, with retreat extraction (Fig. 9.1.3/b). This slicing method was developed in France (Blanty Coal Field) for mining of thick and pitching coal seams, where fill is placed in the excavated void.[3] Each face has three working aisles: one in the process of stowing, a second for transportation and a third for face advancement. The system of coal advancing parallel to the strike allows more efficient transportation of the coal from the face than advancing perpendicular to the strike.

3. *Order of slicing* is the most important parameter of coal extraction, particularly with respect to strata mechanics and ground control. For example, if a thick coal seam is

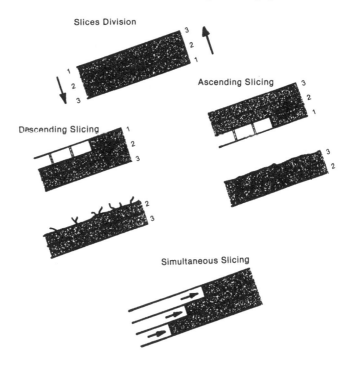

Fig. 9.1.4. Order of slicing
of thick coal seam.

divided in three slices, the order of mining could be carried out by three possible layouts
(Fig. 9.1.4), which are individually discussed below:

a) Ascending slicing: the first slice will be extracted under similar conditions to the
panel mining of normal, thick coal seams. However, extraction of the second slice is
impossible due to subsequent coal deformation in the upper adjacent slices. It is
obvious that ascending slicing excludes caving and requires mining with stowing.[4]
When the lowest slice is extracted, it must be properly filled, such that the fill serves as a
floor for mining the second slice. The success of this method depends on the compaction
of fill and on the strength of the coal material.

b) Descending slicing: the first slice will be extracted also in similar conditions to
medium thick seam. After the roof is caved and has settled, then the second slice will be
extracted. It should be mentioned that an artificial roof is often laid down on the floor of
the extracted slice by cords or wire mesh (5 cm × 5 cm), with the purpose of holding the
gob during extraction of the next slice. Descending slicing is possible but not, however,
without difficulties, particularly if the second slice has a roof of caved rock material and
a floor of coal material. To stabilize conditions of coal extraction the slicing is, in some
cases, done with stowing.

c) Simultaneous slicing: could be assumed as a method of full face mining, where
extraction of all slices advances simultaneously. The cross-sectional position of the
extraction faces is in a zig-zag pattern, where the first slice (most upper one) has an
advanced position in relation to the slice which follows. Success of this method depends
on the homogeneity and strength of coal seams as well as on the readiness of the roof
strata to cave. The simultaneous slice method may have variations, for example, leaving
a protective layer of coal towards the caved gob, which could be in the caving process

together with the gob or recovered as a separate slice by caving, as discussed in some paragraphs of this chapter.

The extraction of individual coal slices is mainly by longwalling (inclined slicing) and long-pillar mining or by shortwall face mining (horizontal slicing). Each method has many variations and peculiarities due to coal seam conditions and prevalent mining technologies in each country. In the past, some local variations of principal room-and-pillar mining were practiced. However, due to low recovery rate, this method is redundant except in some small coal mine operations.

9.1.2. *Longwalling with gob caving*

The most common method for extraction of tabular or gently dipping thick coal seams is longwall mining with ground control by gob caving. The technology of longwalling of thick coal seams follows the panel slice extraction (two or more slices), similar to panel extraction of a single slice. Most longwalling of thick coal seams with gob caving relates to slicing parallel to the plane of stratification. Several variations, due to the order of slicing and the method of roofing, are discussed as follows:

1. Simultaneous slicing with artificial roofing considers the longwall extraction of thick coal seams as practiced in Europe and Japan for many years.[5] The system is based on advancing two or more longwall faces at the same time. Support of the longwall slice faces is with self-advancing supports. As the upper face progresses, an artificial floor is laid between the props to form an artificial roof for the lower face. An artificial roof is made by flexible matts, rolled steel sheet stock, or a combination of both. The flexible mat should be recovered in sequence as extraction with the artificial roof progresses. The mat can be used until worn out. It is very important that caved roof rocks are sufficiently disintegrated, because a large caved individual block may prove extremely difficult to hold above the slice under extraction.

A practical example of this mining system which is illustrated in Figure 9.1.5, is employed in the Miike Colliery in Japan.[1] Two longwall faces in simultaneous extraction are supported by conventional self-advancing supports and a steel matting (metal wire netting, up to 5×5 cm) is laid on the floor in the area between the conveyor and the leg of the support.

In the Japanese system there is a need for extra space for the matting. From Figure 9.1.5, it can be seen that this system is best suited to the one web back support technique.

Simultaneous slicing by longwall mining is the most common method for extraction of coal seams, of up to 10 m thickness i.e. in which a maximum of three slices can be delineated. Some other mining method of longwalling is usually employed for the thicker coal seams. The best productivity and lowest cost of production is achieved for simultaneous slicing, for example, for the coal mines below 300 m, where entry drainage and maintenance problems usually are serious. However, if only two longwall slices are in extraction a single set of development entries can be driven, which should service both the upper and lower slices, as illustrated in Figure 9.1.5. The single entry shown acts as both the head gate for the coal panel at the left and the tail gate for the right coal panel. The entries are lined with steel arches to withstand the high pressure. The entry is packed to seal it from the resulting gob and to provide additional roof support. It is

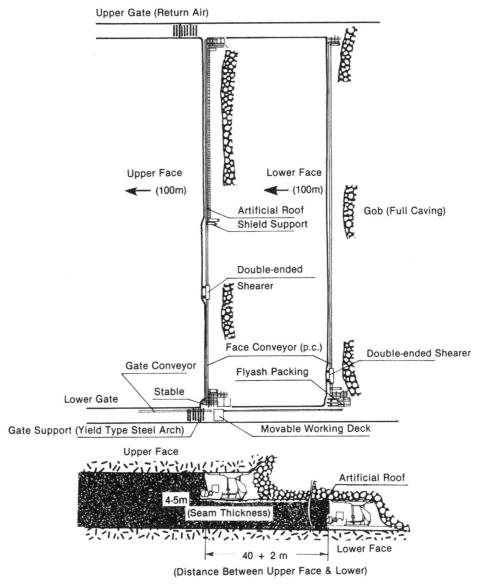

Fig. 9.1.5. Layout of the longwall slicing method at Miike Colliery, Japan.

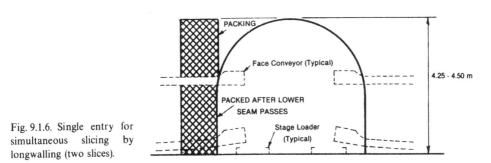

Fig. 9.1.6. Single entry for simultaneous slicing by longwalling (two slices).

obvious that great economies of development cost can be achieved by using single entries for not only two slices of mining in one panel but also for two adjacent panels (Fig. 9.1.6).

2. Simultaneous slicing with waste parting as an intermediate roof consider a coal mine which has a thick coal seam with waste partings which can be left in place during extraction of coal slices. Under these conditions it is not necessary to make an artificial roof between longwall slices within a panel.[6]

For example, use of waste partings as an intermediate roof for simultaneous longwall slicing is used to depths of 600 m below sea level in Japan. The upper coal seam-slice varies between 2.8 and 3 m thick and the lower coal seam-slice is between 2 and 2.5 m.

The details of this mining technique are similar to those for simultaneous slicing with an artificial roof, with the exception that the role of artificial roofing has been taken by rock partings.

3. Simultaneous longwall slicing with coal layers as an intermediate roof, is usually applied for thicker coal seams, where more than two coal slices are in extraction. The thickness of coal layers left as an artificial roof is between 0.5 and 1.0 m. However, by this practice coal recovery is decreased and remnant coal layers in the gob area are a potential source for underground fire, particularly for coal that is very susceptible to self-ignition.[7] The risk of coal self-ignition is increased as the remnant coal layers are fractured during slice extraction.

Descending slicing by longwalling with a coal layer as an intermediate roof, with self-advancing supports, has been practiced in the U.S.S.R. It has been reported that with this system it had been possible to achieve a satisfactory production where three headings simultaneously had an average individual face productivity of between 15 and 18 tonnes per man shift (UGOL', October, 1975).

4. Non-simultaneous descending slicing has been practiced in several mining companies in Europe and Japan by using a method in which extraction of the lower slice comes into effect when the caved material from the roof strata of the upper extracted slice has fallen and consolidated (about 2 years). For this method, it is necessary to have a separate entry for each slice, which increases development costs. However, the productivity should be higher because each slice is extracted independently and each can advance at its own pace.

Non-simultaneous descending slicing by longwalling can have the roof support for the lower face very similar to its simultaneous counterparts (artificial roof; waste partings or coal layers), with the exception that both Japanese and European mines are using techniques wherein the artificial roof is laid by the upper set of supports as they advance. The support is usually by using one-web back chock type supports with a wire mesh strip unrolled behind the shearer. This method is successful but extra crew is required to attach the wire mesh in an area of unsupported roof (Fig. 9.1.7). The requirement for the wire mesh or rock parting, as an artificial roof for the lower slice, is necessitated by the use of chock type roof supports which do not permit spills of caved material.

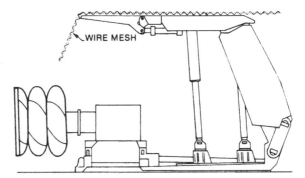

Fig. 9.1.7. Non-simultaneous longwall slicing with an artificial roof support of wire mesh.

However, by introduction of the full canopy chock-shield support (briefly described in simultaneous slicing with a coal layer as an intermediate roof) for non-simultaneous descending slicing, where the gob of the caved roof is settled, it is not necessary to leave coal layers or wire mesh as an artificial roof for the lower slice face. The success of this method is very dependent on providing immediate forward support of the roof after the passage of the shearer and on having skin-to-skin roof coverage to prevent spill of caved rock material.

Studies of the possibility of using consolidated gob as a lower slice face artificial roof have been of particular interest. These studies in European mines suggest that the subsidence and compaction of gob material causes the lower portion of the settled gobs to be very dense and tightly packed. Based on these results, it has been shown in Poland that this packed gob provides an adequate immediate roof for the lower slice without the need for additional artificial or natural roof partings.

5. Non-simultaneous ascending slicing has been used in England, where thick coal seams are divided into two slices waste parting.[2] The lower slice is taken first. In this operation, an artificial roof is formed by laying a wire mesh on top of the powered supports, thus forming a base to prevent caving of the upper slices which then subsides onto the floor of the extracted bottom slices. The intention here is to prevent fracture of the interwaste rock as it bends in the subsidence phase. However, this order of slice extraction is not very efficient and current working faces are being phased out, the main reasons being the inconsistence of output, due to the artificial floor of the waste parting collapsing, and difficulties to support this roof.

Ascending extraction of coal slices is always followed by sagging and fracturing of upper waste or coal layers (dense fracturing particulalrly along joint planes). This phenomenon results in spontaneous heating of the upper coal slices and coal self-ignition can occur within a short period of time (in some cases within six months). Also, by ascending order of extraction, the floor of the upper slice is difficult to work, because it is formed on the extracted slice of coal. At the present time, ascending slicing of thick coal seams is directed toward another system, in which the bottom longwall slice represents an undercut for the coal above, which is then extracted by induced (blasting) or natural (gravity) caving. In this case extraction of the complete thickness of the coal seam is simultaneous.

9.1.3.   *Slicing with stowing*

Hydraulic stowing is mainly in effect at the present time because pneumatic filling is seldom practiced. The source of stowing material in coal mining are quarries and gravel pits, from which material is handled to the mine collars for underground transportation. Fill transportation and placement is hydraulic by pipes. Stowed material is discharged from the lowest point in the filling area, retreating upwards along dip. Drainage is by a combination of decantation, exudation and percolation down-dip to a collecting sump. A drainage line could be installed along the strata floor or on the floor of the stowed area to reduce maximum flow path lengths. In the case of gently pitching and pitching coal seams, hydraulically placed stowing allows a tight filling, right up to the roof of the void. Dewatering of the fill is of paramount importance because water-solid mixtures, with no shear strength, change to a granular solid with considerable shear strength. Retention of some fill water results in apparent cohesion, probably adequate to allow an unsupported vertical face up to 1 to 2 m, depending on the size distribution of fill material.[8]

The main cost of stowing is the construction of the barriers of the placed fill and pumping of drained water back to the surface (up to 1 tonne of water per tonne of solid).

Two principal mining methods are considered, slicing along stratification planes and horizontal slicing of thick pitching seams which are discussed individually.

1. Non-simultaneous slicing with stowing. While coal extraction is by the longwall technique, and may be in ascending or descending order as is the general case with all

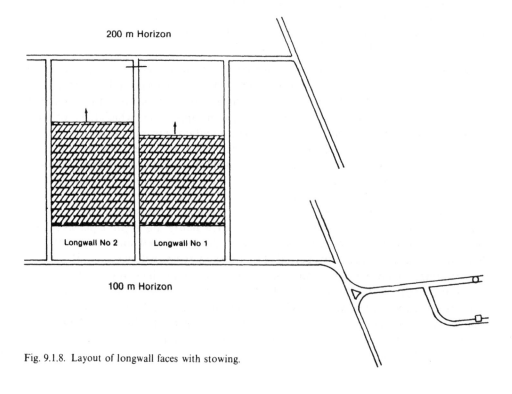

Fig. 9.1.8. Layout of longwall faces with stowing.

mining methods by slicing, the general conclusion is that descending slice extraction has advantages over the ascending one, particularly for extremely thick and pitching coal seams.[9] However, much of the Polish experience has been with a technique of ascending slicing, where the lower slice extracted area has been supported by prop and bar sets, filled by a sand slurry and the upper roof (floor of the next slice) allowed to settle on backfill. A practical example of this method is briefly represented for extraction of a coal seam of 7.3 m thickness and a 25 to 30° pitch at a coal mine in Sudamdih Colliery (India).[4] The extraction is at an average depth of 260 m, where the roof of the coal seam is sandstone and the floor is a sandy shale and sandstone. The direction of extraction is up-dip to allow better compaction and drainage of hydraulic fill. The length of the longwall face is up to 110 m, and the height is 3.5 m (Fig. 9.1.8).

The extraction begins by blasting faces, which are provided by driving 6 m deep stables in the longwall. Face support is by a prop and bar arrangement on a spacing of 1.2 m. Coal handling is by chain conveyors along the face, which is shifted after every 6 to 8 m of face advancement. The extraction has been on two panels, which advance 25 m apart (zig-zag pattern). The stowing cycle starts when the extraction cycle is completed and filled as shown in Figure 9.1.9. Coal extraction and stowing are carried

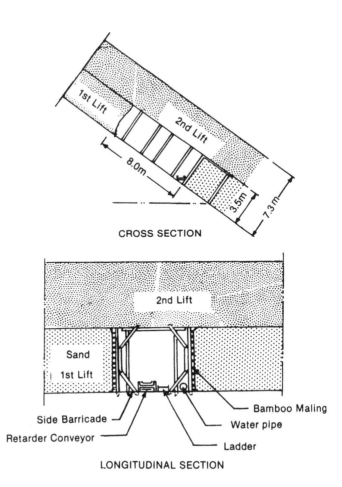

Fig. 9.1.9. Non-simultaneous slicing with stowing in ascending order of extraction (Polish system).

out simultaneously in two panels, one crew works on drilling, blasting, supporting and coal handling; while another crew is engaged in stowing.

The face productivity by this method is up to OMS 5 tonnes and the total production from the longwall face is rather low.

2. Double winged by slicing and stowing is used in France and the U.S.S.R. It was originated in France and was called the Lorraine method, according to the coal field where it originated.[10] The principal elements of this method are that a panel is developed with two raises, which are utilized for material supply and air return. The panel is worked by slices in ascending order, each being 2.5 m thick. Steel tubing is used for coal transportation and air intakes and it rises as stowing advances. They are located at the middle of the panel, dividing it into two wings (Fig. 9.1.10). When a slice is mined-out in one wing, this wing is hydraulically stowed to 1.5 or 2.0 m of coal back. The armoured conveyor, fitted with floats, rises during the stowing operation. The winning machine is meanwhile parked in the other wing. Two types of extraction machines are in use: ANF and SAGEM, the application of which depends on the thickness of the coal seam. The machines are equipped with floats and rise during stowing.

The production and productivity are low, and the only consolation is the possibility of simultaneously working several faces. Success of this method depends on the suitability of stowing material, and its homogeneity. Set-backs are caused by the impossibility of having the conveyor float during the stowing and the hydrostatic pressure crushing the steel tubing.

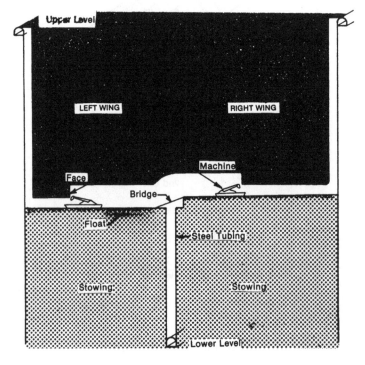

Fig. 9.1.10. Double wing rise slicing with stowing (French system).

It should be mentioned that the double wing rising method in Kusbass (U.S.S.R.), replaced, in some instances, shield mining with caving and, in all instances, descending horizontal slicing, where extensive timbering and formation of artificial roofing for succeeding slices caused certain difficulties, particularly in coordination of the coal extraction and stowing sequence.

9.2. STABILITY OF SLICE-FACE STRUCTURE

The division of a thick coal seam, on the basis of the principal structure could be a single slice, which offers partial or total seam extraction (full face mining), or by multiple slices. A single slice in full face mining is a function of the mining method applied and it could have various relationships to the stope and pillar structure, as discussed in the corresponding chapters. However, a single slice-face structure has similar stability characteristics to medium thick coal seams, with instability increasing with enlarging working height. At present, single slice mining of thicknesses up to 5 m (U.S.S.R.) whilst implementing face reinforcement, sprags, face plates, and/or a battered upper section of the face (shaped face) to offset instability of the stope structure, is practiced.

Primary consideration is given in this chapter to the multiple-slice mining methods where, with the aspect of stability of slice-face structure, three principal stope configurations are considered. There is no limit of slices, for example in some mines in Eastern Europe, up to ten slices are extracted.

9.2.1. *Ascending structure*

This type of structure considers the coal extraction by slicing in ascending order with hydraulic stowing. This order of slicing is usually applied in cases where the roof strata do not cave readily. The analysis of slice-face structure is briefly summarized as follows:

1. Configuration of the face is related to the formation of the next slice when extraction of the first bottom slice has been completed and stowed.[11] Under these circumstances, the upper slice exhibits a displacement, the magnitude of which depends on the degree of fill compaction. There is a general opinion that hydraulic fill compaction is in the range of 15 per cent, which is sufficient to induce displacement with the development of fissures which are more or less normal to the stratification planes, i.e. direction of maximum vertical displacement.

The mining consideration described relates to the formation of an unstable coal block with both a free face and frictional boundary at the roof strata and to the fill in addition to the contact failure at the main body of the slice in extraction. The stability of the critical working face block could be analyzed, with reference to Figure 9.2.1, as follows:

a) For the weight of the coal block:

$$G = \gamma_c H$$

b) For the vertical component of weight:

$$W' = G \cos \alpha$$

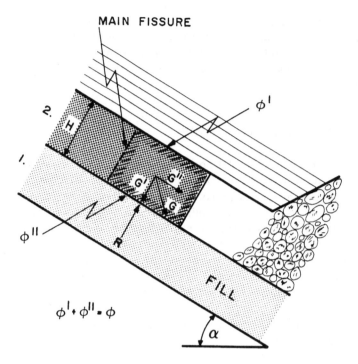

MAIN FISSURE

Fig. 9.2.1. Critical stability of ascending slicing.

c) For the horizontal component of weight:

$$W'' = G \sin \alpha$$

where

γ_c = unit weight of coal;
H = height of slice;
α = angle of inclination of coal seam.

The criteria of stability are expressed by

$$\gamma_c H \sin \alpha < \phi' + \phi'' = \phi \tag{4}$$

$$P_p \geq \frac{\gamma_c h(\sin \alpha + \phi \cos \alpha)}{\phi} \tag{5}$$

where

ϕ' = friction along roof strata;
ϕ'' = friction along fill;
ϕ = total friction;
P_p = resistance pressure of the fill (passive earth pressure).

Sliding of the critical working face is prevented when a passive earth pressure of the fill is developed and shearing resistance along stratification contacts is in effect.

The mine configuration stability obviously depends on the interaction mechanics between the mine roof, the coal slice, and the stowed material and on the magnitude of stress-strain states at the coal block boundary. The slice thickness, the structural-geological features of the seam and the coal's mechanical properties, rather than the dead load of the loose block structure, are the critical stability parameters.

2. The mine floor and roof are of particular importance in ascending slicing with stowing because there is a difficult floor on which to work heavy mechanized face equipment. If supports sink, due to an inadequate bearing capacity of the floor, then the roof will cave on the face. However, in some countries, as for example in Poland, specially adapted face equipment has been developed for such conditions.[12] This equipment is also designed to avoid coal dilution by sand fill during coal face excavation. Hard coal mining experience confirms that this equipment operates better on a hard floor. Hence, an artificial floor mat may be placed during the supporting and stowing stages of the lower slice to obtain better floor conditions.

The most successful ascending coal extraction is achieved if a layer of coal is left in the floor during extraction of the preceding slice (Fig. 9.2.2). The coal, in many instances, may prove a better floor material than some rock types. Under these circumstances a high efficiency operation of longwall face sets is obtained. According to theoretical calculations confirmed in practice, the thickness of the coal layer left in the mine floor should be from 0.3 to 0.7 m, depending on the coal strength properties.[13] Floor reinforcement with coal layers results in lower coal recoveries and some hazard of spontaneous combustion in the gob if the fissured coal is liable to self-ignition.

If a bench of coal (layer) is left on top of the fill then a conventional chock support is used, as is common in the single slice longwall mining of medium thick coal seams.

3. Coal pillars of any kind have a particular consideration in multiple slice mining of thick coal seams. However, in the case of slicing with stowing, the pillar in the extraction structure exerts a lesser stability role because the load is taken by fill. In some

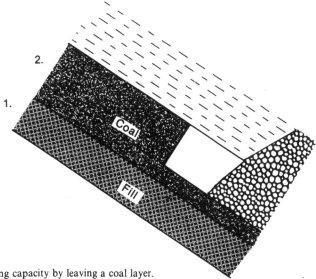

Fig. 9.2.2. Increase of floor bearing capacity by leaving a coal layer.

instances, the stowing is reinforced by artificial pillars, so-called coal-crete, with a strength commensurate with that of the extracted coal.[12] This type of pillar is composed of washings from plant waste or run-of-face coal and portland cement.

Ascending slicing with stowing is generally considered as pillarless mining, where remnants of coal pillar structures are not left in the gob and thus results in a low risk of spontaneous combustion. However, productivity is approximately one half that of slicing with caving. Slicing with stowing offers stability of mine structures and a decrease in the large amount of subsidence normally associated with extraction of thick seams by strata control with caving. Finally, it should be pointed out if thick coal seam mining is with complete filling, the caving and jointing of surrounding strata is insignificant and the limited displacement of overlying strata would be uniform and smooth, without disruption of the continuity of seams above.

9.2.2. *Descending structure*

Descending, or top slicing with caving, is the most common method of mining thick coal seams. After taking the first slice, the roof strata are caved and the fragmented rocks are compressed by the dead load of overlying strata. The caved and consolidated gob is a key factor in analyses of the slice-face structure, as further discussed.

1. The configuration of the face is related to extraction of the top slice, which is mined-out in a similar manner as for the case of single slice mining. The caved roof strata fill the excavated slice void with fragmented rocks resting on the top of the next slice to be mined.

After extraction of the first slice, the bottom of the seam is subjected to stress relief by displacement along, and perpendicular to the bedding planes. It should be mentioned that thick coal seams, which are prone to coal and gas outbursts, are often de-stressed

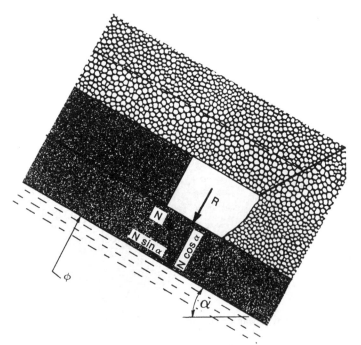

Fig. 9.2.3. Critical stability of descending slicing.

by taking the top slice from under the roof strata and further extraction is continued by full face mining.[14]

With respect to coal face stability, principal consideration should be given to the bottom slice, which represents the mine floor and bears the longwall face self-advancing equipment. The redistributed stress in this slice, due to normal stresses induced by the support load and shear along bedding planes, are as illustrated in Figure 9.2.3. The stability of the working slice face will depend on the stress state of the lower slice exposed for roof support as given by,

$$N \sin \alpha \geq \phi \qquad (6)$$

$$F \phi \geq N(\sin \alpha - \phi \cos \alpha) \qquad (7)$$

$$F \geq \frac{N(\sin \alpha - \phi \cos \alpha)}{\phi} \qquad (8)$$

where

F = roof preload support;
N = floor reaction force (opposite to gravity);
ϕ = friction along coal and floor strata;
α = angle of inclination of coal seam.

From this analysis it could be inferred that the slice face is stable, with a capacity to control the immediate roof of caved and compacted rocks, if the upward reaction from the lower coal slice is in equilibrium with the installed supports.

Transverse instability of the extraction structure is exhibited immediately after completion of the extraction lift. At this point there is a maximum movement upwards, which might activate lateral motion. If lateral motion is in progress, it is very difficult to set the advanced support, the preload pressure of which should hold the floor and roof in place and resist lateral movement.

2. The mine floor and roof also have certain peculiarities in top slicing. For example, in ascending slicing the mine floor is the most difficult to work because it is artificial (fill) but in descending slicing, the roof is most difficult due to its composition of caved rock.

Principal consideration in countries where thick coal seam mining by slicing exists is given to reinforcement of the mine roof and prevention of the caved waste from one slice entering the working space of the next slice.

Three concepts of roof control have been introduced, as described individually below:

a) The most common technique is to lay a mat on the floor of the top slice under the initially caving roof. This mat then moves down with successive slices, preventing fragments of the caved roof material flushing into the current working space, as illustrated in Figure 9.2.5 of the previous section. It should be pointed out that the mat also provides some degree of support to hold the caved rock mass together. As reported previously, only minimal face support has been used when mining slices below the first, relying substantially on the mat to hold open a safe working area at the face. With respect to strata control, the mat characteristics of an artificial roof have little influence because key factors of stability are cavibility, good fragmentation of caved rock and good flushing characteristics of the roof strata.

b) The second common technique is to leave an interslice of coal or rock layers.

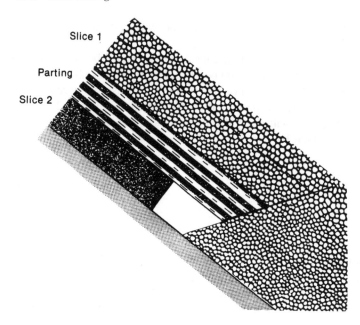

Slice 1

Parting

Slice 2

Fig. 9.2.4. Slice division by waste/coal parting (Zenica Colliery, Yugoslavia).

Some coal seams may conveniently be subdivided into several distinct intervals of economic quality coal, suitable for slice extraction, separated by consistent thickness of waste partings or inferior coal as shown in Figure 9.2.4. The critical parameter of roof stability is that the thickness, geology and mechanical property of the inter-slice layer should remain coherent on the coal face and act as a beam above the supports when mining a slice beneath.[15] The interface must cave readily behind the supports and ensure continuous downward movement of the gob. If thick coal seams do not consist of waste partings and inferior coal layers, the alternative is to leave an inter-slice layer of coal, at a predetermined thickness to given the required roof stability. In this case, the drilling of short holes should be applied to maintain a constant thickness of the coal inter-slice. As with any mining system that leaves coal behind in the gob, one important requirement is that the coal is not liable to spontaneous combustion.

c) The most recent technique requires that the caved rock be compacted by overlying strata after taking the first slice and that it be supported by the intrinsic ability of shield type supports. This type of support permits mining of the slice directly under the caved debris of overlying slices. No artificial mats are required. This technique is satisfactory for operations with a limited number of slices. The properties and compressibility of caved rock strata are the main factors influencing the degree of rock reconsolidation, which might be complete or not. Reconsolidated rock improves the strength conditions of the roof and allows it to be treated as a natural roof, because it overcomes rock flushing and provides some self-supporting capability. Attempts have been made to artificially consolidate caved rock by the addition of portland cement and/or various pozzolanic products. Excessive cost has severely limited application of this concept. It is observed that some rock types (shales, siltstone and others) show a tendency for natural consolidation due to existing gob conditions. From an engineering point of view, the best natural reconsolidation is achieved if the caved rocks have a high clay content (up to 40 per cent) consisting of 4 to 5 per cent moisture and if they are loaded by overburden pressure for a sufficient time, which might vary between 3 months and 5 years.[13] It should be pointed out that, if there is a lack of sufficient natural moisture or natural water inflow, it is necessary to supply 1000 litres per minute per 100 m of longwall face to the caved rocks. Under slice mining conditions, with a reconsolidated gob rock, the use of shield type powered supports is necessary to utilize the advantages of this method.

The analysis in the discussion above considers only the mine roof as a critical parameter of structural stability, by assuming that the floor of the coal slice is of sufficient strength and bearing capacity to withstand coal excavation.[16]

3. Coal pillars in descending slicing could cause certain difficulties during successive coal extraction. It is a well known fact that the success of mining lower coal slices depends also on the degree of de-stressing. The degree of de-stressing is a function of the mining method and the ability for total coal recovery. However, coal losses in the stumps, pillars and others, will produce an extremely unfavourable effect on the stability of coal faces and roadways in the subjacent coal bed, due to transmitting a concentrated strata stress, as illustrated in Figure 9.2.5.

It is obvious that mining should be by continuous faces, where coal losses are less frequent, to avoid adverse effects on subjacent slices. Also, in continuous mining, roof strata subsidence is more rapid and more complete, thus permitting a shorter waiting time for extraction of lower beds and giving more flexibility in mine layout.

High coal recovery, as an aim of good conservation practice, is also a factor in reducing the hazards of underground fires due to spontaneous combustion of coal.

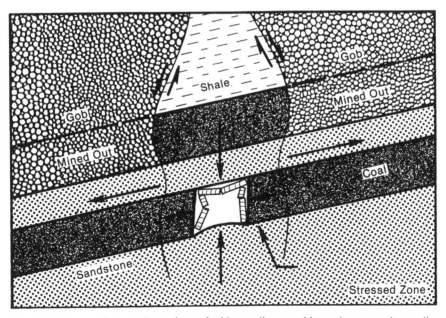

Fig. 9.2.5. Adverse effects on the roadway of subjacent slice caused by coal remnants in top slice.

Crushed and loose coal in the remnants of the gob is always liable to self-ignition. Complete coal extraction not only eliminates a source of fire, but also, due to a process of quick caving and subsidence, the gob is isolated from the inflow of air. This phenomena is equally important for coal and for carbonaceous roof strata, which is also subject to spontaneous combustion.

9.2.3.   *Simultaneous structure*

Simultaneous, or multi slice, systems exhibit parallel faces following one another at close spacings and they represent some variations of full face mining of thick coal seams but without implementing the single pass method. Simultaneous slicing offers the obvious advantage of high productivity but suffers from the potential disadvantages of high capital outlay (multiple sets of equipment) and interdependence of advance rates between the faces. Synchronized slice mining is discussed as follows, in the same manner as single slice mining;

1. Configuration of the face is strongly influenced by the layout of individual slices as well as by coal seam properties. The peculiarity of configuration for multiple slice mining is the zig-zag geometry, which is governed by the two-dimensional parameters illustrated in Figure 9.2.6.

From analyses of deformation and failure in multiple slice mining, it could be concluded that the stability effects of single slicing in both ascending and descending order exist. As illustrated in Figure 9.2.6, stability of the top slice is dependent mostly on the magnitude of component forces perpendicular to the seam, as written:

$$W \cos \alpha \leq N \cos \alpha \tag{9}$$

It may be seen that the stability of the lower slice is of a similar context but

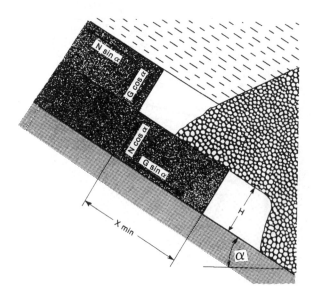

Fig. 9.2.6. Critical distance between multiple coal slice faces.

consideration has to be given to resisting force components parallel to the bedding plane, as written:

$$W \sin\alpha \leq N \sin\alpha \qquad (10)$$

The important feature is that the magnitude of forces acting on the coal faces also depend on the distance between slice faces and the height or thickness of individual slices in addition to the mine depth and angle of inclination of the coal seam as further discussed.

a) The minimum distance between multiple slice faces (X_{min}) is the most important design parameter in this mining method. The stability of individual working faces exists only if an adverse interaction is not developed between them. To eliminate such a possibility, it is necessary to space slice faces at satisfactory distances, if not, then their deformation and failures are eminent. For example, close spacing of slice faces result in severe fracturing and displacement of the coal where broken upper slices exhibit downward movement in the direction of maximum shear stress. Broken lower slices also exhibit movement but along a bedding plane. It is a point of interest that underground observations showed (Zenica Colliery, Yugoslavia) fracturing in top slices had been propagated to the point where interaction effects with a lower slice were eliminated. On the basis of underground observations in Japan's coal mines, practical parameters for spacing between faces of the individual slices have been developed. For example, in the case of synchronized slicing with middle plates left as partings, Nakajima suggested:

– The top slice should start first, when basic cuts are made (Fig. 9.2.7/a).

– Advancement of the top slice should be about 5 m and the lower slice about 3 m (Fig. 9.2.7/b). The lower slice is already under the top slice gob.

– Only the top slice should then be advanced (20 to 30 m) until the first stress concentration occurs (Fig. 9.2.7/c).

– When the distance between lifts equals the safety distance, the lower slice may start (Fig. 9.2.7/d).

– The safest distance between top and lower slices during synchronized coal extraction is in the range of 30 to 40 m (Fig. 9.2.7/e).

Stability of the slice faces and elimination of interaction between them, is of particular importance in adopting powered self-advancing support and coal extraction machines for synchronized slice mining.

b) Optimal thickness of the slices (h) is an important factor for structural stability. In the first place, it is obvious that a designer could choose an optimal slice height which should best accommodate mining equipment of multi lifts. However, the thickness or height of the slice should be taken into consideration for design of the minimum distance between slice faces. For example, on the basis of investigations in underground coal mines (Donets Coal Basin), an empirical equation for minimum distance (X_{min}) between slice faces was derived as a function of their height (h) as given below:[18]

$$X_{min} = 12h + 3.5h^2 \qquad (11)$$

It should be noted that the equation above is in agreement Nakajima's proposed distance of 30 and 40 m between individual coal faces.

The configuration of the coal face of thick coal seams in which mining is in progress by multi slice extraction is simple, but maintenance of constant geometrical pro-

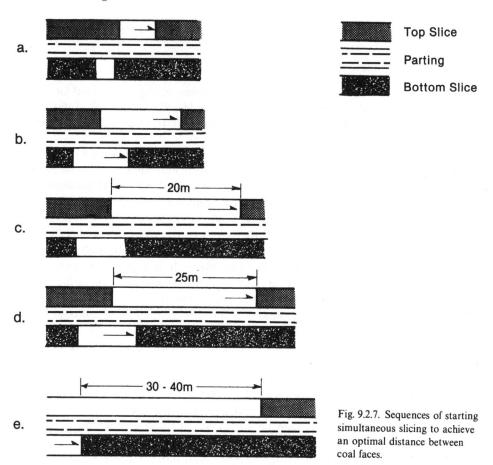

Fig. 9.2.7. Sequences of starting simultaneous slicing to achieve an optimal distance between coal faces.

portions is difficult. Very often it is necessary to stop production of one or more slices until the faces of one or more following slices are lined up in the required position. This phenomenon, as well as temporary stopage of individual slices due to the necessity of maintenance, are main setbacks of synchronized multi slice mining.

Large coal mines with high production rates in Eastern Europe are more favourable for multi slice synchronized mining because complicated and expensive stowing is eliminated by production face concentration and strata caving ground control, regardless of some limitations in structure stability.

2. The mine floor and roof in simultaneous slicing might be of a complex composition. Further complexity of roof and floor units in individual extraction slice structures could be exhibited by leaving coal or waste partings as protective layers between individual lifts.

With respect to the stability of slice-face structures in all slicing methods, particularly in synchronized slicing, the main consideration should be given to the lithology of roof strata, which reflects particular mechanical behaviour and properties, as briefly listed below:

a) Plastic and soft strata (mudstone shale, silty-shale and others) would cave rapidly with good flushing characteristics. The displacement and subsurface subsidence is also rapid with appreciable vertical extension, which in a short period of time could progress to the ground surface. The rate of subsidence propagation to the surface is directly proportional to the mineable thickness of the coal seam. This type of rock strata is optimal for ultimate compaction in a shorter period of time. Longwall mining with shield support is carried out without varying artificial mats laid on the mine floor of the coal to erect an artificial roof during extraction of following slices.

b) Elastic and hard rock strata (silty sandstone, sandstone, conglomerate and others) would not cave rapidly and might even have a delayed caving which often results in roof falls of large area (air blast), where individual rock layers above tend to fall one after another. The hard rock strata with delayed and uncomplete mechanics of caving, could be a very negative factor for synchronized multiple slice mining because sudden caving then crushes the powered support on impact. The solution of this problem often is found by leaving layers of rock of coal between extraction layers. Under these circumstances, with solid roof layers, a chock-shield support is usually applied which could sustain the impact force during large roof falls.

Generally speaking, similar conditions exist for all three types of slicing except for a specific variation of synchronized mining in which several slices could be implemented with coal caving between them during simultaneous advancement. In this case, for each increment, a slice is mined at the base. The mechanization and roof support is represented by two types of self-advancing support. For example, the slice under the strata roof (top slice) is supported by AMS complex but the slice under the coal (bottom slice) is supported by KTU shield complex (Fig. 9.2.8), which is also used in sublevel caving methods. This shield support also includes linked rear beams supported by an inclined prop (so-called banana prop). This support is used under a mesh set at the face. This mesh allows a broken coal layer-roof to be held and also permits caving and drawing through openings (windows) in the canopy. It should be pointed out that dilution could be avoided to certain degree by this variant.

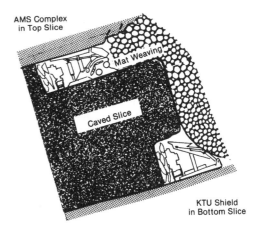

Fig. 9.2.8. Integrated simultaneous slicing and slice caving (Blanzy coal field, France)

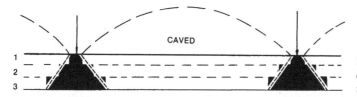

Fig. 9.2.9. Abutment pillar of simultaneous slice mining.

3. Coal pillars in simultaneous slicing are usually exhibited as abutment structures which make a boundary of thick extraction panels. The zig-zag method for full face advancement of thick coal seams is a main factor in the formation of a pyramidal pillar morphology (Fig. 9.2.9).

With respect to mine stability, it should be noted that a large pillar has a limited bearing capacity. A particular disadvantage of this pillar is its capturing of a large coal reserve. For example, a coal pillar in a 1 m thick seam with a least width of 5 m (W/H ratio of 5) per unit length will capture only 5 m³ of coal. However, in a 10 m thick coal seam, a pillar with the same ratio $W/H = 5$, would capture 500 m³ of coal. It is obvious that strata control by pillar in thin coal seams (single slice) is satisfactory but is irrelevant in thick coal seams (multiple slicing).

A general conclusion might be that coal pillars in slice mining should be eliminated as supporting structures as much as possible and that ground control should be effected by artificial support and complete caving of roof strata.

9.3.   STRATA MECHANICS

The strata mechanics of thick coal mining by slicing have a number of characteristics which depend on the choice of ground control methods and mining technology.

Several topics are of concern to the stratum mechanics of this particular mining method, the most important of which being discussed as follows:

– Displacement of the coal roof has been analyzed on the basis of data obtained by measurements of convergence (vertical displacement), lateral movement (horizontal displacement) and expansion above the face. The particular strata mechanics relate ascending coal slicing or integrated slicing to coal caving systems.

– Abutment stresses of the gob roof have been considered for cases of unconsolidated caved rocks and for consolidated gob rocks. The strata mechanics mainly correspond to descending slicing and multiple slice systems.

– Interaction between the coal structure and fill material has been analyzed by the principal mechanics of their deformation and also their common interrelation, although interaction with rock strata has also been examined.

– The influence of geological anomalies on strata mechanics is well exhibited in thick seam mining, particularly with slicing and stowing. Displacement caused by mining factors and geological anomalies have been considered.

Stratum mechanics are based mostly on field observations and on only limited theoretical analyses because many analyses are irrelevant to actual underground mine situations.

9.3.1. *Displacement of the coal roof*

The classic example of strata displacement is in the French coal mines, where extraction of very thick seams with soft coal resulted in appreciable movement and a high abutment stress. The mining methods practiced are either slicing or sublevel mining. Irregular thicknesses and gradients of the coal seams are negative factors for the division of coal seam slices, which explains the development of hybrid methods of undercut (one slice) and sublevel caving (extension of the slice height). For example, the original slice height was 3 m, which was then increased to 4 m and finally to 10 m by integrated caving. This method of mining is discussed in chapter 11, where sublevel caving methods are presented.

Displacement analysis in the French coal mines gave consideration to several components of strata movement as further presented for the case of ascending slicing.

1. Convergence measurements had been of particular interest to investigators because of their relationship to the stability of powered support (Fig. 9.3.1). The convergence of the face for Darcy and Ricard coal fields follows an equation which has been

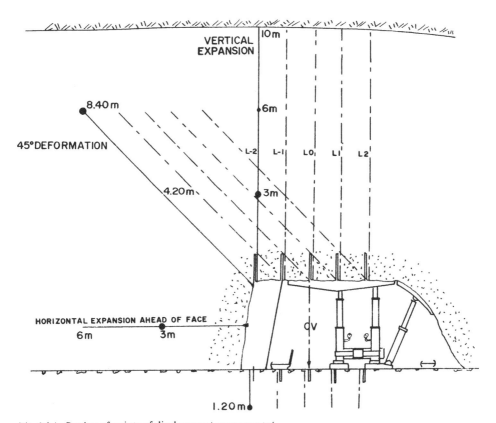

Fig. 9.3.1. Section of points of displacement measurement.

established from statistical analyses of 140 longwall faces in thin seams, as given below:[1]

$$C = 0.2(q \cdot t)^{3/4} h^{-1/4} \qquad (1)$$

where

C = convergence;
q = parameter as a function of gob formation (stowing q ≃ 0.5; caving q ≃ 1.0),
t = total seam thickness;
h = mine depth.

Application of this equation for calculation of convergence has been extended to the Rozeley coal field, with a correction factor to represent the thickness of the lower slice under the parting. Which corresponds to the true amount of mined coal,[19] rather than the total seam thickness.

Measurements of convergence in the top coal slice and the roof rock strata showed the seam mechanics of ground movement as convergence at the extraction face, which is governed by two parameters the working face height and the coal seam thickness. It was also established that the rate of convergence over the roof span held by self-advancing supports is smaller than that ahead of the supports. This indicates a possiblility that convergence may be reduced by increasing the bearing capacity of the support (but this reduction would probably not exceed 15 per cent). The convergence exhibits three principal stages as given below:[20]

a) During the short period in which the roof is unsupported;
b) During coal excavation on the face, due to the nature of roof weakening;
c) During caving, most likely due to incomplete separation of advancing and caving phases.

However, it should be appreciated that the magnitude of convergence is also influenced by time. It is known that increasing the speed of face advance reduces convergence but the magnitude of this reduction is quite small (probably in the range of few per cent).

2. *Lateral movement* indicates that the supported overhead coal is fragmented and progressively tilted toward the gob by the thrust developed from expansion at the unsupported area of the front. For example, the horizontal displacement in the vicinity of the canopies of powered supports, as measured by the relative displacement between the roof and floor anchors, showed that the coal overhead flowed continuously toward the gob area. The mean overall displacement between the face and the caved area reached 550 mm. This horizontal displacement accelerated during advance of the powered supports. Figure 9.3.2. summarizes the percentage horizontal expansion determined from these tests.[20]

For example, underground monitoring of displacement in the Darcy Coal Mine, where a point 6 m ahead of the face was taken as the reference, showed particular expansion movement. Initially, the relative expansion at the 3 m point was 1.2 per cent (the coal most likely was still coherent) and at the face the expansion was 5.5 per cent (the coal was fragmented). These data clearly suggest that a significant horizontal expansion was induced with advancement of the working face, which resulted in de-stressing of coal on the face.

The overhead coal requires a wire mesh above the supports, due to its fragmentation

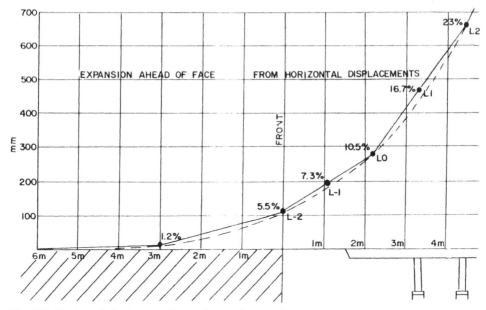

Fig. 9.3.2. Horizontal displacement due to distance from the face.

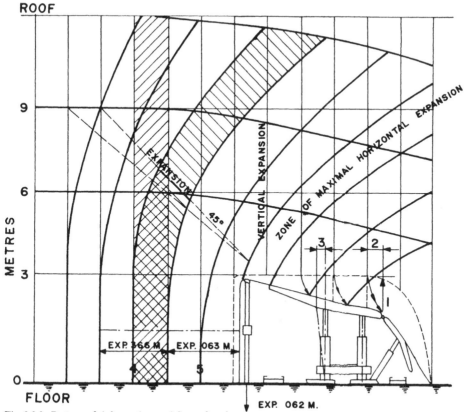

Fig. 9.3.3. Pattern of deformation and flow of coal.

and unconsolidated state, as discussed under previous headings. Absence of the artificial reinforcement by wire mesh may result in coal falls and creation of caving ahead of the face.

3. Expansion above the face has been measured between the roof rock strata (natural roof) and the coal face (artificial roof), and at an angle of 45° from the face; as well as ahead of the face. From all displacement measurements, a pattern of deformation and flow of coal around face structure has been constructed (Fig. 9.3.3). However, it should be pointed out that construction of coal displacement figures involves some assumptions, but nonetheless it gives a good general picture of the flow of coal towards the gob area. These expansion measurements clearly illustrate the strata movement of thick coal seams. It is obvious that the degree of deformation and failure of the coal might cause instability of the working face in progress as well as in the coal slice above the face. To achieve ground control to a certain degree therefore, the upper slice is caved behind the advancing longwall face rather than extracted as a separate structure.[21]

It is necessary to point out that ascending slicing causes an appreciable movement of roof strata (geological roof) which is transmitted to the coal below with a high impact (shock). Investigation showed that the amplitude of movements and the disintegration of coal by the pressure in ascending slicing, exhibits an essential difference from the descending (top) slicing, which suppresses the effects of roof strata pressure (i.e. absorbs shocks).

9.3.2.   *Abutment stress of the gob roof*

The stress distribution and concentration, due to interaction between the longwall face and the mine roof of caved rock, should be considered from aspects corresponding to related mining systems. In the first case, consideration is given to the unconsolidated gob, which is exhibited in simultaneous multi-slice mining, and in the second case consideration is given to a consolidated gob, which is exhibited for individual slice mining in descending order. Each case is separately discussed below:

1. Unconsolidated mine roofs have been studied in thick coal seam mining in Yugoslavia.[22] The investigations showed a particular abutment stress distribution over the whole extraction structure which has been considered as one structural unit. As illustrated in Figure 9.3.4, the diagram of stress distribution shows three zones of various stress concentration, as briefly noted below:

a) Front stress zone, where the abutment stress is induced on solid coal below the immediate roof strata. The maximum stress concentration is approximately 6 m ahead of the face of the top slice.

b) A de-stressed zone, located at the face gob area with minimum stress concentration in the vicinity of the support of longwall faces. Actually, stress in this area relates to the dead load of immediate roof strata which are hanging on supports.

c) Back stress zone formation starts at the point where gob consolidation begins. It reaches a maximum approximately 17 m behind the bottom slice face. The maximum stress concentration is close to 3/4 of the maximum front abutment stress. However, the back abutment stress is wider spread because, appreciably farther back, it reaches virgin stress levels to the point where gob rock is completely reconsolidated.

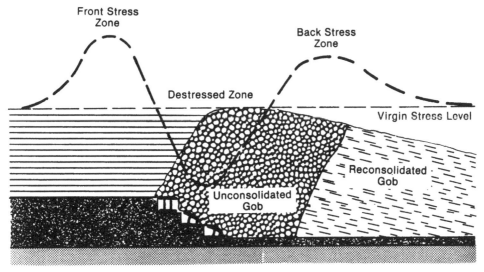

Fig. 9.3.4. Distribution of the abutment stress for multiple slice mining.

The stress pattern described considers a regular stope layout with satisfactory coal extraction under safe roof conditions and without coal losses. The mechanics of abutment stress distribution are in agreement with strata stability because the maximum stress concentrations have been shifted ahead or behind the coal faces. The seam of hard coal without soft partings and caving of roof strata with good rock fragmentation satisfies this stress condition.

Interesting investigations have been carried out in the U.S.S.R. for the case of simultaneous slice mining to establish a stress concentration on frictional props and canopy (top slice) and on the shield supports (middle and bottom slices), with the objective of supporting longwall faces by prevention of spills of caved rock into the working place.[23] However, the investigations suggested that it was necessary to leave a coal layer of at least a 0.35 m thickness to provide protection for the next slice below. Actually, the coal parting has a role to support that part of the exposed roof between the shield and longwall faces, as illustrated in Figure 9.3.5. The measurements of pressure on powered supports indicated:

a) Frictional support of the top slice experienced the highest stress magnitude with stress concentrations at props. For example, a prop to the face had a stress 23 kg/cm², and a prop to the gob had a stress 28.3 kg/cm².

b) Shield supports experienced the main stress concentration at the prop to the coal face. For example, a support at the middle slice exhibited a stress of 22.3 kg/cm², and at the bottom slice of 24.2 kg/cm².

It should be pointed out that in this reported case, the slicing was carried out in hard coal, which offered a solid floor of satisfactory bearing capacity for the self-advancing powered support under a superimposed pressure.

The stress distribution and formation of abutment stress is a key factor for successful synchronized slicing in the case of an unconsolidated gob.

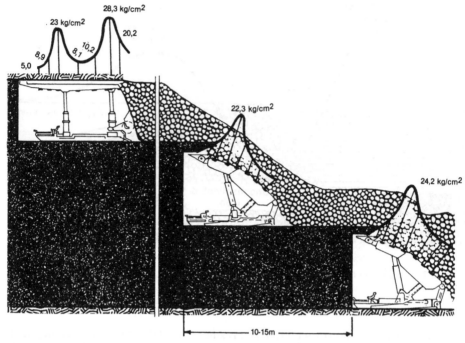

Fig. 9.3.5. Stress concentration on the support for multiple slice mining.

2. A reconsolidated mine roof and its self-supporting abilities were studied in Poland. For old types of longwalling with chock support, the time for gob reconsolidation was estimated between 3 and 5 years. However, this duration is shortened by implementing a shield support. This requires that the reconsolidated gob have sufficient cohesion to ensure the stability of the exposed roof up to 1.0 m wide and several metres long, during that period of time necessary for pushing the successive support set ahead to the new face. The time of reconsolidation of the caved rock could be decreased if adequate gob saturation and sufficient overburden pressure are exhibited.[24]

The main factor for face stability is a stress concentration for the exposed portion of reconsolidated roof. The approximate range of equilibrium could be calculated by the Protodyakonow theory of ellipse stress:

$$f = \frac{1}{2\tan m} \tag{2}$$

where

f = length of longer semi-axis of ellipse;
$\tan m$ = cohesion index of roof material (assumed as 1);
1 = width of face (assumed as 1).

For the assumed conditions, the maximum vertical stress concentration is at $f = 0.5$. This is denoted with a number 5 in Figure 9.3.6. Moreover, the occurrence of both shallower and deeper stress concentrations with ellipse failure are anticipated. Values of more than 1.9 (cohesion index 0.5) at the assumed roof exposure (width 1.0 m) are rather improbable.

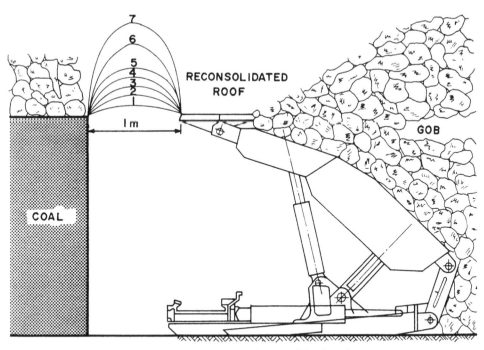

Fig. 9.3.6. Possible stress concentration zone of exposed reconsolidated roof.

J. Lojas et al., performed a simple calculation to establish a stability criterion of the exposed roof as a function of the dead weight of a critical arch and the strength of a reconsolidated gob, as follows:[25]

$$A = sl \tag{3}$$

where

A = area of the vault;
s = width of arch;
l = length of the considered roof section parallel to the face.

Active forces in a roof fall i.e. weights of rocks contained within the assumed ranges of partings, were calculated from the formula:

$$F_a = V\gamma_{sr} \tag{4}$$

where

F_a = active force;
V = volume of rocks occurring under the vault with an area P;
γ_{sr} = unit weight of the caved rock.

The value of tensile stresses, σ_T, for particular vaults, was calculated from the formula:

$$\sigma_T = \frac{F_a}{A} \tag{5}$$

For theoretically assumed possible vaults involved in a roof fall, the forces of resistance

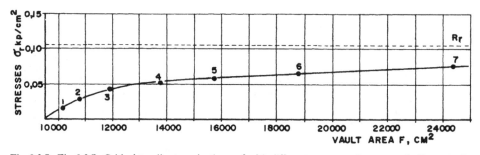

Fig. 9.3.7. Fig. 9.3.7. Critical tensile stress in the roof with different concentration zones (1–7) versus the strength of reconsolidated caved rock.

to collapse (N) were calculated as a product of tensile strength and area of the vault.

$$N = \sigma_T A \qquad (6)$$

By comparing the calculated values of active forces $(P_1 - P_7)$ with the forces of resistance to collapse $(N_1 - N_7)$, it is possible to determine a safety factor for the analyzed roof using the principles of strata mechanics. The diagram in Figure 9.3.7 illustrates the comparison between stresses in the roof and the strength of reconsolidated caved debris. From this figure, it can also be seen that the considered part of the roof will be self-supporting.

To verify the theoretical analysis of stability for a roof made of reconsolidated caved rocks, measurements were made at the longwall face in seam 405 whilst working the next underlaying slice.[25]

Verification of roof stability was based on observations of the roof behaviour, comparison of caved debris compositions, determination of rock humidity and measurements of roof cohesion.

The exposed roof of the second coal slice was comprised of a conglomerate of shale debris and fine coal, with a prevailing part of the former or of the latter component.

Underground measurements of the roof behaviour showed that with a roof rock humidity of more than 4 per cent, the roof sag in the height of the vault varied within the range $\eta = 10 - 30$ mm $(\eta_{sr} = 22$ mm). The width of exposed roof was different and was $1 = 55 - 100$ cm. Roof falls occurred occasionally when the reconsolidated roof was composed of unwatered (dry) rock material $(\eta \geq 19$ mm).

Measurements of roof rock cohesion were made by means of a static penetrometer and consisted of pressing a cone, with taper angle $\alpha = 30°$, into the roof and recording the penetration depth and applied force. On the basis of these measurements, cohesion indices 'R_c' for particular parts of roof, were calculated according to the formula:

$$R_c = \frac{F}{h^2} \qquad (7)$$

where
 F = applied force;
 h = penetration depth.

The measure of stability of the exposed part of the roof was a value of index S determined as:

$$S = \frac{l}{W} \tag{8}$$

where
W = height of vault sag of roof fall;
l = length of the exposed roof span.

Figure 9.3.8 shows the roof stability index 'S' as a function of the cohesion index 'R_c' with separate curves for the roof with either coal as the prevailing component or shale as the prevailing component. As can be seen from the diagram, increments in the stability index 'S' for the coal curve are higher than those for the shale curve at the same values of cohesion index 'R_c'. Though the differences between two types of roof rocks are not significant, a roof composed of conglomerate with prevalence of fine coal is more stable. Figure 9.3.9 shows the roof stability index 'S' as a function of the roof rock water content. In this case separate curves were also drawn. It can be seen from the

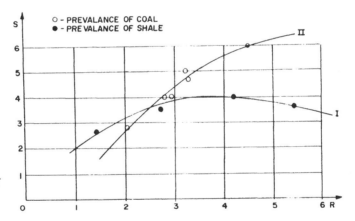

Fig. 9.3.9. Stability index of the reconsolidated rock as a function of humidity.

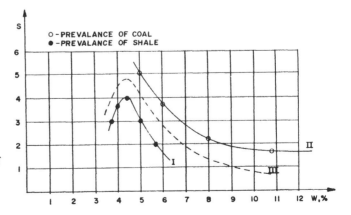

Fig. 9.3.8. Stability index of the reconsolidated caved rocks as a function of cohesion.

diagram that the optimum water content of the rock ($\simeq 4.5$ per cent) corresponds to the maximum roof stability.

9.3.3.   *Interaction between coal structure and fill*

In some paragraphs of this volume the principles of interaction between rock layers and coal seams, with an abutment stress concentration at the boundary of coal seams, have been discussed. Digital models and physical models of equivalent materials, as well as in situ monitoring, showed that coal cannot withstand abutment stresses. Under these circumstances the stressed coal undergoes a transition to its limiting yielding state of stress, which results in the transfer of abutment stresses further into a seam.

However, in slice mining with stowing, as well as in other stowing methods, the interaction of elastic rock strata with inelastic fill material and coal both of which demonstrate nonlinear behaviour under compression, is of particular interest.

V. A. Gogoling and Y. A. Ryzhkov[26] discussed the interaction mechanics as follows:

1. Coal seam structure (pillars) has been analyzed by the stress-strain state due to uniaxial compression, as illustrated in Figure 9.3.10. The diagram of progressive coal loading is close to linear for a stress $\sigma < \sigma_e$ but further loading resulted in yielding with partial loss of supporting capacity. Further inelastic deformation was accompanied by a decrease in load until total failure, which is exhibited by maximum deformation ε_p at $\sigma = 0$. Under this circumstance the coal structure was supporting only a normal compressive load (Fig. 9.3.10), which can be exhibited by:

a) Elastic deformation ($\varepsilon < \varepsilon_e$) with the Young's modulus of elasticity E_{11}, and

b) Inelastic deformation ($\varepsilon_e < \varepsilon < \varepsilon_p$) with the deformation modulus E_{12} and the elastic limit σ_e, which corresponds to the compressive strength which is further represented by the equation of the idealized $\sigma - \varepsilon$ diagram as given below:

$$\varepsilon = \begin{cases} -\sigma/E_{11}, \varepsilon < \varepsilon_e \\ (\sigma_e - \sigma)/E_{12} - \sigma_e E_{11}, \varepsilon_e < \varepsilon < \varepsilon_p \end{cases} \tag{9}$$

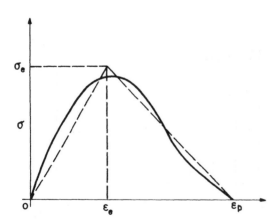

Fig. 9.3.10. Stress-strain diagram of the coal structure.

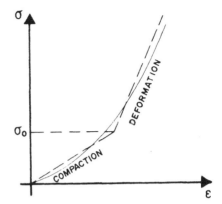

Fig. 9.3.11. Stress-strain diagram of the fill.

From Equation (1) could now be rewritten as

$$\varepsilon_p = \sigma_e \left(\frac{1}{E_{12}} - \frac{1}{E_{11}} \right) \tag{10}$$

It should be noted that if $E_{12} \to 0$, there is a plastic flow for $\varepsilon > \varepsilon_e$ but, if $E_{12} \to \infty$, then there is a violent structure collapse by brittle fracture.

The parameters E_{11}, E_{12} and σ_e of the model could be obtained directly by approximation from $\sigma - \varepsilon$ diagrams determined during uniaxial loading of coal structures, or by in situ instrumentation of the seam boundary displacement and the related magnitude of abutment pressure.

2. Fill material analysis was carried out by investigation of its behaviour under compression (Fig. 9.3.11). The compression curves have a special characteristic which enables approximation of the fill as a linearly deforming body within the same range of deformation, without incurring excessive error. Three stress fields could be delineated regarding deformation of the fill during its progressive compressive loading:

a) When stress $\sigma < \sigma_0$, which is called strengthening, deformation increases linearly and rapidly with stress. The phenomenon of the stress-strain relationship could be explained by a sharp change in the porosity of the fill due to the chipping of sharp corners and grain boundaries and their reconsolidation.

b) When $\sigma = \sigma_0$, particles of fill are approaching close compaction and minimum porosity.

c) When $\sigma > \sigma_0$ the increment of deformation sharply decreases but its dependence on the stress is still almost linear. Therefore, due to further loading, the deformation is usually small and it is determined by the strength and deformation properties of the rock fill.

Under the conditions of fill loading exhibited on the compression diagram (Fig. 9.3.11), with two proportionality coefficients E_{21}, E_{22} and a strengthening stress σ_0, the following equation could be written:

$$\varepsilon = \begin{cases} -\sigma/E_{21} & \sigma < \sigma_0, \\ (\sigma_0 - \sigma/E_{22} - \sigma_0/E_{21}), & \sigma > \sigma_0 \end{cases} \tag{11}$$

In contrast to the previous case, it cannot be assumed that $E_{22} \to \infty$ (the state of total hardening), because $E_{22} \gg E_{21}$ but is still comparable with the modulus of elasticity of the seam E_{11} in the problem of the interaction of adjacent rock strata-coal seam fill material.

3. Interaction mechanics are primarily governed by the properties and behaviour of the coal structure or fill material under compressive loading to total deformation.

The state of interaction stresses could be illustrated by the equation derived by V. A. Gogoling and Y. A. Ryzhkov as given below:

$$\eta_{ij} = \frac{\pi \, w \, E}{2(1 - v^2) a \, E_{ij}} \tag{12}$$

where

η_{ij} = coefficient of stress concentration;

E and v = elastic constants of adjacent rock;

w = width of opening;

a = displacement of opening;

E_{ij} = elastic parameters either of coal or fill (i, j = 1, 2).

In slice mining, the interaction with adjacent rock strata is of lesser importance than the interaction between coal and stowing, where several possibilities exist, as for example:

a) Increase in the rigidity of fill (increase of E_{21} and E_{22} or decrease of σ_0) results in transfer of stress from the coal structure on stowing, as illustrated in Figure 9.3.12/A. Under this circumstance, the abutment stress of the coal structure is pushed closer to contact with the fill due to an increase in inelastic deformation.

b) Increase of the coal elasticity, with very limited inelastic region (E_{12}), will also result in pushing abutment stresses closer to contact with the fill (Figure 9.3.12/B).

It should be pointed out that if there is a reduction of the coal strength or fill rigidity, then the opposite phenomenon is exhibited by pushing of abutment stresses further from the contact between fill and coal.

Interaction mechanics in a three phase system, particularly coal and fill, is of paramount importance for mine stability. The analyses presented show that an increase

a)

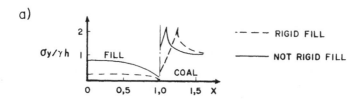

b)

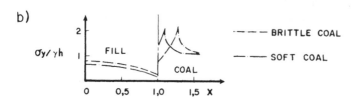

Fig. 9.3.12. Position of the abutment stress due to rigidity of fill (a) and elasticity of coal (b).

in fill strength will increase the stability of a mine structure because a rigid fill will take one part of the stress from the coal structure. It is obvious that it is not a matter of the engineering of stable mine structures in coal slice mining, but of what the cost will be. For this reason, the more realistic analysis for engineering of mine stability is not stowing but consolidation of the gob rock as the mine roof.

9.3.4.   *Influence of geological anomalies on strata mechanics*

Coal falls and rock falls in working areas of the thick coal seams, particularly in the case of slice mining, might also be due to the influence of geological factors. For example, ascending slice mining with stowing at Kuzbass indicated that 70 per cent of coal and rock falls were due to geological anomalies.[27] The geological anomalies are represented by folds, faults, shears, fissures and other structures, which are discussed in the chapters on geological events. It is necessary to point out that instability of geological anomalies is aggravated by the transfer of abutment stresses, particularly along the strike and long pillars left either in working or adjacent coal seams.

To visualize the influence of geological anomalies on strata stability, a layout of the mine workings in the Gorelyi seam is given (Fig. 9.3.13) showing the coal and rock falls which occurred on the working faces during coal extraction by an inclined slicing from the floor of the seam with hydraulic stowing (coal getting by D-11 cutter-loader). During the three year period of up-dip extraction of a 100 m coal seam it was shown that the majority of coal falls occurred during extraction of bottom slices, which were on opening a virgin seam. The coal falls were sudden, with formation of dome-shaped cavities, and in 72 recorded cases their volume was in a range between 5 and 500 m^3. In this case 96 per cent of coal falls occurred during or immediately after coal extraction by the cutter-loader.[27]

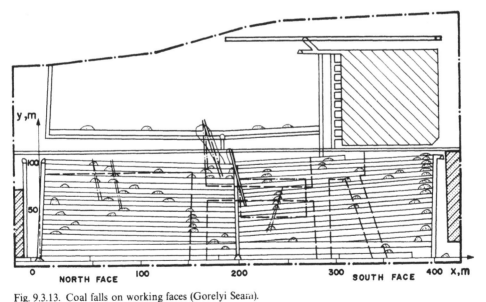

Fig. 9.3.13. Coal falls on working faces (Gorelyi Seam).

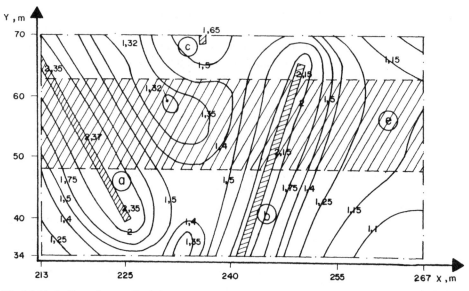

Fig. 9.3.14. Isolines of anomalies influence (Gorelyi Seam).

The strata mechanics could be analyzed simply by the stress-strain (stress-deformation) state of the rock as a function of coal extraction and the influence of geological anomalies. The rate of advancement of the coal face controls the dynamics of stress transfer, which are related to the increments of roof displacement of the extracted area. This concept of strata mechanics offers analytical description of maximum displacement in relation to mined-out areas, which could then be related to mine stability. Knowing the mine layout, mining technology of coal extraction, geological anomalies, and the characteristics of the field, it is possible to define the magnitude of the displacement. The displacement surfaces can be analytically represented and evaluated for improvement of mine stability by changes of layout or mine technology.

The analytical stability analyses do not separate the displacements which are induced by mining factors or geological anomalies. However, field monitoring of rock displacements and analyses of geological defects in the mine workings of the Gorelyi seam, suggested that the action of anomalies on rock dynamics is similar. As the extraction face approaches a geological anomaly, there is an increase in the roof displacement increments, which results in maximum displacement when the extraction face passes on anomaly followed by a subsequent decrease in displacement. From field monitoring data, calculations were made to graphically represent the influence of anomalies in the extraction field. A portion of this representation is illustrated in Figure 9.3.14.

During work in this section, field observations indicated that the mechanics of the strata are influenced by geological faults (a, b), fracturing of the coal seam (c) and pillar left in the overlying coal seam (d), as illustrated in Figure 9.3.14. The isolines of the density of roof fall coefficient show the degree of increase of the strata mechanics parameters at each point in the panel.[28]

The above described conditions enable a quantitative representation of the influence of geological and mining factors on strata mechanics. It might be suggested that systematic field monitoring of the displacement and its cumulative representation, could be used as a parameter of the influence functions of the anomalies with the aim of prediction of rock strata pressure.[27]

9.4. CAVING MECHANICS

The caving mechanics of multiple slice mining have certain peculiarities which are not common in single slice mining. Discussion of caving mechanics is, therefore, related to the specific phenomena of this system.
 – Caving of intact rock strata has been analyzed with two particular aspects: strata flow and strata cavibility, where principal consideration has been given to rock burst phenomena.
 – Caving of fragmented gob rocks has been analyzed with respect to the behaviour of consolidated rock, although brief consideration is given to the failure criteria and flow characteristics of the gob.
 – Coal slice caving, as a means of extraction in integrated mining methods was very briefly discussed, primarily on the basis of physical model investigations.
 – Subsidence phenomena have been presented on the basis of the Japanese experience at Kushiro Colliery. Consideration has been given to both subsurface and surface subsidence.
 Caving mechanics of slice mining to a certain degree share some similarity with those of sublevel mining. Some caving mechanical phenomena in this mining method are thus considered applicable to slice mining.

9.4.1. *Caving of intact rock strata*

The roof cavibility and good flushing characteristics of intact rocks are key factors for successful underground mining. Slice mining could be favourable for required roof strata caving or it could be unfavourable and result in roof strata flow. However, as it has been stated in many paragraphs of various chapters of this volume, the cavibility of roof strata also depends on the lithology or roof strata, which is not considered here because concern is given only to mining conditions.
 Further discussion is related to extreme cases of roof strata deformation and failure as follows.

1. Roof strata flow is typically exhibited in the case of top slicing with stowing. In Figure 9.4.1, the backfill did not permit caving of the roof strata but it only allowed convergence and resting of the strata on the fill material. A great amount of strain is induced in rock layers which are deflected and only partially fissured. In the case of sandstone or conglomerate strata, the development of rock bursts during extraction of the next slice under all an uncaved roof is facilitated. A similar development could be expected in the case of top slice extraction in highly stressed thick seams, with respect to their de-stressing. The convergence of roof rock strata and the coal floor heave could be rapid, such that caving could not be initiated except for limited fissuring.

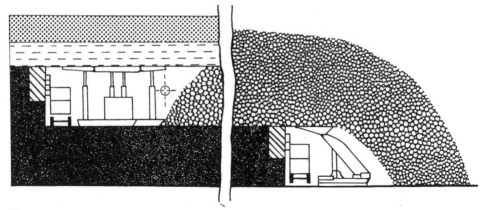

Fig. 9.4.1. Top slicing with stowing and incomplete roof strata caving.

If extraction of the next slice without stowing is commenced, then fill from the top slice will cave in the excavated area but caving of the roof strata is usually incomplete and violent, due to the release of concentrated strain. The stress in front of the elastic wave exerts a thrust on the supports on its arrival at the contact surfaces between the supports, floor and roof. The most common mechanics of caving are that the immediate roof with a weight denoted as W_1 is first fissured and separated from the main roof and then caved. The main roof might subsequently fail violently and cave, with the weight of caved rock donated as W_2. It is evident that

$$W_2 > W_1 \tag{1}$$

Underground investigations in the Ostrava Karvina coal field indicated that the most unfavourable situation, with respect to mine stability, is seen when the main roof strata is strong and undergoes delayed caving, which is expressed by:

$$W_2 \simeq 2.5\,W_1 \tag{2}$$

Rock bursts are most common under this circumstance. This type of dynamic loading of the supports causes rapid transfer of abutment stresses in overlying strata further into the seam from the coal face. The mechanics of impact loading of the supports during violent caving, suggests an increase of convergence by the increase in load borne by the supports (Fig. 9.4.2). The research on the interaction between support and load, has been summarized by B. Zamarski and J. Franek as:[30]

a) The abutment stress redistribution leading to rock burst situations is difficult to control;

b) The maximum stress concentration affects the face support in width of approximately 0.3 m;

c) Maximum speed of loading might be expressed in the order of KN per second, but rarely exceeds one KN/sec. There are some data which suggest that actual values are lower.

One of the most difficult problems in ground control, at the Ostrava Karvina coal field, is manifested by violent rock strata caving by means of rock bursts. The elimination, or rather a decrease in number and intensity of the rock bursts, in this

STATIC LOADING (NO CAVING)

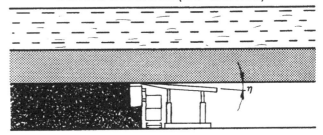

DYNAMIC LOADING (VIOLENT CAVING)

Fig. 9.4.2. Convergences due to static loading (no caving) and dynamic loading (violent caving).

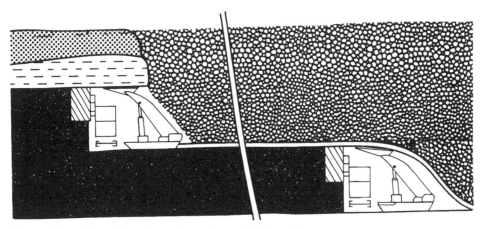

Fig. 9.4.3. Top slicing with caving and complete roof strata caving.

particular case is based on both theoretical research and practical experience. It is most likely that hydraulic fracturing could control the rock burst mechanics, due to successful de-stressing of the roof rock strata.

2. Roof strata caving is typically exhibited in top slicing without stowing or in simultaneous slicing (Fig. 9.4.3). Under this circumstance the possibility of inducing rapid roof strata convergence and flow, as described in the previous paragraph, is very small. The gob of caved rock substituted for the stowing structure and took the role of roof for the next extractive slice, as discussed to a certain extent in previous chapters.

There is nothing particular in the mechanics of roof strata caving other than that which is stated in other subheadings of this volume, except that slicing does not facilitate rockburst because the rock strata is caved before dangerous strain concentrations can occur. Regular roof strata caving, with good rock fragmentation, could be achieved by optimization of mining extraction sequences in space and time. This requirement is of importance for either descending or simultaneous slicing without stowing, because the gob of caved rock is an integral part of the extraction structure. The caved rock will be exposed to caving again when the slice below is extracted and, in this case, an adequate texture of the unconsolidated or consolidated gob rock is important for successful caving.

9.4.2.   *Caving of gob rocks*

One of the characteristics of slice mining (descending and simultaneous) is that fragmented rocks in the gob, situated in caved and mined-out areas, will cave again when mining of the next slice is in effect.

The study of caving mechanics for caved rock could be carried out on the basis of knowledge concerning bulk solids behaviour in bins and hoppers, because of the similarities between the two situations. In this respect several phenomena could be analyzed.[31]

1. The behaviour of caved rocks could be approximated with sublevel caving mechanics in hard rock mining. Under this circumstance it could be assumed that the behaviour of caved rock during drawing of the next slice is exhibited by ready flow without mechanical stimulation. It could also be assumed that the volume and configuration of

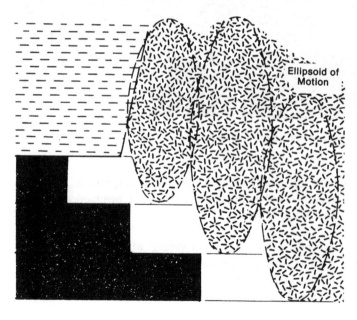

Fig. 9.4.4. Diagrammatic caving flow pattern of gob rock.

flow of caved rock coincide with characteristic sublevel caving patterns (Fig. 9.4.4). In the case of slice mining, the flow of caved rock in the lower part is in contact with the steel shield support as is solid or caved rock in the upper part.

Three cases of approximation of different flow of gob rock could be analyzed on the basis of bulk solids flow in bins and hoppers.

a) Mass flow of rock with 60° mild steel walls (both walls are mild steel inclined at 60° to horizontal).

b) Flow of rock in comparison to flat bottomed bins with 60° slip lines (when both walls are fragmented rock inclined at 60° to the horizontal).

c) Comparison with flat-bottomed bins with 70° slip lines (when both walls are fragmented rock inclined 70° to the horizontal).

Consideration of the angle of inclination of slip lines are given between 60° and 70° as being the most likely to occur in actual mine situations. Hopper flow tables are available for determination of flow characteristics of caved material for 60° (France) and 60° to 70° (Yugoslavia).

2. Failure criteria could be analyzed by the principles of soil mechanics as for example by Mohr-Coulomb postulation (Fig. 9.4.5). The criterion is described by the angle of internal friction ϕ and cohesion c. The tangent A-A' is referred to as the yield locus (YL) and represents those combinations of principal stress σ and shear stress τ at which yielding, or flow, begins. In a real bulk solid ϕ is often not a constant and may vary for low values of σ. The line A-A' will then be curved and assume a shape approximating a parabola. For simplicity, it is generally assumed that A-A' is a straight line and that ϕ is constant. It should be noted that the stress σ is compressive because bulk solids will not support tension.

The significance of the yield locus is seen when the combined state of stress, represented by Mohr's circle of stress, becomes large enough to contact the yield locus, with the material then yielding within itself. This movement occurs along slip planes at an angle β:

$$\beta = \theta \pm \left(\frac{\pi}{4} - \frac{\phi}{2} \right) \tag{3}$$

where

θ = angle between the direction of the major principal stress (σ_1) and the reference x axis, as shown in Figure 9.4.6.

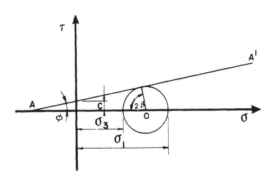

Fig. 9.4.5. Coulomb failure criteria for soils and bulk solid or caved rocks.

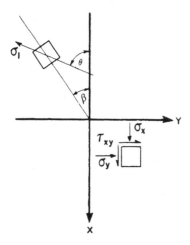

Fig. 9.4.6. Relationship of slip lines to principal stress and reference axis.

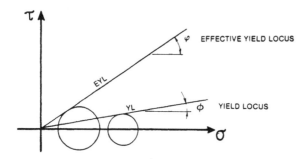

Fig. 9.4.7. Simplified yield locus and effective yield locus.

In a gravity flow problem, the x axis is defined as being parallel to the gravity vector. The principal stress σ_1, due to gravity is approximately parallel to the gravity vector and $\theta = 0$, therefore:

$$\beta = \pm \left(\frac{\pi}{4} - \frac{\phi}{2} \right) \qquad (4)$$

And for a bulk failure due to pure compression:

$\phi = 0$ and

$\beta = \pm 45°$

And for a bulk failure due to pure shear:

$\phi = \pi/2$ and

$\beta = 0$

which indicates that the slip planes will lie between the x axis and a pair of lines inclined 45° from the x axis.

It is of interest to examine the effect of consolidation. When the caved rocks stand for a period of time under pressure, as is usually the case, they become consolidated, as discussed in the previous section. Under this circumstance, the yield locus shown in Figure 9.4.5 is replaced by an effective angle of friction,[32] as illustrated in Figure 9.4.7.

3. Flow characteristics were studied as a functions of the size distribution of fragmented rocks, the area of draw sufficient to prevent arching, the relation between elongation of rock fragments and direction of gravity, and others. For example, a width required for free draw without arching could be calculated by the equation:

$$b \geq \frac{\sigma_c}{\gamma}(1 + m) \tag{5}$$

where

$\quad$ b = width of opening behind support;
$\quad$ σ_c = unconfined yield strength;
$\quad$ γ = unit weight;
$\quad$ m = 1 for circular; m = 0 for rectangular draw openings.

That the draw rate of caved and compacted rock behind the supports is sufficient to permit caving of intact rock strata above is of particular interest in caving mechanics. Ahead of the supports, the draw rate of consolidated gob should be zero for the period of extraction (Fig. 9.4.8). In both cases, the draw rate could be regulated by the size of the draw area, as may be expressed by the following equation:

$$Q_m = \gamma BD \frac{Bg}{2\tan\beta}\left(1 - \frac{ff}{ff_a}\right) \tag{6}$$

where

$\quad$ Q_m = maximum weight of discharge (10 N/sec);
$\quad$ g = gravitational constant (10 m/sec^2);
$\quad$ ff_a = actual flow factor = σ_1/σ_c;
$\quad$ D = length of draw area;
$\quad$ B = width of draw area.

For very free flowing rocks, the slope of σ_c locus is > 4, and

$$\frac{ff}{ff_a} = 0$$

The significance of the ratio above is that it reflects the cohesiveness of the flowing material.

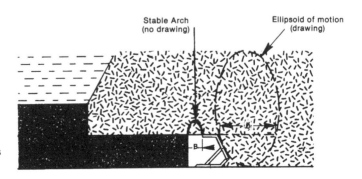

Fig. 9.4.8. Interaction between length of openings and draw of caved rocks.

The time required to draw caved rock can be expressed by the velocity of flow at any time in relation to the maximum velocity by the following equation:

$$V = V_m \frac{e^{t/s} - 1}{e^{t/s} + 1} \tag{7}$$

where

$$V_m = \frac{Qm}{BD};$$

$t = $ time elapsed;

$$s = \frac{Qm}{2g\,BD} \frac{1}{1 - ff/ff_a}. \tag{8}$$

The elapsed time can be evaluated from the plotted function between the dimensionless velocity ratio V/V_m and the dimensionless time ratio t/s. For example, if V/V_m is approximated as 0.5 then $t/s = 3$.

9.4.3.   *Caving of coal slices*

The phenomena apply for the cases of either ascending slicing, where the bottom slice is extracted by longwalling and upper slice by coal caving, or simultaneous slicing, where, the middle slice is extracted by caving. As mentioned before, this method is transitional to sublevel mining, which was discussed in chapter 11. Remarks made in this section could thus be considered equally valuable for sublevel coal extraction.

The mechanics of coal caving, as an integral process of mining extraction in many countries, have been studied primarily by models of equivalent materials. The physical model investigations in the U.S.S.R., suggest that coal drawing mechanics are functions of the thickness of the top coal slice and the nature of its cavibility. These laboratory investigations, in some instances, were followed by full-scale underground tests, which confirmed the mechanisms obtained in model tests, as briefly listed below:[31]

   a) The thickness of the top slice of coal should be 6 m instead of 2 or 4 m.

   b) The strength and cavibility of the top slice of coal, and of the immediate roof rock, should be markedly different.

   c) The coal drawing increment should be no less than 1.2 to 1.5 m.

   d) Continuity of coal caving and drawing is necessary and requires the technology of continuous loading and handling of coal.

   e) If the thickness of the top coal slice is smaller, then dilution of the coal by the caved rock above during the drawing process is greater and the percentage output from the extracted area is smaller because of noticeable coal losses. This mechanism is dictated by ellipsoids of motion because, if the top slice of coal is thin, it would be difficult to 'fill' the flowing structure with coal. Under this circumstance a great part of the ellipsoid of motion would be field with mixed rock and coal.

   f) Dilution of the drawn coal by caved rock occurs primarily along the contact of the caving of the roof rocks and the top slice of coal as it is drawn, following advance of the support.

It should be pointed out that coal caving mechanisms have also been studied in France, particularly with the aspect of coal drawing to avoid those effects which lead to dilution and loss of coal, as discussed in great detail in the CANMET's Technical Bulletin.[33]

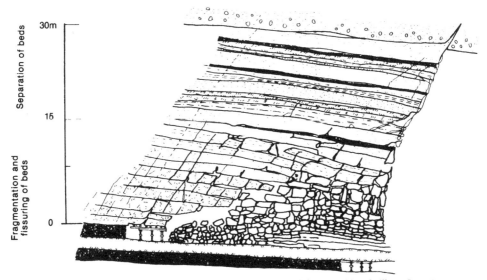

Fig. 9.4.9. Caving and displacement of roof strata due to slice mining (Kushiro Colliery, Japan).

9.4.4.   *Subsidence phenomena*

S. Nakajima studied subsidence phenomena at Kushiro Colliery as result of slice mining. The mining is carried out by two slices, between which a slice of the waste partings is left. The author gave first consideration to subsurface and then to surface subsidence.[34]

1. Subsurface subsidence has been related to the cavibility of roof strata and compaction of fragmented rock in the gob area. The state of the caved roof and subsurface subsidence is represented in Figure 9.4.9. As illustrated, the roof strata from the top slice, up to 15 m, are either crushed and fragmented or fissured both vertically and horizontally. However, the roof strata from 15 to 30 m is affected by the separation of the beds and is resting on the crushed and fragmented rock.

The portion of strata which will cave when the bottom slice is brought into extraction is, at first, represented mostly by fragmented rock and not by consolidated gob.

The magnitude of subsurface subsidence was measured in the upper coal seam above the gob area of the mined-out top slice (Fig. 9.4.10). The central part of the extracted area, about 25 m above the gob, subsided to the same height as the mined-out thickness of the seam. This represents a subsurface subsidence of 100 per cent, which is related to the maximum compaction of the gob and its solidification. However, at the same time, the subsidence near the unmined coal face was not complete.

2. Surface subsidence at the Kushiro Colliery was measured from 1926 to 1950. The data of subsidence from this period are illustrated in Figure 9.4.11. It is interesting to notice that two periods of subsidence have been established:[35]

a) Main subsidence, which exhibits a general pattern characteristic of other coal

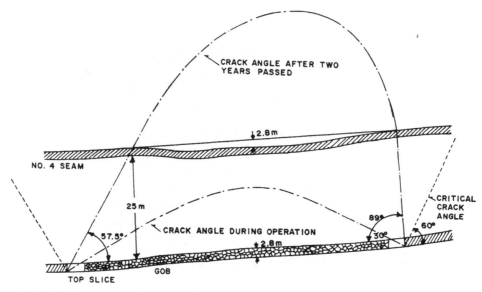

Fig. 9.4.10. Subsurface subsidence of the Number 5 Seam, after two years of mining of the top slice of the lower seam (Kushiro Colliery, Japan).

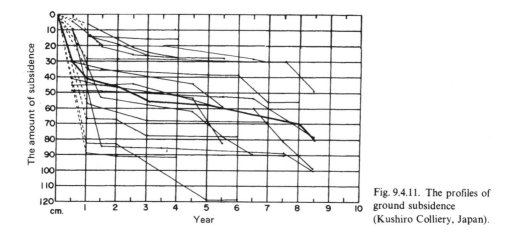

Fig. 9.4.11. The profiles of ground subsidence (Kushiro Colliery, Japan).

fields over the world. This subsidence ceases after two years of mining the top coal slice.

b) Minor subsidence, which came in the effect 7 or 8 years after mining the top coal slice.

The reason for cessation of main subsidence after two years and initiation of minor subsidence after seven or eight years is not clear. However, the probable explanation is that creep deformation of remnant coal structure and pillars occurs with a noticeable displacement after this period and is reflected in ground surface subsidence.[36]

The surface instrumentation showed that the average main subsidence was 35 per cent of the mined-out height of the seam and the minor subsidence was only a further 5 per cent, totalling 40 per cent. A maximum subsidence was recorded at 70 per cent of the mined-out height of the seam. In this case, it is obvious that the magnitude of ground surface subsidence is influenced by two principal factors: gob compaction and consolidation and the coal recovery factor.

CHAPTER 10

Long pillar mining

A common mining system used for extraction of pitching thin and moderate thick seams is long pillar mining. The analyses of this mining system have been carried out using the same topics as in the previous chapters:

– Principal long pillar systems mainly considering drilling-and-reaming (thin seam), cyclic and continuous (medium and thick seams) technologies.

– Stability of stope-pillar structure analyzed with respect to long and narrow pillars between extraction blocks (rib and crown pillars) and their effect on stability of mined-out area.

– Strata mechanics discussed in a manner similar to that of sublevel mining, since there is similarity between these two mining methods.

– Caving mechanics including studies of physical models, since this technique offers visualization of the phenomenon.

The future of long pillar mining, particularly of thin and steep pitching coal seams is very promising, due to possible application of hydraulic mining technology.

10.1. PRINCIPAL MINING SYSTEMS

Long pillar coal mining has been derived from hard rock mining systems such as sublevel mining and vertical crater retreat mining. The consideration of long pillar coal mining is presented under the following topics:

– Principal elements of long pillar mining by open stope structure.

– Conventional long pillar mining, particularly for extraction of thin and steep coal seams.

– Hydraulic long pillar mining by drilling-and-reaming with water jets.

– Shield mining primarily for extraction of medium thick coal seams.

– Bench mining for extraction of steep pitching to vertical coal seams of varying thickness.   .

– Other mining systems based on the principles of open stope structure.

The main technological characteristics of these long pillar mining systems and their variations are briefly presented under individual subheadings.

It is important to point out that in addition to the described mining systems, there is a wide range of variations in long pillar mining which have certain similarity, but they are known under different titles.

If there is strata instability (particularly mine roof cavibility), the long pillar mining concept must be changed to coal mining with stowing. Under these circumstances, mining is carried out in a fashion similar to the extraction of thick coal seams by slicing, but due to the limited thickness of the coal seams, it would be by coal strips or lifts rather than slices.

10.1.1. *Long pillar structure*

The long pillar is the extraction structure and it is usually delineated by rib and crown pillars. The shape, dimension, and orientation of long pillars depends on the mining technology, strata stability, mechanical properties, the behaviour of coal seams and other factors.

Generally speaking, long pillars are categorized in relation to the strike and dip of the coal seams as indicated below:
– Pillars along seam dip;
– Pillars along seam strike;
– Pillars oblique to seam dip and strike.

This is illustrated in Figure 10.1.1. The extraction of long pillars could be by lifts (Fig. 10.1.1 e, f, h, i) by benches (Fig. 10.1.1 g), or by strips (Fig. 10.1.1. a, b, c, d). The choice of technology for coal getting and the direction of extraction are influenced by the cavibility of the roof strata. For example, up-dip extractions are permitted when roof caving does not occur immediately after the removal of coal. This is in contrast to down-dip mining, where it is necessary to synchronize roof caving with a face

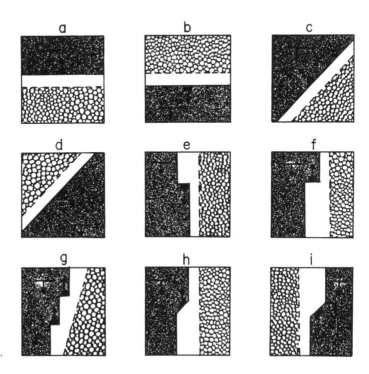

Fig. 10.1.1. Direction of extraction of long pillars.

advance. However, mining in a direction diagonal to the seam is usually most convenient because offers a maximal roof stability on working faces.

10.1.2.   *Conventional borehole mining*

This mining method has a long tradition which started in Great Britain, where it is known as 'auger mining'. Further development of this method continued in the U.S.S.R. where large coal reserves located in thin pitching coal seams were very suitable for extraction by this method. Generally speaking conventional borehole mining is applicable to coal seams with inclination between 35° and 90° and a thickness between 0.5 to 5.0 m. A brief description of this method is as follows:[1]

a) Development of coal panels by horizons at a distance of 80 m. Long holes with large diameters ranging from 0.45 to 0.90 m, depending upon the thickness of the seam, are drilled at 3 to 6 m intervals along the panel strike. The large diameter drillholes are utilized as coal passes to the grizzly level, from which coal is loaded onto a belt conveyor installed in the main haulage level. The excavation of development workings in steeper and thinner coal seams is performed by roadheaders.

b) Coal winning is carried out by the drilling and blasting of long-holes from one pillar to the next. The diameter of drilling holes is 42 mm with charges of 38 mm diameter and the holes are spaced at distances of 1.5 to 2.0 m in one or two rows, depending upon the seam thickness. One drillhole is sufficient to break 30 to 70 tonnes of coal. The long pillars have a length of 80 m and width of 40 m. Between them there is an 8 m wide rib pillar. Each extracting pillar is excavated by three 12 metre

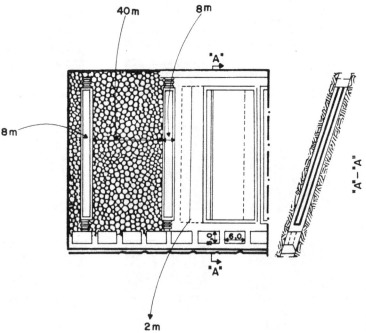

Fig. 10.1.2. Long hole mining.

wide long sub-pillars interested with 'skinny' ribs 2 m in width which are allowed to collapse after stope completion (Fig. 10.1.2).

c) Support of the main level drifts is by timber frames, where partial excavation is carried out in roof rock strata if the thickness of coal seam is not sufficient for in-seam excavation. Support of the stopping area is limited, because this mining method is carried out mainly in thin seams where roof and floor strata consist of hard rocks which could stand unsupported for a certain period of time.

d) Proper ventilation requires separate air intakes for each level, with high air velocity to decrease the danger of methane and coal dust explosion. Fresh air from the haulage level is distributed through the working places and exhaust air is returned by the upper ventilation level.

e) Productivity of this method for coal seams of 1.5 to 2.5 m thickness at angles of inclination of 77° to 90° have been of OMS 18 tonnes, where daily production of the mine is 5,000 tonnes.

From this basic mining system a certain number of variations were developed, some of which did not survive due to their complexity, while others did survive due to their simplicity and high productivity. Examples are hydraulic mining by drilling-and-reaming and shield mining both of which are discussed in the next subheadings.

10.1.3. *Hydraulic drilling-and-reaming*

The philosophy of hydraulic drilling-and-reaming mining should be related to conventional mining which uses the boring technique to gain access to deposits and extract useful minerals. Due to the novelty of this mining method, further discussion is presented under the subheadings of the two principal countries where the method was developed.[2]

1. West Germany, implemented hydraulic mining by drilling and boring in the Hansa Mine, with production of coal from a depth of 850 m. The brief characteristics of this method are listed below.[3]

a) Development of the seams incorporate the mining of multiple seams whose thickness is between 0.6 and 1.2 m. In this case the level drifts are layed out at a gradient greater than 4° so that a mixture of coal and water can flow by gravity. The pilot holes are drilled with rotating water jets from level drifts through the centre of the pillar (diameter 0.3 m) either in up-rise or down-rise direction (Fig. 10.1.3). The level drifts are excavated by small roadheaders particularly designed for driving in thin coal seams. The panel on one level is in extraction by several long pillars, while another is in development.

b) Extraction is achieved by reaming pilot holes with rotating water jets, which break the coal and flush it in the direction of gravity. The pillar is reamed out starting from the top or bottom of the pillar. The pilot holes are used as coal passes for the broken coal, which is pitted around them at a depth ranging from 6 m (hard coal) to 15 m (soft and friable coal). The width of the stopping rib ("skinny") pillar between cavities is a function of the self-supporting capacity of roof rock. The drilling of the pilot hole is achieved by a bit of 5 nozzles (D = 7 mm) and the reaming is done by a bit of 4 nozzles (D = 10 mm) while pumps generate a water flow rate of 45 m³/h.

c) Support is carried out only on the horizon levels, where drifts are supported

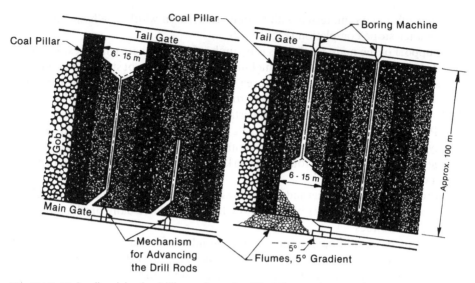

Fig. 10.1.3. Hydraulic mining by drilling and reaming (West Germany's concept).

usually by timber sets centered at 1 m. Long pillar stopes do not require support as they are extracted only by water jet power.

d) The ventilation scheme is very similar to sublevel mining, where each level panel requires two entries: one for intake air (haulage road) and another for return air. These are connected by raises driven in the seam.

e) The productivity objective of Hansa Mine is 120 tonnes/h monitor row coal and of OMS 25 tonnes, with production of 5,000 tonnes per day and coal recovery 70 to 75 per cent from level panel.

2. The U.S.S.R. implemented this technology in a dozen hydraulic mines using conventional borehole mining or shield mining methods. The following brief comments are significant:[4]

a) The development concept is similar to the German hydraulic long-hole mining layout. If geological irregularities permit, the level entries are spaced by 100 m and are driven up 200 m. They are connected every 100 m by raises. The pilot holes are drilled up the rise at centers 3 to 8 m by rotating water jets from the flume level drift (Fig. 10.1.4). The maximum diameter of the hole should be up to 75 per cent of the seam thickness. Level entries are excavated in-seam if the seam is thicker than 1.5 m or partially in the immediate roof if it is thinner. The excavation is by continuous mining machine with hydro/mechanical cutting. Level entries are connected by hydraulic raise boring, which is an integral part of coal extraction.

b) Extraction of coal is by hydraulic reaming and flushing and by natural coal caving when reaming is completed. The reaming is accomplished by various techniques. One of the most popular methods of reaming uses a bit with three nozzles which is either suspended on a hose or mounted on the pipe. The jet pressure at the nozzle is 15 MN/m² and the water discharge rate is 150–250 m³/h.

c) Support is identical to that seen in the German concept of this mining system.

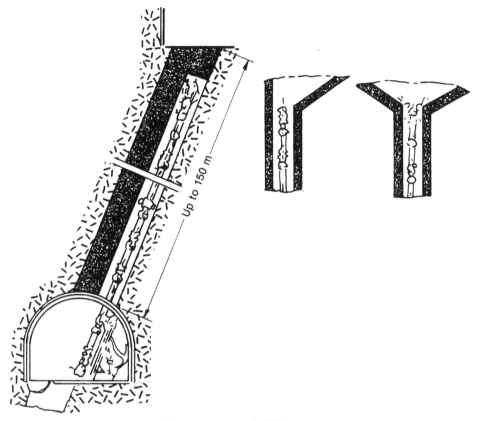

Fig. 10.1.4. Scheme of hydraulic drilling and reaming (U.S.S.R.)

d) The ventilation scheme is also very similar to sublevel mining.

e) Productivity is in the order of OMS 36 tonnes, with an average of 300 tonnes per monitor-shift. The rate of hydraulic drilling of pilot holes is 10 to 20 m/h depending mostly on coal hardness.

More extensive coal extraction by this method should be expected in the near future, because there is little development in relation to produced coal, especially in the case of multiple seam mining.

This technology is particularly important for the mining of thin and steep pitching seams. It could be applicable to a majority of coal reserves which still remain due to their unmineability by conventional mining methods.

10.1.4. *Shield mining*

Shield mining is a hybrid of the long pillar and the caving mining systems. Development in the shield mining method approximates that employed in the long pillar systems while the method of extraction employed parallels that used in the caving systems. However in this method coal is won by drilling and blasting under the guard of a shield. Broken coal falls into a coal pass and gravitates to the haulage road. With the lost support of coal, the shield lowers down under the weight of caved rocks above it and under its own weight. Thus the shield descends through the whole height of the level from the ventilation road to the haulage road. The principles of the shield mining method are described by the following elements:[5]

a) Development divides coal seams into levels. Each level is, in its turn, divided into individual stope blocks (long pillars). The access to the level is achieved by cross-strata drift. Actually only one cross-strata drift has to be driven because the cross-strata drifts of the upper worked-out level are used for ventilation in seam drifts connected by twin raises, between which are drilled chutes 30 to 40 m apart. Those chutes are later widened with drilling and blasting or with pneumatic picks, downward from the ventilation road.

b) Extraction of coal is by full face. Coal is won below the shield support by drilling and blasting; the shield is lowered 0.9 to 1.5 m per cycle. A longitudinal trench is formed under the shield by blasting to allow free movement of men, and to provide straight face ventilation (Fig. 10.1.5). While the shield is being lowered, attention should be directed to prevent the wall rock from caving into the shield face. This is achieved by following a certain pattern of shot holes and a certain order of their firing.

c) Support of development workings is the same as in the case of sublevel caving. However, the stope support by shield could be of several types according to their construction:
– Flexible sectionless shield;
– Concrete reinforced shield;
– Hydraulic operated steel shield;
– Shield on rolls;
– Hydraulically moved shield.
Each type has advantages and disadvantages but hydraulically moved shields are the most efficient for this type of mining.

d) Ventilation is simple, whereby intake air circulates from the main haulage level throughout the coal passes to working faces from where it travels to ventilation levels as exhaust air.

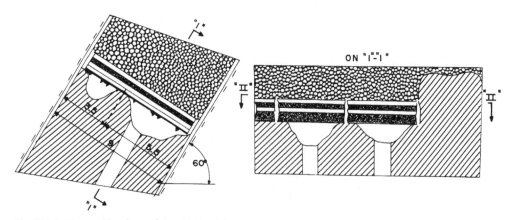

Fig. 10.1.5. The working faces of the shield mining system.

e) Productivity is high because the development of sublevel blocks is not necessary. The coal extraction is performed by only two men, and it has been recorded at rates of up to OMS 20 tonnes.

The advantages of shield mining are efficiency and safety at the production faces. Coal is delivered to the lower entry by gravity, requiring no sublevel haulages. However, the main disadvantage is high coal losses particularly in thick coal seams. If the shield comes into dissaray before it reaches the lower portion of the level, the effected part must be extracted by another method (sublevel caving).

10.1.5. *Bench mining*

This mining method is used in pitching and steep pitching coal seams. The coal extraction by this method is simple. The brief description of this method is as follows:[5]

a) Development is the same as in previous listed methods, where the upper level and lower level drifts are connected by raises, which are in this particular case driven somewhat oblique to the strike in order to decrease the high pitch of the coal seam at the working faces. The level drifts are excavated either by continuous machines adopted for steep seams, or by drill-and-blast technique, which is also used for raise excavation.

b) Extraction starts by coal breaking at the top corner of the bench in order to protect the miners working on the lower bench from any possible hazard. The length of benches is between 6 and 15 m, depending on strength of the coal and thickness of the seam (Fig. 10.1.6). The most difficult operation of this system is cutting corners of

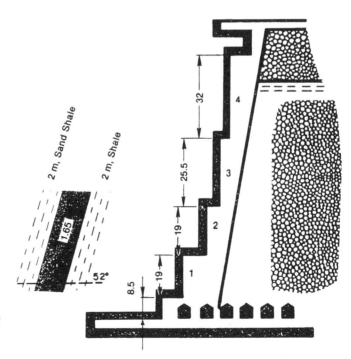

Fig. 10.1.6. Stope layout of bench mining.

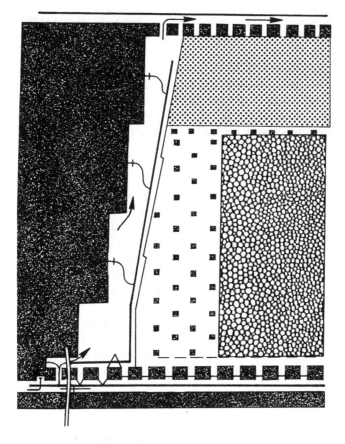

Fig. 10.1.7. Bench mining with partial stowing.

benches manually with pneumatic hammers. The coal from the bench is broken downward. In the Spain (Asturia) this mining method is predominant for extraction of thin pitching coal seams.

c) Support of working faces is intensive by timber props, whose density depends on the roof and floor conditions, as well as on the hardness of coal. This method is convenient for soft and fissured coal, but with respect to the support of the working place, it could impose serious limitations.

d) The ventilation scheme is similar to previously listed methods of long pillar mining.

e) The productivity can range between OMS 1 to 3 tonnes, depending upon the hardness of the coal.

For very steep pitch and heavy ground conditions, the stowing is partially used to support the mined-out area below the upper level drift (Fig. 10.1.7). The fill material plays the role of the crown pillar between the upper extracted level and the lower level being extracted. However, a barrier which could retain the fill material must be built for this part of the mined-out coal. In the Spain the excavated area of some mines is stowed completely (St. Antonio Mine).

Finally, it should be pointed out that in some countries such as the U.S.S.R. and others the bench mining system has been replaced with a cutter-loader technology.

Advantage of the change over is the replacement of costly timber with the self-advancing shield supports, and continuous coal production.

10.2.   STABILITY OF STOPE-PILLAR STRUCTURE

Long pillar mining offers more stability problems than any of the principal mining systems, due to both the mining configurations and the structural features of coal seams. Under these circumstances the matter of concern is as outlined below:

– The concept of stope-pillar structure based on the principal layouts suggests certain relationships between long pillars and rib pillars.

– The criteria of loading of rib and crown pillars relate to their bearing capacity.

– The stability of principal structures is given as a function of the degree of coal extraction as well as the pillars' geometry.

The stability considerations are given primarily with the aspect of satisfactory coal recovery and continuity of coal production.

10.2.1.   *Concept of stope-pillar structure*

The geometrical approximation of the open stope structures formed by long pillar extraction could be represented by an inclined plate with the elongated slots along the dip, along the strike, or oblique to the strike. Generally speaking, two principal mine structures for open stope mining could be suggested (Fig. 10.2.1), as briefly described below:

a) Alternately long stope-rib pillar layout is common for conventional or hydraulic long-pillar mining, particularly of thinner and steeper coal seams. It should be noted that the majority of present layouts has a width of long pillars (W) of up to 20 m which is approximately one third of the stope width (w) where W = 1/3. Unfortunately this rule does not consider mine depth and coal strength, so that in practice the width of rib pillars might be either over or underdesigned. The stability of coal faces during long pillar extraction depends on the bearing capacity of rib pillars. For example, if the rib pillars are in a yielding state, the load could be bounced from them back onto the coal face structure. If the mining is up-dip, then the strata control could be carried out by keeping broken coal in the mined-out area of the stope, which would take one part of the load off the yielding pillars. Under these circumstances the open stope mining is converted to shrinkage mining. This system has been extensively used in metal mining for the extraction of very steep and moderately thick veins.

b) Alternately, long stope – 'skinny' pillar is common for very steep and medium thick coal seams which are extracted by long pillars as in the shield mining system. For this particular layout, the 'skinny' pillars, usually of 2 m thickness, do not have any function of ground support. Their main role is to separate a mined-out stope from a stope in extraction thereby eliminating dilution by caved rocks from the gob. When stope extraction is completed, the 'skinny' pillar toward the gob is completely crushed. However, the battery of several stopes is layed out between actual rib pillars, which accept stress transferred from the working area in a similar manner to the abutment pillar concept for room-and-pillar mining with a rib pillar width of approximately one fifth of the total mined-out area between pillars, as dictated by the

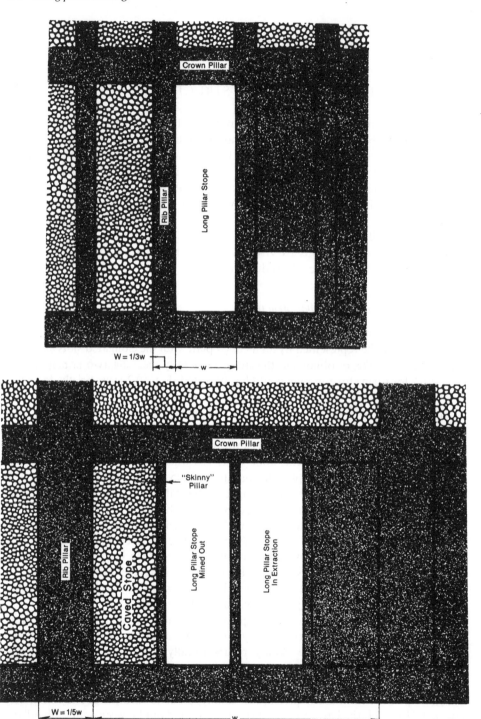

Fig. 10.2.1. Principal layouts of the open stopes.

'rule of thumb'). The justification for this approximation is the same as it is in the previous paragraph for rib pillar width defined by the 'rule of thumb'. Ground stability could be controlled in a satisfactory manner if previously mined-out stopes have been caved so that the stress is not transferred from them to the mining area in progress.

The conceptual representation of long pillar layouts integrates both open stope and rib pillar concepts as the principal structure of coal extraction. From represented layouts (Fig. 10.2.1), it can be concluded that crown pillars might be of a lesser influence on the principal structure of excavation and of greater influence on the load-supporting downdip, particularly between horizons.

10.2.2. *Criteria of pillars loading*

Generally speaking, the rib and crown pillars coupled together represent the side walls of the open stope structure supporting the superimposed dead load.[6] Both rib and crown pillars have common geometric characteristics in that:

    – They are inclined at the seam pitch angle;

    – They have less height due to the limited thickness of coal seams;

    – Their width is in the same range as the pillar width in room-and-pillar mining systems.

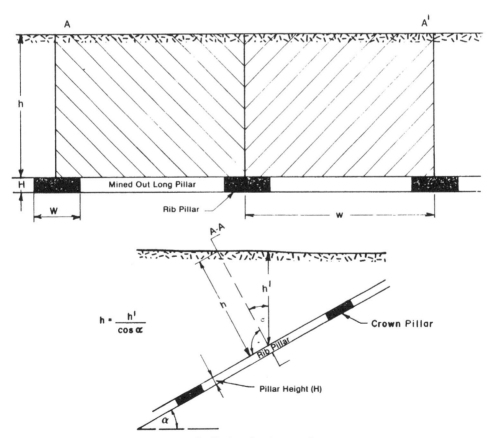

Fig. 10.2.2. Rib pillar loading along longitudinal section (seam strike).

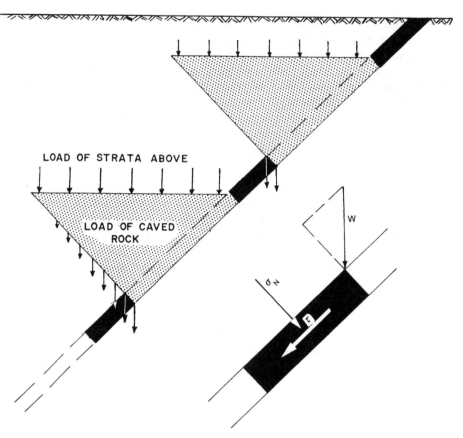

Fig. 10.2.3. Crown pillar loading along cross-section (seam dip).

However, the length of crown pillars is much greater than that of rib pillars, which is delineated by mine layout.

The analysis of pillar loading is carried out in a biaxial plane per unit length as discussed below:

a) Rib pillars are designed along the strike of coal seams. It is inferred that any rib uniformly supports the overburden weight above it (Fig. 10.2.2).[7, 8]

b) Crown pillar loading shows a preference towards point loading of a three-sided prism of caved rock upon which is applied the load imposed by the overburden strata.[9] The magnitude of the total load at its point of application could be calculated in the cross-sectional plane (Fig. 10.2.3) as the sum of the weight of broken rock within the prism boundary and the weight of the rock strata from the prism to the ground surface. The point load induces a shear stress in the crown pillar which constitutes a main factor in its stability. The existence of this shear stress causes deformation and failure of the crown pillar in a manner similar to that discussed in the chapter on sublevel mining. Due to these deformations, the crown pillar is the weakest link in the open stope structure. Increasing the pillar width to achieve higher bearing capacity is not a solution to this problem because activated shear stresses, regardless of pillar size, could cause deformation and shear failure in the pillar, particularly along the structural weaknesses in the geology of the coal. The activation of shear stress depends upon the magnitude of the point load, which increases with mine depth. The compressive stress is of secondary importance except in a case where the pressure from the footwall is of sufficient magnitude to cause compression of the crown pillar, resulting in coal fissuring.

The represented criteria of loading corresponds to geostatic stresses, where dynamic stresses due to mining effects are not considered. The analysis of dynamic loading in open stope structures by underground instrumentation and by experimental analytical methods has been investigated in some underground mines of various countries.[11] However, the results obtained by these investigations could not be generalized because geological structural features of individual coal deposits vary a great deal from basin to basin.

10.2.3. *Stability of principle structures*

The open stopes are supported by four principle structures whose stability is described as follows:

1. A long pillar represents a structure which could be extracted in directions as illustrated in Figure 10.1.1 in the previous section. At this point, only one layout of long pillar extraction will be considered, namely the 'down-dip direction' which exhibits particular characteristics as follows.

At the onset of stope extraction marked by the excavation of a primary slot, the long pillars are going to be in a highly stressed zone, as seen in Figure 10.2.4. This stress concentration is due to the transfer of loads from the caved prism to the footwall block perpendicular to coal seam stratification planes. The minimum load transferred is at the top of the prism (at the base) and it increases toward the center (progressive load transfer), so that it reaches a maximum at the lower tip of the prism. By enlarging the size of the prism, the load on the prism will also be increased and, consequently, so will the magnitude of the load transferred to the footwall. If mining is commenced up-rise, the initial extraction slot would be excavated under virgin stress conditions of the coal seam, and therefore, a high stress concentration would be avoided. However, with further extraction and shrinking of the long pillar, the stress state would depend on the magnitude of transferred abutment pressure and on the size of the pillar remnant currently in extraction.

2. Rib pillar experience variable stress concentrations and distributions due to the decreasing size of long pillars and to the relevant increase to its length during extraction. The stability of the rib pillar depends on the slenderness ratio (H/W) which is progressively decreasing as the extraction of the coal face advances (Fig. 10.2.5). The rib pillars formed by the initial stope excavation have a slenderness ratio $H/W < 1$, and they have sufficient bearing capacity to support a high stress concentration. As the stope extraction gradually advances, the bearing capacity of the rib decreases in accordance with the slenderness ratio. At a certain rib length, the slenderness ratio becomes critical with respect to stress concentrations, thereby decreasing pillar strength and initiating yielding and failure of the pillar. Under these circumstances, a

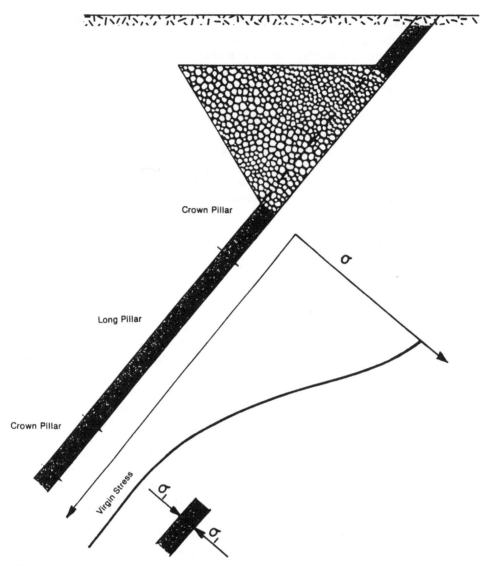

Fig. 10.2.4. Compressive stress distribution within long coal pillar.

rib looses its bearing capacity resulting in the need for additional artificial stope support.

3. Crown pillar weakness as discussed in the previous heading represents a great problem to mine operators mainly because of two aspects: firstly, due to the pillar's inability to support abutment loading, the stresses are transferred to working area, which affects strata stability and ground control; and secondly, the roadways in crown pillars (return airways) are in a state of progressive deformation and it is very difficult to maintain them in good standing condition, particularly when the extraction face

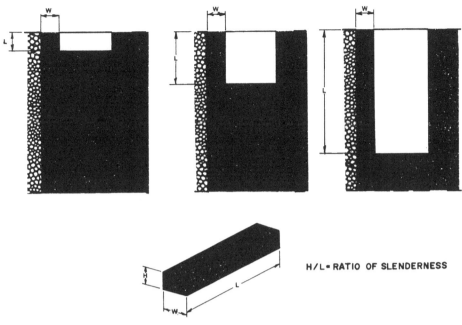

H/L = RATIO OF SLENDERNESS

Fig. 10.2.5. Changes of slenderness ratio (H/W) of rib pillar during stope extraction.

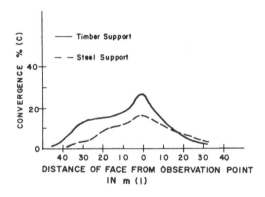

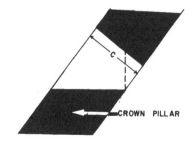

Fig. 10.2.6. Convergence of the crown pillar's roadway reinforced by timber and steel support.

advances towards the vicinity where maximum deformations are developed. For example, in some coal mines in the U.S.S.R., the convergence movement between hanging wall and footwall strata could be as much as 25 per cent for yielding steel support (Fig. 10.2.6).[12] The matters of particular consideration are crown pillar fissuring, and the closure of the ventilation roadways.

Fissuring constitutes a potential fire hazard by facilitating the self-ignition of coal. If such conditions are encountered, the only prevention is the extraction of crown pillars and the replacement by stowing in certain zones of the resulting void as an integral part of mine operations. In some European mines, the waste rock excavated from development workings has been used as a backfill to build up artificial crown pillars which would support part of the load from the surrounding area and consequently would relieve high stress concentrations on the working faces.

4. *The intersection of pillars* is the most important element in underground mine stability because this area bears the maximum stress concentration. Investigations carried out on models and in underground mines have led to the suggestion of several main stability characteristics of this particular structural element:

a) The stress pattern observed on models indicated that the intersection between the rib pillar and crown pillar constitutes a stress concentration several time greater than that found in the central part of the crown pillar. This observation is in agreement with other data reported in literature which were based also on two-dimensional photoelastic analysis of celuloid model plates.[6] The vertical stress magnitude at the middle part of the crown pillar is less than one half of the overburden load ($< 0.5\,\gamma$h), while at the intersection with the rib pillar, it reaches more than three times the overburden load ($< 3\,\gamma$h), as illustrated in Figure 10.2.7.

b) Underground observations have also shown great stress concentration at the corners between the crown pillars and rib pillars. In some places the peak corner stresses were reduced by coal falls which tended to round corners.

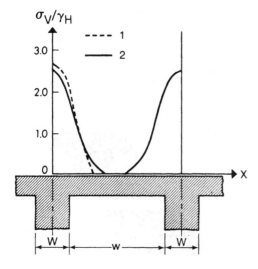

Fig. 10.2.7. Major principal stress distribution at rigid intersection between rib pillar and crown pillar (1. stress defined by model analyses; 2. stress defined by theoretical analyses).

c) The theoretical analysis based on the photoelastic model investigations has prompted the derivation of an equation for calculating shear stress at the rigid interaction between the rib and crown pillars as given below:

$$\tau = \frac{2W\,w\,\sigma\,c\,\tan\phi}{1 - \sin\phi} \qquad (1)$$

where
    W = width of rib pillar;
    w = width of the stope;
    σ = acting stress (1.6 γh);
    ϕ = angle of internal friction of coal;
    c = cohesion.

In the Tuzla Basin of Yugoslavia, the calculated shear stress at pillar intersections of hard and tough coal was 20.0 MN/m^2 and the tested shear strength of the coal material was 22.0 MN/m^2. This suggested stable conditions of the rigid connection at pillar intersections. However, underground mine observations showed only a semistable state under these conditions.

The photoelastic analysis shows that with an increasing stope span and a decreasing width of rib pillars, stress concentration is rapidly increased at pillar intersections.

The stability investigations of principal structures in open stope mines for various mining configurations could be successfully investigated by digital models, provided the limitations are known.

10.3. STRATA MECHANICS

The analysis of strata mechanics of long pillar mining systems has been based on model studies, underground instrumentation and visual observations of deformation and on actual failure of mine structures. On the basis of these diverse investigations, three principal subheadings can be considered:
– A stress redistribution of the cross-sectional mining configuration due to open stope extraction studied by photoelastic model analyses, in cases similar to metal mining.
– A longitudinal stress redistribution discussed on the basis of observation and instrumentation of long pillar mining in the open stope structure of some coal mines in Europe.
– Closure deformation of a mined-out area and strata stability individually discussed, from the aspect of actual underground situations.

It would be a fair statement that there are, to a certain degree, some similarities between strata mechanics in long pillar mining systems and strata mechanics in sublevel caving systems. Under these circumstances, the represented strata mechanics in these two chapters actually compliment each other.

10.3.1. *Cross-sectional stress analysis*

A stress redistribution in a mine's cross-sectional configuration has been demonstrated by the stress analysis of photoelastic models. Until the development of the numerical

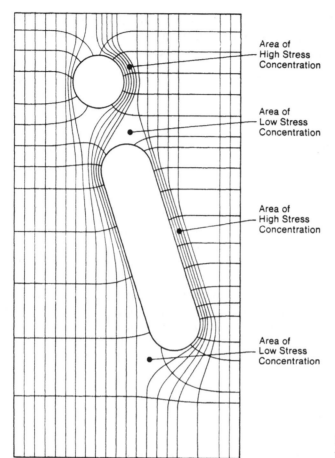

Area of High Stress Concentration

Area of Low Stress Concentration

Area of High Stress Concentration

Area of Low Stress Concentration

Fig. 10.3.1. Stress trajectories of the mine openings.

finite element technique, photoelastic modeling was the most common method used for stress analysis. Nevertheless, for the investigation of long pillar structure, photoelastic analysis might remain a valid method. The stress concentration around a level drift and the open structure between a crown pillar has been studied. The configuration openings were slotted in a $30 \, \text{cm}^2$ celluloid sheet with a thickness of $0.6 \, \text{cm}$ (CR-39). The level drift and stope structure were cut circularly to prevent infinite stress concentration. The angle of inclination was $70°$ to simulate a coal seam of this pitch and the selected thickness was $3 \, \text{m}$.

A brief description of model investigations and the interpretation of obtained data is presented below:

a) Isoclines were observed in polarized light as a colored interference of fringe patterns. They were interpreted in terms of direction of the principal stress.

b) A principal stress illustrated in Figure 10.3.1 shows distribution in the form of an orthogonal system of curves located around mine openings and within the crown pillar. The critical stress concentration within a crown pillar was found to be oblique to the pillar, from the tip of the undercut hanging wall to the tip of stope footwall. On the sides above and below the point of high stress concentration, the crown pillar consisted of areas of low stress concentration, which indicated its suitability for drift layout.

The elementary photoelastic model study of ideal materials illustrates some events found to occur in underground mines and raises some ideas for consideration. One drawback is that model analyses cannot distinguish between slow failure or violent fracturing, because whichever does take place may well be primarily due to the property of the material "enmasse" in a particular stope.[14]

10.3.2. *Longitudinal stress distribution*

Longitudinal stress redistribution due to long pillar mining has been studied to a lesser extent in the mining industry than a cross-sectional stress analysis. In this subheading, consideration has been given to the general representation and orientation of the principal stresses with respect to the strike and dip of coal deposits as well as to the particular analysis of abutment stress redistribution along the longitudinal profile.

1. Principal stresses and particularly their longitudinal distribution are important factors in the mechanics of deformation and failure of the open stope structure. The underground stress measurements indicated that the direction and orientation of the principal stresses is controlled by the strike and dip of deposits.[15] The long pillars, stressed due to the effect of nearby mining, have the stress distribution shown in Figure 10.3.2 and described below:

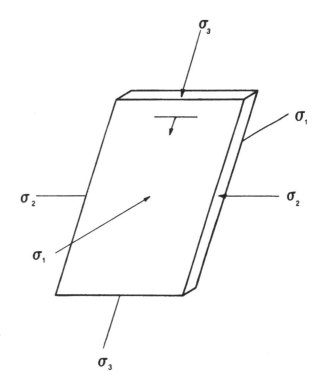

Fig. 10.3.2. The principal stress distribution of the long coal pillar structure.

a) The major principal stress (σ_1) is perpendicular to the bedding planes. This is the stress which exhibits abutment stress fields, and which is primarily dependent upon the virgin stress (composed by geostatic and geological stresses) and mining stresses transferred from the excavated area. The major principal stress is always compressive in mining configurations. The most common case of occurrence of major principal stress is in cross-sectional mine layouts. However, for strata mechanics evaluation, the longitudinal stress analysis along open stope faces is of equal importance as will be discussed shortly for the Kuzbass coal mine in the U.S.S.R.

b) Intermediate principal stress (σ_2) is in the direction of the coal seam strike for the majority of the mining configuration. It is not present in the cross-sectional stress analyses because only major and minor principal stresses can be considered. The intermediate stress in the longitudinal direction should be taken into acccount during stability analyses of the pillar-roof structure and it can be simply approximated by beam structures (Fig. 10.3.3). The intermediate principal stress (σ_2) is compressive and confines the pillar in a lateral direction. At the pillar core, the magnitude of the principal stresses can be redistributed due to their tendency for equalization and formation of a hydrostatic stress state. However, σ_2 is tensile on the y axis along the boundary of the coal pillar, as discussed in previous chapters. The intermediate stress in roof strata or on a rock beam of short length is usually compressive for a majority of mining configurations. With an increase in the length of the rock beam, the stress becomes tensile and it is transformed in minor principal stress (σ_3). This phenomena has also been discussed in previous chapters which consider mine roof failures.

c) Minor principal stress (σ_3) is compressive and is in line with the dip of the strata. However, when coal faces are parallel to the dip of coal seams, stress dynamics caused by nearby mining can cause a minor principal stress to become an intermediate (σ_2) or even a major principal stress (σ_1). Major attention in this principal stress is shear stress analyses which are carried out along the bedding planes of the long pillars. This has also been discussed in previous chapters.

By application of continuum mechanics the stability analyses by principal stresses are most suitable for long pillar structures. The principal stress analysis has a particular importance for the evaluation of stability of the roof span in a mined-out area of a long pillar mine. However, with respect to sublevel caving, principal stress analyses, to

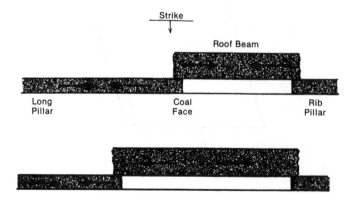

Fig. 10.3.3. Intermediate principal stress redistribution in the long coal pillar and roof strata along a strike of deposit.

certain extent, exert an importance on coal pillar stability analyses, as will be briefly discussed in the next chapter.

2. Abutment stresses distributed longitudinally along the mine structure are particularly important with respect to the roof stability of long pillar structures. This particular element of stress mechanics was studied in the coal mines of Kuzbass (U.S.S.R). In this region there is an appreciable number of pitching and steep pitching thin and medium thick coal seams. The most common mining method is by mining long pillars whose longer axis could be either along the dip or strike of coal seams. Coal winning is mechanical in the lifts (Tempo Cutter-loader) without supports in the mined-out area because mining is done under a stable roof span. A large number of mining methods start extraction from a raise slot with retreats along the strike. The underground measurements of stress and roof convergence have been carried out to evaluate the stress-strain state at the coal faces (abutment zone) and the behaviour of the rock strata of the mine roof. Figure 10.3.4. illustrates the abutment stress distribution which exists ahead of the coal face as a function of the length of roof span. A wide range of measurements, together with visual observations, provided a base for the assessment of the mechanics of formation of abutment pressure zones for different progressive stages of long pillar extraction as discussed below:[17]

a) Face advances up to 15 m are still influenced by stress transfer due to the widening of the raise slot, and stress relief can be observed on the coal face. However, a

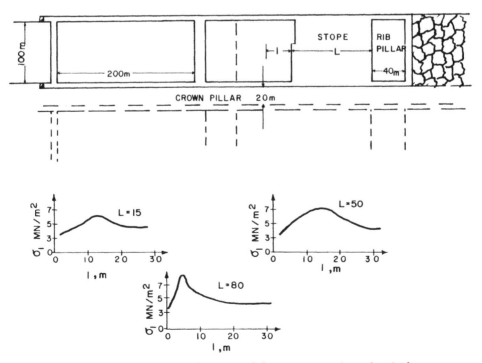

Fig. 10.3.4. Long pillar mining system with patterns of abutment stresses due to length of stope span.

coal face ahead indicated almost entirely elastic recovery, because small exposure of the mined-out area and rapid coal extraction do not allow displacement of the surrounding rock strata. Instrumentation indicates that the abutment stress gradually increases toward the solid pillar and that there is a maximum push 10 to 15 m ahead of the coal face. The maximum abutment stress is up to 1.5 of the virgin strata stress (Fig. 10.3.4). For this initial phase of coal extraction, the behaviour of roof strata can be approximated as a plate with its edges supported by pillars. Due to a fairly strong bonding between the rock layers above the mined-out area, as well as to the existence of elastic deformation recovery only, the vertical and horizontal displacements of the roof span are very small.

b) Face advances up to 50 m still have a roof span which remains standing, but bed separation and deflection are in progress. Vertical fissures can be observed and the bonding between beds of the mined-out area and the strata above the coal pillars decreases. Under these circumstances, the roof strata could be approximated as a beam on two supports. The increased deformation is followed by the increase of the abutment stress in the order of 2.9 times the virgin strata pressure, as well as by the formation of larger regions of abutment stresses up to 15 m in length. It should be noted that the peak of the abutment stress is not yet exhibited under these particular stress mechanics (Fig. 10.3.4).

c) Face advances over 50 m increase the length of the roof span to a critical state where intense displacement and roof caving takes place close to the working face. The situation can be approximated by a cantilever beam representation. Under these circumstances the load on the immediate area of the working face is increased, which results in coal fracturing and a simultaneous decrease in its bearing capacity. The relieved stress is transferred to the adjoining solid coal, where a peak abutment stress is formed, whose concentration is up to 2.5 times the virgin strata pressure. With the progressive extraction of a long pillar, the abutment stress is increasing, reaching a peak at a distance of up to 5 m from the coal face.

An equal consideration has been given to mine roof convergence, particularly with respect to roof caving. Investigations have established that the length of a roof span, before it starts to cave, is a dominant variable because of the similarity of the critical rate of convergence (Fig. 10.3.5). The instrumentation of roof convergence indicates that a majority of rock falls had been at a convergence of the range of 400 mm/m, regardless of the length of the stope span.

V. I. Murashev et al. developed an empirical equation for the determination of the length of the stable span in long pillar mining, when advancing in the direction of the strike of coal seams with a thickness of less than 3 m. This equation, containing geological and engineering factors, is given below:[18]

$$l = 300 \left[1 - e^{-8.32 \times 10^{-2} \left(\frac{F \sqrt{V_{adv}}}{\sqrt{D_d/D_w}} + 0.35 \right)} \right] \tag{1}$$

where

F = Protodyakonov factor of rock hardness;
V_{adv} = rate of face advance (m/day);
D_d = length of working face along dip (m);
D_w = width of working face (for this method 3.5 m).

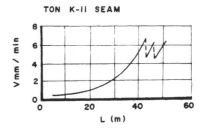

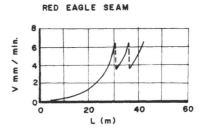

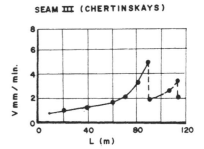

Fig. 10.3.5. Diagrams of rates of roof convergence before and at failure, for various coal seams of Kuzbass Basin (after V. I. Murashev et al.).

The proposed equation for the determination of stable roof spans should be used with great caution when applied to coal fields other than the Kuzbass, where different geological and mining parameters are in effect. The represented equation has the purpose of illustrating the sophistication of open stope engineering elements rather than providing a strata control parameter.

10.3.3. *Closure deformations*

Closure are the product of two types of deformations: roof sagging and floor heave. The intensity and character of closure of the long pillar structure depend on several factors, for example:

a) The stratification feature of the surrounding strata as well as the lithology of individual rock layers along a transverse profile of the stope structure.

b) Strength properties of individual rock layers as well as their behaviour characteristics within the roof span.

c) Geological structural defects such as: shear strength of stratification planes due to either original bounding of rock layers or their orogenic shearing, and shear strength

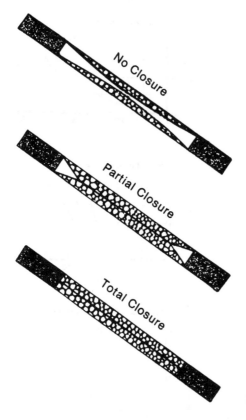

Fig. 10.3.6. Models of the types of open stope closure.

of structural discontinuities (joints), their density, character of bridging, etc.

d) Other factors, such as working thickness of coal seam, length and width of the stope area, mining with or without external roof support, rate of coal extraction, technology of coal winning (cyclic, continuous or hydraulic) and other mining parameters.

All these factors have been discussed in greater detail in the previous chapter. However, at this point it should be noted that they are instrumental in the origination of the various types of closure deformation, which can be classified in three groups using Berry's models of surface subsidence (Fig. 10.3.6).[18] Each category of the closure deformation of the open stopes of the pitching and steep pitching thin to moderate thick coal seams is briefly discussed below:

a) Closure mechanics do not apply to the case where roof sag is usually up to 50 per cent of the working height, and where floor heave is not significant. Differential slip along bedding planes and the lack of geological structural defects are the predominant factors which maintain the roof strata intact during sagging deformations. Two types of deformation can be considered when approximating roof strata with the beam: firstly, deflection in a short period of time (elastic deformation), secondly, sagging with time (viscous deformation) until failure occurs (Fig. 10.3.7). Viscous deformation has the same effect as a slow reduction in the modulus of elasticity of the rock, causing the rock beam to yield with time. This mechanism of roof span deformation could be

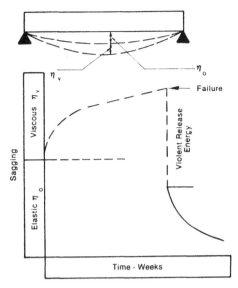

Fig. 10.3.7. Model of rock beam deformation until failure.

represented by asymptotic creep until the critical point as given below:

$$\sigma = E\eta_0 + E_1 \frac{\eta_1}{dt_1} + \frac{d\eta_1}{dt_1} + E_2 \frac{d\eta_2}{dt_2} \tag{2}$$

where

E = original coefficient of elasticity;
E_1, E_2 = deteriorated coefficients of elasticity;
η_0 = elastic beam deflection;
η_1 = viscous deflection due to time t_1 at constant speed sagging;
η_2 = critical deflection at time t_2, due to frictional displacement of mineral grains;
t = time.

The underground observation suggests that the rock beam undergoes mine roof popping before it collapses.

b) Partial closure mechanics exhibit a roof sag by the same principles as described in the previous heading, but deformation in the open stope area is supplemented by floor heave. The floor heave also could reach 50 per cent or more of the height of the opening, so that the roof and floor squeeze together for a major portion of the mine opening. Under these circumstances, due to bed separation, the rock beam is supporting only its own load, and the rest of the load must be transferred to the long pillar adjoining the pillar in extraction (Fig. 10.3.8). The induced mining stresses due to closure of mined-out area could cause severe ground conditions in the long pillars which are to be extracted next. It should be pointed out that closure deformation causes a cessation of progressive roof caving. Due to discontinuity of roof strata caving, a further build-up of induced stresses should be expected. Under these circumstances, ground de-stressing and artificial roof strata caving might become integral part of mine operation and could result in a burden on coal production and cost.

c) Total closure mechanics exhibit a complete sealing of the mined-out void. These

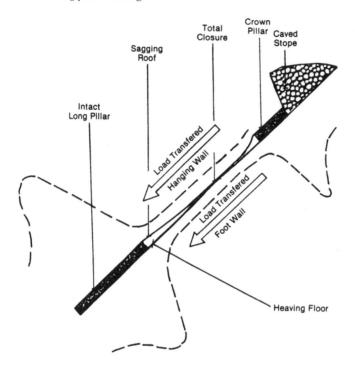

Fig. 10.3.8. Schematic sketch of transfer of induced mining stresses as result of partial stope closure.

mechanics belong to a case where the floor strata is composed of mudstone and/or shale layers, which have great ability towards swelling deformation when exposed to moisture. The total closure of mined-out stopes of thin coal seams (thickness 1.0 to 2.5 m) has been recorded for period of time ranging from 6 weeks to 6 months. It is a matter of interest that in the case of total closure, particularly a rapid one, there is no high stress concentration in the long pillars under development because the mining stresses are partially induced in the surrounding strata. In fact, after completion of the total closure, the transferred stresses can bounce back on the area of the former extraction structure. After completion of total closure of the mined-out stope, the stress state of the virgin strata could be reconstituted in a short period of time. However, the mechanics of total closure for open stope structures is not a common case because successful mining of long pillars requires equally strong and stable rock in both the mine roof and floor.

At the present time, the most common underground instrumentation is applied in the measuring of convergence in both development and extraction openings. These measurements are usually related to two factors: time versus deformation (progress of viscous deformations), and effect of nearby mining (intensified deformations due to superimposing additional load).

10.4. CAVING MECHANICS

The caving mechanics of the long pillar stopes are very similar to the caving mechanics of the sublevel caving system. The principal difference is that in the case of long pillar mining, the roof structures is approximated as a 'rock beam' but in sublevel caving, the roof structure is approximated as a 'rock cantilever'. In long pillar mining, the 'roof beam' will frequently be subjected to viscous deflection which progresses to closure deformation as opposed to roof failure and caving, particularly in case of thin seams. This is not often the case encountered with the 'rock cantilever' in sublevel caving of thick coal seams.

Consideration to caving mechanics of the long pillar mining system is given in terms of the following aspects:

– Caving of the roof strata has been studied by physical models using a base friction testing frame.

– The angle of break at the working faces has been investigated particularly in relation with seam pitch.

– Subsidence of the ground surface has been considered on the basis of the theory of elasticity, which produces more or less idealized solutions.

It is a point of interest that caving mechanics of long pillar mining considers a roof strata as a unit 'rock beam'.

10.4.1. *Roof caving dynamics*

The general case of long pillar mining has been investigated for an inclined coal seam, with coal extraction down-dip. The caving mechanics have been analyzed with the aid of the base friction model. The loading frame and the friction belt as well as the model material have been already discussed under the heading of the multiple slice (seam) mining.

The model was built to simulate the interaction between coal extraction sequences and roof strata caving. A crown pillar was left at the surface according to a layout of open stope structure. The model material was cut so as to simulate dipping strata at 50°. The jointing was also cut so that there were continuous bedding. The jointed blocks had jointing perpendicular to bedding. The jointed blocks have roughly a 2:1 length to height ratio. The coal seam itself was 4 times the thickness of the surrounding strata layers. The friction base model was allowed to run after each extraction sequence until equilibrium was reached.[20]

Mechanics of caving have been followed by the photographing of individual events which are grouped in four principal phases. It should be pointed out that this analysis considered only cross-sectional deformations and failure of the stope structure, as briefly described below:

1. Phase I represents commencing of the coal extraction by retreat lifts. The length of one retreat lift was approximated to 3/4 thickness of the coal seams. In this phase three of the six lifts have been extracted. Two photographs are presented, which are most representative for this phase of coal extraction (Fig. 10.4.1). From this it could be inferred that:

a) Very little displacement occurred in the hanging wall during extraction of these three lifts.

b) Upon extraction of the second lift, some loosening of the joints did appear but there was no caving activity.

The simulated mining conditions exhibited satisfactory stability of the stope

Fig. 10.4.1. Slight roof deformation during progressive coal extraction (lift two and three are completed).

structure which lasted with certain degree of degradation until completion of the final (sixth) lift. It is obvious that the existence of the crown pillar influences roof strata stability through the extraction of all lifts until the final one is completed.

2. *Phase II* represents the mining situation immediately after coal extraction and the formation of the mined-out stope structure. The following deformations of the roof beam were observed (Fig. 10.4.2) and are briefly commented on below:

a) The convergence of roof span by the exhibition of a maximum deflection in the area of the third lift might suggest that deformation in this area had been induced during coal extraction.

b) The bending of the 'rock beam' or convergence of the mine roof continued at a

Fig. 10.4.2. Deflection and tensile cracks initiation in roof strata after completion of coal extraction (six lifts).

progressive rate until it reached a maximum. The maximum convergence was approximately 1/3 of the original height of the mine opening.

c) The model indicated that maximum convergence was followed by formation of individual tensile cracks for each simulated sandstone layer with the extension in roof strata for a thickness of approximately one seam.

The simulated mining conditions of this phase clearly indicate that the critical length of the roof span is 4 times the height of the original stope opening, although under these circumstances the visible deformations started immediately after the completion of the last lift of coal extraction.

3. Phase III is a continuation of the progressive deformation and failure from the

Fig. 10.4.3. Progressive
caving roof strata with
kink bands development.

second phase, which was exhibited by roof caving dynamics (Fig. 10.4.3) as described
below:

a) Tension cracks had developed for the classic case of the beam deformation. Three
sets of failure were clearly exhibited, of which one was a tensile failure located in the
middle of the rock beam which was approximately perpendicular to plane stratification. The other two were shear failure and were located close to the clamped edges of
the beam at an angle between 70° and 75° to the bedding plane.

b) The rock beam of the mine roof was clearly separated from the upper rock beam.
The maximum separation corresponded to the area of tensile failure (i.e. to the central
part of the beam). The separation progressed as the caving of the lower beam advanced.

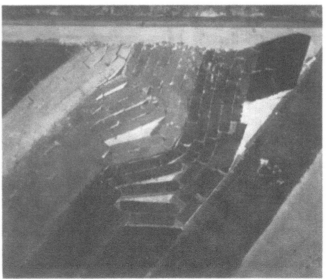

Fig. 10.4.4. Completion of caving of the roof strata before gob settlement.

c) The failed rock beam exhibited fragmentation by individual rock block which were delineated by a stratification plane, a failure plane, and to certain degree of jointing.

d) Caving of the fragmented block represents kink bands, where rotation of the block is toward a mined-out area. The block rotation in the lower part was clockwise whereas in the upper part, it was opposite.[21] As caving progressed the degree of block rotation was increased.

This prototype of caving of the roof strata provided a good simulation of the actual mining situation, because visual observations of caved material showed reverse rotation of the caved blocks. Also, these observations showed that boundaries of rock fragments are delineated by structural continuities, discontinuities and failure planes.

4. *Phase IV* represents completion of the caving cycle which resulted in several phenomena (Fig. 10.4.4), which are discussed briefly below:

a) A greater dilatation between caved blocks, particularly in the upper roof strata, had been exhibited. The strata fracturing in the blocks is emphasized in a downward rotation.

b) It was noticed as caving progressed, that tension failure between blocks rather than a complete displacement of the entire roof strata became the main contributing factor. Actually, tension fractures occurred in the blocks, thus permitting better caving of the roof strata and filling of the excavation.

c) The shear failure propagation at both ends of the collapsed beam formed the arch (trapezoid shape), which divided the caved and deformed strata from the intact strata unaffected by the caving process except for the displacement due to stress redistributions.

d) After caving the model was run (friction belt) for a few hours to simulate gravity loading on the caved material to produce compaction and observe the surface subsidence. The degree of compaction observed was much less than expected.

The described mechanics of caving are influenced by structural elements of the extraction face whose roof was approximated with the beam. The key element of this structure was a crown pillar, which also had been affected by caving dynamics. The model exhibited that the crown pillar also was rotated and fractured, which is quite possible because of its outcropping on the ground surface. Crown pillars which are located within underground mine are displaced and crushed as discussed in the second heading of this chapter.

10.4.2. *Breakage of roof strata at the face*

If the roof strata does not demonstrate self-sustaining abilities, then it is going to break when the roof span is formed, regardless of its length. Long pillar mining in this case, has to be carried out by installing external support on the working face. Under these circumstances the direction of mining becomes an important parameter. There will be appreciable differences in the mechanics of caving and in the loading conditions depending upon the mining direction being employed.

1. *Up-dip advance* results in roof caving behind the support, and the caved gob has very limited interaction with the support. Actually, there is only the interaction between the uncaved strata above the coal face and the support. This intersection is represented by the development of shear stresses between the mine roof and floor supports. The shear stress will depend on the dead weight of the fissured block of rock superimposed on the support, which is defined by the following parameters (Fig. 10.4.5):

δ – angle of caving,
β – angle of break,
α – angle of inclination of coal seam.

If the frictional resistance between the support and the roof and/or floor is exceeded

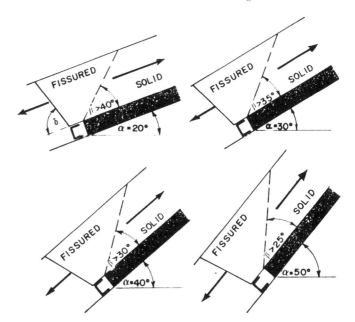

Fig. 10.4.5. Formation of break plane in relation to angle of inclination of coal seam.

by acting shear stress then the working area will collapse due to premature roof caving. The prevention of such a type of deformation of the working face requires a change-over of the mining system as, for example, the implementation of stowing of the mined-out void. However, it would be a fair statement that under these circumstances, long pillar mining should not have been implemented in the first place.

2. Down-dip advance results in roof caving directly on artificial supports. The bearing capacity of the support depends on its interaction with roof strata cavibility and compaction of caved material.

The compaction of caved material depends on the angle of coal seam inclination, the extent of rock caving above the seam and the lithology of the roof strata. The investigations showed that when coal seam inclination is over 45°, caved rock moves immediately beyond the canopy, which leads to dilatation of the roof strata above. This could promote additional caving, which would result in compaction of roof strata. The required pressure on the support is essential for stope face stability. For example, incomplete caving relating to insufficient pressure on the support could cause a cease in production until the required pressure has been achieved by induced caving.

Research has been carried out in the U.S.S.R. on the interaction between the lithology of roof strata and its cavibility behind supported working faces as a function of thickness and strength of lithological units. The following characteristics of roof caving and of pressure on supports would be suggested:[22]

a) The roof caving has been evident in three intervals: firstly, caving of a false roof (minimum pressure); secondly, caving of the immediate roof (average pressure). An idealized oscillogram of the pressures on the support for these roof strata caving mechanics and related loading is shown in Figure 10.4.6/a.

b) If the immediate and main roofs are both weak, then probably only the maximum

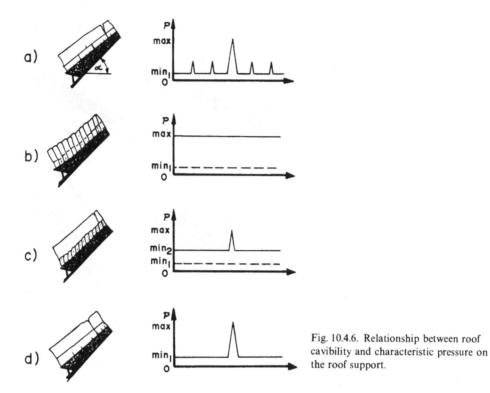

Fig. 10.4.6. Relationship between roof cavibility and characteristic pressure on the roof support.

pressure on the support will be exhibited, without a gradual load setting. An idealized oscillogram of the pressures on the support is illustrated in Figure 10.4.6/b.

c) If the immediate roof is weak (shale) and the main roof is strong (sandstone), the characteristic pressure on the support should be a transitional one between the first and second cases as illustrated in Figure 10.4.6/c.

d) If the immediate and main roof strata are of the same strength (medium or high), caving of the roof strata will superimpose loading on the support as an integrated pressure of the first and second case, but of higher magnitude.

Thus, for inclined coal seams the four models of characteristic pressure on the support are proposed due to roof strata cavibility, all other conditions being equal.

The calculation of the load on the support for inclined coal seams is based on the anticipation of the existence of two zones of roof caving: immediate and main roof. Some authors consider that caving of immediate roof has been orderly (block rotation) and that of the main roof, disorderly (block crushing). This is discussed in the next chapter in greater detail

10.4.3. *Subsidence of ground surface*

The subsidence of surface due to long pillar mining has been investigated analytically. The theoretical analyses of this type subsidence have been carried out based on the principles of soil and rock mechanics.[23] In the latter case, the subsidence mechanics have been investigated by applying a theory of elasticity, and a theory of elasto-plasticity, in

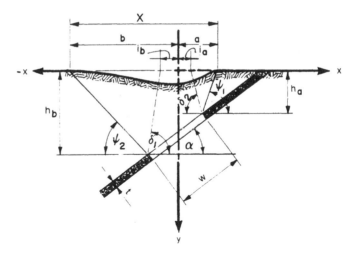

δ_1 = ANGLE OF CAVING (UP DIP)

δ_2 = ANGLE OF CAVING (DOWN DIP)

ψ_1 = ANGLE OF LIMIT OF SUBSIDENCE (UP DIP)

ψ_2 = ANGLE OF LIMIT OF SUBSIDENCE (DOWN DIP)

α = ANGLE OF INCLINATION OF THE SEAM

Fig. 10.4.7. Analytic model of subsidence for open stope mining of pitching coal seam.

a very similar manner to that used in other cases of subsidence analyses. However in this case, subsidence has been considered on the basis of the theory of elasticity, where trough subsidence is induced by the extraction of an inclined seam where displacement is toward the deeper edge of a minedout area, depending on the seam inclination (α).

The general characteristics of the subsidence are illustrated in Figure 10.4.7 and are described below:

a) The caving of the hanging wall is commenced from the up-dip area and continues down-dip with progressive coal extraction.

b) Subsidence trough of the inclined coal seam is accompanied by asymmetrical vertical movement, and it occurs as a maximum along the ordinate, y.

c) The angles of limit for subsidence, ψ_1 and ψ_2, and the angles of caving δ_1 and δ_2, with their corresponding planes depend on rock type and on coal seam inclination (α).

Considering the general case of subsidence for an inclined coal seam with its cross-section placed in the x-y coordinate system, the curve of vertical displacement can be expressed by the following equation:

$$v = v_{max} \exp\left(-\frac{x^2}{(i' + i'')^2} \right)\left[1 \exp\left(-\frac{t}{(T-t)^n} \right) \right] \tag{1}$$

where

v = vertical displacement;

v_{max} = maximum vertical displacement;

i' and i'' = abscissa of zone of compression;

x = length of subsidence influence (along axis x);

T = total time of subsidence;

t = subsidence at given time;

from which can be obtained a general expression for slope and curvature of the subsidence profile as follows:

$$g = \frac{\Delta v}{\Delta x} \tag{2}$$

Taking into account a final time of subsidence $(t + T)$, and variable substitution, the following expression is obtained:

$$g = \frac{dv}{dx} = -\frac{x}{(i' + i'')^2} v \tag{3}$$

where curvature radius of the subsidence is approximated as:

$$\rho = \left(\frac{\Delta^2 v}{\Delta x^2}\right)^{-1} \tag{4}$$

$$\rho = \left(\frac{d^2 v}{dx^2}\right)^{-1} = \frac{(i' + i'')^4}{[(i' + i'')^2 + x^2]v} \tag{5}$$

Furthermore, the degree of intensity of subsidence trough can be defined by the following relationships:

$$N_u = K \frac{g}{\rho} = K \frac{x[(i' + i'')^2 - x^2]v}{(i' + i'')^6} \tag{6}$$

where

K = constant, determined from practical experience 10^6, if value g and ρ are given in metres;

N_u = intensity of subsidence.

The same values for the slope (g) and radius of curvature (ρ) of subsidence do not occur at the same place because the maximum slope (g_{max}) is in the point between the zone of extension and the zone of contraction. However, minimum radius of curvature occurs between points of variations along the axis and boundary of subsidence.

It should be noted that these same analyses carried out for some coal mines of Southern Europe showed a discrepancy between analytically defined curves of subsidence and curves of subsidence obtained by measuring horizontal and vertical displacement in field.[24]

CHAPTER 11

Sublevel caving

The most widely used mining method for the extraction of thick and pitching coal seams is sublevel caving where the roof and floor rocks are strong and stable. This chapter deals with the following topics:
- Current sublevel caving systems which should have wider use in the near future.
- Stability of stope-pillar structures analyzed for various coal seam inclinations, where shear strength along bedding planes, as well as, pillar size must be considered.
- Strata mechanics of sublevel caving analyzed for ground stability, the key to successful underground operations.
- The use of physical models to investigate ground stability and comparisons of model studies with real mining situations.

The understanding and appreciation of the topics are of paramount importance for safe and economic coal production. The particular importance of sublevel caving systems is the potential for increased production and productivity through rapid coal extraction on a large scale. An example of this is the application of hydraulic sublevel caving.

11.1. PRINCIPAL MINING SYSTEMS

Sublevel caving for coal extraction has been borrowed from hard rock mining and modified to suit soft rock deposits. Generally speaking the variations in the layouts for sublevel caving of the coal are similar to the variations found in hard rock mining, and they can be broken down into the following catagories:
- Conventional sublevel caving, which uses the cyclic technology of drilling and blasting to fragment the coal, along with mechanized transportation.
- Hydraulic sublevel caving, which is particularly suited to the extraction of pitching coal seams. Water is used both to fragment and transport the coal.
- Gravity caving, to a certain degree, represents a transition from open stope mining. This technology is based on the formation of an undercut followed by gravity caving and drawing of the coal. Additional help by drilling and blasting may be required to achieve satisfactory caving.

The principles of the individual sublevel caving methods are briefly discussed below.

11.1.1. Cyclic sublevel caving

This mining method originated in hard rock mining, but during recent years has been introduced with significant success into the coal industry. The main requirement of the geological-structural conditions of the coal seams is that they are thick and pitching with strong and stable wall rocks. Seams with harder coal and limited tectonic disturbance are particularly suited to this type of mining. The following are typical applications of cyclical sublevel caving.

1. Coal seam thickness in the order of 6 m. A single entry development system is used, with the entry width governed by the horizontal seam width. The angle of inclination of the coal seam should be 30 degrees or greater. The system involves long-hole drilling and blasting in a retreat sequence. A typical layout consists of:

a) The development of the sublevel blocks with drifts driven 600 m long, 8.5 m wide and 3 m high (Fig. 11.1.1) and spaced at 20 to 25 m intervals measured down the slope of the seam. The drift height is governed by the drilling equipment. The drift is partitioned into an intake and return airway by cribs and sealant-sprayed metal panels. This partition permits a return of exhaust air. The excavation of the drifts is by a continuous miner fitted with side mounted roof bolters. A step system of advance is used where the left side of the drift is advanced a specified distance, then, after cleaning the face, the miner backs out of the cut and begins excavating the right side. A modular belt unit transports the coal from the face to the feeder breaker, where it is sized for further transportation by belt conveyor.

b) The full width of the coal to the sublevel above is drilled off with 44.5 mm diameter auger holes in rings spaced at 1.2 m intervals (Fig. 11.1.2). All holes should be kept under 20 m in length to limit hole deviation and to maintain an acceptable penetration rate. The coal is blasted in 6 m (5 ring) blasts using a low density permissible explosive. The broken coal is loaded by a mechanical loader to a feeder breaker for

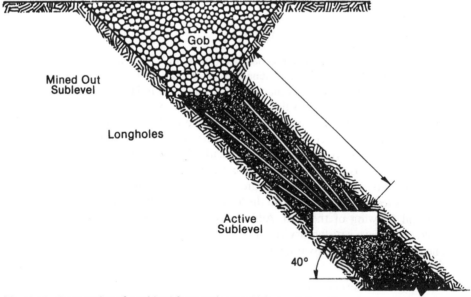

Fig. 11.1.1. Cross-section of a sublevel for a coal seam thickness of about 6 m.

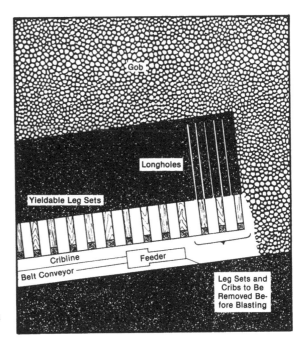

Fig. 11.1.2. Plan view of the extraction face of the section in Figure 11.1.1.

transport by belt conveyor. The feeder breaker and belt are retracted at the end of the loading operation and the cycle repeated.

c) Drift support is provided by roof bolts, mats and cribs. Extra roof support is required during retreat blasting, and this is normally provided by yielding leg sets placed on the intake side of the crib line. The sets are positioned between each pair of cribs at approximately one metre centers.

d) Ventilation of the face is conventional using brattice cloth, tubing and auxiliary fans.

e) productivity averages OMS 10 tonnes with seven men for development and four men for extraction. Production from underground mines of up to a half million tonnes per year has been achieved. This mining system has been successfully practiced for years in the extraction of coal seams with dips between 30° and 80°.

2. Coal seams up to 15 m thick and dipping steeper than 25° are usually developed with double entries.

a) Development is by arched sublevel drifts driven by a boom type miner. The fresh air sublevel drift is driven along the footwall and the exhaust air-drift along the hanging wall (Fig. 11.1.3). The drifts are at 11 m centers with crosscuts driven periodically to connect the intake and return drifts. A shuttle car transports the mined coal from the advancing face to the feeder breaker where it is sized for belt conveyor transport.

b) Two rubber tired mobile drill rigs are used to drill off the coal in rings spaced 1.2 m apart. The five yielding arch sets nearest the retreat line are removed from both drifts and five rings of holes in each drift are loaded with low density permissible explosive and blasted together (Fig. 11.1.4). When broken coal loading to the feeder

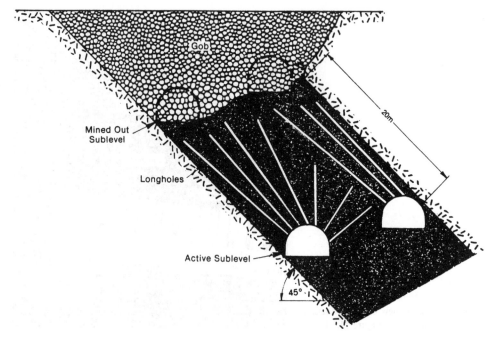

Fig. 11.1.3. Cross-section of a sublevel for a coal seam thickness in the range of 15 m.

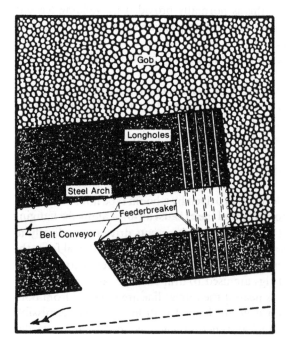

Fig. 11.1.4. Plan view of the extraction face of the section in Figure 11.1.3.

breaker is complete, the next five ring lift is drilled through the lagging of the arches to start the next mining cycle.

c) Support in the drifts crosscuts is provided by yieldable arches, chosen for their ability to resist intensive strata displacement. The cave brow is supported with timber sets which are left in the mined out area.

d) Ventilation is again supplied conventionally with brattice cloth, duct and fans.

e) Productivity averages OMS 15 tonnes with seven men for development and six men for extraction. Its increase is a function of the increased coal seam thickness. Production from underground mines of up to 1 million tonnes per year has been achieved.

The sublevel mining system with drilling and blasting is primarily applicable to hard coal blasting required to fragment the coal. For soft seams, hydraulic sublevel caving offers improved productivity.

3. Remote sublevel caving is under development in the U.S.S.R. for seams of a thickness between 2.6 and 6.0 m and with dips from 40° to 80°. The technology is based on the complete mechanization of all operations (KSBD complex) as summarized below:[2]

a) Sublevel development is done by an integrated system of roadheader with powered support, rock bolter, rolling-stage roller and monorail.

b) Extraction on retreat is provided by a remotely controlled drilling-and-shot firing unit with automatic coal loading under the protection of the roadway support. The retreat of the powered supports keeps pace with the progress of coal extraction. There is no need for men at the face.

c) Support is provided by powered hydraulic props supplemented with rock bolts and a timber barrier toward the gob.

d) Ventilation is the same as in conventional sublevel mining, a methane drainage system is required in the gaseous coal seams.

e) Productivity of the coal face should be two to three times greater than for conventional sublevel mining. However, difficulties in controlling the more active effects of gravity forces such as the development of floor failures could cause unpredictable, long and severe production shortfalls because of jammed power supports. Seam offsetting by faulting could also seriously restrict productivity.

It would be fair to say that this system belongs more to the drawing board than to practical applications. Further refinements to this technology, with greater equipment flexibility, are required.

11.1.2. *Hydraulic sublevel caving*

Hydraulic mining consists of coal cutting by powerful jets of water from hydraulic monitors operated by remote control. The broken coal is crushed and mixed with water and then transported from the face to surface using gravity wherever possible. This is a simple, safe and productive method. It will easily satisfy North American productivity standards provided the structural features and coal properties are favourable for water jet cutting technology as, for example, at the B.C. Coal hydraulic mine.

Sublevel caving with hydraulic technology is similar to sublevel caving with drill-and-blast techniques. The basic difference between them is in the method of coal extraction and transport. A hydraulic sublevel layout has to accommodate the continuous extraction of coal by monitor jets and the transportation of a coal slurry by gravity. The layouts for hydraulic sublevel caving can be grouped according to the thickness of the coal seam.

1. Medium thick coal seams normally inclined at more than 30° has been successfully mined by hydraulic sublevel caving in Japan and the U.S.S.R.[4,5,6]

a) Development of the seams is along strike (drifts) and along dip (raises). All development is in the coal seams. The sublevel drifts are spaced between 5 and 15 m apart, depending on the coal hardness and corresponding monitor jet effectiveness at impact. The drifts are driven upgrade at slopes between $2\frac{1}{2}°$ and 5° for gravity transportation of the coalwater mixture in flumes installed on the drift floor. The excavation of drifts is by drilling and blasting, Japan, or by monitor jets, U.S.S.R. The hydromonitors are mounted on a crawler, whose advance as well as the operation of the monitors is remotely controlled. Sublevel drifts are connected by twin raises, one utilized for slurry transport by pipes and the other for material and men movement.

b) The coal is mined in a retreat sequence in slices ranging in length between 2.5 and 5.0 m, depending on the coal hardness. The harder the coal, the smaller the sublevel block. The hydromonitors are positioned to break coal from the seam on the rising side so that the mixture of coal and water can flow by gravity along the mine floor (Fig. 11.1.5). The support of the retreating lift is not removed because the water jet impact on the coal face is between the timber posts and caps. The caving of the coal is

Fig. 11.1.5. The coal breaking by hydromonitor in Sunagawa Mine (courtesy of Takaji Kato, General Manager).

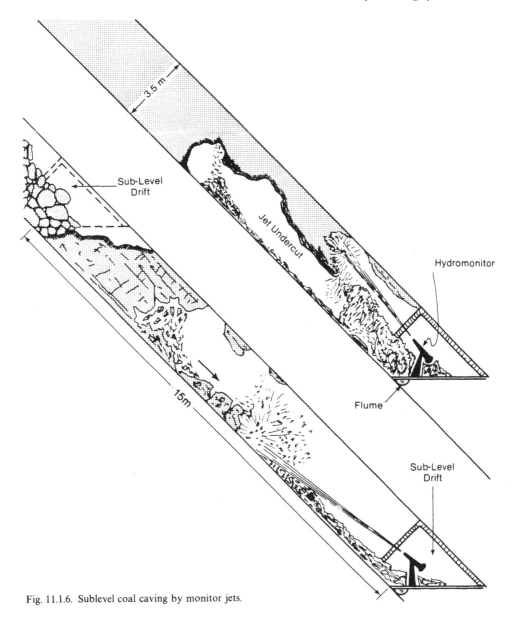

Fig. 11.1.6. Sublevel coal caving by monitor jets.

started close to the floor by making an undercut with the water jet so that the rest of coal can be easily fragmented and washed away (Fig. 11.1.6). In the sublevel drifts, the ccal-water mixture is transported in open flumes. Slurry transport in raises is in pipes, because the gradients are too steep for flume transportation. From the raises the slurry is transported by main haulage way flumes to the slurry collecting stations.

c) Timber is usually used to support the drifts and raises. The framing of the timber sets varies to accommodate the thickness and pitch of the coal seam. Timber frame sets are spaced at centers of 1 m or less.

d) Ventilation is conventional for a sublevel layout. However, the gas content in coal seams superimposes certain limitations on mine layouts. For example, Japan's coal seams are oversaturated with gas. They have small sublevel blocks delineated by pairs of raises spaced 35 to 55 m apart to achieve a satisfactory ventilation of methane gas emissions. In gaseous coal seams in the U.S.S.R., the distance between twin raise pairs is up to 150 m.

e) Productivity based on twelve men for single development and production face is limited by the seam thickness to range OMS up to 12 tonnes. An underground mine could achieve a coal production of up to 1 million tonnes per year.

2. *Thick coal seams*, 5 to 20 m, with dip angles in excess of 25° have been successfully mined in Canada and Russia.[6,7]

a) The seam is normally divided into panels for development. Each panel is opened by twin main entries, one for the flume roads and the other for the intake road. From the flume road, sublevel drifts are driven at the same slope angle as the main entries, but in the opposite direction (Fig. 11.1.7). The main entries and sublevel drifts are excavated by a continuous miner. The oblique position of the main entries and sublevel drifts to the strike and dip of the coal seam results in the optimum (2° to 7°) slope for coal-water mixture flow in flumes.

b) Extraction in a retreat sequence starts with the removal of the steel support arches for the length of a lift for soft coal, this can be up to 12.5 m. The hydromonitor is usually mounted on a feeder breaker, which is submerged in the water flow. Coal produced by hydraulic jet cutting is highly variable in both sizing and production rate and the feeder breaker is needed to control the coal size and feed rate to the flume to minimize blockages. The direction of the monitor jets is remotely controlled by an operator stationed in a steel cabin 10 to 20 m from the stope face (Fig. 11.1.8). The water

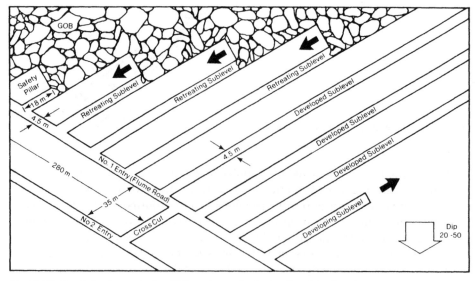

Fig. 11.1.7. Panel development of a thick coal seam for hydraulic sublevel caving.

Fig. 11.1.8. Operator's cabin, feeder breaker and hydromonitor.

pressure at the monitor jet nozzle ranges from 10 to 20 MN/m² and the water discharge from 200 to 500 m³/h, depending on local conditions and the coal hardness.

c) Support is similar to that used for the conventional sublevel mining of thick coal seams with the exception that stiff steel arches are used instead of yielding arches. The support is salvaged as the face retreats and this is made possible by the remote operation of the monitors and the rapid coal caving and flushing out power of the water jets.

d) The ventilation system is the conventional one used for sublevel mining by the drill-and-blast technique in thick coal seams.

e) Productivity based on a seven men development crew using a continuous miner and shuttle cars and two men per monitor can be in the range of OMS 30 tonnes. Coal recovery is 70 per cent from the sublevel block as is the case for the sublevel methods previously listed. The production from an underground mine can range between 1 and 3 million tonnes per year.

Sublevel caving by the monitor jet technology is one of the most successful hydraulic mining systems. The main advantage of this system is its simplicity and the minimization of equipment break downs. The integration of coal excavation and transport operations is achieved through the use of the relatively simple technology of fluid flow.

11.1.3.   *Gravity caving of coal*

The undercutting of the coal blocks and caving by gravity or with the aid of drilling and blasting has been practiced in several countries: France, U.S.S.R., and Yugoslavia.[8,9,10] Each country developed systems to suit local geological-structural conditions and the coal hardness. The three most diverse examples of caving systems are briefly discussed below:

1. A method using wire mesh to control dilution is practiced in the U.S.S.R., for dips greater than 60° and widths over 5 m.

   a) The layout is similar to sublevel caving systems except for a development of a grizzly level, to draw the broken coal for loading into mine cars on the main haulage

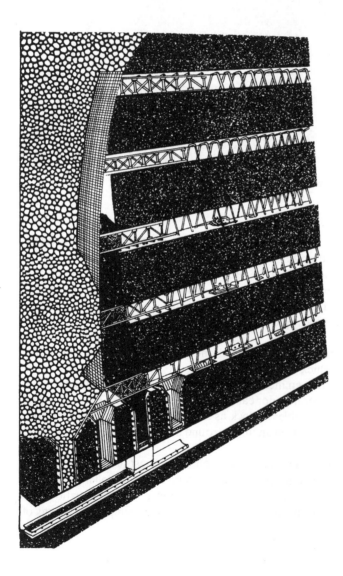

Fig. 11.1.9. Sublevel mining of steep and thick coal seams with the wire mesh method.

(Fig. 11.1.9). The size of the sublevel blocks is the same as for conventional sublevel caving and depends on the coal hardness and cleat development.

b) Extraction is based on gravity caving and fragmentation of the coal. The sublevel faces are separated from the gob by wire mesh to hold back dilution. Each sublevel contains moveable frame constructions placed next to the production face with extensions to the gob on which the wire mesh is tightly spread and attached. The steel frame supports are pulled back in retreat as coal caving progresses. The success of coal extraction depends on coal fragmentation and breakage at impact on the grizzly level.

c) Sublevel support is provided by two leg frames, timber props and steel beam caps and the independent mobile frame structure which can be easily pulled back by the whinch when each extraction lift is completed.

d) Ventilation is provided conventionally as for the other sublevel methods; however, the velocity of the air current has to be increased to ventilate the increased dust produced during caving.

e) Productivity by this method has been below OMS 10 tonnes in contrast to the expected figure of OMS 30 tonnes. The reasons for this are, the irregularities in coal caving due to gravity and the limited practibility of containing a large area of caved gob rock with flexible wire mesh.

This particular mining method is a hybrid type of open stope and sublevel mining. It is presented in this section because of a similarity to sublevel caving at the Kiruna Iron Mine in Sweden where all sublevels are integrated into one stope structure which is filled with broken ore by removing only the swell ore on the sublevel horizons. In the final extraction phase the ore is drawn at drawpoints on the main haulage level.

2. 'Soutirage' method, developed in France for mining thick and pitching coal seams. This method is applied where the coal seam cross-sectional horizontal width is greater than 12 m.[8,9] A brief description follows:

a) Panel development consists of two inclined entries, one in the hanging wall and one in the footwall (Fig. 11.1.10). The hanging wall entry is used for coal transportation, and the footwall entry for service. At each level a crosscut connects the two entries. From this crosscut airways are driven in both directions along the hanging wall contact. At the ends of these drifts two crosscuts are driven in the coal, the full undercut width, which can range between 6 and 9 m, to the footwall contact.

b) Mining retreats in short faces to the central crosscut. The height of the undercut is approximately 2.5 m. The coal above the undercuts may be from 3.5 to 12 m thick and is extracted by caving and drawing (Fig. 11.1.11). Mining is done with a self-advancing shield support, with banana props, that jacks against the caving coal. The coal draws

Fig. 11.1.10. Plan view of development for the 'Soutirage' method.

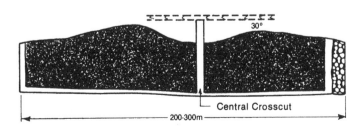

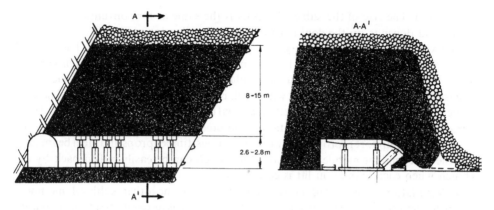

Fig. 11.1.11. Sections of 'Soutirage' method sublevels: A. a transverse section; B. a longitudinal section showing the shortwall and caving face.

through windows in the shield canopy and loads on a belt conveyor below. The undercut face is mined by a frontal single drum shearer (Eickhoft) which is integrated with a chain conveyor and the shield support.

c) Support of the entries and crosscuts is provided by yielding steel arches. The shortwall face support of the undercut is with 'Marrel FB 21–305 Soutirage Shields', at an operating pressure of 35 MN/m^2 and yielding pressure 45 MN/m^2. Each shield unit covers a 4 m^2 roof contact area and weights 12.75 tonnes.

d) Ventilation is laid out such that the intake is the entry in the hanging wall and the return is the entry in the footwall. Ventilation is controlled to provide a fresh air flow to all working places.

e) Productivity of the integrated, advancing, coal caving face is OMS 40 tonnes but for the entire mine output is only OMS 6 tonnes. Annual production is up to a half million tonnes.

In France it is considered the 'Soutirage' method, under the right circumstance of a thick regular coal seam, might be very productive with good coal recoveries. However, under these circumstances hydraulic mining might be simpler and more productive, with production costs of less than one half of the 'Soutirage' method.

3. The 'Velenje' method, developed in Yugoslavia, similar to the 'Soutirage' method.[10]

a) The 16 m thick coal seam with dips of between 30° and 60° is developed by a main crosscut from which two roadways are driven, one in the hanging wall and other in the footwall. These roadways develop blocks for shortwall mining using longwall equipment. The coal is hard and tough.

b) Mining retreats along the seam strike in sublevel blocks, 10 to 13 m high. Three sublevel blocks are simultaneously mined. Each sublevel block is undercut by a retreating short face and the coal above is drilled and blasted (Fig. 11.1.12). The coal undercutting, loading and transport system is similar to the French system. The caved and blasted coal is loaded in part by gravity and in part manually.

c) The short face is supported by OKP powered supports, and the caving face by hydraulic props and beams. Previously timber support was used as illustrated in Figure

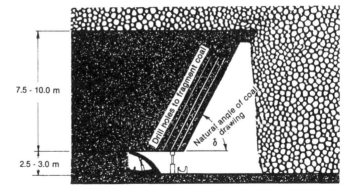

Fig. 11.1.12. Transverse section of the 'Velenje' system with two faces, the shortwall-mechanized face and the drill-and-blast caving face.

7.5 - 10.0 m

2.5 - 3.0 m

Fig. 11.1.13. Loading of caved coal in the late 50's (Kreka Coal Mine, Yugoslavia).

11.1.13. The supports of the roadways is by yielding arches lagged with timber halves.

d) Ventilation scheme uses the hanging wall roadway for intake air and the footwall roadway for return air.

e) Productivity for the system is OMS between 15 and 45 tonnes and is very dependent on manual loading of the caved coal. Coal recovery is in the range of 60 to 70 per cent.

The success of this mining system depends on the stability of the cave brow. Further improvement of the system requires some form of a mechanical loader to load the caved coal onto the chain conveyor.

11.2. STABILITY OF STOPE-PILLAR STRUCTURE

The stability of sublevel extraction pillars have certain peculiarities which are related to the thickness and inclination of the coal seam. The three elements essential for

stability analysis of stope-pillar structures in sublevel mining are listed below:
  1. The influence of the inclination of the seams on the strength properties of coal pillars.
  2. The average stress across sublevel pillars as controlled by local field stresses, the ratio of extraction and the seam pitch.
  3. The stress distribution within sublevel pillars relative to their shape and inclination.

The satisfactory bearing capacity of sublevel blocks is a key factor for safe and efficient operations. It is a point of interest that pillar-stope structures are in a progressive state of loading. Examples studied should be considered as snapshots of the mining operation at various points in time during the full mining cycle.

11.2.1.   *Pillar strength due to seam inclination*

The influence of anisotropy on coal pillar stability in relation to either the angle of seam inclination or bedding has been discussed to some degree under other subheadings. At this point the main concern is the strength variation of coal relative to the angular difference between the principal stress direction and the stratification planes in the coal.

1. Uniaxial compressive strengths have been investigated for mini coal pillars made into cubes with edge lengths of 7.5 cm. The cubes have been cut from coal lumps taken from various coal seams in Western Canada.[11] The testing results showed that the uniaxial compressive strength of mini coal pillars varied by a factor of approximately 4, and depended on the orientation of the stratification planes to the principal stress direction. Figure 11.2.1 is a plot of uniaxial strength versus the bedding orientation. From Figure 11.2.1 it is possible to predict the strength of a mini coal pillar which could be scaled to a mine pillar. Knowing this, it should be possible to evaluate the bearing capacity of a sublevel block and its stability, provided the average load is known.

2. Triaxial compressive strengths have been investigated for cylindrical coal samples with a diameter of 2.5 cm, and a height of approximately two diameters. Testing has been carried out on an INSTRON vertical loading system (TT-D), using an improvised

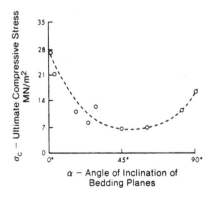

Fig. 11.2.1. The anisotropy effect (inclination of stratification plane) on the uniaxial compressive strength of mini coal pillars.

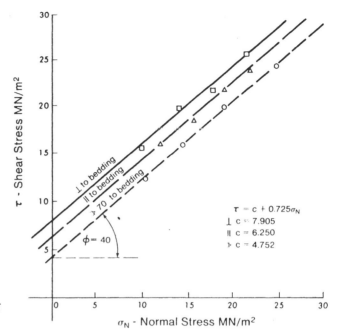

Fig. 11.2.2. The anisotropy effect (inclination of stratification plane) on the triaxial compressive strength of coal samples.

apparatus for lateral stress and a Soiltest triaxial cell. Figure 11.2.2, is a plot of shear strength versus normal stress for three different bedding orientations.[12]

a) The maximum confined strength of the coal occurs where the principal stress direction is perpendicular to the bedding planes, as was the case for the uniaxial compressive strength. For this loading orientation, coal has also a maximum cohesion intercept ($c = 7.905$ MN/m^2) and shear strength.

b) The confined strength of the coal where the principal stress direction is parallel with the bedding planes, is somewhat lower. This coincides also with the uniaxial compressive strength test results. The cohesion intercept is also lower ($c = 6.250$ MN/m^2) as is the shear strength.

c) The minimum confined strength occurs where the principal stress direction is at the same angle to the bedding as for the lowest uniaxial compressive strength. The cohesion intercept decreases approximately 40 per cent ($c = 4.752$ MN/m^2) reducing the shear strength. It should be noted that the angle of friction, regardless, of the orientation of the principal stress direction to the bedding planes is constant, $\phi = 40°$. The coal samples fractured during the triaxial tests exhibited an angle of failure to the vertical axis of between $10°$ and $25°$, average $\beta = 17\frac{1}{2}°$.

The observations of coal pillars in the Plains Region showed that between 80 and 90 per cent of their volume was confined, suggesting that for evaluating their bearing capacity, one should take into account the triaxial compressive strength of coal. However, lateral confinement can vanish when sublevel blocks near the completion of mining and there is a significant decrease in pillar size.

3. Shear strength of the coal samples relative to the angle of bedding inclination is shown in the previous paragraph and also the subheading of sliding deformation in the

Chapter on Deformation and failure of coal structures. The shear strength of coal decreases where the stratification planes are inclined.

Model studies and underground phenomenological investigations also indicated that the shear stress of coal due to the ratio between the vertical and horizontal stresses, expressed by parameter K (Fig. 11.2.3).

The pillar core, which is commonly under hydrostatic pressure has a shear stress close to zero (i.e. $\sigma_1 - \sigma_3/2 \simeq 0$). Shear stress in this case is independent of coal seam inclination. The outer zones of inclined pillars are under non-hydrostatic stress ($\sigma_3 \ll \sigma_1$) and the shear stress is of much greater intensity. This is particularly true for seam inclinations ranging from 30° to 60°, where the shear stress may approach one-half the applied vertical stress.[13]

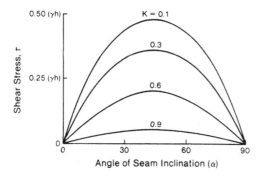

Fig. 11.2.3. Shear stress concentration as a function of the overburden load (γh), inclination of stratification plane (α) and the ratio between vertical and horizontal stresses (K).

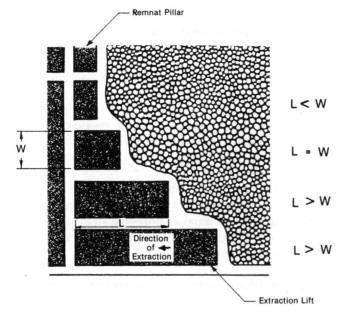

Fig. 11.2.4. The principal elements of sublevel pillars.

11.2.2. *Concept of sublevel pillar structures*

The sublevel extraction blocks are developed with sublevel drifts along strike and by raises up-dip. Several aspects relative to the stability of pillar-stope structures are briefly discussed below:

1. Geometry of sublevel blocks is similar to the pillars in room-and-pillar mining except that sublevel blocks are of greater length. However, with respect to pillar geometry the following parameters are important (Fig. 11.2.4):

a) The least pillar width (W) is a constant and is measured along the dip of the coal seam. The least width of sublevel blocks is chosen on the basis of coal recoveries, mine ventilation and other production factors. The relationship of the least pillar width to the strength of the sublevel blocks is not considered. W, ranges between 5 and 15 m and this is the main factor for limiting the mining depth of this method, to between 700 and 800 m.

b) Pillar length (L) is between 50 and 250 m, and is primarily governed by the gas content in the coal seam and the corresponding ventilation requirements. With a retreat extraction along strike, the pillar length is progressively decreased. At a certain point in time it becomes smaller than the least width. The balance of the block is then left unmined as a remnant pillar to protect the raise and drift structure.

c) Pillar height (H) corresponds to the working thickness of the coal seam, which in the majority of cases corresponds to the distance between hard and strong roof and floor stratum which can sustain load after coal extraction. In thick coal seams, the appreciable height of the sublevel block is a factor that reduces block strengths because of the unfavourable W/H pillar ratio.

Hard mining experience suggests that some consideration should be given to include stability analysis in sublevel block design. Unfortunately at the present time there is no simple method to carry out such a requirement. The following is a procedure for stability analysis which includes a number of simplifying assumptions which are shown in Figure 11.2.5:

a) A sublevel block is developed by two parallel drifts with a slope spacing of 3 times the thickness of the coal seam. The upper drift is in close vicinity to the gob and is surrounded by fissured coal remnants. It is assumed the lower drift is also surrounded by fissured coal as result of the excavation effects of coal breaking by drilling and blasting and of stress concentrations. The resulting approximation of a complex configuration is as shown in Figure 11.2.5(b).

b) The second assumption is that the resulting sublevel block is loaded in a similar manner to room-and-pillar mining, where the percentage of extraction is a main factor for load calculations.

2. The average pillar stress in a sublevel block is calculated by the tributary area theory as is the case for room-and-pillar mining. The average pillar stress of sublevel block is a function of the overburden weight, mine depth, the effective extraction ratio and the angle of seam inclination. The normal stresses on the pillars along with the shear stresses at the pillar roof and floor line are calculated in sequences:

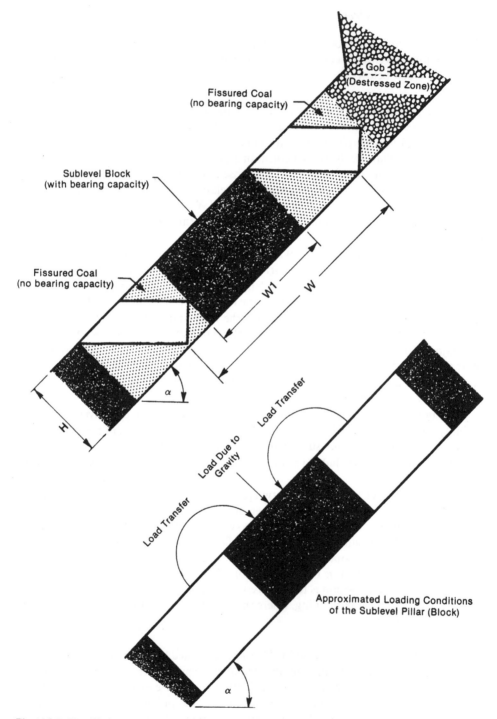

Fig. 11.2.5. Simplified static loading of sublevel pillars.

a) Calculate vertical and horizontal stresses on the unmined seam by the following equations:

$$\sigma_V = \gamma h \tag{1}$$

$$\sigma_H = K \sigma_V \tag{2}$$

where

γ = unit weight;
h = depth to seam;
K = ratio of the horizontal to vertical in situ stress (0.3–3.0).

b) Resolve vertical and horizontal stresses of the unmined seam into components normal to bedding and in shear parallel with the bedding (Fig. 11.2.6). The following equations can be used to calculate the normal and shear stress components

$$\sigma_N = \sigma_V \cos^2 \alpha + \sigma_H \sin 2\alpha \tag{3}$$

$$\tau_{seam} = \frac{\sigma_V - \sigma_H}{2} \sin 2\alpha \tag{4}$$

where

α = the seam dip.

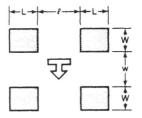

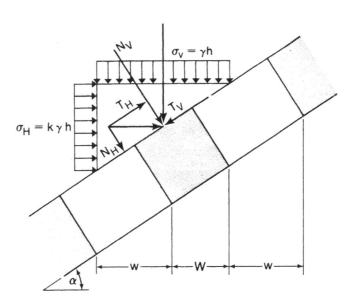

Fig. 11.2.6. Average pillar stress of the inclined block (calculated by Tributary area theory).

c) Define tributary and pillar areas as follows:

$$A_t = (W + w) \times (L + 1) \tag{5}$$

$$A_p = W \times L \tag{6}$$

where

A_t = tributary area;
A_p = pillar area;
W = least pillar width;
w = width of mined out stope;
L = length of the pillar;
l = length of the mined out stope.

d) Solve for resultant pillar normal and shear stresses on the tributary area

$$\sigma_{\text{pillar}} = \sigma_N \times \frac{A_t}{A_p} \tag{7}$$

$$\tau_{\text{pillar}} = \tau_{\text{seam}} \times \frac{A_t}{A_p} \tag{8}$$

For σ_{pillar} positive seam ride (shear) is down dip and for σ_{pillar} negative seam ride is up dip.

The above equations determine average stress conditions and do not include dynamic loading conditions during coal extraction.

3. Dynamic stresses distributions in sublevel blocks are of primary importance in the stability of the pillar-stope structure. The static calculation of the average pillar stress gives only the parameters for the design of a stable sublevel block least pillar width (W), prior to the start of mining. However, this dynamic stress pattern is directly related to ground control during mining operations. The particulars of this stress state are discussed below (Fig. 11.2.7):

a) Stress transfers during mining operations are primarily down-dip and are the

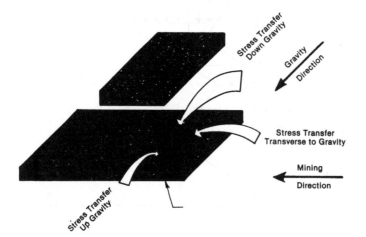

Fig. 11.2.7. Model of the dynamic loading on sublevel pillars.

main factor in pillar fracturing. These transferred stresses are high and they concentrate in the narrow abutment area in the solid coal below. As the coal fractures the abutment area broadens and unit stress values reduce. This fracture process has been observed underground and is further discussed under the heading of Stress mechanics in sublevel caving.

b) Stress transfer along strike, are of lower magnitude. They spread over a broader area of the working face because of earlier down-dip fracturing. Underground instrumentation in coal mines in the U.S.S.R. showed maximum values for stress transfers along strike to be approximately one-half of the values for down-dip stress transfers.[13]

c) Stress transfers up-dip are of the smallest magnitude because the area up-dip is soft, relative to the area down-dip. These phenomena show that sublevel block loading relating average stresses to tributary area theory are somewhat idealized.

Dynamic loading of the sublevel blocks during extraction is a most complex problem because such changing stresses are difficult to monitor in the field. Due to this reality we are directed towards the use of models to learn more about changing stress patterns.

11.2.3. *Model pillar stress distributions*

Visual observations of stress distributions within sublevel blocks in relation to their angle of inclination and shape can be carried out by photoelastic model analyses. The modeling of pillars is done in the same manner as briefly discussed in the chapter on room-and-pillar mining.[13, 14]

The study of inclined and trapezoidal shaped pillars is very convenient with photoelastic modeling. This type of pillar is a common structure in mining operations where the dip of the coal seam is over 20°. The studies show that the variations of the principal stresses across the pillar depends on the angle of inclination of the coal seam.[15] For example, compressive and shear stress concentrations occur at the pillar-floor contact on the down-dip side with a similar pattern existing at the pillar-roof contact on the up-dip side of the structure (Fig. 11.2.8). Mining observations of pillars in inclined seams show they fail differently during coal extraction than those in flat seams. The stability of inclined pillars have been studied as function of their shape. Photoelastic model studies carried out on inclined pillars indicate increased stability

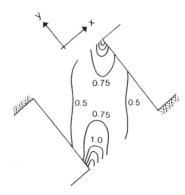

Fig. 11.2.8. Shear stress concentration in an inclined rectangular mine pillar where $\tau = (0.5 - 1.0)\gamma h$.

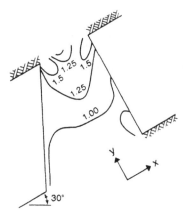

Fig. 11.2.9. Shear stress concentration in an inclined trapezoidal pillar where $\tau = (0.5 - 1.0)\gamma h$.

with the down-dip side of the pillar vertical and up-dip side normal to the walls. Trapezoidal pillars require subdrifts with trapezoidal cross-sections. The countours of equal maximum shear stress are shown as multiples of the vertical overburden stress γh, as illustrated in Figure 11.2.9. It was found that both the shear stresses and the compressive stresses were greatest at the top corners of the pillar. Maximum compressive stresses occur on the down-dip side and they are smaller across the pillar at mid height. The same model studies provided a common basis for comparing stress concentrations in the specially shaped pillars to those in rectangular pillars having faces perpendicular to the roof and floor. The stress distribution in trapezoidal pillars is more favourable than in rectangular pillars where the seam is inclined. Trapezoidal shaped pillars with a reduced cross-sectional area at the top of the pillar have stress concentrations 50 per cent greater than in the rectangular pillars. This could be a critical factor for the stability of trapezoidal pillars where there is a sheared stratification plane at the roof contact. The opinion of many researchers is that specially shaped pillars have little or no advantage over the rectangular shaped pillars. Also, it is obvious that it is difficult to develop sublevel headings with a constant trapezoidal shape and this could result in undersized pillars.

It should be pointed out that control of the pillar shape and size during the extraction phase of hydraulic mining is out of the question and any utilization of these elements to increase the stability of the stope structure is irrelevant. This fact does not contradict to philosophy of hydraulic mining because men are not exposed at the extraction face. If the stope span fails the coal can be still produced because the monitor jet has only sufficient power to flush out the coal but not the higher density rock.[16]

Further consideration of the stability of the pillar stope structure is given in the next subheading of strata mechanics.

11.3.   STRATA MECHANICS

The analysis of the strata mechanics of sublevel caving are based on monitoring and observations in underground mines, and by means of digital and physical models. On the basis of these investigations three principal phenomena have been evaluated.

– Coal strata are displaced by roof movements along planes of stratification by differential shearing in the coal pillars, and by heave deformations in the floor strata.

– Sublevel caved structures influence the formation of stress-concentration zones and stress-relaxed zones. High abutment stress occur for up to 50 m, away from mining, and influence of mining induced stresses extends for 150 to 300 m.

– The mechanics of deformation and the failure of mine structures due to coal extraction are well demonstrated by physical models, which produce failures comparable to actual mining situations.

Strata mechanics here are limited to sublevel caving systems. However, some of the information may be pertinent to other mining methods with open stopes.

It is a point of interest that the primary consideration of the coal caving studies is given to hydraulic mining where strata mechanics is most complex and still not well understood.

11.3.1. *Methods of investigations*

The investigations of strata deformation and the failure of sublevel caving structures in thick and steep coal seams in the Canadian Rockies have been carried out by three rock mechanics procedures:

1. Underground phenomenological studies have been conducted in the hydraulic coal mine of B.C. Coal Limited by Fisecki et al.[17, 18] for comparison to similar studies in the conventional coal mines of Coleman Collieries Limited and McIntyre Mines Limited by Jeremic.[19, 20] The main objective of the hydraulic studies was to understand strata movements due to the rapid breaking and washing out of coal by water jets to form large unsupported cavities, Figure 11.3.1. Field instrumentation was installed to monitor loads on sublevel and main entries, pillar deformations, strata movement above and below the seam, and roof to floor convergence in the entries.

The underground studies at the conventional mines were directed towards a better understanding of the deformation and failure of pillars due to the mechanics of loading and unloading, the influence of seam thickness and depth on the deformation of sublevel entries, and the caving of the roof strata.

2. Digital model studies were carried out for several different sublevel caving configurations in thick, 15 m, and, steep 45°, seams sandwiched between sandstone and siltstone strata. The stress analyses were done using two-dimensional finite element programs. The programs perform linear elastic analyses for plane strain problems using a constant strain triangle and provide displacements, element stresses, element principal stresses, maximum principal stresses, and average nodal stresses.

The finite element mesh used in the model studies and illustrated in Figure 11.3.2 has 812 triangular elements with 432 nodal points. This mesh divides the stope and pillar structures into idealized elements which are connected at the nodes to form a continuous structure. The forces and displacements have to be compatible at each node. A high density of elements is used in the pillars and the immediate hanging wall and footwall for detail. Also, the element nodes are so arranged that different mining configurations can be modelled with the same mesh by assigning different element

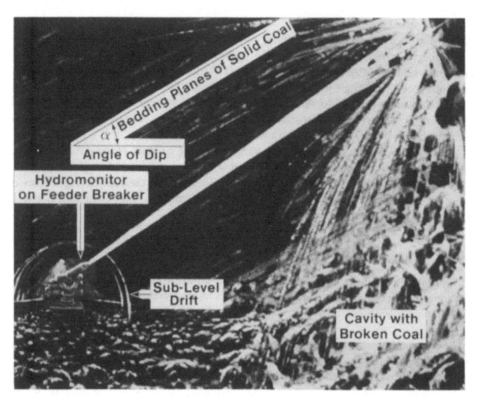

Fig. 11.3.1. Stope cavity with broken and washed-out coal.

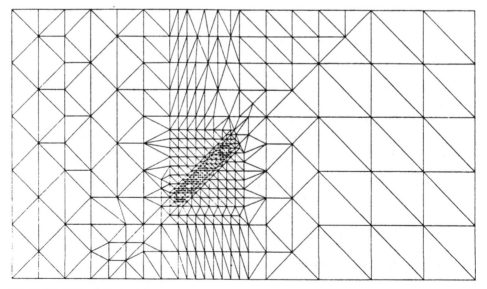

Fig. 11.3.2. Finite element mesh.

material properties and loading conditions. The mesh is extended to boundaries beyond the limits of mining induced stress changes and displacements. Planes of symmetry are utilized where possible to reduce the number of elements and computer costs. In the sample case, a high number of elements are placed at the mineral deposit sandstone interface in the mined region. This is to prevent displacements in the coal seam being overly restrained by the relatively stiff sandstone which has a modulus of elasticity of up to 30 times that of coal.

For excavated zones, stopes and drifts, the elements are assigned near zero stiffness.

The actual value used for the Young's Modulus of the sandstone was 10.0 GPa and that for Poisson's Ratio, 0.25. These values were selected to allow deformation into these regions to occur readily.[21] The caved material was assigned a Young's Modulus of 1.2 GPa and a Poisson's Ratio of 0.25. No specific values were available for caved material so the moduli ratio of coal to the caved sandstone was chosen as 2 to 1.

The initial vertical stress distribution is assumed to be due to the weight of overlaying rock material. Therefore, the vertical stress at point Z metres below the surface is given by:

$$\sigma_V = \gamma Z \tag{1}$$

where

γ = unit weight of overlying rocks.

If the overlaying formations are made up of layers of material with different unit weights, the vertical stress value can be calculated as:

$$\sigma_V = \gamma_1 Z_1 + \gamma_2 Z_2 \ldots \gamma_n Z_n \tag{2}$$

where Z_i is the overall thickness of material with an unit weight, γ_i and $\Sigma Z_i = Z$.

The initial horizontal stress magnitude is calculated from:

$$\sigma_H = K \sigma_V \tag{3}$$

where

K = Ratio $\sigma_H / \sigma_V (0.3 - 3.0)$.

The element mesh boundary conditions are selected such that the nodes on the boundaries are fixed in both the horizontal and vertical directions. The mesh is sufficiently large that the boundary conditions are not influenced by the behaviour of the stope structure.

3. Physical model studies have been conducted in the Department of Mineral Engineering of the University of Alberta using a base friction model.[22]

The base friction system, however, has been modified to confine a modelled structure in the third dimension, out of the plane of the friction model by a sheet of plexiglass which is subsequently loaded with evenly spread lead weights. Additional loading of the model is achieved by inclining the belt so that a gravitational component is added to the shear stress induced by the belt, and by adding lead weights along the top edge of the model to add a further load component (Figure 11.3.3).

The unconfined base friction model required the use of a very low strength modeling material, without the possibility of changing the properties of individual rock beds. However, under the modified loading system it was possible to use model materials of higher strengths. In the case studies, two model materials were investigated and

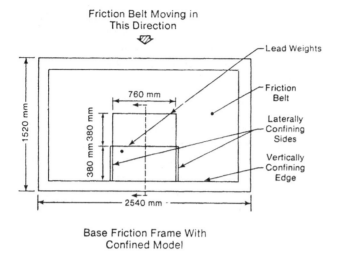

Base Friction Frame With
Confined Model

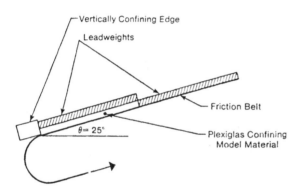

Fig. 11.3.3. Plane and section
of integrated base friction frame
with confined model.

selected to simulate the strength properties and behaviour of coal and sandstone. Coal was modelled with a mixture of plaster, coarse-grained sand, vermiculite, and water. It was compacted in 0.5-cm layers to simulate bedding, with the vermiculite grains producing a cleated texture. It is a low density material (1.11 to 1.28 t/m^3) with a low uniaxial compressive strength between 0.05 and 0.1 MN/m^2. The strength scaling factor of the model material to coal is 1:160 for coal samples taken from the Number 11 Underground Mine at Grande Cache. The uniaxial strength of this moderately hard coal is between 10.0 and 17.2 MN/m^2. The stress-strain diagrams for both coal and the model material were found to be sufficiently similar to make this a valid modeling material; both exhibited elasto-plastic failure (Fig. 11.3.4).

Sandstone was modelled with a mixture of fine-grained sand, plaster, and water, compacted in layers to simulate bedding. This fine-grained material had a density of 1.56–1.67 t/m^3, and a uniaxial compressive strength of 0.58–1.03 MN/m^2. When multiplied by the scaling factor of 1:160 this strength falls within the range of 96–200 MN/m^2, for the sandstone. The stress-strain behaviour to failure of the model material is similar to sandstone in that both are linear and exhibit brittle failure (Fig.

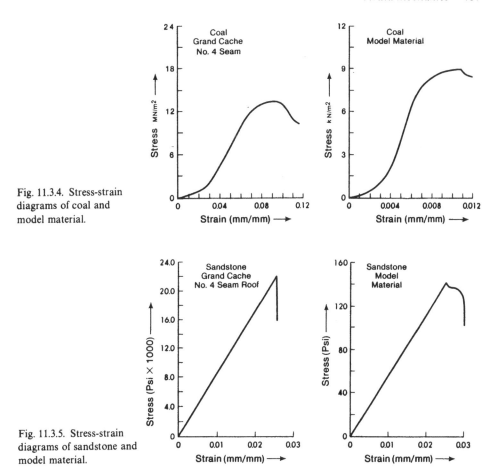

Fig. 11.3.4. Stress-strain diagrams of coal and model material.

Fig. 11.3.5. Stress-strain diagrams of sandstone and model material.

11.3.5). Laboratory tests of both sandstone and its model material show axial strains, at failure, about one-quarter of those for coal and its model material.

The studied model represents a coal seam 10 m thick at an inclination of 30°, sandwiched between sandstone strata. It simulates mining by a sublevel system, where sublevel pillars 15 m wide are delineated by sublevel entries. The model investigated simulated rapid coal extraction and the deformation and failure of sublevel drifts.[23]

11.3.2. *Displacement of the coal strata*

The main factor in the initiation of strata displacement is the development of shear stresses in excess of shear strengths. The mechanism of displacement depends on the geometry of the planes of weakness, their frictional resistance, and lithological changes along the seam profile. The behaviour of the strata should be analyzed for three horizons; the roof, seam and floor.

1. Roof strata mostly consist of layers of hard sandstone and siltstone and exhibit two types of displacement:

a) Shear displacement with slip failure occurs preferably between orogenically-sheared bedding planes which represent de-stressed zones[24] with shear movements of up to 0.5.

b) Stick slip failures along the rough non-planar surfaces of joints. These are sometimes severe enough to be classified as bumps.

In situ monitoring has shown that the displacement of roof strata can occur for up to three or four seam thicknesses above the seam.[17] The measurements also indicate integral roof strata movements toward the mined-out coal with a gradual increase of displacement closer to the coal seam. The same mechanical phenomena have been observed in the physical model where the maximum displacement of the roof adjacent to the coal seam has been recorded and scaled to 150 mm. The investigators of strata control at B. C. Coal's hydraulic coal mine expressed surprise that the area of influence of even a small excavation tends to be much greater than would be expected from the conventional mining of relatively thin seams.[17, 18]

2. Coal pillars in the vicinity of mine excavations have differential movements along bedding and thrust fault planes which can be attributed to the existence of differential stresses. This phenomenon is strongly exhibited in thick and steep seams in Western Canada, both in conventional and hydraulic mining. In general, the displacement patterns suggest three distinctive zones of deformation.[19, 20, 25] as illustrated in Figure 11.3.6. Coal layers close to the sandstone roof may be crushed and move down-dip and exhibit elasto-brittle deformations. Some layers at the middle of the seam, however, may have a tendency to move in opposite directions due to the different elastic constants between hard and soft coal layers and related frictional effects in direct-shear deformations. Finally, coal layers in the vicinity of the shale floor may flow towards the mine openings and exhibit large floor heaves due to viscous-flow deformations.

Monitoring of pillar deformations in relation to coal extraction was carried out at B. C. Coal's hydraulic mine. Only the middle part of the coal seam was monitored with borehole extensometres. The maximum displacement was 43 mm for a pillar 25 m wide.[18] This limited displacement corresponds to underground observations where the

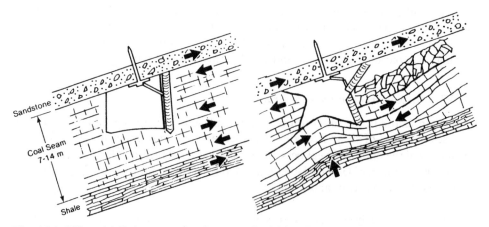

Fig. 11.3.6. Differential displacement of coal structure in vicinity of sublevel drifts.

middle layers of the seam are constrained and there is limited differential movement of the type noted above.

3. *Floor strata* are mostly shales which have a tendency to flow upwards rather than be displaced in a manner similar to roof strata. This type of deformation depends mainly on the magnitude of compressive stresses in the footwall.

Underground observations at the Vicary Creek Mine of Coleman Collieries have shown that displacements of floor strata can be approximately equal in both vertical and horizontal directions.[24]

The displacement mechanics of coal strata greatly influences mine stability which is directly related to the critical length of roof spans and the bearing capacity of pillars.[26, 27]

11.3.3. *Stress patterns in sublevel structures*

Stress mechanics in sublevel caving in hard rock mining have been studied extensively by Jeremic[27] and Hoek.[26] Hydraulic sublevel mines, however, have unique features which have been evaluated by Lin and Pang in Chinese mines[28] and by Fisecki et al. in Western Canadian mines.[28] These features have been further studied using digital and physical models by Jeremic and Lutley.[23]

The stress analyses discussed in this chapter consider two particular mine layouts and have been carried out mainly with digital models. It was assumed the seam had been mined to a depth of 150 m by underground methods and that the sandstone roof strata would cave into the mined-out area in the manner shown in Figure 11.3.7.

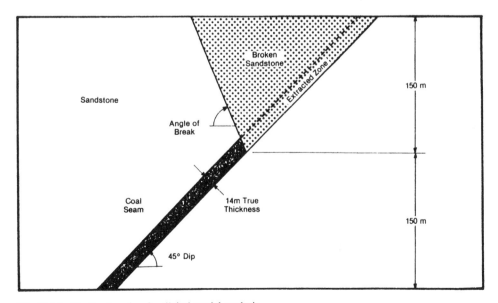

Fig. 11.3.7. Idealized section for digital model analysis.

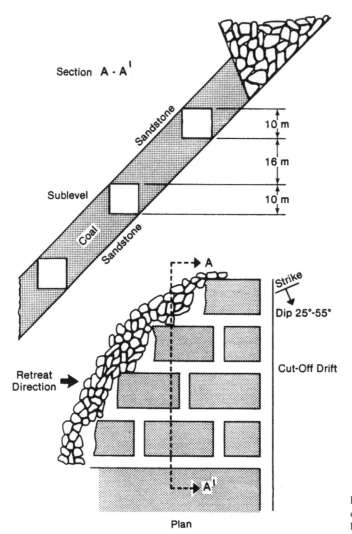

Fig. 11.3.8. Plan and section of sublevel mining analyzed by the digital model.

1. Sublevel block structures are delineated by the gob and sublevel drifts, and further subdivided along strike by raises to improve access and ventilation as illustrated in Figure 11.3.8. Loading conditions for the modelled section through the workings are illustrated in Figure 11.3.9.

The stress distributions in the model are gradually presented to show minor principal stress (σ_3) contours, major principal stress (σ_1) contours and the directions and the magnitudes of σ_1 and σ_3 at the mesh nodes.

a) The minor principal stress contours show a steady increase with depth; in the seam hanging wall adjacent to pillars, it exhibits a small peak, which is mainly due to stress transfers from each sublevel drift (Fig. 11.3.10). Underground phenomenological studies, however, show greater stress peaks adjacent to the pillars due to some load transfer from the extraction faces and the action of lateral tectonic stress.

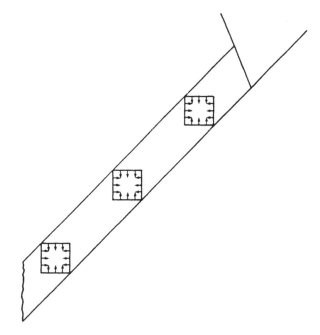

Fig. 11.3.9. Loading conditions of sublevel configuration (digital model).

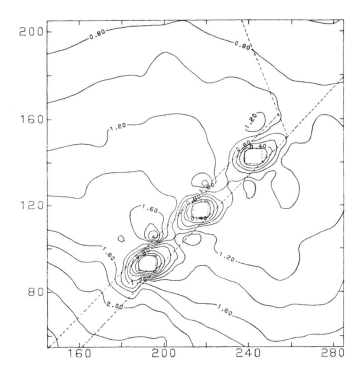

Fig. 11.3.10. Least stress distribution for a sublevel configuration.

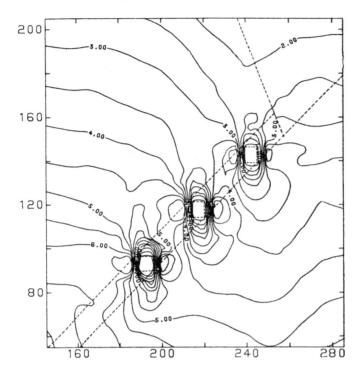

Fig. 11.3.11. Principal stress distribution for a sublevel configuration.

b) The major principal stress contours show a pattern of stress concentrations and relaxations in the hanging wall and footwall, around the sublevel openings. The pillars show the classic stress redistribution pattern with peak stresses at the abutments and the average pillar stress increase with depth (Fig. 11.3.11). Underground studies indicated much greater differences in stress concentrations between upper and lower pillars than the digital model suggests. This phenomenon should be attributed to the effect of yield in the upper pillars causing a transfer of the main abutment stress front to pillars below.[21]

c) The stress vector diagram illustrates the stress trajectories within pillars, which are influenced by the geometry of mining and geological structures (Fig. 11.3.12). An anomaly in the roof and floor of the excavations is caused by the way in which stresses are calculated at nodes. Stresses at nodes are calculated by averaging the stresses in the elements surrounding the node. On the boundary of the excavations, stresses are therefore averaged between the elements at zero in the excavation and the stressed elements in the surrounding rock. Above and below the excavations, vertical de-stressing has occurred but the magnitude of the average stress in the vertical direction is greater than the horizontal stress which is contrary to the true case. This results in false stress values at the excavation boundaries.[21]

The significance of the finite element stress analysis for this particular layout is obvious. It indicates the widths of the pillars should be increased to decrease average pillar stress magnitudes as well as to decrease concentrations of abutment stresses in the vicinity of the sublevel walls, the main cause of structured instability.

It is necessary to point out that the section analyzed passes through three pillars and

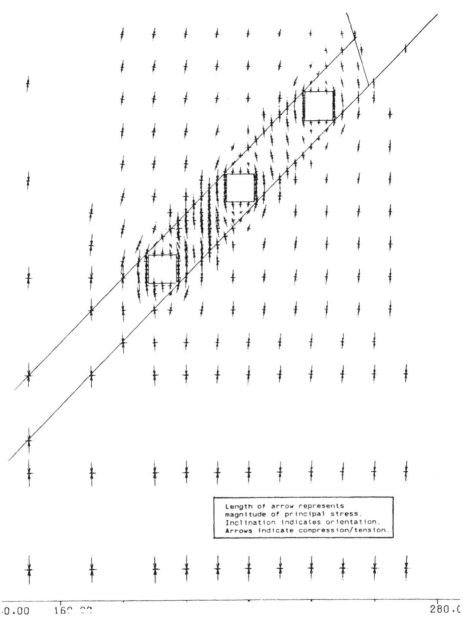

Fig. 11.3.12. Stress vectors for a sublevel configuration.

no allowance has been made for the transfer of stresses from the approaching retreat line. In fact, along the longitudinal section there exists a large unsupported excavation area and the transfer of stress from this area has not been modelled. Mining stresses from nearby mining retreats as each face retreats. As the excavation front approaches the section analyzed, stress transfers will arise in the third dimension. The evaluation of integrated mining stress transfers from cross- and longitudinal sections is, however, beyond the scope of this particular finite element program and any other two-dimensional program. This statement should be clearly understood or the stress analyses will be misleading.

2. The sublevel structure represented by the sectional mining configuration (Fig. 11.3.13) could be related to any system of sublevel caving where loading conditions are the same. The digital model analyses were carried out and the results compared to similar mining operations where stress and structure stability were instrumented.[21, 28, 29, 30] Particular consideration is given to the patterns of stress concentration zones as briefly represented below:[27]

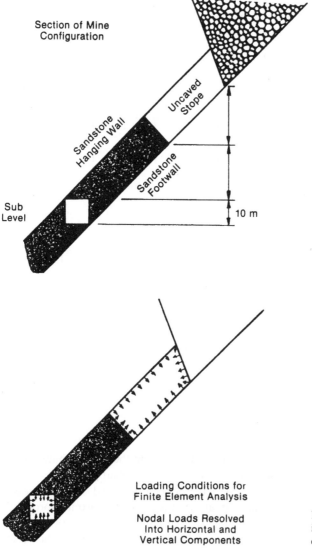

Section of Mine
Configuration

Uncaved
Stope

Sandstone
Hanging Wall

Sandstone
Footwall

Sub
Level

10 m

Loading Conditions for
Finite Element Analysis

Nodal Loads Resolved
Into Horizontal and
Vertical Components

Fig. 11.3.13. Excavation and loading conditions for an open stope configuration (digital model).

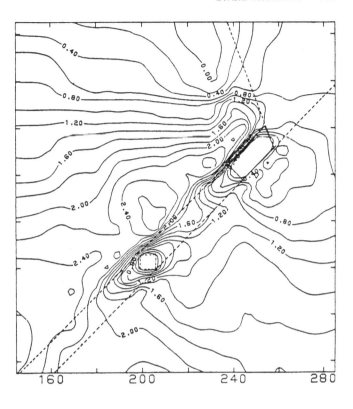

Fig. 11.3.14. Least stress
distribution for an open
stope configuration.

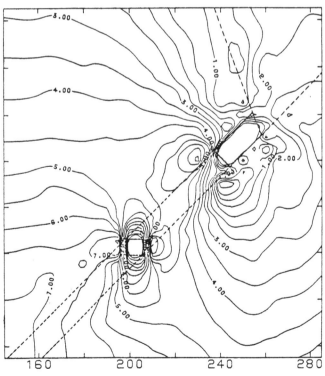

Fig. 11.3.15. Principal
stress distribution for an
open stope configuration.

a) There are stress concentration zones in the sublevel structure around mined openings where solid coal supports the transferred load. The stress concentration increases proportionally with an increase in the size of the mined stope. Patterns of the stress concentrations around mine openings are illustrated in Figure 11.3.14 for the minor principal stress and in Figure 11.3.15 for the major principal stress. Both figures show the large areas of influence of mining on the stress pattern which are coincidental with underground measurements of strata displacement. Also, the diagrams of stress concentrations suggest that the absence of roof caving in the mined-out area is due to an insufficient span to generate tensile caving conditions.

b) The stress concentration zone is the hanging wall of the mined out area shows the alignment of major principal stress to be approximately parallel to the roof strata of sublevel stope. The digital model studies showed that this stress is compressive (Fig. 11.3.16). This stress state might influence roof span stability and delays caving as

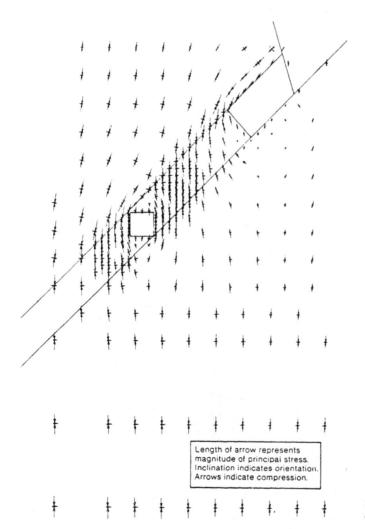

Length of arrow represents
magnitude of principal stress.
Inclination indicates orientation.
Arrows indicate compression.

Fig. 11.3.16. Stress vectors
for an open stope
configuration.

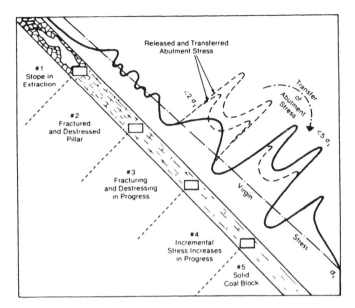

Fig. 11.3.17. Mechanics of
abutment stress transfer
and distribution.

discussed in the next subheading. However, a most dangerous situation occurs when a
large area of roof strata, after coal removal, still stands up. The stress concentration in
the roof strata, particularly in sandstone, can reach a critical magnitude which is
released suddenly by a violent roof collapse. This phenomena in underground coal
mining is known as an air blast and it affects the stability of working places and the
safety of mine personnel.

Of particular interest in sublevel caving, regardless of the extraction technology, is
the strength of the coal and the bearing capacity of coal pillars. If transferred stresses
from nearby mining can be supported by the immediate coal pillars without their
fracturing and yielding, then strata mechanics follows the stress-strain state extensively
discussed in literature for flat moderately thick coal seam mining. However, if the
abutment pillars cannot support the new mining and yield stresses the front abutment
is pushed further into solid coal until it has been extended to the point where a new
equilibrium condition exists (Fig. 11.3.17). The interaction between advancing
abutment fronts and progressive coal fracturing is a key factor in the formation of large
de-stressed zones. Peak mining stresses are moved large distances from the active
mining area and the magnitudes of the induced stresses are reduced. Underground
investigations in coal mines of Western Canada have shown that progressive coal
fracturing is also assisted by cleating. These investigations have shown that abutment
front stress can be transferred more than 50 m from the stoping area and that mining
can influence the stress field as far away as 150 to 300 m.

11.3.4. *Deformation and failure of structures*

The deformation and failure of sublevel structures have been monitored underground
and studied in physical models. Both investigations show similarities.

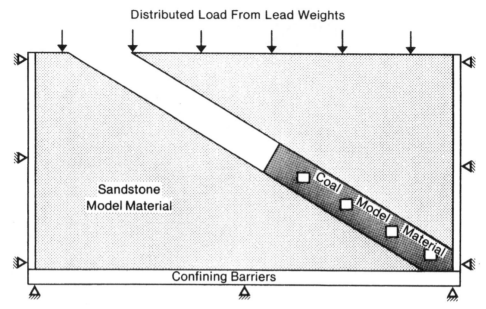

Fig. 11.3.18. Loading conditions for a sublevel configuration (physical model).

The loading conditions and the mine layout of a physical model are illustrated in Figure 11.3.18. The physical model was confined horizontally and along the bottom edge by rigid barriers to prevent displacements in these directions. The mining configuration is similar to the digital model of sublevel caving.

The results of the investigations of the physical model can only be used qualitatively to identify areas of potential instability and modes of failure, but the modeling technique does provide one of the best visual representations of pillar deformation induced by mining (Fig. 11.3.19).

a) In the uppermost sublevel, caving of the hanging wall progresses into the stoping area and crushes the pillar body and the sublevel drift. This phenomenon of collapse of the stope-pillar structure was observed at B. C. Coal's hydraulic mine, where the roof caved into the drift and buried a hydromonitor and a feeder breaker. These mechanics of the failure confirm the need for sublevel support here because the pillars are not capable of controlling the surrounding strata during extraction operations.

b) A pillar adjacent to a sublevel in extraction may fracture due to high abutment stress transfers from above. The model shows deformation by differential shear displacement along bedding planes and subsequent fracturing of the pillar accompanied by floor uplift. The same phenomenon has been observed in a steep thick coal seam at Coleman Collieries Limited, where room-and-pillar mining was practised.

c) The next pillar is in a state of stress concentration because the abutment front is transferred onto this sublevel from the pillar above by progressive fracturing. The existence of high stress concentrations was observed in the model by cracking and chipping of the sublevel pillar.

d) The last pillar also experiences the same stress concentration due to stress transfers from the pillar above. This phenomenon corresponds to underground

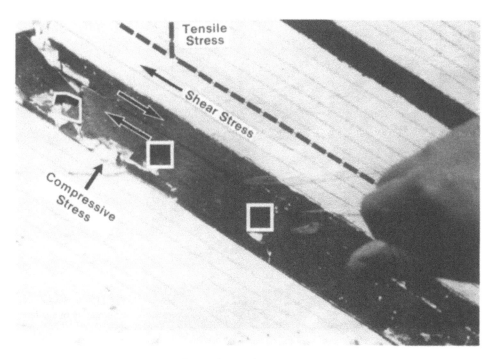

Fig. 11.3.19. Model deformation and failures in a sublevel structure.

phenomenological studies of the transfer of abutment stress as previously described.

These model investigations permit study of pillar instability from the uppermost level to the last sublevel. Besides the visual manifestations of pillar deformation and failure for each individual sublevel, stress mechanics can also be inferred. For example, the uppermost sublevel experiences stress relief, and stress is transferred to the next lower pillar where it is manifested by the concentrations of shear stresses along bedding planes. Also, vertical stresses in the floor go to zero at the excavation boundary but stress transfers around the opening increase the horizontal stress component in the floor so that the floor is effectively in uniaxial compression. This leads to the compressive buckling failures which cause floor heave.

Results were confirmed by underground observations. For example, sublevel drifts supported by timber sets without lagging exhibit severe failure and deformation close to the mining area. The severe deformations are caused by intensive displacements and stress concentration (Fig. 11.3.20). However, sublevel drifts supported by yieldable steel arches (5 m × 3.8 m) with tight lagging to form a shield around the opening did not experience appreciable deformations, and convergence measurements between two predetermined points on the roof and floor showed that the maximum roadway deformation during sublevel extraction was 60 mm. Load cells placed either between the arch leg and floor or between the wooden post and the hanging wall measured maximum support loads of 21 tonnes.[18] This phenomenon corresponds to the mechanics of deformation of coal pillars where a shielded drift flows together with the coal seam. The existence of the large yield zone around the mining area is an optimal

Fig. 11.3.20. Severe deformation of a sublevel drift.

condition for strata de-stressing to avoid sudden and violent failures. For this situation, the design of the sublevel drift support has to permit the synchronized flow of the entry with the coal pillar. Such a support is easily engineered, but its cost may be prohibitive.

In hydraulic mines the failure patterns around mined-out structures are altered by high humidity. Hydraulic working faces are very foggy and the moisture content in the strata increases as coal extraction progresses. This phenomenon is accompanied by a decrease in the rock strength. Laboratory testing of the triaxial compressive strength of sandstone samples from Grande Cache with structural defects showed strengths of saturated samples appreciably lower than those of dry samples. The difference in strength was about 48 MN/m². This permits a scaling factor for the reduction in strength due to saturation to be estimated:

$$\frac{\text{Saturated Rock Strength}}{\text{Dry Rock Strength}} = \frac{68}{116} = 1/1.7$$

The investigation of the influence of moisture on the compressive strength and the elastic properties of the rock as well as swelling properties should be integrated in studies of the stress patterns around sublevel structures.

11.4. CAVING MECHANICS

The presentation of the caving mechanics of roof strata (hanging wall) is based on practical investigations in producing mines (B. C. Coal's hydraulic mine, Western Canada; and Mindola Mine, Central Africa) and the physical model investigations described in detail in the previous subheading. With sublevel mining, progressive hanging wall caving is essential for stress relief on working faces and drifts where stress concentration exists due to abutment stress transfers. Caving mechanics are briefly analyzed in the following order:

– Caving of the immediate mine roof, considered to be a problem of a rigidly clamped cantilever loaded by overburden and its own weight.

– Caving of the main mine roof, considered to be structurally similar to the immediate mine roof but with a different mode of caving.

– Overhang break, considered to be a block of cantilever structure which fails close to the clamped edge.

– Ground surface subsidence, with particular attention paid to the cave line.

11.4.1. *Immediate roof caving*

The immediate mine roof can be approximated in the majority of cases as a sandstone-siltstone rock cantilever with one clamped edge. The critical width of the roof span depends on the strength of the rock material and its deterioration during mining, the volume of mined coal, geological structural defects, the magnitude and pattern of principal stresses, the geometry of the stope-pillar structure and others. The several basic elements can be related to the mechanics of immediate roof caving into the open stope.

Fig. 11.4.1. The initial deformation of the immediate roof by deflection and shearing (physical model).

1. Displacement of the rock cantilever starts with support removal by coal extraction. The rock cantilever is loaded by gravity, which produces a component of stress normal to the bedding (σ_N) and a shear component along the stratification planes as illustrated in Figure 11.4.1. The rock beam displacement is expressed by shearing along planes of weakness.

a) The first stage is dilation between the rock cantilever of the main strata.

b) The shear displacement of the immediate roof rock cantilever is toward the mined out area. Displacement begins when the shear force exceeds the static shear strength.

The mechanics of displacement fits the Bi-Linear shear failure envelope suggested by Patton.

The development of a shear plane along structural continuities parallel with the coal seam is a common case in open stope mining. The sliding plane is normally located above the mine roof a distance of between 0.75 and 1.5 times the mined seam thickness. This sliding plane divides the rock cantilever of the immediate roof from that of the main roof. However, progressive block sliding along the shear plane becomes difficult because the intact block generates mechanical asperities. Once sliding starts, the next stage of deformation, as discussed below, begins.

2. Sagging of the rock cantilever is the main deformation and is expressed by both vertical and horizontal displacements. The rock cantilever develops a buckling instability due to axial and transverse loading.

However, rock cantilever sagging in the rapid coal extraction by monitor jets in hydraulic mining may be delayed by freezing the original stresses in the roof strata for a short period of time. The redistribution of axial and transverse stresses in the roof strata due to excavation may have a reinforcing influence on stability. Transverse stresses effect the stability of the rock cantilever in several ways:

a) By closing joints perpendicular to bedding planes, the resistance to strata deflection is increased.

b) The reduced rate of strata deflection may delay failure, such that the continuation of bed sagging may be by viscous deformation.

c) Creep deformation of the rock cantilever may bring the roof span to complete closure, without the need for stress relief in adjacent areas.

It is obvious, that if sagging of the roof strata is carried to full closure, rock failure is eliminated as well as roof caving.

2. Failure and caving of the rock cantilever is induced by a combination of axial and transverse stresses, where the rock lacks tensile strength. Under these circumstances the rock cantilever deflects and fails into seperate blocks which rotate into the mined-out area.

The mechanism of multiple rock strata failure with block rotation is the dominant mode for immediate roof caving.[31]

Of interest is the investigation of the dynamic caving of individual blocks, where size usually depends on the joint spacing. This phenomena has been studied by several investigators who observed that vertical block displacement is followed by rotation for a certain angle. Horizontal block displacement, on the other hand, is followed by an in phase rotation where each block rotates the same angle.[32] Similar dynamics of block

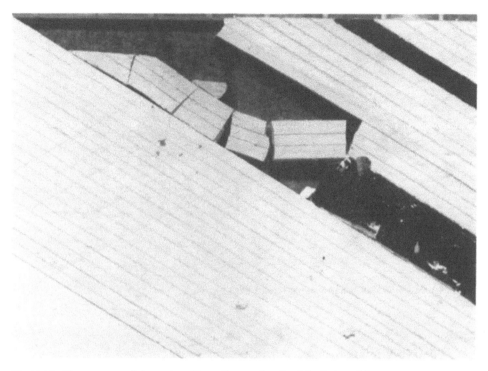

Fig. 11.4.2. The pattern and dynamics of immediate roof caving (physical model).

caving have been observed in physical models, but the angle of rotation between individual members has been different (Fig. 11.4.2).

Hammett[33] developed a mathematical relationship between the driving force required for the rotation ratio between shear and normal stress, and the aspect ratio, ratio of block height to width, of individual blocks. He established that for a single layer of joint blocks there is a limiting length of the failure plane, nb, over which rotation can occur:

$$nb = 2\,a/b \qquad (1)$$

where

      n = number of blocks;
      b = block width;
      a = block height.

The driving force of rotation is represented by the following equation:

$$\frac{\tau}{\sigma_N} = \frac{n(a\cos\theta - b\sin\theta) + \dfrac{(1+2\ldots+(n-1)(b\tan\phi)(\tan\theta + \tan\theta)}{(1+\tan^2\phi)\cos\theta}}{n(b\cos\theta + a\sin\theta) + \dfrac{(1+2\ldots+(n-1)(b)(\tan\theta + \tan\phi)}{(1+\tan^2\phi)\cos\theta}} \qquad (2)$$

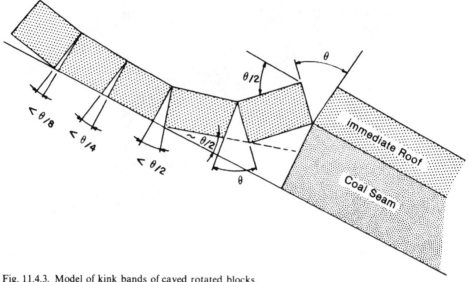

Fig. 11.4.3. Model of kink bands of caved rotated blocks.

where

θ = block rotation angle;
ϕ = friction angle;
τ = shear force;
σ_N = normal force;

if $\tau/\sigma_N < \tan \phi$ – there is rotation, if $\tau/\sigma_N > \tan \phi$ – there is no shearing and related rotation.

However, the model studies as well as those of Ladanyi and Archambault,[31] Baumgartner[34] and others have shown that kink bands are formed with high aspect ratio composite block rotations on the plane of excavation (Fig. 11.4.3.). The physical model studies have shown:

a) That there is no contact between the kink band columns (i.e. there is no lateral dilation force along the failure plane).

b) Kink band hinges indicate that blocks rotate at various angles. If the first block rotates θ-degrees, then the adjoining up-dip block rotates up to $\theta/2$-degrees, and next block rotates up to $\theta/4$-degrees, etc.

c) In order for the kink band columns to remain in-phase, the overlying joint block must make up the difference in rotation and should rotate up to $\theta/2$-degrees also.

It would be fair to say, the dynamics of block rotation and their functional relationships are not well understood. This refers particularly to underground conditions where the heterogenity, structural and lithological, of the roof strata causes deviations from the described mechanisms. Block dynamics does, however, play an important role in the mechanics of strata movement around mine excavations.

11.4.2. *Main roof caving*

The mechanisms of caving of the main roof strata is represented by two concepts:

1. A general concept[31, 34, 35] of main roof strata caving is based on the physical model studies described previously. The results have been compared to actual mining situations, B. C. Coal's hydraulic mine and Mindola Mine, and experimental data not in agreement with actual mine situations are not considered.

a) Block fragmentation is governed by the action of normal stress, because most of the shear stresses have been relieved. The pattern of strata fracturing is in blocks and to the point where large normal forces are transmitted by the side of one block to a face of an adjacent block. There are two different mechanisms to develop the tensile stresses which cause fragmentation of the blocks. One is related to flexural loading and the other to direct point loading as shown in Figure 11.4.4. Block fragmentation is dependent on the tensile strength of the strata and the intensity of the acting stresses.

b) Block crushing is a typical mode of deformation of the upper rock strata, where vertical displacements approach zero because the mined out void is filled with caved rock from the lower rock strata. In this case, the rock hinges are affected by limited or zero displacements and their corners deform due to high stress concentration followed by local rock crushing. The physical model investigations also produced crushing of the blocks of the main mine roof as illustrated in Figure 11.4.5.

It is a point of interest that caving of the main mine roof may not occur in high strength rock with wide joint spacings supported by the large blocks of caved immediate roof. Under these circumstances stress relief is only partial and nearby working places experience high stress concentrations. Our investigations suggest that the important factor for successful caving is the degree of rock fragmentation which, in turn, depends on the rock lithology and the patterns of geological structural defects.

**Bending Tensile Stress
(Simulate Beam Test)**

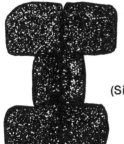

**Direct Tensile Stress
(Simulate Point load Test)**

Fig. 11.4.4. Model of caved blocks fracturing due to tensile stresses.

Fig. 11.4.5. Caving patterns of the immediate roof (rotation of blocks) and the main roof (block crushing).

2. The Chirkov et al. concepts are also based on physical model investigations of the mining of thick and steeply pitching coal deposits.[36] These authors delineated the roof strata into three different zones of caving instead of the two considered above. By their concept, caving of lower strata corresponds to the immediate roof. The caving middle and upper strata corresponds to the main roof (Fig. 11.4.6).

a) Zone I, area of primary caving, is considered to be a rock cantilever of the lower roof strata. Caving starts when the support of the coal is removed by mining. The mechanics of bed separation, failure and rock caving follows the principles described for immediate roof caving.

b) Zone II, area of massive caving, is considered to be a rock cantilever in the central part of the roof strata. Displacement in this part of the mine roof, begins when the lower limits start to rotate towards the void partially filled with caved rock of the lower roof. The degree of displacement is a function of the compressibility of the primary caved rock and the degree of void filling. These authors suggested very similar mechanics of failure and caving as for the immediate roof, especially the development of kink bands.

c) Zone III, area of internal caving, which covers the upper rock strata of the mine roof and corresponds to the general concept of caving mechanics. Roof strata fragmentation consists of bed seperations at the bottom and fracturing at the top by

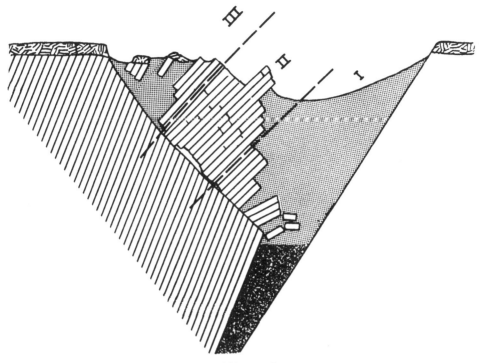

Fig. 11.4.6. Hanging wall cave model of Y. I. Chirkov et al.

block splitting. Physical model investigations have not been successful in simulating zones of internal caving, and the knowledge of this phenomena is obtained from underground observations.

There is an obvious similarity in the caving mechanics of the zones of preliminary caving and massive caving. In our opinion this distinction is due more to the time pause between these two caving cycles than their mechanics. It is common in underground mines of sedimentary deposits, for the roof to cave gradually and in several cycles until the void is filled with compacted caved rocks. Massive roof caving seldom occurs.

11.4.3. *Overhang break*

The overhang break of the hanging wall during sublevel mining of thick and pitching coal seams has been investigated in B. C. Coal's hydraulic mine as well as in physical models with similar mining configurations to this mine. The construction loading and analyses of data obtained from physical models is as described earlier.

The investigations at the hydraulic mine have been carried out for Panel Number 5 to a maximum depth of coal extraction of 250 m. The underground observations showed that the overhang formed by coal removal usually failed after the extraction of each sublevel. Failure, as a tension rupture propagated to the ground surface. The phenomena of successive roof breaking caused de-stressing of the stope-pillar structure to a certain degree and offered greater mine stability.

a. Deflection Deformation

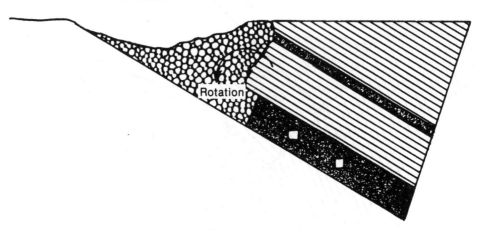

b. Failure Deformation

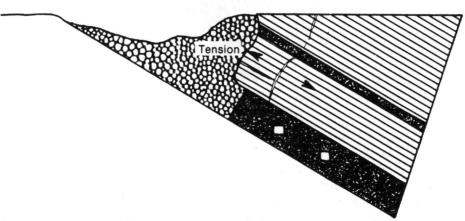

Fig. 11.4.7. Deformation and failure of the overhang (physical model).

In another part of the mine, bumps were recorded in strata where a very strong thick-bedded sandstone of compact fabric was present. In this case bumps occur only where the rock is appreciably stronger than the coal near the pillar-stope structures. This leads to poor caving conditions and concentrations of shear stress. Bumps in strata above uncaved stope areas were followed by the coal seams absorbing the shocks.

However, the overhang deflection and failure primarily could be related to dead weight of the rocks within overhang boundaries. This phenomenon has been studied by a physical model, which indicated mechanics of deformation and failure.

a) Uncaved overhang immediately after coal removal starts to rotate toward the mined out area. As a result of this motion the cantilever rock beam exhibits a bed separation (Fig. 11.4.7/a).

b) Due to time elapse and the coal extraction, the overhang in the area of potential

a. Overhang Break and Caving

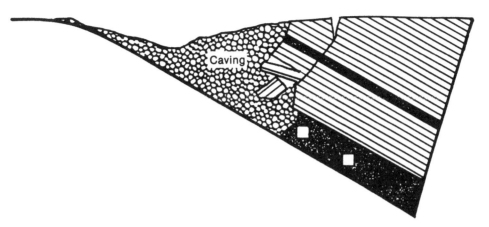

b. Formation of New Overhang

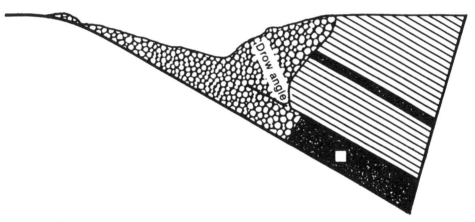

Fig. 11.4.8. Break and caving of overhang (physical model).

break exhibits the initiation of tensile failure (Fig. 11.4.7/b).

 c) When mining of the new sublevel block progressed, a failure is propagated to the ground surface (Fig. 11.4.8/a).

 d) When mining of a sublevel block is finished, a new overhang is formed similar to the one described in paragraph (a) and illustrated in Figure 11.4.8/b.

 The cycle repeats with the start of mining of the next sublevel block.

 At B. C. Coal's hydraulic mine each break of the overhang is accompanied by the development of tensile fissures at the ground surface.

11.4.4. *Subsidence of ground surface*

Subsidence in sublevel caving has not been studied to the same extent as far other mining methods because of the limited potential for subsidence control. The main

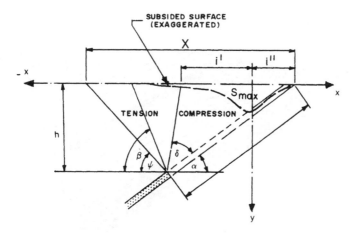

SUBSIDED SURFACE
(EXAGGERATED)

ψ - ANGLE OF LIMIT OF SUBSIDENCE

δ - ANGLE OF CAVING

β - ANGLE OF BREAK

α - ANGLE OF SEAM INCLINATION

Fig. 11.4.9. Analytical model of subsidence from sublevel mining of pitching coal seams.

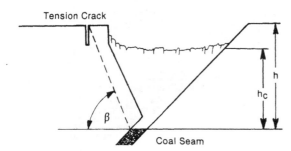

Fig. 11.4.10. Illustration of the angle of break (after E. Hoek).

consideration of subsidence relates to stability studies on main accesses to coal seams as provided by shafts and main haulage ways. Here the potential adverse effects of subsidence due to sublevel caving have been controlled by the proper design and placement of protective pillars.

The theoretical analyses of the subsidence of the ground surface with a regular depression curves, Figure 11.4.9, as discussed in the previous chapter have a certain limitations.[37, 38] This is caused by two factors: First, the elastic analyses of an idealized solid body does not consider the heterogeneity of coal strata which is particularly evident in thick and pitching coal seams. Secondly, there are mining factors, such as partial coal recovery, incomplete caving, fragmenting of the roof strata in large rock blocks and others which can cause deviations in subsidence behaviour.

However, for practical mining purposes the main concern is to determine the break planes along which the sliding deformation of subsided ground occurs.

The break plane with the mine floor delineates a prism of caved hanging wall rocks.

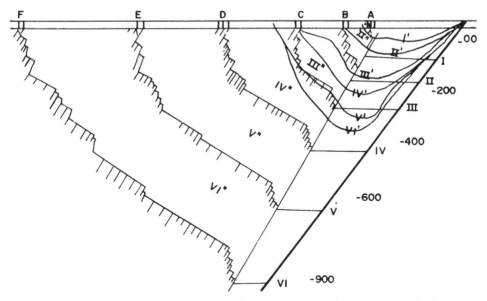

Fig. 11.4.11. Relationship between caving and subsidence: I–VI mine levels; I'–VI' surface subsidence; I"–VI" break line, A–F surface fissures (after Y. I. Chirkov et al.).

Fig. 11.4.12. Sink hole, Tuzla Coal Basin (Yugoslavia).

E. Hoek[39] presented a complete limiting equilibrium analyses of angle of break (Fig. 11.4.10). His analysis shows that the angle of break is not a constant, but alters with mining depth. However, many designers use the angle of break to be known from past experience or inferred from the rock type.

Russian studies on vertical and lateral subsidence displacements during the caving of thick and steeply pitching coal seams show them to be dependant on the thickness and angle of inclination of the coal seam, the strength properties of the rock strata and particularly to the mine depth (Fig. 11.4.11). It should be noted that this profile of surface subsidence shows significant vertical movements which can form sink holes. The maximum vertical displacement of a sink hole in the Tuzla Basin was measured at 15 m at the last reading (Fig. 11.4.12).[38] It is obvious that this degree of subsidence is devastating, and that sublevel caving cannot be applied in industrial and populated areas. Alternative mining systems are required which use stowing.

CHAPTER 12

Coal pillar structure

Coal pillar structures are represented with various aspects of strata mechanics, from laboratory stability investigations to design implication. Areas of interest are:

1. Mini pillars investigations examine the applicability of coal as a prototype model material. The coal pillar models show the possibility of the solving a certain number of problems relating to pillar deformation and failure.

2. Time dependent deformation of coal pillars is an important parameter of mine stability. This was examined with the creep properties of coal materials of different rank.

3. Strength of coal pillar representations utilize knowledge from many fields and suggests a need for adequate pillar design formulas for various coal types with different properties and behaviour.

4. Protective pillars are discussed with the solid coal or which is partially recovered. Final comments are made in regard to shaft mining.

It is important to realize that all theoretical and practical pillar data and design parameters must be varified by solid mining experience.

12.1 MINI COAL PILLARS

Accepting the possibility that there is a direct analogy between the strength of a coal sample and an underground pillar, between the testing loading frame and mine loading, the research was carried out on mini coal pillars. Several aspects of mini coal pillars have been investigated:

– The relationship between sample size and its strength.

– The relationship between dimension ratio and pillar strength.

– The mechanics of the interaction between deformation and the aspect ratio of model pillars.

– Pattern of model pillar fracturing as a function of aspect ratio.

– The stability of mini coal pillars in the simulation of the mining of thick coal seams.

The coal for the preparation of model pillars was taken from underground mines in Western Canada.

Design of the satisfactory size of coal pillars is the major problem in underground mining, particularly for the extraction of deeper coal seams. The laboratory investigations of the mining coal pillars showed that their data can be utilized for analysis of mine pillar stability.

12.1.1.   *Mini pillar size-strength relationship*

Since the end of the last century, a number of investigators have studied the relationship
between compressive strength and sample size. Their investigations suggested a
number of seemingly different predictive equations. The differences are actually in
format and style because they all follow the general exponential equation

$$\sigma_c = \alpha H^{-d} \tag{1}$$

where

σ_c = compressive strength of the sample cube;
H = length of cube edge (varies with sample size);
α = strength constant for a particular coal;
d = constant, varies inversely with root of the length of cube edge.

The investigation of the relationship between the uniaxial compressive strength and the
size of the coal cubes was carried out in the laboratory for various coal ranks in Western
Canada. For example, tests of the uniaxial compressive strength of the coal samples of
the sub-bituminous coal of the Plains region (W. Canada) showed that besides their
size, the weathering factor relates to their strength. The test results are represented by
linear logarithmic function as illustrated in Figure 12.1.1. However, the size-strength
relationship for both weathered and unweathered coal is expressed by the equation

$$\sigma_c = k H^{-1/2} \tag{2}$$

where

H = length of cube edge;
k = strength constant for each type of coal.

W. A. Hustrulid studied the relationship between cube size and cube strength for
various length of the cube edge. He concluded that the relevant literature concerning
the relationship between compressive strength and specimen size of coal samples

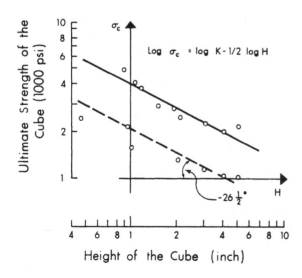

Fig. 12.1.1. Strength of coal cubes of
different sizes (Star-Key Mine, Alberta).

suggested the equations given below, which describe this relationship rather well.[1]

$$\sigma_c = k\, H^{-1/2} < 91.44 \, \text{cm} \tag{3}$$

$$\sigma_c = 0.1 k\, H > 91.44 \, \text{cm} \tag{4}$$

There is general agreement that these equations could be used for prediction of actual pillar strengths if the smaller cubes on which the uniaxial compressive tests were performed are representative of the coal seam. The larger cubes show smaller differences in uniaxial compressive strength. For example, a strength difference between cube lengths of 91.44 cm and 182.88 cm is below 10 per cent, so that equation 4 is acceptable.

Testing the ultimate compressive strength of 2.54 cm length of the cube edge is quite difficult due to the structural nature of the coal. C. Holland in his investigations arrived at the conclusion that constant strength value – k should be determined on the basis of strength of a cube of 7.62 cm length and suggested the equation

$$k = \sigma_c \sqrt{7.62} \quad \text{or} \quad k = 2.76\,\sigma_c \tag{5}$$

which correspond to equation 3, suggested by W. A. Hustrulid.

12.1.2. *Relationship between dimensional ratio and strength of sample*

Parallel with investigations of strength of coal cubes due to variation in their size, there was also an investigation into the relationship between the compressive strength and dimension ratio of coal samples. This test was carried out on coal from several seams in Western Canada. However, the data represented here considers only the Number 11 coal seam (Grande Cache). The tested samples were drilled out perpendicular to bedding planes from one large coal lump. Two sets of cylindrical samples were prepared: one with constant height and variable diameter the other with constant diameter and variable height.

The strength values of all coal samples were recalculated in the so-called normalized form, which was suggested by J. Baushinger in 1876 as a basis for representation of test results performed on sandstone prisms from Switzerland.[3] The testing and representation of the relationship between coal cylinders of ratio H = D and coal cylinders of ratio H ≠ D strengths followed W. A. Hustrulid's procedure.[1] Two sets of coal samples with different dimensional parameters produced different functions of the normalized form equation as described briefly below:

1. Constant height and variable diameter of cylindrical coal samples produced a strength versus dimension ratio which relation can be interpolated as a linear function (Fig. 12.1.2). The best fit of this function corresponds to the equation of normalized form as given:

$$\sigma_p / \sigma_c = 0.778 + 0.222 D/H \tag{6}$$

where
σ_p = compressive strength of coal sample $D/H \ne 1$;
σ_c = compressive strength of coal sample $D/H = 1$;
D = diameter of sample;
H = height of sample.

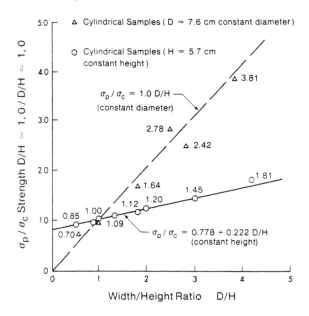

Fig. 12.1.2. Relationship between dimension ration and strength of coal samples, for parameters of constant diameter and constant height.

It should be noted that Equation 6 is in agreement with most formulae for determining the relationship between the strength of coal and dimensional ratio of sample. The represented linear function follows an equation suggested by C. Holland which stated that the strength of the coal sample varies as the square root of the sample width or diameter.

2. Constant diameter and variable height of cylindrical coal samples produced a relation of strength versus sample dimension which is inverse to sample height (Fig. 12.1.2). The coal sample strength rapidly increases with decrease in its height, which can be expressed by the following linear equation:

$$\sigma_p/\sigma_c = 1.0 \, D/H \tag{7}$$

It is obvious that the above equation does not follow the generally accepted normalized form (Equation 6) of the relationship between sample dimension and strength. Under these circumstances, the investigation of the strength of the mini coal pillar in relation to mine pillar design is not valid, and for this reason it is necessary to use a sample with constant height and variable width or diameter.

The acceptance of the modeling of the constant height of the mini coal pillar having variable width corresponds to the natural condition of coal seams. For example, the individual coal seams have more or less a constant thickness, and thus constant pillar height and variable pillar width due variable load.

12.1.3.   *Interaction between deformation and mini pillar dimension ratio*

The stress-strain diagrams are different as a result of relation of the effective loads and the resulting deformation of the mini coal pillars. Early testing results suggested a difference due to variations in dimension ration of model pillars. On the basis of testing

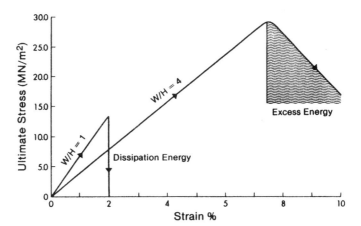

Fig. 12.1.3. Stress-strain diagram for different ratio W/H of mini coal pillars.

of the mini coal pillars of semi-anthracite rank, the following mechanical phenomena are indicated:

a) In the first stage of deformation, model pillars are generally similar in mechanical properties to linear elastic bodies which follows the state represented by Hook's law. The angle of inclination of slope of the stress-strain curve is inversely proportional to pillar dimension ratio (Fig. 12.1.3). Under these circumstances the variation of tangent modulus of elasticity ($d\sigma/d\varepsilon$) is controlled by pillar dimension ratio. However, the tangent modulus is also influenced by the discontinuous nature of the coal (fissure, bedding and other) which at the same time controls the stress formation within the mini pillar. When the shear stress is mobilized due to vertical load, the cracks of mini pillars are opened (tension) or closed (compression) and the tangent modulus of elasticity will depend on the relation between the actual area of shear to total area of the pillar section. With increasing pillar dimension ratio (W/H), the model pillars increase the area of confinement followed by closing the fissures and defects, and forming constant shear resistance. It was established that the magnitude of the tangent modulus of elasticity increases with the increase of the size of the confined area. Further investigations suggested that increase of the tangent modulus of elasticity due to the dimension ratio is different for semi-anthracite coal and medium-volatile coal (illustrated in Figure 12.1.4). This difference can be explained by the nature of coal material.

It could be concluded that mini coal pillars with larger tangent modulus of elasticity and lower Poisson's coefficient can achieve the same resistance with a smaller dimension ratio than model pillars of lower tangent modulus of elasticity and larger Poisson's coefficient.

b) The changes in the relationship between stress and strain close to the ultimate stress is in great deal influenced by the model pillar dimension ratio. For example, laboratory investigations showed that model pillars of ratio W/H < 1.5 have an increase of deformation and strength of the sample until failure which is sudden and brittle with complete loss of cohesion across the failure plane. However, mini coal pillars of ratio W/H > 1.5 exhibited linear stress-strain relations beyond the deformation limit (ultimate stress), where a decrease in strength corresponded to an increase

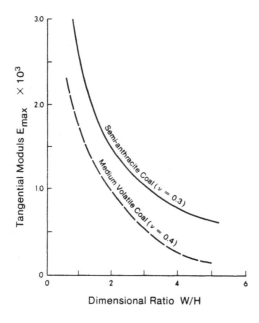

Fig. 12.1.4. Relationship between elastic constants and dimension ratio of mini coal pillars.

in deformation. The mini coal pillars of ratio $W/H > 3$ exhibited stabilization of the sample strength in the region beyond the deformation limit which is characterized by the residual strength. This can be explained by an increase in the angle of internal friction.[4]

c) The relationship between stress and strain of model pillars and their dimension ratio has a great effect on the storing and releasing of strain energy. It has been established that with an increase of the W/H ratio of the model pillars there is an increase in the capacity to store strain energy. The stress-strain diagrams to the ultimate strength of model pillars showed that at failure for dimension ratio $W/H = 1$ where is violent dissipation of energy but for dimension ratio $W/H = 4$ there is a gradual release of excess energy (Fig. 12.1.4). When mini coal pillars reached a dimension ratio $W/H \geq 6$, the majority of mini coal pillars of semi-anthracite rank failed violently.[5] This mechanical phenomenon of interaction between dimension ratio of pillars and the mechanics of failure is utilized for mine pillar design, as discussed under the heading of design considerations.

12.1.4.   *Mini pillar failure in relation to dimension ratio*

The mechanics of deformation and failure of the model pillars in relation to dimension ratio is influenced by the particular stress state of the sample and the geological structural defects of the coal.[6,7] The investigations carried out on the coal samples of low-volatile bituminous coal showed also that the influence of the coefficient of friction between grains and structural defects is different for different dimension ratios of model pillars. The mini coal pillars were represented by cylindrical coal samples of various diameter-to-height ratio. The testing results are briefly discussed below.

a) Ratio $D/H \leq 0.5$. The mini coal pillars tested to failure showed a single shear

Fig. 12.1.5. Shear failure of model pillar with ratio D/H = 0.5.

fracture at an angle to the sample vertical axis of approximately 25° (Fig. 12.1.5). The shear stress at failure was:

$$\tau = 0.5\sigma_1 \cos 2\phi \tag{8}$$

However, due to differential shear movement and the interlocking of asperities on the shear plane, tensile stresses are produced, which overcame the tensile strength of the coal and tension cracks appear in the upper half of the specimen (Fig. 12.1.5).

b) Ratio D/H ≤ 1.5. The model pillars after failure in initial compression showed the formation of double shear fractures which intersect each other. The fractures were formed due to conjugate shear stresses, which acted in the inner part of cylindrical coal samples (Fig. 12.1.6). However, due to the formation of confining stress at the center of the sample the shear stress should be represented by the following equation:

$$\tau = (\sigma_1 - \sigma_3) \sin 2\phi \tag{9}$$

A tensile stress was developed at the free boundary of the mini coal pillars, which resulted in the formation of tensile cracks and coal chipping.

c) Ratio D/H = 1.5 − 5.0. The cylindrical mini coal pillar exhibited three zones of strain. At the free boundary there is a tensile stress and at pillar core hydrostatic stress. However, between these two stress states developed a critical shear stress where shear slips were formed (Fig. 12.1.7). The core of the model pillar remained intact while the boundary exhibited extension slabbing, which progressed as the load increased. Progressive fracturing and disintegration of the periphery of the mini coal pillars gradually reduced the cross-section of the pillar core, until complete collapse.

Fig. 12.1.6. Conjugate shear failures of model pillar with ratio D/H = 1.5.

Fig. 12.1.7. Development of tensile fracture-periphery and shear failure-inner body of the model pillar with ratio D/H = 3.

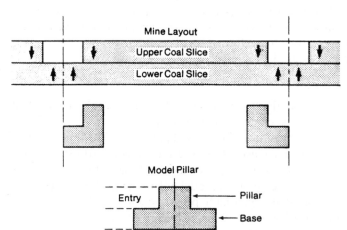

Fig. 12.1.8. Room-and-pillar layout of thick coal seam, with derivation of model pillar-base structure.

Fig. 12.1.9. Set up for testing of the model of a pillar-base structure.

The experimental investigations of mini coal pillars confirmed existence of large changes in strength and changes in angle of internal friction of coal material, as a function of dimension ratio.

Also, the relationship between dimension ratio and mini coal pillar strength could be expressed as:

$$\sigma_c = f\!\left(\frac{D}{H}\right) \tag{10}$$

It should be pointed out that with the changes of the strength of the coal pillars due to dimension ratio, there are also changes in the fracture pattern.

12.1.5.   *Failure of mini pillars of thick seams*

For these investigations particular attention was paid to the cutting of the coal models. Two sets of mini coal pillars were machined, one cylindrical and the other prismatic. In both cases the pillar body was intact with the base which simulated the unmined slice of a thick coal seam, as illustrated in Figure 12.1.8. After formation of the coal pillars strain gauges were mounted on the model sides to record displacement of structure, which could be compared with displacement in actual mine conditions. The setting of mini coal pillar loading by an INSTRON loading frame is shown in Figure 12.1.9.

The general conclusion of all the tests of the mini coal pillars simulating mining conditions in thick coal seams are summarized below:

a) The monitoring of deformation of the pillar model, showed that, in the majority of cases, there was vertical contraction of the pillar body and vertical and horizontal extension of the pillar base, as illustrated in Figure 12.1.10. These mechanics of the model deformation corresponds to displacement of pillar-entry structure, that had been observed in Number 2 Mine at Grande Cache. However, the intensity of model deformation is very low compared with the actual mining closure. The reason for this, is

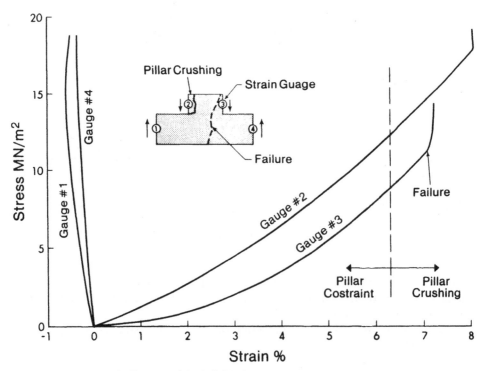

Fig. 12.1.10. Stress-strain diagrams of the individual components of the model of a pillar-base structure.

Fig. 12.1.11. Failure of the
model of a pillar-base
structure.

that the model structure is made from harder coal, which cannot simulate seam profile
of the soft coal in the underground mine. Due to this fact, the model structure could not
exhibit a floor heave of the entries that is so much characteristic for underground
conditions. The trend of the model displacement coincided with the closure of the
underground mine structure, which can be a valuable parameter for investigations of
pillar mechanics for thick coal seams.

b) The fractures of model pillar failures are of two types. Firstly, a shear failure
which is propagated throughout a pillar body and its base, with preference for shearing
along planes of geological structural defects (Fig. 12.1.11). Secondly, tensile failure in
the form of progressive slabbing and crushing which is limited only to periphery of the
model pillar body (Fig. 12.1.12). The mechanics of extension and failure of the model
pillar at free boundary and constraint of the pillar core corresponds to the fracture
pattern of mini coal pillars which simulate a single seam thickness, as described in the
previous subsection.

c) The final pillar collapse is exhibited by fracturing of the pillar-base structure in
individual blocks, which still could support load to a certain degree. After removing the
load, the fractured pillar fragments showed that strain accumulated during loading had
been dissipated usually in two ways, as illustrated in Figure 12.1.13.

The loading conditions of the mini coal pillar simulating thick coal seam conditions
was slow and gradual, where strain was linear with time. During failure propagation
one portion of the accumulated strain was instantly released and followed by a short
exponential strain release. In this case 25 per cent of the total strain was released. For
the case of fragmenting of a model pillar in two blocks larger and smaller, the strain
release was as described previously. In a smaller block, about 30 per cent of the total
strain energy was stored, with decay not too fast, but definite. The strain release was
linear for the duration of one hour. However, a larger block stored about 40 per cent of

Fig. 12.1.12. Pillar periphery crushing and slabbing.

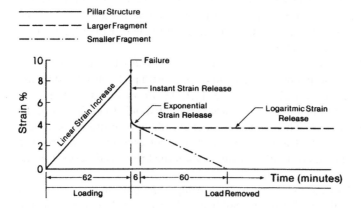

Fig. 12.1.13. Diagram of strain-time concentration and decay.

the total strain energy where the decay was very slow (infinite) following a logarithmic function.

12.2.   TIME DEPENDENT DEFORMATIONS OF PILLARS

The time dependent nature of the behaviour of mine openings is normally evidenced by observations such as the increase in closure with time, heaving floors and sagging roofs, delayed roof falls, collapse of pillars, and others. For safe mine operations, it is essential to maintain the integrity of the pillars and, hence, it is of primary importance that both time dependent deformation patterns as well as modes of failure be understood in order to improve our design procedures.

The time dependent investigations in the laboratory and underground mines have shown certain dissimilarity in creep deformations of the different coal ranks and petrology, particularly in mode of failure as given below:

– Coal of higher rank prone to outbursts exhibits all three creep stages when tertiary creep is ceased by violent and sudden failure.

– Coal of lower and higher rank prone to yielding deformation shows a great similarity with idealized creep curves oblique shear or double cone shear fracture combined with tensile failure.

– Coal of higher rank but not sufficiently 'aged' (mechanically unstable) exhibits only later stage of creep deformation with coal extrusion without any particular failure.

Each of these coal types is seperately discussed including an introductory representation of sampling and investigations as well as final representation of digital modeling of pillar creep deformations.

12.2.1. *Investigation procedures*

The investigation procedures of the creep properties and behaviour of coal structure has been carried out in the rock mechanics laboratory of the University of Alberta, Edmonton and in Western Canadian underground coal mines, as briefly given below:

1. Laboratory investigations have been based on a certain number of coal samples of different rank and petrology taken from various coal seams. These samples could be considered as representative for coal deposits of Western Canada.

The samples in question were brought in drums from mine sites to the laboratory to protect them from disintegration in transit. From large coal lumps both cylindrical and cubic samples were cut. No attempt was made to orient the samples underground and also the structure of the coal was not measured.

For testing creep properties of the coal samples, a simple creep machine has been designed and constructed. A hydraulic system with a bladder-type accumulator to maintain the necessary load constant was chosen. The loading frame consists of two 30.5 cm × 30.5 cm × 3.75 cm steel plates spaced 30.5 cm apart by four 1.9 cm high tensile steel bolts giving a load capacity of 222 KN. The hydraulic ram is ENERPAC RC 256 25 tonnes cylinder with 15.25 cm stroke. An ENERPAC P-39 single speed hand pump drives the ram and pumps up to accumulator (Fig. 12.2.1). The accumulator made by American Bosh is 130 cm^3 in volume and limits the system pressure to 51.51 MN/m^2. A 34.47 MN/m^2 Marcsh pressure gauge of 0.25 per cent accuracy, and 0.00025 cm dial indicator completes the required instrumentation.

A preload of 222 N is applied and the dial indicator is attached and set at 0. The pump is then activated to bring the load to the required level. The accumulator will compensate for fluid loss through the seals.

The creep testing of coal samples is carried out in the dark room, without draft, no windows, at room temperature and humidity. All coal samples have been loaded perpendicular to the bedding planes.

The test results in this heading consider samples given in Table 12.2.1 with the stress levels at which each test was carried out.

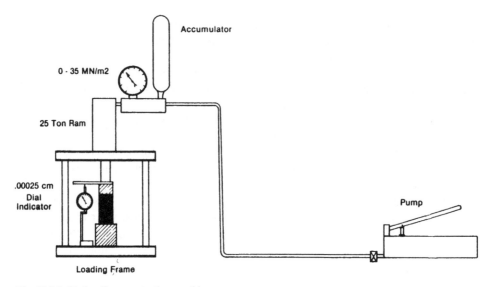

Fig. 12.2.1. Hydraulic creep testing machine.

For each test, a total strain versus time curve and a strain rate versus time curve was plotted. The estimate of the strain rate from the set of data total strain, ε_i, and associated time, t_i, was made as discussed by Cruden.[8]

2. Underground investigations of coal pillar deformations and failures in Western Canada have been carried out in the past[9, 10, 11] and recently.[12, 13] However, time dependent deformations in this heading have been considered only for coal pillars stressed by approximately constant overburden load, its magnitude having been estimated by mine depth and excavations size. The pillar deformations were measured by several techniques: multi-wire extensometers (Newcastle Mk II Extensometer); Davis recorder to measure cumulative convergence; and convergence rods to monitor total closure of entries. The monitoring of pillar deformations has been supplemented by visual investigations of failure mechanics.

Table 12.2.2 represents configurational characteristics, loading levels and failure mechanics of the three coal pillars discussed in this heading. The pillar creep deformations have been represented by plots of comulative convergence versus time. These diagrams have been compared with the corresponding set of data given by plots of total strain versus time obtained during laboratory investigations of coal samples.

12.2.2.   Creep of hard coal

The characteristics of the hard coal or coal of high strength are discussed in several paragraphs throughout this volume. However, the representative creep characteristics are derived on the basis of laboratory and underground investigations of the semi-anthracite coal from Canmore Mine, Western Canada. The coal is of amorphous fabric without individual lithotypes present. The structureless coal is low in ash and moisture and generally is mechanically stable, hard coal.

1. Laboratory investigations considered only samples of solid and unfissured coal which exhibited elasto-brittle behaviour during progressive uniaxial loading.[14] The creep testing of the coal sample (Test 1, Table 12.2.1) gave the following results:

a) Total strain versus time curve has been obtained from 'mini-coal pillar' loading for 95 days at a constant stress level which is approximately equal to 75 per cent of the uniaxial compressive strength. Figure 12.2.2 exhibiting all the three stages of an idealized creep curve is typical. Note the increase of cumulative strain and the second stage shows a sudden increase of total strain that differs from idealized secondary creep. This is followed by an accelerated stage of creep before sudden and violent failure. The laboratory testing showed that duration of the tests is to a great extent affected by stress level.

b) Strain rate versus time is represented by log–log scale curves (Fig. 12.2.3) and can be described by the following:

– At the early stage of the test, for approximately 10 per cent of the total duration, the strain rate decreased continuously. Small deformation has been observed in this stage.

– For the next 20 to 30 per cent of total duration tests there was a transition zone with scattered data and for which a trend was not very clear. If the plot is interpolated as constant strain rate, then secondary creep is indicated.

– For the rest of duration of the test, the strain rate increased continuously until its

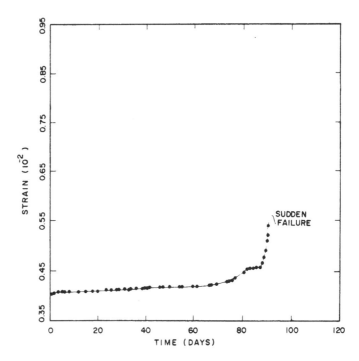

Fig. 12.2.2. Curve of creep deformation.

Table 12.2.1 Sample characteristics and the stress levels of loading

Test No.	Petrological type	Rank	Location	Sample size (cm)	Axial stress (MN/m²)	Stress level (%unc)**	Remarks
1	Amorph	Semi-anthracite coal	Canmore Mine	W = 7.32 L = 7.68 H = 4.42	17.2	.75	Test carried out up to failure
2	Mostly Durain	Coking coal	Cardinal River	D = 7.39 H = 12.97	10.3	< .50 < 1.00	Test carried out for t = 45 days before stress was increased to 20.6 MN/m²
3	Mostly Vitrain	Coking coal	Grande Cache No 4 seam	D = 7.45 H = 5.87	7.2	> .95	Test carried out up to failure
4	Composite	Sub-bituminous coal	Star-Key Mine	W = 4.62 L = 4.67 H = 4.85	19.4	.94	Test carried out up to failure

* Section square;
** Uniaxial compressive strength; average of minimum of two values determined from samples with same shape

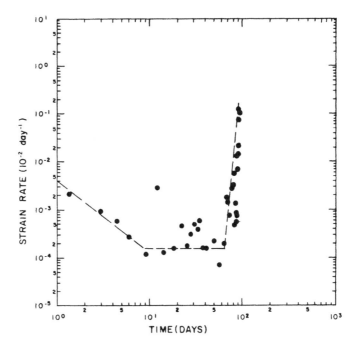

Fig. 12.2.3. Plot of strain rate versus time of semi-anthracite coal (Canmore Mine).

termination by sudden failure. Most of the deformation occurred in this stage. The scatter of strain rate values can be related to the internal fabric of the coal as well as a constant load adjustment during test because the creep curve is sensitive to stress.[9]

c) Mode of failure of the coal samples can be described as sudden and violent with total sample disintegration where coal particles were scattered all over the room. The same mode of failure has been observed in uniaxial compression with progressive loading except that failure of the samples in creep tests showed a greater degree of coal disintegration (Fig. 12.2.4). This is quite an unusual mode of failure in creep and it may be related to the brittleness of the coal. Additional tests on the same rank of coal have to be carried out before a more substantial explanation is given to this rather interesting mechanics of failure.

2. *Underground investigations* were conducted on pillars of hard coal, which is prone to coal and gas outbursts. The underground instrumentations of the pillar deformation in Canmore coal mine have been carried out for three decades before the mine was closed down.

In this presentation particular consideration is given to the entry pillar loaded by overburden dead weight which is not affected by nearby mining. The investigations have been carried out by monitoring convergence which has been plotted against time as illustrated in Figure 12.2.5 (Pillar number 1, Table 12.2.2). From the plot of the convergence three zones of pillar behaviour is inferred as follows:

a) Continuous convergence without definite change in deformation rate, for 65 to 75 per cent of the total duration of pillar instrumentation.

b) Decrease of convergence rate for a shorter duration. The existence of least

Fig. 12.2.4. Mini coal pillar
before and after creep
failure (Canmore Mine).

convergence may be a criterion for detecting a pre-burst state of stress of the coal
pillars.

c) Accelerated convergence which coincides with coal and gas outburst. Due to
violent deformation in a very short of period of time, the deformation-time relationship
is not well defined.[14] If it is part of the deformation accepted as tertiary creep, then it
should be considered as a rapid one.

The convergence at Canmore Mine exhibits a low intensity of total strain and strain
rate, most likely due to the strength and competence of the roof and floor strata and
possibly to the brittle nature of the coal.[10] This creep deformation is similar to
laboratory observations of time dependent deformation of the coal from the same mine.

The similarity between creep characteristics of mini coal pillars tested in the
laboratory and mine coal pillar instrumented underground is an important phenomenon for the study of pillar mechanics and mine stability.[14]

Finally, it should be pointed out that outburst pillar failures are unusual for creep

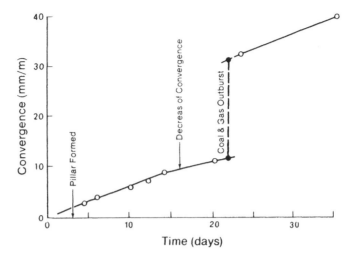

Fig. 12.2.5. Diagram of creep deformation of the coal pillar, with outburst failure (Canmore Mine).

deformations, and this mechanism might be attributed to the inability of the coal to undergo creep deformation.

12.2.3. *Creep of moderately hard coal*

The moderately hard coal strength characteristics have been frequently considered in this volume under various paragraphs. The creep behaviour of such a coal is discussed for sub-bituminous coal, Star-Key Mine (laboratory investigations)[12] and medium-volatile bituminous coal, Coleman Colliery (underground investigations). Both ranks of investigated coal exhibits cleat very well developed perpendicular to the mountain range axes with same strength deterioration either by orogenic movement or weathering.

1. Laboratory investigations of a sub-bituminous mini coal pillar, by the creep rig (Test number 4, Table 12.2.1), exhibited the following characteristics:

a) Total strain versus time plot (Fig. 12.2.6) suggests the existence of all three stages of creep deformation and they, to a certain degree, correspond to a quantitative description of generalized creep given by the following equation:

$$\varepsilon(t) = \varepsilon_1(t) + \varepsilon_2(t) + \varepsilon_3(t) \tag{1}$$

where

$\varepsilon_1(t) = $ primary creep;
$\varepsilon_2(t) = $ secondary creep;
$\varepsilon_3(t) = $ tertiary creep.

It is important to consider the relative magnitudes of qualities t_1, t_2 and t_3; ε_1, ε_2 and ε_3. The test represented here has shown that ε_3 was relatively large compared to ε_1 and ε_2.

b) Strain rate versus time plot is scattered (Fig. 12.2.7) but can be interpreted as follows:

Table 12.2.2. Characteristics of the coal pillars and failure mechanics

Pillar no.	Coal mine	Petrological composition	Rank	Least width (m)	Mineable seam thickness (m)	W/H ratio	Stress* level MN/m^2	Pillar type	Failure
1	Canmore	Not exhibited	Semi-anthracite	14.00	2.00	7	16.00	Rib	Outburst (violent fracturing)
2	Coleman Collieries	All lithotypes	Medium-volatile bituminous	20.00	3.30	6	18.10	Rib	Progressive fracturing (slabing)
3	Grande Cache	Mostly vitrain	Low-volatile bituminous	24.00	3.00	8	6.90	Chain	Extrusion (closure of entry)

* Average pillar stress calculated by equation $\dfrac{\gamma h}{1-e}$;

Pillars in flat and gently dipping coal seams (2–20 degrees)

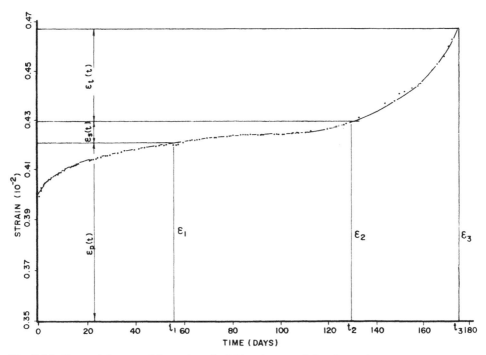

Fig. 12.2.6. Curve of the creep deformation of sub-bituminous coal (Star-Key Mine).

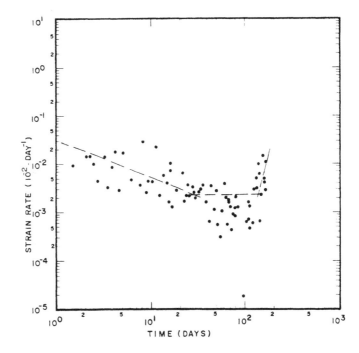

Fig. 12.2.7. Plot of strain rate versus time of sub-bituminous coal (Star-Key Mine).

– Continuous strain rate decreases and relatively small deformation has been observed;

– Constant strain rate was accompanied by observable sample deformation;

– Continuous increase of strain rate, for a large part of the duration, is followed by coal slabbing until ultimate failure.

The similar type of deformation has been shown in instrumented coal pillars in the Lethbridge area where continuous increase of creep rate accelerated before pillar failure.[11]

c) Failure of the coal was located on the sides of the sample where shear stress and tension were developed. Oblique shear or double cone shear failure at an angle acutely inclined to the direction of compressive stress was observed. This is the most common failure of the coal with yielding properties and behaviour. The rupturing pattern is a combined shear and tensile failure producing wedging cones (Fig. 12.2.8). A similar type

Fig. 12.2.8. Mini coal pillar at the beginning and at the end of a creep test of sub-bituminous coal (Star-Key Mine).

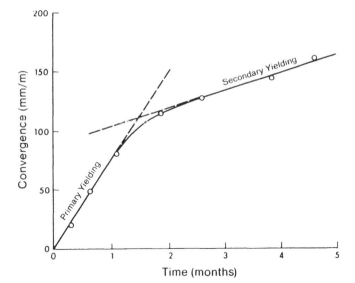

Fig. 12.2.9. Diagram of creep deformation of the pillar with coal of yielding behaviour (Coleman Collieries).

of failure has been observed during progressive uniaxial compressive loading of weathered coal samples from the same mine,[12] and bituminous coal from Cardinal River Mine.[15]

2. Underground investigations have been carried out on the pillars of the medium-volatile coal which has somewhat lower strength then semi-anthracite and higher strength than low-volatile bituminous coal. The instrumentation of coal pillars under constant loading and longer time loading showed a great similarity with the deformation of represented pillar creep characteristics (Pillar Number 2, Table 12.2.2), which is illustrated in Figure 12.2.9 and described below:

a) Primary yielding is clearly exhibited by exponential convergence and very limited pillar fracturing. The convergence of the pillar could be easy observed by timber posts compression and its break close to the mine floor. The time of primary creep and the magnitude of convergence varies from mine to mine. The pillar in question had a rather high convergence, for example in range 100 mm/m and a longer duration, between one and two months. However, the monitoring of pillar convergence in Michael Mine showed primary creep with a lower magnitude of deformation, by nearby 50 per cent, and for a shorter duration.

b) Secondary yielding has shown steady-state deformation, with cumulative convergence linearly increasing for a period close to three months. The coal pillar has not shown any significant degradation, except slabbing of the pillar ribs and slight yielding of nearby timber posts.

c) Tertiary yielding exhibited a rapid increase in convergence, followed by progressive pillar deterioration and collapse of the broken timber posts. Progressive pillar slabbing resulted in total pillar failure over a period of one month.

The results of laboratory investigations of mini coal pillars and underground phenomenological studies of mini pillars showed greater similarity to the creep deformation and failure than in the case of the other two coal types. It seems most likely that creep properties of moderate strength coal, which are evaluated in the laboratory, could be strongly related to underground mine pillars and their stability. Actually, the only difference observed in the laboratory and underground is that post-failure coal pillars still stand and probably support loads directly imposed on them.[16]

Finally, it should be pointed out that creep diagrams either of mini coal pillars or mine coal pillars are very similar to creep plots for rock salt and potash deposits. In the absence of digital model creep representations for coal pillars in last subheading, creep analysis is given for potash pillars principal elements of which could be equally applied to coal pillars.

12.2.4. *Creep of soft coal*

The majority of coal seams of low volatile bituminous rank consist of vitrain as the main lithotype component. They exhibit pillar creep deformations which are easy observable in underground mines but very difficult to simulate in the laboratory. However, the following observations were made.

1. Laboratory investigations first faced a great of difficulty in preparing a sample for creep testing, because of a high degree of coal fracturing and deterioration. Only one mini pillar from Grande Cache (Test Number 3, Table 12.2.1) was available for a creep testing and showed:

a) Total strain versus time curve described coal characteristics and high relative stress levels (approximately 95 per cent of the short term strength) which probably influenced the particular creep behaviour of the sample as illustrated in Figure 12.2.10 and suggested only two stages of progressive creep:

Stage I – moderate creep deformation (about 50 per cent of the total test duration);

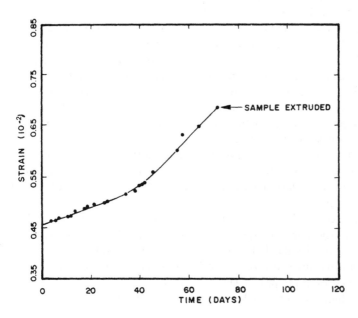

Fig. 12.2.10. Curve of the creep deformation of vitreous coking coal (Grande Cache).

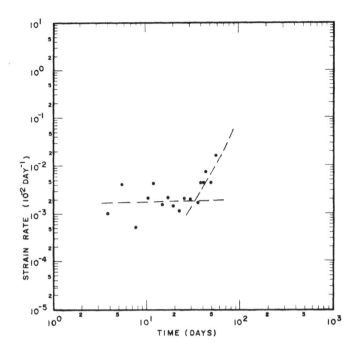

Fig. 12.2.11. Plot of strain
rate versus time of vitreous
coking coal (Grande
Cache).

Stage II – accelerated creep followed by coal extrusion until the sample collapsed.
It seems most likely that coal within its natural setting was already in a state of creep so
that after unloading (taking out sample) and reloading (putting sample in creep rig), the
advanced creep of coal has been resumed and continued until total failure.

b) Strain rate versus time is represented by the diagram in Figure 12.2.11. This plot
also might suggest the existence of only the third stage of creep deformation which
supports the possibility of such unusual type of creep deformation as outlined by the
following possibilities:
– The scatter might suggest constant strain rate so that secondary creep may exist;
– The continuous strain rate increase suggests tertiary creep.
The strain rate versus time plot is in agreement with the total strain-time diagram
which supports the posibility of such an unusual type of creep deformation.

c) Mode of failure exhibited the upper part of the sample which was crumbled and
with time extruded, that finally leaving the lower part in the cone shape. This test, to a
certain degree, has similarity with the mine pillars which exhibit squeeze-out
deformation which can result in total closure of the entries.
It is necessary to develop a technique for creep testing of soft coal in laboratory
conditions and achieve a better understanding of the failure mechanics.

2. Underground investigations have been limited to convergence monitoring of coal
pillars in thick coal seams (Number 4 seam, Grande Cache) where the pillar body was
located in the upper slice, because the lower slice was left unmined.[16] The following
discussion is based on data obtained by instrumentation of the coal pillar in Number 2

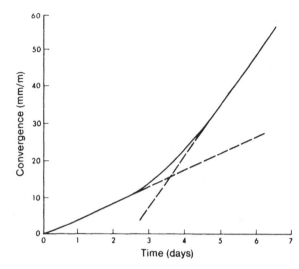

Fig. 12.2.12. Diagram of creep deformation of a soft coal pillar (Grande Cache).

Mine (Pillar Number 3, Table 12.2.2), as illustrated in Figure 12.2.12, and discussed below:

a) Convergence at lower rate, caused a partial closure of entry, within several days of pillar formation. The underground observations suggested that the intact coal seam was in state of primary creep, and when pillar was created, the duration of primary yielding was very short and ceased before convergence instrumentation started.[17] It is most likely that partial closure corresponds to secondary rather than primary creep.

b) Convergence at high rate suggests the presence of tertiary creep. Almost a total entry closure has been observed for the period between 7 days and 7 months. Acceleration rate of the closure dependent on the coal strength and the stress level. When this magnitude of the entry closure is reached, the pillar failure is completed.

The high yielding properties of coal material are the cause of mechanical instability for both mini coal pillars and mine pillars. It is very difficult to maintain the entry opening in the room-and-pillar configurations regardless of either extraction in advance or retreat.

In general, the investigations carried out suggest that creep properties are different for different coal strengths, related to rank and petrology. The investigations have shown that creep behaviour of coal is controlled by the stress level (it is depicted as the ratio between constant axial stress and the uniaxial compressive strength) which agrees with experimental evidence recorded in literature.

The mode of creep failure is of particular interest in the evaluation of mine stability which requires some adjustment due to coal rank particularly in case of semi-permanent and permanent coal pillars with more or less constant stress level.

12.2.5.   *Digital modeling of creep*

A practical example of application of this modeling technique is given by C. Fairhurst et al.[18] for stress analyses of potash room-and-pillar mining. They modelled the room-and-pillar structure on the computer using the STEALTH explicit finite difference

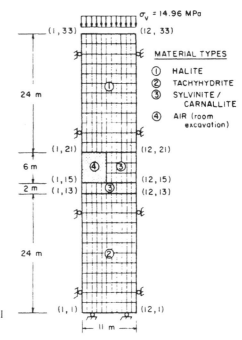

Fig. 12.2.13. Finite different grid for numerical simulations.

program for the conditions of plane-strain deformation for the geometry and loading conditions as illustrated in Figure 12.2.13. The finite different grid exhibits three different layers of halite and sylvinite/carnallite of the same creep behaviour, as well as tachyhydrite of different creep behaviour. The finite difference analyses of the mine structure indicated several important points with the respect to mine stability:

a) The stress distribution within pillars can be defined even after considerable creep has taken place. For example Figure 12.2.14 shows the pattern of stress distribution after creep for one year. It is obvious that this numerical technique is of particular importance for the evaluation of mine pillar stability over time.

b) The elastic stress redistribution due to the relationship between geometry and size of the room and pillar, can be represented by the stress concentrations both normal and tangential (parallel and perpendicular to the cross-section) in the mine pillar. The finite difference model stress analysis suggests that stress relief occurs both in the halite roof and in the tachydrite floor. Also, most of the roof-floor convergence will result from the high stress concentration in the pillars, which are represented by sylvinite.

c) The computer analyses were carried out to obtain the creep behaviour of the respective evaporite materials. A strain hardening form of creep (i.e. creep strain rate) decreasing under constant stress was used in the computer analysis. It was shown that the closure continues at a more or less constant rate, with some acceleration towards the end of the period (Fig. 12.2.15). This result suggests that there is a constant rate of room closure, as observed in underground mines,[19] and which is not inconsistent with the observed strain hardening creep behaviour of evaporite samples during laboratory testing.

The study demonstrated the possibility of the evaluation of a time dependent stress

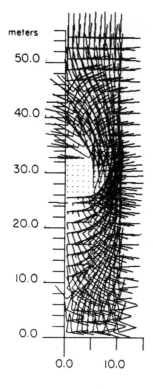

Fig. 12.2.14. Plot of stress distribution after creep for one year.

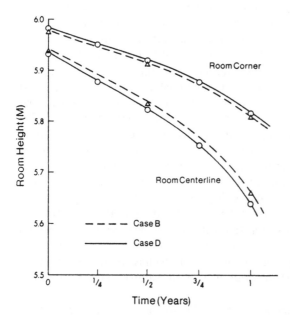

Fig. 12.2.15. Room closure one year after
excavation.

distribution above and below the rooms and pillars, which is a key factor for strata control in underground mines. The authors suggested that this investigation is only an aid in assessing the stability of an actual mine situation. The validity of the data is in question until they can be verified by underground observations.

12.3. STRENGTH OF COAL PILLARS

The basic function of coal pillars is to maintain the stability of an exposed mine roof for a particular duration. The time can vary considerably. For example, if a coal pillar has to protect the ground surface, which should not be damaged by underground excavation, then the design time duration of the pillar might be forever. However, if coal pillars are protecting shafts, haulage or ventilation ways their duration is over some number of years, mostly for the life of the mine. In retreat mining operations, pillar support is required for a short period of time e.g. for some months only. The seam thickness is relatively constant for anyone mine. In this circumstance, the height of the pillars is going to also be constant, but its least with is going to be variable. This depends on mine depth and also on the coal strength. Thus consideration has been given to the following:
- Principal pillar formulas by different authors,
- Pillar formula for coal of high strength,
- Pillar formula for coal of moderate strength,
- Pillar formula for coal of very low strength,
- Philosophy of coal pillar stability.

The detailed consideration of coal pillar strengths is given, because the principal objective of a producing mine is to achieve the maximum coal extraction under safe conditions. To achieve maximum extraction of the coal, there is a tendency to leave pillar as small as possible. Overdesigned pillars could result in severe roof conditions by failing to yield in such a way as to maintain the structural integrity of the mine.

12.3.1. *Principal pillar design formulas*

The first consideration of mine pillar design was given at beginning of this century in relation to studies of the compressive strength of coal and rock. From this period until the present the abundance of design formulas have been developed and most of them represent the same basic phenomena; which is one of the reasons for their lack of acceptance in mine design.

The coal deposits in various locations around the world exhibit great variations in geological and structural composition. With respect to underground mining particular interest is given to coal rank and related mechanical stability of the seam, degree of orogenic deformations and deterioration of coal strength, different strata lithology and the corresponding interaction with the coal seam, intensity and extent of geological structural defects as well as coal weathering and other elements related to mine stability.[20] As a result, of the variation of these natural factors, several authors have suggested different equations for the evaluation of coal pillar strength:

Bieniawski formula[21]

$$\sigma_p = 400 + 220 \, W/H \, \text{psi} \tag{1}$$

where
> W = least width of pillar;
> H = height of pillar.

Holland's formula[22]

$$\sigma_p = \sigma_c (W/H)^{1/2} \text{ psi}, \ \sigma_c = \frac{\kappa}{\sqrt{3}} \tag{2}$$

where
> κ = strength of 7.62 cm cube edge length.

Salamon's formula[23]

$$\sigma_p = 1320 \frac{W^{0.46}}{H^{0.36}} \text{ psi} \tag{3}$$

Surprisingly, all three formulas gave approximately the same results for coal pillar strength.

The assessment of quantitative and qualitative properties and the behaviour of coal pillars showed that their strength and supporting capacity can vary over a wide range.[24, 25] From a practical engineering point of view, the coal pillars, could be classified in four principal strength groups: pillars of excessive strength, moderate strength, low strength, and minimal or no strength.

Pillar loading is a time continuous process during which progressive pillar fracturing could be underway and which will result in a transition of strength from one group to another until ultimate collapse. Under these circumstances, the suggested four types of pillar strength could be snapshots of a continuous process of pillar loading and continuous strength deterioration. However, at present, there is no more acceptable static formulae for pillar strength.

12.3.2. *Pillars of high strength*

Under this category coal pillars of greater bearing capacity are considered which, from their formation to ultimate collapse, may not deteriorate appreciably in strength.

1. Characteristics of coal profile exhibit coal seams of moderate thickness at various angles of inclination, where hard adjacent rock strata have a tendency to constrain coal pillars against their lateral expansion. To this profile belongs coal of higher rank (anthracite, semi-anthracite, some coking coal) which exhibit excessive strength.[25] Since the coal fabric is amorphorous with a limited development of cleat, coal strength deterioration during progressive loading either underground or under laboratory conditions is negligible due to its elasto-brittle properties.

2. Tests of coal pillar strength have been carried out on laboratory scale, on the semi-anthracite coal from the Canmore Mine, Western Canada (M. L. Jeremic)[25] and anthracite coal from Donbas, U.S.S.R. (Y. A. Gramatikov).[26] Both of these tests suggest that there is a critical width-to-height ratio (W/H) at which the pillar collapses violently accompanied by coal outbursts (Fig. 12.3.1).

The coal outburst phenomenon experienced in the laboratory during uniaxial

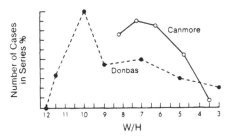

Fig. 12.3.1. Frequency of coal pillars outbursts in relation to W/H ratio.

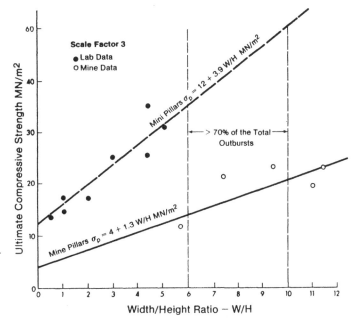

Fig. 12.3.2. The strength of mini coal pillars and mine pillars of differing W/H ratio (coal of excessive strength).

compressive testing of mini coal pillars has also been observed in situ. For example, M. A. Gramatikov[26] suggested, on the basis of 29 years of observations of pillar failures, that over 80 per cent of coal pillar outbursts occurred for the ratio W/H = 8 − 10. This parameter of critical pillar size is in agreement with the underground observations in Canada (M. L. Jeremic)[25, 27] and U.S.A. (C. T. Holland).[28]

3. The coal pillar strength formula is the result of laboratory testing of model pillars to estimate the relationship between pillar strength (σ_p) and W/H ratio. Underground investigations of mine pillars have been carried out to evaluate a scaling factor for these data.

The test results of strength of model pillars in relation to the corresponding width-to-height ratio suggest a linear function between the two which can be expressed by the following equation (Fig. 12.3.2):

$$\sigma_p = 12 + 3.9 \, W/H \, [MN/m^2] \qquad (4)$$

The linear relationship between pillar strength and size obtained for underground mine pillars (Fig. 12.3.2) is:

$$\sigma_p = 4 + 1.3\,W/H\,[MN/m^2] \tag{5}$$

In this case the load at pillar failure (ultimate strength) has been calculated from overburden weight and extraction ratio. The width-to-height ratio has been defined on the basis of underground measurements of each pillar in question (Canmore Mine).

The scaling factor for strength from laboratory data to mine data is 3 which is close to the scaling factor of 2.5 for anthracite coal pillars (U.S.A.) given by D. Bunting and which are similar in strength and composition to pillars in the Canmore Mine.[29]

The suggested formula for pillar design of coal of excessive strength should be considered sufficient if the mine design permits pillar sizes other than in the critical range at which coal and gas outbursts might occur.[25]

12.3.3. *Pillars of moderate strength*

This strength of pillar is optimal for mine stability, but coal strength deterioration over a period of time may be associated with ultimate pillar collapse.

1. The characteristics of the coal profile concerns coal seams of various thicknesses and inclinations. If the adjacent rock strata have a higher strength and constraint than the

Fig. 12.3.3. Progressive fracturing of coal pillars.

coal pillars could fail as an elasto-brittle body. However, if bounding rock strata have a very low constraint effect, the pillars might experience elasto-plastic failure (Fig. 12.3.3). This type of coal can be found in all ranks, however the majority of pillars under consideration belong to the lower rank where the coal fabric is laminated by beds of more or less uniform thickness with a well developed cleat.[24] The coal strength is adequate but can be significantly deteriorated due to either fracturing or weathering as laboratory investigations suggested.

2. Tests of coal pillar strength have been related to the results of the uniaxial compressive investigations for both weathered and unweathered coal as discussed in heading of 'Mini Coal Pillars'. Besides the relationship between coal cube size and strength, it is also important to know how weathered and unweathered coal material behaves under load. For example, different load-displacement curves are obtained from laboratory tests of weathered and unweathered coal cubes.[30] Unweathered coal samples produce a linear loading curve with a high tangent modulus and with an easily distinguished peak strength at which there is a sudden failure. Weathered coal samples do not give a linear load-displacement curve and the approximate tangent modulus is low. Peak strengths are poorly defined and yielding is observed over a range of loads, as the material is more plastic.[25]

Clearly, unweathered coal can support relatively high loads in the mine, but may fail suddenly whereas weathered coal will yield slowly under lower loads.

3. Coal pillar strength formulae were obtained from the model pillar investigations to estimate the relationship between strength, σ_p, and width/height ratio (W/H).[26, 31] The test results, shown in Figure 12.3.4 are described by the equations:

$$\sigma_p = 13.51 + 4.55\,\text{W/H}\,[\text{MN/m}^2] \tag{6}$$

$$\sigma_{pw} = 5.17 + 4.55\,\text{W/H}\,[\text{MN/m}^2] \tag{7}$$

where σ_{pw} is the strength of a model pillar in weathered coal. Again, the functional relationship has the same form for weathered and unweathered coal, the strength difference being about 8 MN/m^2.

However, there is a difference in material behaviour during tests. The unweathered coal pillars fractured into large blocks while the weathred coal pillars produced considerably fine material.

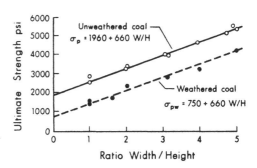

Fig. 12.3.4. The strength of mini coal pillars of differing W/H ratio (coal of moderate strength).

If the model pillars were scaled to the strength of mine pillars, Equation (6) and (7) could be used in mine design. Following the procedure suggested by W. A. Hustrulid[32] we normalize equation (6) and (7) to give:

$$\sigma_p/\sigma_c = 0.75 + 0.25\,W/H \tag{8}$$

$$\sigma_{pw}/\sigma_{cw} = 0.53 + 0.47\,W/H \tag{9}$$

where σ_c and σ_{cw} are the strengths of unaltered and altered 2.5 cm cubes of the coal. Then we substitute for σ_c and σ_{cw} the strength of cubes of the height of the pillars. Cube strength can be estimated as one sixth of the strength of 2.5 cm cubes, giving pillars strengths σ_p and σ_{pw}:

$$\sigma_p = 2.24 + 0.79\,W/H \; [\mathrm{MN/m^2}] \tag{10}$$

$$\sigma_{pw} = 0.86 + 0.79\,W/H \; [\mathrm{MN/m^2}] \tag{11}$$

Equations (10) and (11) indicate substantial differences in the design properties of coal from within the same seam in the same mine (Star-Key Mine). The principal observable difference in the two types of coal is the considerably higher intensity of cracks in the coal we have termed altered.[30]

12.3.4.   *Pillars of low strength*

The pillars of this category have limited stability which last for shorter periods of time, and after that they become unstable what is followed by their ultimate collapse.

1. The characteristics of the coal profile are represented by coal seams of various thicknesses which can be increased due to tectonic deformations. Adjacent rock strata might have the tendency to constrain coal pillars (sandstone and siltstone) or to restrain coal pillars (shale and mudstone).[23] The coal rank is mainly coking coal with limited 'aging' processes belonging to younger geological formations (Jurassic/Cretaceous). Examples are the coal deposits of the Rocky Mountain Belt (North America) and Kuznetsk Basin (U.S.S.R.).[27] The coal fabric has marked laminations identifying the coal beds of different petrology and strength.[33] However the coal strength is deteriorated by orogenic movement of weathering processes. This coal, during progressive compressive loading, exhibits extrusion deformations which are represented by a plastic curve in stress-strain diagrams.[25]

2. Tests of coal pillar strengths are also based on the laboratory investigations. The coal samples were cut by a circular saw from large coal lumps taken from three underground mines (Number 4 Mine–McIntyre, Number 11 Mine–McIntyre and Vicary Mine–Coleman Colliery).[34] The model pillars had a constant height of 2.5 cm with variable widths from 2.0 to 15.0 cm.

The laboratory testing showed a squeezing out of model pillars before ultimate failure, similar to underground observations, but at a lower deformation rate. However, the mini coal pillars tested in laboratory have shown much greater strength than coal pillars underground. The most notable strength differences between model and mine pillars were magnified by the sample model pillar size. These differences can be attributed to the fact that model pillars were cut from the stronger portions of coal

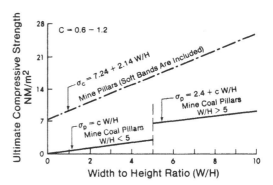

Fig. 12.3.5. The strength of mini coal pillars and mine pillars of differing W/H ratio (coal of low strength).

lumps containing only thin bands of soft coal. Any attempt to cut model pillars from soft coal failed because it broke up during preparation.[34]

3. Coal pillar strength formulae have been related to the underground phenomenological studies which recorded immediate pillar squeeze out (up to 5 cm) after its formation. For a period of between 6 and 12 months complete pillar squeeze out and roadway closure were experienced. This process was accompanied by progressive pillar fracturing combined with a decrease in bearing capacity.

Mining experience showed that in the majority of cases, pillars designed by present pillar strength formulae were underdesigned because their bearing capacity lessened after only a short period of time. Seven collapsing pillars in three underground mines were examined. In all cases, pillar fracturing had been initiated in the early stages and progressed to ultimate pillar collapse.[34, 35]

The model pillars also have been used to estimate the relationship between pillar strength and width-to-height ratio. The test results shown in Figure 12.3.5 correspond to equation (12):

$$\sigma_p = 7.24 + 2.14 W/H \, [MN/m^2] \tag{12}$$

However, from underground mines, data on the relationship between ultimate pillar strength (calculated from overburden weight and extraction ratio) and width-to-height ratio have been used to construct equations for pillar strengths. These do not quite correspond to the function for model pillars. The pillar strength evaluation is suggested by two equations as given below:

$$\sigma_p = CW/H \, [MN/m^2] \quad \text{when } W/H < 5 \tag{13}$$

$$\sigma_p = 2.4 + W/H \, [MN/m^2] \quad \text{when } W/H < 5 \tag{14}$$

The ratio W/H = 5 as a conditional boundary between two pillar sizes and these equations were arbitrarily suggested on the basis of underground determination of pillar stability. Also, on the basis of underground measurements of the percentage of soft coal layers along a workable height, we were able to infer pillar stability values for factor 'C' as given below: where 75 per cent or greater of the seam height is made up of soft layers

$$C = 0.6$$

where 50 per cent or less of the seam height is made up of soft layers

C = 1.2

It is a fair suggestion that factor 'C' should be determined separately for each coal seam in which a pillar design is contemplated.

12.3.5. *Pillars of minimal or no strength*

Under this category coal pillars of very limited bearing capacity are considered which could support only a light uniaxial load before total disintegration.

1. The characteristics of the coal profile are: the coal seams consist of weathered coal or tectonically fractured coal. Adjacent rock strata have minimal influence on pillar constraint due to the weakness of coal material. To this group belongs coal of any rank where its strength has appreciably deteriorated.[20] Internal seam texture is represented by laminated thick and thin coal beds mostly fractured and disintegrated with almost no strength.[35] The samples of this type of coal under uniaxial progressive loading exhibited in the beginning cracks, compaction and after that total viscous flow where flow velocity depended on coal viscosity.[35]

2. Tests of coal pillar strength in all cases for uniaxial compressive strength of the coal failed. Also, studies of the model pillars made of weak material could not produce pillar deformations similar to those observed underground. The only valuable data from an engineering point of view have been obtained underground from severely tectonically disturbed parts of coal seams where the coal pillars could not support their own total load.[25] Present mining practices suggest an elimination of coal pillars from the mine layout for this section of the coal seam.[34]

3. Coal strength formulae for these coal properties and behaviour should not be considered as an engineering tool because the pillars of this type as underground supporting structures do not exist, and any derivation of a formulae for their strength is invalid for practical mining.[25] For these reasons, any further discussion of the stability elements of coal pillars with minimal or no strength is concluded at this point.

12.3.6. *Philosophy of coal pillar stability*

To achieve the required mine stability it is necessary to apply an adequate equation for pillar design which will not provoke adverse effects. It is obvious that the existence of coal pillars with significant differences in both strength and stability also require different concepts of mine design.[24, 25]

1. Coal pillars of excessive strength require the limitation of stress build up. This can be achieved by:
   a) Hydraulic fracturing which should be included as an integral part of the pillar design, with an aim to change the property of the coal from elasto-brittle to elasto-plastic. Therefore, instead of sudden and violent fracturing, yielding deformations are induced, which can be successfully controlled.

b) Restricting of the pillar size so as to avoid sudden and violent collapse. For example coal pillars with width-to-height ratios between 8 and 10 are the most susceptible to coal and gas outbursts.[26]

2. Coal pillars of moderate strength have the most optimal behaviour for successful strata control because they do not violently fail and also they have satisfactory bearing abilities for reasonable lengths of time.[24, 25] The principles of design are:

a) Determination of pillar sizes in relation to coal strength could be satisfactorily done by existing pillar formulae where room-and-pillar mining can be a successful method of coal extraction.

b) For semi-permanent or permanent pillars it might be necessary to use safer formulae for the design, because of the possibility of either coal weathering or progressive coal fracturing due to nearby mining.

c) Local irregularities due to geological conditions or coal petrology has to be taken into account for pillar design.

3. Coal pillars of low strength include mainly bituminous coal seams with heterogeneous transverse anisotropy which greatly influence pillar instability. For this reason, greater concern has been devoted to design parameters as listed below:

a) The coal extraction should be rapid to prevent the progression of coal extrusion deformation and deterioration of the coal's self-supporting ability in the pillar-stope structures.[34] For example, development in advance (entries) and mining pillars in retreat should be for a period of three to six months maximum.

b) The mining concept should involve pillarless extraction of panels or blocks, thus eliminating the need for natural support during extraction procedures.

c) The mining method should minimize the length of development workings if they are located in the unstable coal media because heavy support is required.

d) Mining, especially in pitching coal seams with thicknesses over 3 m, should be performed in such a way that it will prevent creeping of coal faces and shear movement along orogenically sheared planes. Regular roof caving will release high stress concentration of the pillars.

e) The size of the coal pillars and their orientation to the strike and dip of the coal seam should be optimized so that the roof will cave behind the coal face as the pillar is extracted. This will help to avoid pressure build up on the coal pillars in which extraction is underway.

4. Coal pillars with minimal or no strength, until the present, have not been considered feasible for an underground mining. The present mining experience suggests that the design of very weak coal pillars is limited or impossible,[23, 24, 34] but several comments are possible:

a) Rib pillars might be reinforced by artificial means using the injection of particular sealants. The constrained pillar with a decreased rate of fracturing can delay total coal pillar extrusion.

b) If coal pillars are used as support under the circumstances suggested above, mining faces should be vacant so that collapsing of the pillar and associated roof falls will not cause a human casualty.

c) Serious consideration should be given in mine layout to design artificial pillars to replace unstable and dangerous natural coal pillars.

Finally, particular attention should be given to the evaluation of pillar strength and bearing capacity as they are the key factors to underground mine stability.

12.4.   PROTECTIVE PILLARS

The bearing capacity of protective pillars is an important factor in the safety of surface and underground structures. The layout of protective pillars is an integral part of design in mine development. Their layout is definite because they are seldom subject to changes.

The protective pillars, by their function, can be classified into three principal groups:
– Solid coal barrier and safety pillars,
– Barrier and safety pillars partially extracted by room-and-pillar mining,
– Shaft pillars.
Of particular consideration is the size of protective pillars in relation to the degree of damage to mine or surface structures.

The approach to the design of protective pillars can be based either on empirical formulae, digital model studies or analytic solutions in relation to permitted ground displacement.

12.4.1.   *Protective pillars of solid coal*

The protective pillars without any excavation (solid coal) have been common place in shallow coal mining, because at shallow depth they have a limited required width.

Several European countries carry out extensive investigations to design solid protective pillars particularly with the respect to permitted displacement of the ground surface.

1. Barrier pillars in underground mining have a role to protect the active mining area from any adverse effects of the surrounding ground. The barrier pillars at outcrops protect underground workings from runoff and other surface effects. The pillars underground protect the extraction structure from gases, water and fire which are accumulated in a gob area so that layouts of barrier pillars might be either at the boundary of the mineral deposit to protect the mine integrity or at the central part of the deposit to separate the mined out sections from the neighboring virgin sections (Fig. 12.4.1). In both cases, proper design of the pillars is a key factor of mine safety because collapse of such pillars usually results in mine catastrophes. For this reason, a barrier pillar of large size is unlikely to fail except for punching into the roof or floor. This could happen only if the average stress on the pillar is three or more times greater than the uniaxial compressive strength of either roof or floor.[36]

At present, there are several design formulae[37] in coal mining and most of them empirical where a rule of thumb guideline was developed, for example:
– The British formula considers that pillars should be one tenth of the overburden height (h) plus 45 feet.[38]

$$W_p = 0.1\,h + 45 \tag{1}$$

The pillars in consideration are long and rectangular in shape.

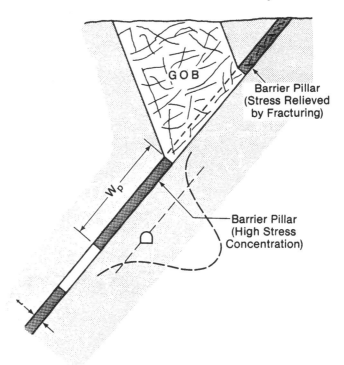

Fig. 12.4.1. Layout of two types of barrier pillars with related stress concentration.

– The Mine Inspectors formula, developed in Pennsylvania by Ashley,[39] is given as:

$$W_p = 20 + 4H + 0.1h \tag{2}$$

where H and h are the height of the pillar and the thickness of overburden, respectively.

– The size of the barrier pillars in European vein mining is based on the depth (h) and thickness (t) of the ore body;

$$W_p = 10t + 0.1h \tag{3}$$

Barrier pillars generate pervasive stress concentrations in the surrounding ground and particular care must be taken in the layout adjacent to them. This case is particularly exhibited in hard rock mining where haulages in the vicinity of barrier pillars can be in high stress concentration zones (Fig. 12.4.1).

2. *Safety pillars* are solid ore or coal blocks left intact to protect the surface structures located directly above them.

The design of safety pillars is based on the intensity of horizontal strain as a criterion of structure deformation (Fig. 12.4.2).[40] The general equation for horizontal strain is given as:

$$\varepsilon = \frac{2\pi v_{max}}{r^2} x e^{-\pi x^2 / r^2} \tag{4}$$

$$r = \frac{h}{\tan \beta'} \tag{5}$$

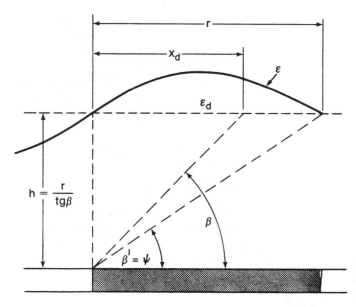

Fig. 12.4.2. Design of safety pillars by the criterion of permitted strain.

where

h = mine depth;

β' = limit angle of subsidence, depends on rock type as discussed under shaft pillars;

r = horizontal distance of subsidence limit;

x = any point on line to subsidence limit;

v = vertical displacement.

The key factor for safety pillar design is in knowing the permitted strain (ε_d) which develops at the point of inflection at a distance from the surface structure. Its value, x_d, can be obtained from the equation:

$$\varepsilon = \frac{2\pi V_{max} X_d}{r^2} e^{-\pi x_d^2/\pi^2} \tag{6}$$

or by graphical construction as illustrated in Figure 12.4.2.

It should be noted that the horizontal distance of subsidence limit (r) depends on the mining depth. Under these circumstances, with increasing mining depth, the length of the safety pillar increases. However, this increase should be accepted to a certain limit because after this the deformation of a surface structure is small and can be neglected.[41]

3. The relation of surface subsidence to pillar width in the case of underground mining usually represented by long and narrow rectangular blocks which are surrounded on both sides by mined out areas. The determination of the pillar size can be done by analyses of subsidence curvature which represents the difference in slope between two neighboring stations. The curvature function is similar to the strain one, but it is of a different scale and its maximum could be written:

$$K_{max} = \pm \frac{2\pi v_{max}}{r^2} e^{-1/2} \tag{7}$$

$$K_{max} = \pm 1.52 \frac{V_{max}}{h^2} \tan^2 \beta' \tag{8}$$

or written as a minimum of curvature radius:

$$\rho_{min} = \pm \frac{1}{K_{max}} \tag{9}$$

$$\rho_{min} = \pm 0.66 \frac{h^2}{V_{max} \tan^2 \beta'} \tag{10}$$

The interaction between two curvature diagrams due to mining on both sides of the pillar is illustrated in Figure 12.4.3 where four possible cases are analyzed:

a) A safety pillar of smaller width (0, 2r) has a total strain of lesser intensity than individual strains on each side.

b) A pillar of increased width (0, 8r) has a total strain as a sum of the individual strains on each side.

c) Further increase of pillar width (2, 0r) results in a decrease in total strain.

d) Large increase of pillar width (4, 0r) will eliminate the influence of the surface subsidence above its central part.

From these analyses it is obvious that safety pillars with insufficient width do not

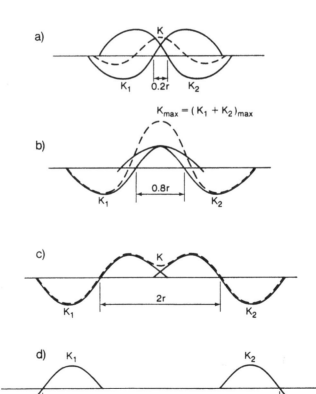

Fig. 12.4.3. Pillar design in relation to the distance between two subsidences (mining on both pillar sides).

protect the surface structures and most likely would initiate their damaging. Adequate attention has not been paid in practice to this problem and in many instances safety pillars have not been effective in protecting surface structures.

4. Surface structure damages can we avoided by controlled underground mining. One of the ways to control surface damage is to design safety pillars of required size.[42]

The most common criterion in Continental Europe is the relation between the angle of the pillar inclination (β) and the permitted strain (ε_d) as illustrated in Figure 12.4.4 for which four categories are given:[43]

Category I: Strain 0.01–1.5 mm/m and angle of pillar $\beta = 54°$. The cracks are invisible. Protection of mine shafts, industrial objects (prep. plants, steel plants, power plants and others), highways, railways, dams, bridges, hospitals, rivers and lakes.

Category II: Strain 1.5–3.0 mm/m and angle of pillar $\beta = 58°$. Cracks are visible (2–5 mm). Protection of drillholes, warehouses, pipelines, apartments up to 2 levels, transformer stations, high voltage electric lines, irrigation objects, agricultural, industrial structures.

Category III: Strain 3.0–6.0 mm/m and angle of pillar $\beta = 62°$. Cracks opened 10–20 mm. Protection of secondary railways and roadways, bungalows, underground workings, airports.

Category IV: Strain 6–12 mm/m and angle of pillar $\beta = 66°$. Protection of temporary buildings like trailers, agricultural land and forest.

If the strain is greater than 12 mm/m, the buildings are unusuable because they are heavily damaged with collapsed walls (Fig. 12.4.5).[44]

It is very important to point out that the critical angle β of the safety pillars, used as a criterion for permitted ground surface damage, is subject to change from one mining region to another. For example, the coal deposits of the Plains Region have to have a

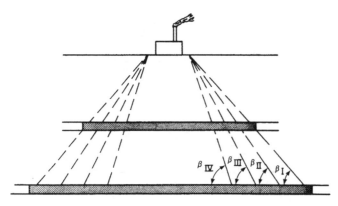

Protective Pillars

$$\beta_I = 54°$$
$$\beta_{II} = 58°$$
$$\beta_{III} = 62°$$
$$\beta_{IV} = 66°$$

Fig. 12.4.4. Design of safety pillar by criterion of critical angle.

Fig. 12.4.5. Heavily damaged building due to surface subsidence (Tuzla, Yugoslavia).

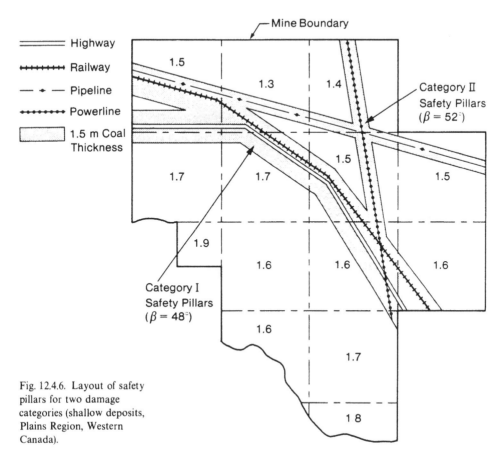

Fig. 12.4.6. Layout of safety pillars for two damage categories (shallow deposits, Plains Region, Western Canada).

lower angle of the safety pillars because coal bearing strata consist of weak rocks. The research in this direction suggests a decrease in angle β by approximately 11 per cent from the European limits. Thus a large mass of coal has to be left underground within safety pillars even in conditions of shallow mining (Fig. 12.4.6).

12.4.2   *Partially extracted protective pillars*

The size of protective pillars increases with increasing mine depth. Large masses of the coal had been contained within protective pillars and there is strong tendency to

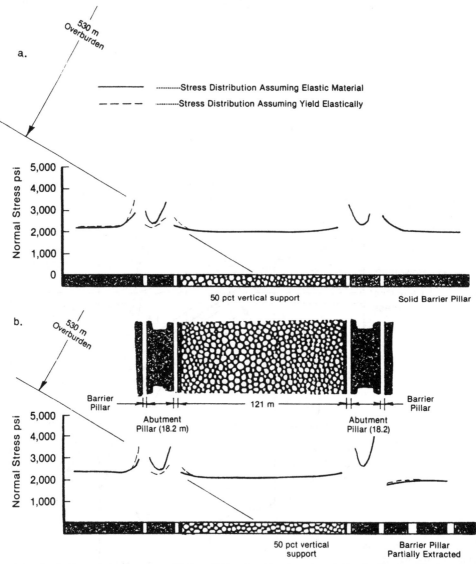

Fig. 12.4.7. Normal elastic stress profile for solid barrier pillar (a) and partially mine barrier pillar by room-and-pillar (b).

recover some coal from them. The partial extraction of the coal from protective pillars (shaft pillars) by room-and-pillar mining started in South Africa some time ago, without endangering mine and surface structures.

Recently a paper has been published called 'Analysis of Pillar Stability on Steeply Pitching Seam Using the Finite Element Method' by N. P. Kripakov to design the abutment and barrier pillars in a coal seem mined by longwal system.[45] The input data for model accounted for the irregular and nonsymmetrical nature of the mine configuration. The digital model consists 791 nodal points, 64 truss elements, and 634 two-dimensional continuum plane strain quadrilateral elements. It has been used as a modeling technique of for imposing applied loads on a structure through boundary displacements.[46]

The finite element analyses by N. P. Kripakov have been carried out for solid barrier pillars as shown in Figure 12.4.7/a, where the normal stress profile is illustrated across the mine configuration for an overburden depth of 530 m acting over the center of the longwall panel. The coal seam was considered to be homogeneous, linear-elastic material (solid line) with a redistribution of stress locally on the coal pillars allowed to yield elastically (doted line) and supporting ability decreased by 50 per cent. Figure 12.4.7/b considers the effect of a partially mined barrier pillar by room-and-pillar mining. Under these circumstances the stress has been increased in the abutment pillar of a tail roadway without effects on the head entry abutment pillar. The subsidence profile is illustrated in Figure 12.4.8, where its maximum lies to the right of a line projected perpendicular to the center of the panel. However, the influence of a partially extracted barrier pillar on subsidence is negligible, what leads credibility to the application of each type of protective pillar. It should be noted that a maximum

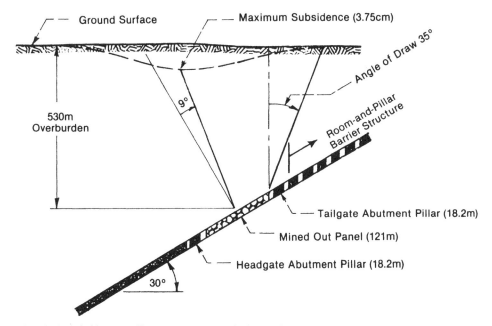

Fig. 12.4.8. Subsidence profile on cross-section of mine configuration.

subsidence of 3.75 cm for an extracted coal seam of 3.3. m does not corresponds to reality. Such a small value of maximum subsidence is misleading and probably due to the high estimate of overburden stiffness of the model. However, the importance of this subsidence profile is as an indication of the qualitatively stress redistribution. For example, it is clearly exhibited that the weight of overburden above the mined out area was being shifted laterally onto the barrier pillars as the face retreated.

The author paid particular consideration to the stability of individual pillars due to their width regardless to their interaction with the longwall panel or the barrier structure. He observed that initial elastic normal stress levels at rib side and extending some distance into the pillar exceeded their respective strength values, in the case of critical pillar widths. Based on the premise that fractured rock still retains some bearing capacity or residual strength and has elastic properties, a new input file was generated with zones of progressive failure represented by different values of effective modulii of elasticity consistent with an inverse value of the safety factor. The analysis was return to obtain a new stress distribution at each interaction. The progressive change in effective stiffness at each interaction is illustrated in Figure 12.4.9. The redistribution of stresses through the pillar section for each interaction: 18.2 m, 24.2 m, 30.3 m and 45.5 m pillar widths. The interactions of stress redistributions were continued until the stress at rib side converged to the theoretical unconfined strength value of the pillar edge. At this point the author assumed that the final steady-state equilibrium conditions existed.[45] The obtained results by this analysis clearly exhibited that the least width of the pillars within barrier structure at the depth of 560 m should be 30.3 m. This would reduce coal recovery to the 25 per cent instead of 50 per cent as in case of a shallow barrier structure. On the basis of failure criteria it could be assumed that although the pillar has fractured to some distance from the rib side, it is still safe.

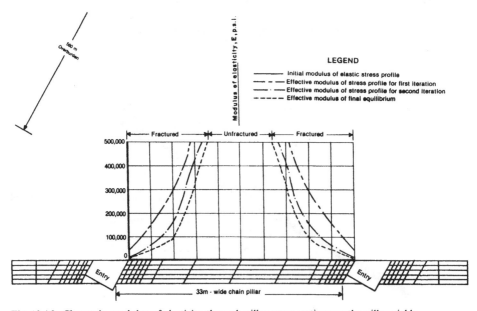

Fig. 12.4.9. Change in modulus of elasticity through pillar cross-sections as the pillar yields.

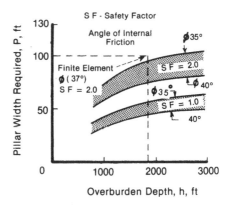

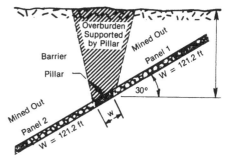

Fig. 12.4.10. Curves for design of protective pillars.

The stability analysis of the barrier pillars in the pitching coal seams where longwall mining is in effect were compared with the confined core pillar design method used in U.K., as illustrated in Figure 12.4.10. Wilson's confined core pillar design method assumes an average hydrostatic virgin stress field in the deposit prior to mining,[38] whereas, the finite element analysis assumed the lateral confinement on the pillar to be approximately equal to the Poisson effect only. Therefore, the finite element results would be expected to be conservative. This pillar design, represents rational analysis which incorporating numerical techniques that consider post-failure behaviour, and its validity should be checked out in actual mining operations.

12.4.3. *Shaft pillars*

There are two concepts used in shaft pillar design; one a civil engineering approach, and the other a mining engineering approach.

The civil engineering approach to shaft pillar design considers:
 – Induced stresses due to the shaft opening,
 – Shaft depth,
 – Strength of the rock mass,
 – Properties of individual beds and their angle of inclination (sedimentary strata only).

Shaft pillar design using this approach is appropriate in a mining engineering context only if the shaft is to be located outside of the excavation perimeter.

When the mining engineer is confronted with the problem of having to site shafts within the excavation perimeter new concepts are required. The influence of narby mining on the shaft's stability is then of paramount importance. There are two approaches which relate the mining extraction pattern to shaft pillar design, one in which extraction proceeds towards the shaft, and the other where extraction proceeds from the shaft.

1. Shaft pillar in a homogeneous competent rock mass is based on the fundamenta' approach to pillar design which depends on:
  – Size of the opening,
  – Shaft depth,
  – Tectonic stresses,
  – Strength properties of the rock mass.
Analyses for shaft pillar size are based on the principles of solid mechanics. It is assumed that the presence of joints and other geological irregularities of the rock mass do not affect the elastic stability of the shaft wall.[46] From the interaction between the intensity of the abutment stress and the strength of the rock mass three stress zones around the shaft opening may be inferred (Fig. 12.4.11).

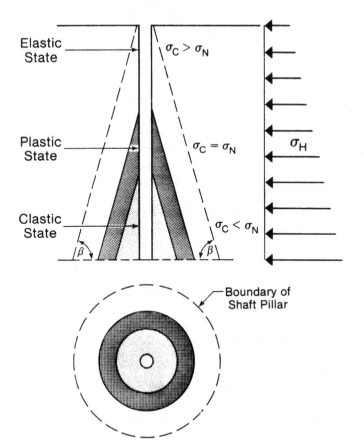

Fig. 12.4.11. Model of stress zones development around shaft opening.

a) Elastic state: The induced stresses around the shaft opening do not exceed the yield point of the rock mass in compression. The radial stress (σ_r) and tangential stress (σ_t) can be expressed as:

$$\sigma_r = \sigma_H\left[1 - \left(\frac{r}{a}\right)^2\right] \quad \sigma_t = \sigma_H\left[1 + \left(\frac{r}{a}\right)^2\right] \tag{1}$$

where

   σ_H = horizontal stress;
   r = radius of shaft;
   a = radius of any point around the shaft opening.

Under this stress state elastic or visco-elastic deformations are present and, theoretically, the shaft wall should be able to withstand the outside pressure without support. In practice, however, shaft walls with concrete lining (up to 15 cm thick) are desirable. The lining is considered a natural extension of the rock mass, which is in an elastic state of stress.

   b) Plastic state of stress occurs with greater shaft depth. The magnitude of the induced stresses due to the shaft opening may exceed the yield point of the rock mass in compression. Shaft pillar and shaft support are required to withstand radial dynamic stresses (rock extrusion – plastic flow). The plastic or visco-plastic stress zone around the shaft opening can be transformed into the clastic zone with increased stress.

   c) Clastic state is usually caused by high induced stress around the shaft opening which creates an excessive shear zone. In this state radial stresses are related to a stress relieved zone. Their determination could be approached in a way similar to failure stress conditions in cohesionless rocks. The final derivation of this approach is:

$$\sigma_r = \sigma_H(1 - \sin\phi)\left(\frac{r}{a}\right)^{e-1} \tag{2}$$

where

   ϕ = coefficient of friction;
   r = radius of the shaft opening;
   a = radius of any point around the shaft opening.

Rock destroyed by shear around the shaft wall will crumble and fail into the shaft opening. Under these circumstances large shaft pillars are required, and concrete lining has to be immediately applied following shaft excavation.

   The uniform concentric zonal distribution of stress states requires a circular shaft pillar which is cone-shaped. The main design parameter used for shaft pillars is the angle of break β. The determination of β is primarily dependent on rock strength and magnitude of the lateral stress at the shaft bottom.

2. Shaft pillar in heterogeneous sedimentary strata should be considered in relation to mineral deposits, because a shaft location might be in either virgin ground or in the mining ground (Fig. 12.4.12). Under identical geological conditions the influence of geological factors on shaft pillar stability is greater in the mining area than in virgin ground.[47] This is particularly exhibited when the strata consist of an interchanging of hard and very weak rock layers. This interaction might lead to dramatic deformations.

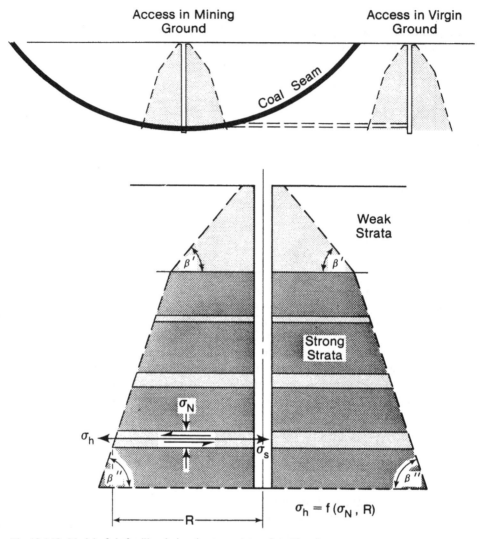

Fig. 12.4.12. Model of shaft pillar design due to geology of coal bearing strata.

The mechanics of deformation can be explained by the compressibility of a visco-plastic layer between two rigid plates. In this process large inelastic deformations (creep) might be accumulated for long periods of time before viscous flow (extrusion) is initiated (Fig. 12.4.12). However, with the increase of compressive stress (for example, with greater shaft depth), and the decrease of shear strength along bedding planes, the deformation may be initiated during the early stages of shaft existence.[48]

The consequences of the compressibility of soft strata layers is vertical displacement of shaft pillars (subsidence), followed by corresponding horizontal movement. Surveying of permanent surveying monuments around the shaft collar in one Russian coal mine have shown[38] that vertical displacements of the shaft pillar were in the range of

15 to 18 mm/year, where the shear strength of nine clay shale layers was $\tau_0 = 14\,\text{kg/cm}^2$ and viscosity was $\eta = 63 \times 10^6\,\text{kg/m}\cdot\text{sec}$. It has been established that intensive vertical displacement velocity should be attributed principallly to the limited size of the shaft pillar ($\beta = 80°$), which is underdesigned for existing geological conditions.

It is a known fact that there is displacement of the rock near shafts due to creep deformation of clay strata. This process is facilitated by the sandwiching of weak, soft strata between thick, hard strata.[38] Then the shaft pillar angle β should correspond to weak rock. Variation due to lithological types may occur according to the following table:

Saturated sand	$\beta = 28°\text{--}35°$
Sand and gravel	$\beta = 40°\text{--}45°$
Clay shale	$\beta = 48°\text{--}55°$
Mudstone	$\beta = 56°\text{--}60°$
Rock salt	$\beta = 58°\text{--}65°$
Shale	$\beta = 60°\text{--}65°$
Siltstone	$\beta = 68°\text{--}70°$

The two limiting angles of shaft pillars should be applied in strata with glacial till of great thickness, as in the Plains Region of Western Canada, where coal seams are located in weak rock strata. For example, under given geological conditions the limiting angle of shaft pillars in glacial till should be $\beta = 45°$ (till depth 25–100 m), and in rock strata $\beta = 60°$. The lithological profile of coal bearing strata favours large shaft pillars which will capture large reserves of coal.

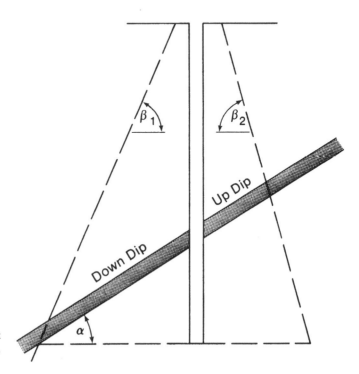

Fig. 12.4.13. Model of shaft pillar of inclined coal seam.

3. Shaft pillars in inclined strata take into consideration pillar relations to coal deposits which depend mainly upon the mechanics of ground surface subsidence and shear stress along bedding planes, which is mobilized by the compressive load.[38] There are various equations based on the theory of elasticity for the design of shaft pillars, as well as an empirical equation.[39] In this case, further analyses are limited to empirical solutions, which are suggested for carboniferous coal strata (Fig. 12.4.13). The equations for shaft pillar design are given for two cases of strata inclinations as follows:

case 1: $\alpha < 45°$

$$\beta_1 = \beta + (90° - \beta)(\alpha/45°) - \text{down-dip} \tag{3}$$

$$\beta_2 = \beta - (90° - \beta)(\alpha/45°) - \text{up-dip} \tag{4}$$

case 2: $\alpha > 45°$

$$\beta_1 = 90° - (90° - \beta)(\alpha - 45°)/45° - \text{down-dip} \tag{5}$$

$$\beta_2 = \alpha - (90° - \beta)(\alpha - 45°)/45° - \text{up-dip} \tag{6}$$

The parameters for calculation of the limiting angles of shaft pillars are: angle of inclination of strata (average value used), and strength of rock strata represented by the limiting angle β (with corresponding values listed in the two previous headings).

4. Interaction between shaft pillars and nearby mining is most common in coal mining, for shafts located within mineable seams. It is a well known fact that shaft pillars experience abutment stresses transferred from nearby mining. High stress concentrations result in pillar displacement along bedding planes and whole pillar movement towards or away from the worked area. With the aspect of shaft pillar deformation, two principal interactions between pillar and gob may be suggested as a function of extraction direction with respect to the shaft.[40] In both cases the main parameter for shaft pillar design might be an angle of break (β) of the rock strata.

Mining from the shaft to the mine boundary is favoured with respect to shaft stability, because the pillar is surrounded by gob where the ground is relaxed and in a state of extension deformation (Fig. 12.4.14). However, mining is preferable in the opposite direction, particularly in coal mining where the gob usually contains water, gas accumulations, and coal remnants liable to self-ignition. Mining from the mine boundaries is the common case, however, it is not favourable for shaft stability. The shaft pillar is surrounded by ground in a state of shortening deformation (Fig. 12.4.15). Under these circumstances shaft pillar design cannot be done only by using the angle of break (β) because the influence of ground subsidence might extend into the safety pillar.

The determination of shaft pillar enlargement is derived from Figure 12.4.15. For example, the horizontal distance of subsidence limit can be written:

$$r = h \cdot \cot \beta' \tag{7}$$

where

      r = radius of subsidence limit;
      h = depth of mining;
      β' = angle of strain limit (ψ).

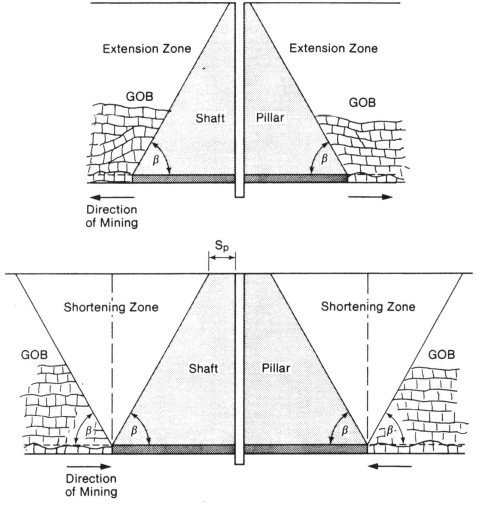

Fig. 12.4.14. Design of the pillar in relation to direction of mining to the shaft.

In the case of an extendent horizontal strain limit the formula should be written as below:

$$r = h \cdot \cot \beta' \tag{8}$$

$$r' = h \cdot \cot(\beta' - \Delta\beta) \tag{9}$$

Enlargement of the strain limit is calculated as:

$$\Delta r = h \cdot \cot(\beta' - \Delta\beta) - h \cdot \cot \beta' \tag{11}$$

$$\Delta r = h \cdot \cot(\beta' - \Delta\beta) - \cot \beta' \tag{12}$$

$$\Delta r = h\cot(\beta' - \Delta\beta) - \cot \beta' \tag{12}$$

$$\Delta r = S_p \tag{13}$$

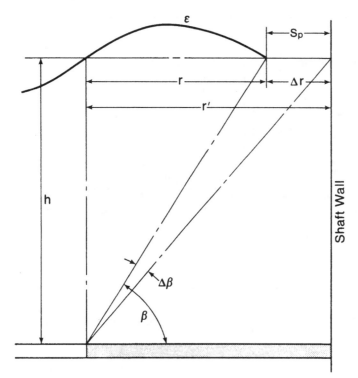

Fig. 12.4.15. The concept of the pillar enlargement.

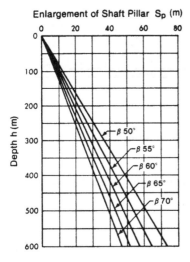

Fig. 12.4.16. Enlargement of the pillar due to mine depth and rock type.

The represented equation of the extension of the strain limit will give a safety factor which should eliminate the damaging influence of possible ground displacement over limiting angles of shaft pillars.

Figure 12.4.16 illustrates a method for the direct reading of values of pillar enlargement as a function of extended angle ($\Delta\beta$) and depth (h). In some European coal fields the enlargement of the shaft pillar is related to shaft diameter, and is expressed by the empirical equation:

$$S_p = (2 - 4)D \qquad (14)$$

where
D = shaft diameter.

Preventative steps to avoid shaft deformation are required because permitted horizontal displacements of the shaft's walls are only up to 1.5 mm/m of shaft circumference.

Shaft stability is the main concern of any underground operation, and if damages are induced they are very difficult to correct. For this reason proper and safe pillar design is of primary importance.[49]

CHAPTER 13

References

CHAPTER 1

1. Aristotle, A., 351 B.C. Meteorology, Central Library, Paris, France, Volume IV, Chapter 9, pp. 36–37.
2. Taylor, R. K., 1975. *Characteristics of Shallow Coal Mine Working and Their Implications in Urban Development Area*, Site Investigations in Areas of Mine Subsidence, Newnes-Butterworths, pp. 125–128.
3. Moore, E. S., 1940. *Coal*, John Wiley & Sons, Inc., New York, Second Edition, pp. 1–2.
4. Lama, R. D., 1977. Principles of Underground Coal Mine Design – An Approach, *Colliery Guardian*, England, May, June & July.
5. Jeremic, M. L., 1957. Ancient and Recent Mining Techniques of the Central Bosnia, *Min. Met. Bull.*, 2: 37–42 (in Yugoslav).
6. Terzaghi, K., 1943. *Theoretical Soil Mechanics*, John Wiley & Sons, New York, pp. 510–511.
7. Jaeger, J. C. & N. G. W. Cook, 1969. *Fundamentals of Rock Mechanics*, Methuen & Co. Ltd., London, pp. 9–25.
8. Boshkov, S. & M. T. Wane, 1969. *Rational Procedures for Underground Mine Design*, Henry Krumb School of Mines, Columbia University, New York (Manuscript).
9. Asszonyi, Cs. & R. Richter, 1979. *The Continium Theory of Rock Mechanics*, Trans Tech Publications.
10. Morrison, R. G. K., 1976. *A Philosophy of Ground Control*, Department of Mining and Mettalurgical Engineering, McGill University, Montreal, pp. 1–27.
11. Woodruff, S. D., 1966. Theory and Application of Rock Mechanics to Roof Control and Support Problems, *Methods of Working Coal and Metal Mines*, Pergamon Press, Vol. I, pp. 138–149.
12. Jeremic, M. L., 1966. Application of Mohr's Circle in Rock Mechanics, *Bull. of Fac. of Min. and Met.*, 123–150 (in Yugoslav).
13. Fiks, I. I., 1972. The Dynamic Aspect of Fracture by Shock Bumps, *Fiz-Tekh. Probl. Raz. Polez. Iskop.* 2: 44–48 (in Russian).
14. Avershin, S. G., 1971. *Some Applied Problems in Rock Mechanics*, Izd-vo Ilim, Frunze (in Russian).
15. Jeremic, M. L., 1966. An Ellipse Stress in Rock Mechanics, *Bull of Fac. of Min. and Met.*, 5: 38–47 (in Yugoslav).
16. Boshkov, S. & M. T. Wane, 1969. *Rational Procedures for Underground Mine Design*, Henry Krumb School of Mines, Columbia University, New York (Manuscript).
17. Jeremic, M. L., 1968. Application of Theory of Plates in Ground Control, *Min. Met. Bull.*, 1. and 2: 57–64 and 221–233 (in Yugoslav).
18. Kuznecov, S. V. et al., 1973. Determination of Stresses in an Uniform Bent Cantilever, *Fiziko-Tekhnicheskie Problemy Razrabotki Poleznykh Iskopaemykh*, September, October, 5: 78–82.
19. Goodman, R. E., 1966. *Methods of Geological Engineering*, West Publishing Company, pp. 227–368.
20. Zienkiewicz, O. C., 1977. *The Finite Element Method*, McGraw-Hill, 3rd ed., pp. 787.
21. Cundall, P. A., 1971. A Computer Model for Simulating Progressive, Large Scale Movements in Block Rock System, *Proc. of Symp. of the Int. Society of Rock Mechanics*, Nancy.
22. Crouch, S. L., 1976. *Analysis of Stresses and Displacements Around Underground Excavations*, An Application of the Displacement Discontinuity Method, Dept. of Civil and Mineral Engineering, University of Minnesota, Minneapolis.

23. Fumagally, E., 1976. Model Simulation of Rock Mechanics Problems, Chapter 11, *Rock Mechanics in Engineering Practice*, pp. 353–383.
24. Hoek, E., 1966. *Rock Mechanics – An Introduction for the Practical Engineer*, Part II – Laboratory Techniques in Rock Mechanics, Mining Engineering, London, pp. 13–16.
25. Jeremic, M. L., 1980. Philosophy of Hydraulic Mining, *CIM Bull.*, September: 81–85.
26. Coon, R. F., 1968. *Correlation of Engineering Behaviour with the Classification of in Situ Rock*, Ph.D. Thesis, University of Illinois, Urbana.
27. Obert, L. & W. I. Duvall, 1967. *Rock Mechanics and Design of Structures in Rock*, John Wiley & Sons, Inc., New York.
28. Jumikis, A. R., 1979. *Rock Mechanics*, Trans Tech Publications.
29. Roberts, A., 1978. The Measurement of Strain and Stress in Rock Masses, *Rock Mechanics* (Editors Stagg, Zienkiewicz), John Wiley & Sons, Inc., pp. 157–202.
30. Peng, S. S., 1978. *Coal Mine Ground Control*, John Wiley & Sons, Inc., New York, pp. 365–398.
31. Vutukuri, V. S. & R. D. Lama, 1974. *Handbook on Mechanical Properties of Rocks*, Trans Tech Publications. Vol. I.
32. Lama, R. D. & V. S. Vutukuri, 1978. *Handbook on Mechanical Properties of Rocks*, Trans Tech Publications. Vol. II, III and IV.
33. Maksimov, A. P., 1963. *Strength Properties of Rock Strata and Stability of Underground Workings*, Gortekhizdat, Moskow, pp. 63–68 (in Russian).
34. Bielenstein, H. V., 1975. Thrust Faults a Problem in Western Canadian Coal Mining, *CIM – A. West. Meet.*, Edmonton, Alberta, Oct. 26–29.

CHAPTER 2

1. Jeremic, M. L., 1979. Influence of Geological Factors on Coal Design in the Plains Region, *West. Min.*, April, 83–89.
2. Holter, M. E. et al., 1975. *Geological and Coal Reserves of the Ardley Coal Zone of Central Alberta*, Report 75-7. Research Council of Alberta, 1975.
3. Locker, J. G., 1973. *Petrographic and Engineering Properties of Fine Grained Rocks of Central Alberta*, Bull. 30. Research Council of Alberta.
4. Anonymous, 1976. *Report of Rock Properties of the Horseshoe Canyon Formation*. Research Council of Alberta.
5. Pearson, G. R., 1959. *Coal Reserves for Strip Mining, Wabamun Lake*, District of Alberta. Prel. Report 60-1. Research Council of Alberta.
6. Anonymous, 1974. Northern Alberta Route (No. 14), *10th Commonwealth Mining and Metallurgical Congress*, September.
7. Jeremic, M. L., 1969. Elastic Deformations of the Rock Mass, *Min. & Met. Bull.*, 7 & 8: 1010–1020 and 1205–1211 (in Yugoslav).
8. Williams, W. A. & P. A. Gray, 1980. The Nature and Properties of Coal and Coal Measure Strata, *The Aus. I.M.M., Roof Support Colloquium*, Illaware Branch, September, 1: 1–12.
9. Holter, M. E. et al., 1975. *Geology and Coal Reserves of the Ardley Zone of Central Alberta*, Report 75-7, Research Council of Alberta.
10. Locker, J. G., 1973. *Petrographic and Engineering Properties of Fine-Grained Rocks of Central Alberta*, Bull. 30, Research Council of Alberta.
11. Jeremic, M. L., 1981. Elements of Rock Weakening of Coal-Bearing Strata of the Canadian Plains Region, *Int. Symposium on Weak Rocks*, Tokyo, Japan, September 21–24.
12. Jeremic, M. L., 1981. Strength Properties of Coal Bearing Strata of the Plains Region, Western Canada, *Min. and Met. Quarterly*, 28, 1: 63–76.
13. Williams, W. A. & R. G. Wilson, 1976. Geological Aspects of Coal Seam Roof Conditions in the Southern Coalfield of New South Wales, *Illaware District Conference*, The Aus. I. M. M., pp. 1–9.
14. Jeremic, M. L., 1981. Elements of Rock Weakening of Coal Bearing Strata of the Canadian Plains Region, *Int. Symp. on Weak Rock*, Tokyo, Japan, September 21–24, pp. 43–49.
15. Pininska, J., 1977. Correlation Between Mechanical and Acoustic Properties of Flysch Sandstones, Int. Symp., *The Geotechnics of Structurally Complex Formations*, Capri, Italy, Vol. 1, pp. 387–394.
16. Jeremic, M. L., 1982. *Elements of Hydraulic Coal Mine Design*, Trans Tech Publications, pp. 25–32.

17. Jeremic, M. L., 1981. Strength Properties of Coal Bearing Strata of the Plains Region, Western Canada, *Min. and Met. Quarterly*, 28, 1: 63–76.

18. Williams, W. A. & P. A. Gray, 1980. *The Nature and Properties of Coal and Coal Measure Strata*, The Aus. I. M. M., Illaware District Conference, pp. 1–9.

19. Mauritsen, S. A., 1975. Introduction to the Genesis of Planetary Structure and Terrestrial Maria, *Geologischen Rundschau*, Band 64, 3: 889–915.

20. Jeremic, M. L., 1980. Rupturing Criteria of Coal Bearing Strata, Western Canada, *Mod. Geol.* 7: 191–199.

21. Bielenstein, H. V., 1975. Thrust Faults a Problem in Western Canadian Coal Mining, *CIM A. Western Meet.*, Edmonton, Alta, October 26–29.

22. Jeremic, M. L., 1978. Tectonic Stresses South Bay Mine, Canada, *Geol. Runds.*, 67, 3: 895–912.

23. Hucka, V. J. & B. Das, 1975. Better Seam-Mining by Evaluating Joints, Cleats, Petrological Profile, *West. Min.*, March: 36–39.

24. Jeremic, M. L., 1977. Primitive Stress in the Underground Mines at Kitwe-Nkana Area, Zambia, Africa, *Geologische Rundschau*, 66.

25. Patton, F. D., 1966. Multiple Modes of Shear Failure in Rock, *1st Int. Cong. Rock Mech.*, Lisbon, Vol. 1.

26. Abramov, B. K. & V. T. Sapozhnikov, 1977. Friction at Contacts Between Hard Rocks, *Fiz-Tekh. Probl. Raz. Polez. Iskop*, March–April, 2 (in Russian).

CHAPTER 3

1. Williamson, I. A., 1967. *Coal Mining Geology*, Oxford Univ. Press, London, pp. 218–236.

2. Clarke, A. M., 1963. A Contribution to the Understanding of Washouts, Swalleys, Splits and other Mining in South Durham, *Trans. Instn. Min. Engrs.*, 122: 567–599.

3. Jeremic, M. L., 1980. Effects and Constraints of Geological Factors on Hydraulic Mining, *Coll. Guard.*, England, December: 559–564.

4. Cook, M. E., 1978. Variability and Anisotrophy of Mechanical Properties of the Pittsburg Coal Seam, *Rock Mechanics*, 11: 3–18.

5. Jeremic, M. L., 1980. Coal Strengths in the Rocky Mountains, *World Coal*, 6, 9: 41–44.

6. Protodyakonov, M. M., 1965. *Investigations on the Brittleness and Viscosity of Coals*, Ugletkhizdat, Moscow (in Russian).

7. Jeremic, M. L., 1979. Hydraulic Mining-Possible Method for Rocky Mountain Coal, *Mining Magazine*, London, England, October: 330–338.

8. Jeremic, M. L., 1980. Characteristics of Western Canadian Coal Seams and Their Effect on Mine Design, *Mining Magazine*, London, England, December: 557–564.

9. Das, B., 1974. Contortional Structures in Coal and Their Effect on the Mechanical Properties of Coal, *Int. J. Rock Mech. Min. Sci. & Geomech.*, 11: 453–457.

10. Pirsson, L. V., 1966. *Rocks and Rock Minerals*, John Wiley & Sons, Inc., pp. 281–286.

11. Williamson, I. A., 1967. *Coal Mining Geology*, Oxford University Press, London, pp. 218–256.

12. Jeremic, M. L., 1980. Coal Strengths in the Rocky Mountains, *World Coal*, 6, 9: 41–44.

13. Jeremic, M. L., 1980. Effect of Coal Seam Stability in Hydraulic Mine Design, *J. of Min. Met. Fuels*, India, May: 93–99.

14. Jeremic, M. L., 1980. Characteristics of Western Canadian Coal Seams and Their Effect on Mine Design, *Mining Magazine*, London, England, December: 557–564.

15. Hucka, V. I. & B. Das, 1975. Better Seam Mining by Evaluating Joints, Cleats, *Petrol. Prof. West. Min.*, March: 35–40.

16. Jeremic, M. L., 1980. Coal Strengths in the Rocky Mountains, *World Coal*, September, 6, 9: 40–43.

17. Jeremic, M. L., 1980. Effect of Coal Seam Stability on Hydraulic Mine Design, *J. of Min. Met. & Fuels*, India, May: 93–99.

18. Jeremic, M. L., 1980. Characteristics of Western Canadian Coal Seam and Their Effect on Mine Design, *Min. Mag.*, London, England, December: 557–564.

19. Das, B., 1974. Contortional Structures in Coal and Their Effect on the Mechanical Properties of Coal, *Int. J. Rock Mech. Min. Sci. & Geomech.*: 453–457.

20. Jeremic, M. L., 1979. Coal Behavior in the Rocky Mountain Belts of Canada, *IV Int. Congress on Rock Mechanics*, Montreux, Switzerland, pp. 189–195.

21. Jeremic, M. L., 1980. Influence of Shear Deformation Structures in Coal Selecting Methods of Mining, *Rock Mechanics*, 13: 23–38.
22. Mjasnikov, U. G. et al., 1975. Determination of Spacing Geological Structural Defects Per Unit Length During the Pillars Development, *UGOL*, Moskow, pp. 62–64 (in Russian).

CHAPTER 4

1. Jeremic, M. L., 1981. Strength Properties of Coal Bearing Strata of the Plains Region, Western Canada, *Min. and Met. Q.*, 28, 1: 64–76.
2. Brown, A. et al., 1958. Ground Stress Studies in Coal Mines of Western Canada, *CIM Bulletin*, November: 686–697.
3. Jeremic, M. L., 1981. Elements of Rock Weakening of Coal Bearing Strata of the Canadian Plains Region, *Int. Symposium on Weak Rock*, Tokyo, Japan, September, Vol. 1, pp. 21–24.
4. Shevyakov, L. D., 1965. *Mining of Mineral Deposits – Long Pillar Mining in the Moscow Coal Fields*, Foreign Languages Publishing House, Moskow, pp. 350–353.
5. Kidybinski, A., 1979. Experience With Hard Rock Penetrometer for Mine Rock Stability Analysis, *IV Int. Cong. on Rock Mechanics*, Montreux, Switzerland, Vol. 1, pp. 293–301.
6. Jeremic, M. L., 1982. *Elements of Hydraulic Coal Mine Design*, Trans Tech Publications, pp. 25–32.
7. Peng, S. S., 1978. *Coal Mine Ground Control*, John Wiley & Sons, pp. 130–174.
8. Bielenstein, H. V. et al., 1974. Multi-Seam Mining at Smoky River, *Strata Control Conference*, Banff, Canada, September.
9. Sikora, W. et al., 1979. Designing of Hard Roof Rock Destressing System for Safe Mining of Rock Burst Prone Coal Seams, *IV Int. Cong. on Rock Mechanics*, Montreux, Switzerland, Vol. 1, pp. 543–549.
10. Whittaker, B. M. & D. R. Hodgkinson, 1971. Reinforcement of Weak Strata, *The Min. Eng.*, London, England, June: 595–609.
11. Whittaker, B. N., 1974. An Appraisal of Strata Control Practice, *The Min. Eng.*, London, England, October: 9–49.
12. Patton, F. D., 1966. Multiple Modes of Shear Failure in Rock, *1st Int. Cong. Rock Mech.*, Lisbon, Vol. 1.
13. Jeremic, M. L., 1980. Rupturing Criteria of Coal Bearing Strata W. Canada, *Modern Geology*, 7: 191–199.
14. Shephard J. & N. I. Fisher, 1978. Faults and Their Effect on Coal Mine Roof Failure and Mining Rate: A Case Study in a New South Wales Colliery, *Min. Engng*. AIM, September: 1325–1334.
15. Kidybinski, A., 1979. Experience with Hard Rock Penetrometers Use for Mine Rock Stability Predictions, *4th Int. Cong. on Rock Mechanics*, Montreux, Switzerland, Vol. 1, pp. 293–301.
16. Jeremic, M. L., 1956. Influence of Geological Factors on Coal Mineability at Zenica Colliery, *Geol. Bull.*, 3: 36–49 (in Yugoslav).
17. Ladanyi, B. & G. Archambault, 1977. Shear Strength and Deformability of Filled Indented Joints, *Int. Symp. The Geotechnics of Structurally Complex Formations*, Capri, Italy, Vol. 1, pp. 317–327.
18. Hucka, V. I. & B. Das, 1975. Better Seam-Mining by Evaluating Joints, Cleats, Petrological Profile, *West. Min.*, March: 36–39.
19. Abramov, B. K. & V. T. Sapozhnikov, 1977. Friction at Contacts Between Hard Rocks. *Fiz.-Techn. Probl. Raz. Polez. Iskop*, March–April, 2: 18–25 (in Russian).
20. Pawlowicz, K., 1967. *Distribution of Planes of Weakness in Rocks; a Method for Determining the Strength and a Project of Classification of Coal Seam Roofs in the Upper Silesian Coalfield*, Prace GIG, Komunikat No. 429, (in Polish).
21. Bilinski, A. et al., 1973. *Criteria of the Choice of Patterns of Single and Multiple Powered Supports for Longwall Workings*, Glowny Instytut Gornictwa, Katowice (in Polish).
22. Jeremic, M. L., 1981. Effect of Subcoal Strata on Coal Pillar Stability, *The Coal Min.*, March: 39–46.
23. Rockway, J. D., 1977. Geotechnical Evaluation of Subcoal Strata for Coal Pillar Support, *Int. Symp. – The Geotechnics of Structurally Complex Formation*, Capri, Italy, Vol. 1, pp. 407–415.
24. Jeremic, M. L., 1980. Influence of Shear Deformation Structures in Coal on Selecting Methods of Mining, *Rock Mechanics*, 13: 23–38.
25. Brown, A. et al., 1958. Ground Stress Studies in Coal Mines of Western Canada, *Can. Min. and Met. Butt.*, November: 32–45.

26. Peng, S. S., 1974. *Roof Control Studies at Olga No. 1 Coal Mine*, Coalwood, W. V., Mining Engineering Transaction, pp. 1–6 (1C3).
27. Woodruff, S. D., 1966. *Methods of Working of Coal and Metal Mines*, Vol. 1, Pergamon Press.
28. Jeremic, M. L., 1963. *Exploration, Development and Evaluation of Mineral Deposits*, Sarajevo University, pp. 474–484 (in Yugoslav).
29. Enever, J. R. et al., 1979. *Geomechanical Considerations Related to the Choice of a Working Floor for a Proposed Longwall Development*, Report No. 14 CSIRO, Australia.
30. Vesic, A. S., 1973. Analysis of Ultimate Loads of Shallow Foundations, *J. of Soil Mech. and Found.*, Div., ASCE Vol. 99, SM1: 37–42.

CHAPTER 5

1. Lama, R. D., 1969. The State of Stress in Virgin Ground, *Met. and Min. Rev.*, 8: 3–12.
2. Jaeger, J. C., 1972. *Rock Mechanics and Engineering*, Cambridge Univ. Press, pp. 38.
3. Denham, D. et al., 1979. Stresses in Australia Crust: Evidence From Earthquakes and In-Situ Stress Measurements, *BMR J. of Aust. Geol. & Geophys.*, 4: 289–295.
4. Kuznecov, S. V. et al., 1977. *Primitive Stress State of a Rock Mass, Reflection of Present-Day Stress Fields and Rock Properties in the State of Ledge Rock Massifs*, Izd. KF Akad. Nauk SSSR, Apatity, pp. 140–145 (in Russian).
5. Tuchaninov, I. A., 1978. Interrelation Between the Stressed State of Rocks, *Fiz-Tekh. Probl. Raz. Polez. Isk.*, March–April, 2: 20–25 (in Russian).
6. Zolotarev, G. S., 1978. Geological Factors Governing the Stressed State of the Rocks, *Fiz-Tekh. Probl. Raz. Polez. Iskop.*, March–April, 2: 26–34 (in Russian).
7. Jeremic, M. L., 1977. Primitive Stress in the Underground Mines at Kitwe-Nkana Area Zambia, Africa, *Geol. Rundschau*, 1, Stuttgart, 66: 237–255.
8. Kidybinski, A., 1979. Experience with Hard Work Penetrometers Used for Fine Rock Stability Predictions, *4th Int. Cong. on Rock Mechanics*, Montreux, Switzerland, Vol. I, pp. 293–301.
9. Campbell, W. F., 1952. Deep Coal Mining in Spring Hill, No. 2 Mine, *Min. Engng.*, AIM, pp. 968–972.
10. Peperakins, J., 1952. Mountain Bumps at Sunny Side Colliery, *Min. Engng.*, AIM, pp. 982.
11. Golecki, J. J., 1977. Stresses in Folded Rocks, *Int. Symp.-Geotechnics of Structurally Comples Formations*, Capri, Italy, Vol. I, September 19–21, pp. 245–254.
12. Otkidach, V. V. et al., 1973. Rise of Temperature in Coal Pillars Due to Deformation by Dynamic Load, *Fiz-Tekh. Probl. Iskop.*, 2: 105–108 (in Russian).
13. Lindner, E. N. & J. A. Halpern, 1976. *In-Sity Stress: Analysis*, AIM Transaction, 4C1/1-4C1–4C1/7.
14. Williams, W. A. & P. A. Gray, 1980. The Nature and Properties of Coal and Coal Measure Strata, The Aus. I.M.M., Illawara Branch, *Roof Support Colloquium*, September, 1/1–1/12.
15. Jeremic, M. L., 1981. Coal Mine Roadway Stability in Relation to Lateral Tectonic Stress, W. Canada, *Mining Engineering*, June: 704–709.
16. Norris, D. K., 1978. *Structural Conditions in Canadian Coal Mines*, Geol. Surv. of Can., Ottawa, Bull. 55.
17. Jeremic, M. L., 1980. Rupturing Criteria of Coal Bearing Strata, W. Canada, *Modern Geology*, 7: 191–199.
18. Jeremic, M. L., 1979. Coal Behavior in the Canadian Rocky Mountain Belt, *4th Int. Cong. on Rock Mechanics*, September 2–8, Montreux, Switzerland, Vol. 1: 189–194.
19. Jeremic, M. L., 1980. Influence of Shear Deformation Structures in Coal on Selecting Methods of Mining, *Rock Mechanics*, 13, 1, pp. 23–38.
20. Barron, K., 1975. *An Injection Technique for Investigating Pillars and Ribs in Coal Mines*, CANMET Report, ERP/MRL (CF) 75–86 (TR). Ottawa.
21. Evans, W. H., 1941. The Strength of Undermined Strata, *Trans. Inst. Mining Met.* 1940–1941: 475–500.
22. Emery, C. L., 1960. The Strain in Rocks in Relation to Mine Opening, *Min. Eng. No. 1*, London Paper, October, No. 3834.
23. Zolotarev, G. S., 1978. Geological Factors Governing the Stresses State of the Rocks, *Fiz-Tekh. Probl. Raz. Polez. Iskop.*, 2: 26–34.
24. Jeremic, M. L., 1981. The State of Geological Stresses in the Athabasca Oil Sand Deposits, *Mod. Geol.*, 8: 43–50.
25. Reiner, M., 1960. *Deformation, Strain and Flow – An Elementary Introduction to Rheology*, Interscience Publishers, New York.

26. Protodyakonov, M. M., 1965. *Investigations on the Brittleness and Viscosity of Coals*, Ugletek-izdat, Moskow (in Russian).
27. Jeremic, M. L., 1980. Influence of Shear Deformation Structure in Coal on Selecting Method of Mining, *Rock Mechanics*, 13, 1: 23–38.
28. Jeremic, M. L., 1979. Coal Behaviour in the Canadian Rocky Mountain Belt, *4th Int. Cong. on Rock Mechanics*, September 2–8, Vol. 1, Montreux, Switzerland, pp. 189–194.
29. Van Der Poel, C., 1960. *Road Asphalt*, Asphalistic Mixtures, Chapter IX, pp. 360–413.
30. Lama, R. D., 1980. Adsorption and Desorption Technique in Predicting Outbursts of Gas and Coal, The Aus. I.M.M. South, Queensland Branch, *Prediction and Control of Outbursts in Coal Mines Symposium*, September, pp. 173–191.
31. Patching, T. H., 1970. The Retention and Release of Gas in Coal – A Review, *CIM Trans.*, 73: 328–334.
32. Puzirev, V. N. et al., 1970. Gas Flow Prognosis in Coal Seams by Drill Holes, *UGOL*, 12: 43–47.
33. Ripu, D. L., 1968. Outbursts of Gas and Coal, *Colliery Engng.*, March: 103–109.
34. Watanabe, Y. et al., 1979. The Application of Accoustic Emission Methods to Forecasting the Rock and Gas Outburst in Coal Mine, *Mem. of the Fac. of Engng.*, Hokkaido University, 15, 2: 187–198.
35. Jeremic, M. L., 1979. Hydraulic Mining – Possible Method for Rocky Mountain Coal, *Min. Mag.*, London, England, October: 330–338.
36. Ricca, V. & M. Hemmerich, 1978. *Underground Mine Drainage Quantity and Quality Generation Model*, SIAMOS, Granada, Spain.
37. Richter, R. & E. Bobok, 1978. *Complex Continuum Model for Description of the Simultaneous Solid–Fluid Movements*, SIAMOS, Granada, Spain.
38. Shubert, J. P., 1978. *Reducing Water Leakage Into Underground Coal Mines by Aquifer Dewatering*, SIAMOS, Granada, Spain, pp. 911–931.
39. Salustowicz, A., 1959. Protective Coal Plate in the Roof of Mine Workings, *Min. Bull.*, 12: 332–340 (in Polish).
40. Zrelec, V., 1966. Determination of Thickness of Protective Plate in Lignite Coal Seams with Quick Sand Strata, *Min. and Met.*, 12: 275–280 (in Yugoslav).

CHAPTER 6

1. Jeremic, M. L., 1966. Application of Mohr's Circle in Rock Mechanics, *Bull. of Fac. of Min. and Met.*, 5: 38–47 (in Yugoslav).
2. Schischka, L. & V. Zamarsky, 1977. Principle of Prevention of Rock Bursts in Ostrawa-Karvin Colliery, CSSR, *UGOL*, September: 71–74 (in Russian).
3. Jeremic, M. L., 1967. Influence of Shear Deformation Structures in Coal on Selecting Method of Mining, *Rock Mechanics*, 13, 1: 23–38.
4. Holland, C. T., 1958. Cause and Occurrence of Coal Mining Bumps, *Mining Engineering, Transaction AIME*, September: 994–1004.
5. Lama, R. D., 1966. Rock Burst – A Comparison of the In-Situ Mechanical Properties of Coal Seams, *Coll. Engng.*, January: 20–25.
6. Lama, R. D., 1968. Outbursts of Gas and Coal, *Coll. Engng.*, March: 103–109.
7. Brown, A. et al., 1958. Ground Stress Studies in Coal Mines of Western Canada, *CIM Bull.*, November: 1–11.
8. Jeremic, M. L., 1980. Characteristics of Western Canadian Coal Seams and Their Effect on Mine Design, *Min. Mag.*, London, December: 557–564.
9. Medvedev, B. I. & V. V. Oxokin, 1973. Mechanics of Gas Outbursts, *Fiz.-Tek. Probl. Raz. Polez. Iskop.*, May–June, 3: 11–17 (in Russian).
10. Makarov, Yu. N. et al., 1977. Camouflet Blasting in Outburst Prone Coal Seams, *Fiz.-Tek. Probl., Raz. Polez. Iskop.*, 1: 114–117 (in Russian).
11. Mauck, H. E., 1958. Coal Mine Bumps can be Eliminated, *Min. Engng. Transaction AIME*, September: 993–994.
12. Jeremic, M. L. & D. M. Cruden, 1979. Strength of Coal From the Star-Key Mine, Near Edmonton, Alberta, *CIM Bull.*, February: 94–99.
13. Hustrulid, W. A., 1976. A Review of Coal Pillar Strength Formulae, *Rock Mechanics*, 8: 115–145.
14. Evans, I., 1961. The Tensile Strength of Coal, *Coll., Engng.*, 38: 428–434.

15. Jeremic, M. L., 1956. Influence of Geological Factors on Coal Mineability at Zenica Colliery, *Geol. Bull.*: 83–90 (in Yugoslav).
16. Jeremic, M. L., 1980. Coal Strength in the Rocky Mountains Belt, *World Coal*, 6, 9: 40–44.
17. Jeremic, M. L., 1982. *Elements of Hydraulic Coal Mine Design*, Trans Tech. Publ., Claustal, pp. 105–123.
18. Fadeev, A. B. & E. K. Abdyldayev, 1979. Elastoplastic Analyses of Stress in Coal Pillars by Finite Element Method, *Rock Mechanics*, 11: 243–251.
19. Zienkiewicz, O. C. et al., 1970. Analysis of Non-Linear Problems in Rock Mechanics With Particular Reference to Jointed Rock Systems, *Proc. 2nd Cong. Int. Soc. Rock Mech.*, Beograd, Vol. 3.
20. Faustov, G. T. & P. A. Abashin, 1974. Computation of Pillars in Elasto-Plastic State, *Fiziko-Tekhnicheskie Problemy Razrabotki Poleznykh Iskopaemykh*, May–June, 3: 27–37 (in Russian).
21. Lutzens, W. W. & L. B. Childress, 1976. *Criteria For Slab Removal at Luisiana Salt Mines*, U.S. MESA REPORT, Informational Report 1047, p. 8.
22. Jeremic, M. L., 1980. Influence of Shear Deformation Structures in Coal on Selecting Methods of Mining, *Rock Mechanics*, 3: 23–38.
23. Lone, P., 1975. Comment previor la deformation des massifs rocheux, *Annls. des Min.* Paris, Fevrier-Mars.
24. Atkinson, R. H. & Hon-Yim Ko, 1977. *Strength Characteristics of U.S. Coals*, Bureau of Mines Report, pp. 2B3–1–2B3–6.
25. Afansea, B. G. & B. K. Abramov, 1975. Determination of the Long-Term Shear Strength Along Rock Contacts, *Fiz-Tek. Probl. Raz. Polez. Iscop.*, 5: 131–134 (in Russian).
26. Bielenstein, H. V., 1975. Thrust Faults – A Problem in Western Canadian Coal Mining, *CIM An. West. Meet.*, Edmonton, Alberta, October, 26–29.
27. Jeremic, M. L., 1980. Rupturing Criteria of Coal Bearing Strata, W. Canada, *Modern Geology*, 7: 191–199.
28. Norris, D. K., 1958. *Structural Conditions in Canadian Coal Mines*, Geol. Surv., Canada, Bull. 44.
29. Patching, T. H., 1958. Part II – Investigations Into Outbursts of Gas and Coal, Ground Stress Studies in Coal Mines of Western Canada – A Progress Report, *CIM Bull.*, Transactions, 61: 364–375.
30. Jeremic, M. L., 1979. Coal Behaviour in the Rocky Mountain Belt of Canada, *4th International Congress of Rock Mechanics*, Montreux, Switzerland, pp. 189–195.
31. Kovari, K., 1977. Micromechanics Models of Progressive Failure in Rock and Rock Like Materials, *Int. Symp. – The Geotechnics of Structurally Complex Formations*, V. 1, Capri, Italy, pp. 307–316.
32. Noonan, D. K. J., 1972. *Fractured Rock Subjected to Direct Shear*, M.Sc. Thesis, Department of Civil Engineering, The University of Alberta, Edmonton.
33. Sakatikjants, S. A. & A. I. Bashkov, 1977. Thin, Level and Steep Seam Equipment Used in the Donets Basin, *World Coal*, July: 40–43.
34. Jeremic, M. L., 1980. Effect of Coal Seam Stability on Hydraulic Mine Design, *J. of Min., Met. & Fuels*, May: 93–99.
35. Holland, C. T., 1967. The Strength of Coal in Mine Pillars, *Symp. on Stress and Failure Around Underground Openings*, Paper 14, University of Sydney.
36. Jeremic, M. L., 1958. Concept of Entry Design for Raspotocje Coal Mine, *Min. Met. Bull.*, 1: 8–15 (in Yugoslav).
37. Wilson, A. H., 1972. An Hypothesis Concerning Pillar Stability, *The Min. Eng.*, 131, June: 409–417.
38. Kryzhanovskayz, T. A., 1960. *An Investigations of Pressure on Horizontal Workings by Visco-Plastic Deformations*, Isd-vo Lib-ry po Gornoma Delu, Moskow, pp. 28–33 (in Russian).
39. Jeremic, M. L., 1980. Influence of Shear Deformation Structures in Coal on Selecting Methods of Mining, *Rock Mechanics*, Springler-Verlag, 13: 23–38.
40. Bielenstein, H. V. et al., 1977. Multi-Seam Mining at Smoky River, *International Conference on Strata Control*, Banff (Canada), September.
41. Afansea, B. G. & B. K. Abramov, 1975. Determination of the Long-Term Shear Strength Along Rock Contacts, *Fiz.-Tekh. Probl. Raz. Polez. Iskop.*, 5: 131–134 (in Russian).
42. Jeremic, M. L., 1979. Hydraulic Mining, Possible Method For Rocky Mountain Coal, *Min. Mag.*, London, October: 330–338.

CHAPTER 7

1. Given, L. A., 1973. Room-and-Pillar Methods, *SME Mining Engineering Handbook*, The Am. Inst. of Min. Met. and Pet. Eng., Vol. 1, Chapt. 12, pp. 45–71.
2. Parking, R. J., 1981. Underground Mining in the Collie Coalfield at Western No. 2 Mine, *I. E. Aust. Sumposium, Western Australia's Energy Resources and Utilization to 2000*.
3. Ellis, S. P. & F. E. Kirstein, 1975. To Strip or Not to Strip Coal, *Gold & Base Minerals of South Africa*, April: 59–75.
4. Shield, J. J., 1976. Pillar Mining Systems – Continuous (Non-Cyclic) Mining, Coal Mining System, SME *Min. Eng.* The Am. Inst. of Min. Met. and Pet Engng.: 109–153.
5. Wright, D. L., 1973. Layout of Continuous Miner Operations in the Smoky River Mines, *CIM Bull.* *March*: 167–171.
6. Shevyakov, L. D., 1965. *Mining of Mineral Deposits*, Foreign Languages Publishing House, Moskow, pp. 338–353.
7. Hargraves, A. J. & R. J. Kininmsath, 1980. Factors Determining the Dimensions of Opening, *The Aus. I.M.M., Illawarra Branch, Roof Sup. Coll.*, September, Paper 2, pp. 1–10.
8. Polowitch, E. R. & T. J. Brisky, 1973. Designing the Hendrix No. 22, Shortwall, *Min. Cong. J.*, 59, 6: 16–22.
9. Green, L. E. & E. R. Polowitch, 1977. *Comparative Shortwall and Room-and-Pillar Mining Costs*, U.S.B.M., No. 8757.
10. Duval, V., 1948. *Stress Analysis Applied to Underground Mining Problems*, Part II, USBM, Re. 4378.
11. Salomon, M. D. G., 1970. Stability, Instability and Design of Pillar Workings, *Int. J. Rock Mech. Min. Sci.*, 7.
12. Salomon, M. D. G. & K. I. Oravecz, 1976. *Rock Mechanics in Coal Mining*, Chambers of Mines of South Africa, PRO Series No. 198, pp. 19–26.
13. Sheorey, P. R. & B. Singh, 1974. Estimation of Pillar Loads in Single and Contiguous Seam Workings, *Int. J. Rock Mech. Min. Sci.*, 11: 97–102.
14. Jeremic, M. L., 1982. Internal Structure of Coal Seams and mechanical Stability, *CIM Bull.*, May: 71–75.
15. Stephansson, O., 1971. Stability of Single Openings in Horizontally Bedded Rock, *Eng. Geol.*, 5: 55–71.
16. Hofer, K. H. & W. Menzel, 1963. *Vergleichende Betrachtungen uber die mathematisch and aus Messungen unter Tage ermitteleten Pfeiler Belastungen im Kalibergau*, Bergakademie, Vol. 15, pp. 326–334.
17. Tincelin, E. & P. Sinou, 1960. Collapse of Areas Worked by the Small Pillar Method, *3rd Int. Conf. on Strata Control*, Paris, pp. 571–589.
18. Jeremic, M. L., 1967. The Application of Ellipse Stress in Rock Mechanics, *Bull. of Fac. of Min. & Met.*, Bor, 5: 191–203 (in Yugoslav).
19. Whittaker, B. M. & M. L. Jeremic, 1979. Longwall Mining Potential of Plains Region Coal Deposits in Western Canada, *Coll. Guard. Coal Int.*, April: 31–39.
20. Jeremic, M. L., 1978. Stress Mechanism at Mindola Mine, Zambia Africa, *CIM Bulletin*, November: 77–84.
21. Shevjakov, L. D., 1941. *Size and Deformability of Mine Pillars*, Izv. ANSSR, No. 7, 8, 9, Moskow (in Russian).
22. Slesarev, V. D., 1948. *Design of the Optimal Mine Pillars*, Mechanika, Gornoe Delo, Ugletehizdat, Moskow, pp. 238–261 (in Russian).
23. Katkov, G. A. et al., 1979. Use of Optically Sensitive Materials to Model Brittle Failure of the Rocks Around Mine Openings, *Fiz-Tekh Probl. Raz. Pol. Iskop.*, January–February, 1: 92–95.
24. Iljstein, A. M. et al., 1964. *Methods of Pillar Design for Room-and-Pillar Mining Systems*, Izdateljstvo Nauka, pp. 173–192 (in Russian).
25. Hoek, E., 1965. The Design of Centrifuge for the Simulation of Gravitational Force Fields in Mine Models, *J. of the S. Afr. Inst. of Min. & Met.*, 65: 455–487.
26. Jeremic, M. L., 1980. Influence of Shear Deformation Structures in Coal on Selecting Methods of Mining, *Rock Mechanics*, 13, 1: 23–38.
27. Gramatikov, Y. A., 1977. Critical Size of Pillars Dangerous for Coal Outbursts, *UGOL'*, September: 56–57 (in Russian).
28. Holland, C. T., 1964. The Strength of Coal in Mine Pillars, *Proceedings of the 6th Symposium on Rock Mechanics*, Univ. Missouri Rolla, pp. 450–460.

29. Yemashchikov, V. S. & S. M. Shikinyavskii, 1978. Qualitative Evaluation of Limiting State of Rocks, *Fiz. Tekhn. Prob. Raz. Polez. Iskop.*, September–October, 5: 28–36 (in Russian).

30. Jeremic, M. L. & D. M. Cruden, 1979. Strength of Coal from the Star-Key Mine, Near Edmonton, Alberta, *CIM Bulletin*, April: 94–99.

31. Szwilski, A. B. & B. N. Whittaker, 1975. Control of Strata Movement Around Face Ends, *The Min. Eng.*, July: 515–525.

32. Korbin, G. & T. Brekka, 1975. *A Model Study of Spiling Reinforcement in Underground Openings*, Technique Report MRD-2-75, University of California, Berkley.

33. Stillborg, B. et al., 1979. Three-Dimensional Physical Model Technology Applicable to the Scaling of Underground Structures, *4th Int. Rock Mechanics Congress*, Montreux, Switzerland, Vol. 2, pp. 655–662.

34. Dudukalov, V. P., 1976. Estimation of the Stress State of the Models of Equivalent Materials for Inclined Strata, *Fiz-Tekh. Probl. Raz. Pol. Iskop.*, January–February, 1: 84–88 (in Russian).

35. Szwilski, A. & M. L. Jeremic, 1979. *Stability of Coal Seam Strata Undermined by Room-and-Pillar Operations*, Report, Dep. of Mineral Engineering, The Univ. of Alberta, Edmonton.

36. Wright, R., 1981. *Stability Analysis in Room-and-Pillar Mining*, M.Sc. Thesis, Dep. of Mineral Engineering, The Univ. of Alberta, Edmonton.

37. Terzaghi, K., 1943. *Theoretical Soil Mechanics*, Wiley, New York, pp. 194–197.

38. Hagraves, A. J. & R. J. Kininmsath, 1980. Factors Determining Dimensions of Openings, *The Aus. I.M.M.*, *Illawara Branch Roof Sup. Colloquium*, September, 2, pp. 1–10.

39. Jeremic, M. L., 1978. Elements of Rock Weakening of Coal-Bearing Strata of the Canadian Plains Region, *Int. Symp. on Weak Rock*, September 21–24, Tokyo, Japan.

40. Jeremic, M. L., 1981. Effect of Sub-Coal Strata on Coal Pillar Stability, *West. Min.*, March: 34–46.

41. Jeremic, M. L., 1982. Stability Analysis of Mine Structure, *Elements of Hydraulic Coal Mine Design*. Trans Tech Publications, Clausthal, pp. 105–123.

42. Hofer, K. A. & W. Menzel, 1976. Comparative Study of Pillar Loads in Potash Mines Established by Calculation and by Measurements Below Ground, *Int. J. Rock Mechanics Min. Sci.*, 1, 2: 181–190.

43. Jeremic, M. L. & R. Wright, 1980. *Stability of Coal Seam Strata Undermined by Room-and-Pillar Operations*, Manuscript, Dept. of Mineral Engineering, The University of Alberta, Edmonton.

44. Markov, G. & S. N. Savchenko, 1975. Stress Distribution in Shallow Rectangular Excavation, *Fiz.-Tekh. Prob. Raz. Polez. Iskop.*, 5: 14–18 (in Russian).

45. Salamon, M. D. G. & K. L. Oravecz, 1976. *Rock Mechanics in Coal Mining*, Chamber of Mines of South Africa, PRD series, 198, pp. 59–65.

46. Jeremic, M. L., 1975. Subsidence Problems Caused by Solution Mining of the Rock Salt Deposits, *Proceedings of the 10th Can. Rock Mechanics Symp.* Vol. 1, Kingston, Ontario, Sept. 2–4, pp. 103–115.

47. Goodman, R. E., 1976. *Methods of Geological Engineering in Discontinuous Rocks*, West Publishing Company, San Francisco.

48. Bray, J. W., 1973. *Similitude in the Base Friction Model*, Unpublished notes, Imperial College, London, England.

49. Spang, R. M., 1977. Possibilities and Limitations of the Base Friction Model, *Rock Mech.*, 12: 185–198.

50. Baumartner, P. et al., 1979. Development of a Tilutable Base Friction Frame for Kinematic Studies of Caving at Various Depths, Technical note, *Int. J. Rock Mech. Min. Sci.*, 16: 265–267.

51. Bielenstein, et al., 1977. Multi-Seam Mining at Smoky River, *6th Int. Strata Control Conference*, Banff, Canada. September.

52. Pepeljev, G. P., 1975. Experiment of Coal Extraction Influence by Undermining, *UGOL'*, Moskow: 53–55 (in Russian).

53. Jeremic, M. L., 1978. *Multiple Coal Seam Mining*, Unpublished notes. Dep. of Mineral Engineering, University of Alberta, Edmonton.

CHAPTER 8

1. Blades, M. J. & B. N. Whittaker, 1974. Longwall Layouts and Roadway Design for Effective Strata Control in Advance and Retreat Mining, *Min. Eng.* April: 277–288.

2. Szwilski, A. B., 1980. Longwall Mining Methods Applies to the Plain Coal Region, *CIM Bulletin*, January: 85–87.

3. Jeremic, M. L. & A. B. Szwilski, 1979. *Elements of Longwall Mine Design in the Plains Region of Western Canada*, Manuscript, The University of Alberta, Edmonton.

4. Hagraves, A. J. & R. J. Kininmsath, 1980. Factors Determining the Dimensions of Openings, *The Aus. I. M. M., Roof Support Colloquim, Illaware Branch*, September, 2/1–2/14.

5. Whittaker, B. N. & M. L. Jeremic, 1979. Longwall Mining Potential of Plains Region Coal Deposits in Western Canada, *Coll. Guard. Coal Int.*, April: 31–39.

6. Bessonov, Y. M. & M. I. Boganov, 1977. Working in the Permafrost of Penshora, Frozen North, *World Coal*, July: 61–64.

7. Ashwin, D. P. et al., 1978. Some Fundamental Aspects of Face Powered Support Design, *The Min. Eng.*, August, London, 119, 129: 659–671.

8. Whittaker, B. M., 1974. An Appraisal of Strata Control Practice, *The Min. Eng.*, London, October: 9–24.

9. Whittaker, B. M., 1977. *A Review of the Contribution Made by Powered Roof Supports to Longwall Mining*, University of Nottingham, pp. 1–13.

10. Whittaker, B. M. & G. J. M. Woodrow, 1977. Design Loads for Gateside Packs and Support Systems, *The Min. Eng.*, February: 263–275.

11. Wilson, A. H., 1964. Conclusions From Recent Strata Control Measurements Made by the Mining Research Establishment, *The Min. Eng.*, April: 367–380.

12. Wilson, A. H., 1975. Support Load Requirements on Longwall Faces, *The Min. Eng.*, June, 173: 479–488.

13. Peng, S. S., 1978. *Coal Mine Ground Control*, Face Powered Support, John Wiley & Sons, pp. 232–268.

14. Sakatikjants, S. A. & A. I. Bashkov, 1977. Thin, Level and Steep Seam Equipment Used in the Donets Basin, *World Coal*, July: 40–43.

15. Gritsko, G. I. et al., 1978. Principles of Operation of Hydraulic Supports Regarding as Rheological Model, *Fiz.-Tekh. Prob. Raz. Polez. Iskop.*, 2: 3–10 (in Russian).

16. Popovic, V., 1965. Mining Stresses of the Excavation in Flat and Gently Dipping Coal Seams, *Min. and Met. Bull.*, Belgrade, 8: 169–175 (in Yugoslav).

17. Everling, G., 1972. Rock Pressure: Its Prediction and Evaluation, *5th Int. Strata Control Conf.*, London, Vol. 8, pp. 1–10.

18. Whittaker, B. N., 1974. An Appraisal of Strata Control Practice, *The Min. Eng.*, October: 1302–1309.

19. Jacobi, O., 1967. Die gebirgsmechanisch gunstige Fuhrung von Abbaustrecken, *Gluckauf*, 103: 1302–1309.

20. Irresberg, H., 1980. Strata Control in Face and Road at Deep Levels, *Gluckauf* (Translation), 116, 5: 88–91.

21. King, H. J. et al., 1972. The Effects of Interaction in Mine Layout, *5th Int. Strata Control Conf.*, London, Vol. 17, pp. 1–8.

22. Zhzlov, N. I. et al., 1974. Stress Evaluation in a Coal Seam with Strata Control by Chocks and Stowing, *Fiz-Tekh. Prob. Raz. Polez. Iskop.*, 2: 125–127 (in Russian).

23. Szwilski, A. B. & B. N. Whittaker, 1975. Control of Strata Movement Around Face-Ends, *The Min. Eng.*, London, England, July: 515–525.

24. Whittaker, B. N. & D. R. Hodgkinson, 1971. Design and Layout of Longwall Workings, *The Min. Eng.*, 134: 79–91.

25. Whittaker, B. M., 1972. Design and Planning of Mine Layouts, *Min. Dep. Mag. Nott. Univ.*, 24: 57–68.

26. Jeremic, M. L., 1982. Stability Analysis of Mine Structures, *Elements of Hydraulic Coal Mine Design*, Trans Tech Publications, 105–123.

27. Gorrie, C. & G. Scott, 1970. Some Aspects of Caving on Powered Support Faces, *The Min. Eng.*, Augustus: 677–691.

28. Kenny, P., 1969. The Caving of the Waste on Longwall Faces, *Int. J. of Rock Mech.*, November: 541–555.

29. Harris, G. W., 1974. A Sandbox Model Used to Examine the Stress Distribution Around a Simulated Longwall Coal Face, *Int. J. of Rock Mech.* 8: 325–335.

30. Baumgartner, P. et al., 1979. Development of a Tiltable Base Friction Frame for Kinematic Studies of Caving at Various Depths, Technical Note, *Int. J. of Rock Mech. Min. Sci.*, 16: 265–267.

31. National Coal Board, 1975. *Subsidence Engineers Handbook*, N.C.B., Mining Department, London, England.

32. Whittaker, B. N. & M. L. Jeremic, 1979. Longwall Mining Potential of Plains Region Coal Deposits in Western Canada, *Coll. Guard. Coal Int.*, April: 31–39.

33. King, H. J. et al., 1974. *Minerals and Environment*, I. M. M. London, Paper 25, p. 26.

34. Whittaker, B. N. & C. D. Breeds, 1977. The Influence of Surface Geology on the Character of Mining Subsidence, *Int. Symp. The Geot. of Struct. Complex Struct.*, Capry, Italy, Vol. 1, pp. 459–469.
35. Jeremic, M. L., 1974. Subsidence Problems Caused by Solution Mining of Rock Salt Deposits, *Proceedings of the 10th CDN. Rock Mech. Symp.*, Kingston, Vol. 1, pp. 203–223.
36. Jeremic, M. L. & Z. Nikolic, 1967. The Caving Influence on Surface Deformation with an Example of Lubnica Coal Mine, *The Fac. of Min. & Met. & Inst. of Copper Bull.*, Bor, 1: 161–177.
37. Kowalczyk, Z., 1966. Effect of Mining Exploitation on the Ground Surface and Structures in Heavily Industrialized and Populated Areas, *CIM Bull.*, October: 1201–1208.
38. Salustowicz, A. & W. Parysiewicz, 1964. Problem of Bamping in Mines, *Przglad Gorniczy*, April: 38–47 (in Polish).
39. Brauner, G., 1973. *Theory and Practices in Predicting Surface Deformation, Subsidence due to Underground Mining I*, U.S.B.M., IC 8471, pp. 1–56.

CHAPTER 9

1. Vemura, T., 1965. Mine Structure and Transportation in Steep and Thick Seams at Akabira Colliery. *4th Int. Min. Cong.*, Paper D3, pp. 1–8.
2. Wilson, J. W., 1975. *The Longwall System of Mining With Particular References to Coal Seams.* Meeting, California University, U.S.A.
3. Barron, K., 1974. *The Mining of Thick, Flat Coal Seams by a Longwall Bottom Slice with Caving and Drawing*, Mining Research Centre, Technical Bulletin, TB 189, Ottawa.
4. Chowdhary, S. K. & P. J. Kozlowski, 1975. Extraction of Thick and Steeply Inclined Seam with Longwall System. *J. of Mine. Met. & Fuels*, India, October 1975: 461–465.
5. Bise, Ch. J. & R. V. Remani, 1975. *An Evaluation of Underground Mining Technology for Western Thick Coal Seam.* AIME Preprint Number 75-F-341 SME Fall Meeting, Salt Lake City, Utah, September 10–12, 1975.
6. Jeremic, M. L., 1962. Investigation of Strata Behaviour, with the Aspect of Underground Mining Extraction. *Exploration, Development and Evaluation of Mineral Deposits.* The University of Sarajevo, Tuzla, Yugoslavia, pp. 474–484 (in Yugoslav).
7. Coates, D. F., T. S. Cochrane & G. Ellie, 1972. Three Mining Methods for Vertical, Inclined and Thick Coal Seams Used in France. *CIM Transactions*, 85: 96–102.
8. Hans, A. H., 1976. A Case for Coal Mining Research in Australia, *Symp. on Thick Seam Mining by U/G Methods*, pp. 50–58.
9. Hesgit, M. E., 1980. Mining Thick and Vertical Seams in Turkey, *World Coal*, September: 30–34.
10. Adam, R., 1976. French Thick Seams Mining Practice, *Symp. on Thick Seam Mining by U/G Methods*, Queensland Branch, Australasian Inst. of Min. and Met., September, pp. 41–50.
11. Enever, J. R. & C. D. Rowlings 1980. A Geomechanical Assessment of Thick Seam Mining with Regard to Their Application to Australian Black Coals, *The Aus. I.M.M. Conference*, New Zealand, May, pp. 135–149.
12. Thomas, E. G., 1976. Thick Seam Underground Coal Mining by a Slice and Fill Method, *Symp. on Thick Seam Mining by Underground Methods*. The Aus. I.M.M., Central Queensland Branch, September, pp. 75–81.
13. Sikora, W., 1976. Methods for Working Thick Seams and Research on Increasing Their Effectiveness, *Symp. on Thick Seam Mining by Underground Methods*. The Aus. I.M.M., Central Queensland Branch, September, pp. 93–107.
14. Zamarski, B., 1976. Strategy of the Rock Bursts Prevention in the Ostrava Karvina Coal Basin, *Uhli*, 6: 220–224 (in Czech).
15. Jeremic, M. L., 1956. Influence of Geological Factors on Coal Mineability at Zenica Colliery, *Geol. Bull.*, 2: 31–39 (in Yugoslav).
16. Jeremic, M. L., 1979. Influence of Shear Deformation Structures in Coal on Selecting Method of Mining, *Rock Mech.*, Springer-Verlag, 13: 23–38.
17. Nakajima, S., 1962. Consideration of Starting Methods for Multi-Slice Mining, *J. of the Min. & Met. Inst. of Hokkaido*, September: 38–43 (in Japanese).
18. Schevjakov, 1964. Mining of Contiguous Beds, *Mining of Mineral Deposits*, Moskow Publishing, pp. 561–585.

19. Adam, R., 1976. French Thick Seams Mining Practice, *Symp. on Thick Seam Mining by Underground Methods*, September, Central Queensland Branch, Australia, pp. 41–50.
20. Barron, K., 1974. *The Mining of Thick, Flat Coal Seams by a Longwall Bottom Slice with Caving and Drawing*, CANMET, Ottawa, TB 189, pp. 196.
21. Coates, D. F., T. S. Cochrane & G. Ellie, 1972. Three Mining Methods for Vertical, Inclined and Thick Coal Seams Used in France, *CIM Trans*. 85: 96–102.
22. Jeremic, M., 1962. Investigation of Strata Behaviour, with the Aspect of Underground Mining Extraction. *Exploration, Development and Evaluation of Mineral Deposits*. The University of Sarajevo, Tuzla, Yugoslavia, pp. 474–485 (in Yugoslav).
23. Chowdhary, S. K. & P. J. Kozlowski, 1975. Extraction of Thick and Steeply Inclined Seam with Longwall System, *J. of Mines, Metal Fuels*, October 1975: 461–465.
24. Lojas, J., 1970. Results of Tests on Rock Movements Caused by Working a Thick Seam. *Przegl. Gorn.* 5: 31–37 (in Polish).
25. Lojas, J. et al., 1978. Working the Lower Lift of a Thick Seam Under the Caved Debris Reconsolidated With Waters from Drainage of Overburden Strata, *Int. Str. Contr. Conf.*, September, Banff, Canada.
26. Gogolin, V. A. & Y. A. Ryzhkov, 1977. A Nonlinear Model of the Interaction Between Adjacent Rock Strata, the Backfill and Coal Seam, *Fiz-Tekh. Prob. Raz. Polez. Iskop.*, 1: 23–27 (in Russian).
27. Vlasenko, B. V. et al., 1979. Influence of Geological and Mining Factors on Rock Strata Pressure, *Fiz-Tekh. Prob. Raz. Polez. Iskop.*, 1: 3–9 (in Russian).
28. Vlasenko, B. V. & A. S. Fleishman, 1977. Dynamic Laws of Roof Displacement Due to Coal Extraction with Stowing, *Aspects of Rock Pressure*, Izd. IGD Sib. Otd., Akad, Nauk SSSR, 35: 37–42 (in Russian).
29. Zamarski, B., 1976. Strategy of the Rock Bursts Prevention in the Ostrava Karvina Coal Field, *Uhli*, 6: 224–229 (in Czech).
30. Zamarski, B. & J. Franek, 1977. Theoretical Principles, Organization Means and Mining Practice in Czechoslovakia, *Int. Symp. on Strata Control*, Banff, Canada, September, p. 12.
31. Wallis, M. G. et al., 1977. *Longwall-Caving Mining Method for Thick Coal Design and Feasibility Study*, U.S.B.M., Report No. R-3621, Vol. 232, March, pp. 38–46.
32. Johanson, J. R., 1965. Method of Calculating Rate of Discharge from Hoppers and Bins, *Transactions of SME*, 232, March: 38–46.
33. Barron, K., 1974. *The Mining of Thick, Flat Coal Seams by a Longwall Bottom Slice with Caving and Drawing*, CANMET, TB 189.
34. Nakajima, S., 1976. Thick Seam Mining Techniques in Japan, *Symp. on Thick Seam Mining by Underground Methods*, Central Queensland Branch, September, Australia, pp. 21–40.
35. Nakajima, S., 1976. Eine Entwicklung des Schreitausbau in Japan, *Gluckauf*, July: 709–712.
36. Jeremic, M. L. & Z. Nikolic, 1967. Caving Mechanics and Surface Subsidence at Lubnica Coal Mine, *Bull. of Fac. of Min. and Met.*, 5: 32–40 (in Yugoslav).

CHAPTER 10

1. Brozovic, T., 1964. Kuzbas Coal Basin, *Min. & Met. Bull.*, 10 & 11: 207–216 (in Yugoslav).
2. Jeremic, M. L., 1981. *Elements of Hydraulic Coal Mine Design*, Trans Tech. Publications, pp. 99–104.
3. Herzer, H. & L. B. Geller, 1978. German Experience in Hydraulic Coal Mining and Its Application to Canadian Conditions, *CIM Bull.*, 71, 788, January: 70–82.
4. Cooley, W. C., 1975. *Survey of Underground Hydraulic Coal Mining Technology*, Vol. III, Appendix B, Terraspace Incorporated, 15 October, Bureau of Mines – Open File Report 28(3)–76.
5. Shevyakov, L., 1966. *Mining of Mineral Deposits*, Foreign Languages Publishing House, Moskow, pp. 329–337.
6. Jeremic, M. L., 1968. Theory of Plates Applied in Rock Mechanics, *Min. and Met. J.*, Belgrade, 1 & 2: 57–64 and 221–233 (in Yugoslav).
7. Schevyakov, L. D., 1941. *Size and Deformation of Mine Pillars*, Izv. Akad. Nauk. U.S.S.R., No. 7, 8, 9, Moskow (in Russian).
8. Iljstein, A. M. et al., 1966. *Methods of Pillar Design for Room-and-Pillar Mining*, Izdateljstvo, 'Nauka', Moscow (in Russian).
9. Jeremic, M. L., 1978. Stress Mechanics at the Mindola Mine Zambia, Africa, *CIM Bull.*, November: 1–7.
10. Agoskov, M. J., 1961. Influence of Geological Factors on the Pillar Sizes, *Izvestija A.N.S.S.S.R.*, 3, Moscow (in Russian).

11. Berthold, E. et al., 1979. Experimental-Analytic Method of Determination of the Three-Dimensional Stress-Strain State During Extraction of East German Deposits, *Fiz. Tekh. Prob. Raz. Polez. Iskop.*, 2: 10–17 (in Russian).
12. Ivanov, I. F. et al., 1970. Support of the Roadways of Steep Pitching Seams, *UGOL'*, 8: 41–44 (in Russian).
13. Barron, K. & H. E. Cross, 1964. *A Photoelastic Model Investigation of Stress Around Cut and Fill Stopes, Based on the McIntyre Configuration*, CANMET, Report FMP64/25 MRL, Ottawa.
14. Cohrane, T. S. et al., 1964. *Ground Behaviour in a Mine in Northern Ontario*, The Trustees of Columbia University, Vol. I, Observations and measurements in mines, pp. 1–16.
15. Jeremic, M. L., 1978. Stress Mechanism at the Mindola Mine, Zambia, Africa, *CIM Bull.*, November: 77–83.
16. Corlett, A. V. & C. L. Emery, 1959. Prestress and Stress Redistribution in Rocks Around Mine Opening, *CIM Bull.*, 52, 566: 372–384.
17. Murashev, A. I. et al., 1973. Stress-Strain State of the Working Faces During Long Pillar Mining, *Fiz. Tekh. Probl. Raz. Polez. Iskop.*, 2: 11–16 (in Russian).
18. Berry, D. S., 1963. Ground Movement Considered as an Elastic Phenomenon, *Min. Engng.*, October: 28–41.
19. Jeremic, M. L., 1969. Elastic Deformation of Rock Mass, Part II, *MGM Bull.*, 8: 163–169 (in Yugoslav).
20. Jeremic, M. L. & P. A. Craig, 1980. *A Physical Model Study of Sublevel Mining*, Manuscript, Dept. of Min. Eng., Univ. of Alberta, Edmonton.
21. Goodman, R. E., 1976. *Methods of Geological Engineering*, West Publishing Company, pp. 277–285.
22. Stazhevskii, S. B., 1976. Load Calculation on Self Advancing Shield Support in Inclined Seams, *Fiz. Tekh. Prob. Razr. Polez. Iskop.*, 2: 3–18 (in Russian).
23. Kalcov, A., 1965. Methodology of Investigation of Subsidence and Protection of Surface Structure, Tuzla Basin, *Archiv for Technology*, Tuzla, 3, 2–3: 77–88 (in Yugoslav).
24. Smailbegovic, F., 1963. *The Results of Measurements of Vertical and Horizontal Displacement With Precise Nivelman in Area of Tuzla City*, Ph.D. Thesis, The University of Zagreb (in Yugoslav).

CHAPTER 11

1. Bise, Ch. J. et al., 1977. Underground Extraction Techniques for Thick Coal Seams, *AIME*, October: 35–40.
2. Dokukin, A. V., 1977. Research in the Coal Industry of the U.S.S.R., *World Coal*, July: 65–68.
3. Shevyakov, L., 1966. *Mining of Mineral Deposits*, Foreign Language Publishing House, Moskow, pp. 368–380.
4. Otsuka, T., 1980. Hydraulic Mining at Sunagawa Coal Mine, *4th Joint Meeting MMIJ-AIME*, Tokoyo, B-4-5, pp. 63–75.
5. Mills, L. J., 1978. Hydraulic Mining in the U.S.S.R., *The Min. Eng.*, London, June: 655–663.
6. Jeremic, M. L., 1979. Hydraulic Mining Possible Method for Rocky Mountain, *Coal, Min. Mag.*, London, October: 330–338.
7. Parkes, D. M. & A. W. T. Grimley, 1975. Hydraulic Mining Coal, *Min. Congr. J.*, May: 26–29.
8. Coates, D. F. et al., 1972. Three Mining Methods for Vertical, Inclined and Thick Coal Seams Used in France, *CIM Transactions*, 85: 96–102.
9. Schneiderman, S. J., 1980. Mining Thick and Irregular Seams in France, *World Coal*, September: 30–34.
10. Ahcan, R., 1977. Mechanization and Concentration of Thick Coal Seams Mining in SFR Yugoslavia, *Proc. of Int. Symp. on Thick Seam Mining*, Dhanbad, India, Paper 04, pp. 1–5.
11. Jeremic, M. L., 1979. Coal Behaviour in the Rocky Mountain Belt of Canada, *4th Int. Congress on Rock Mechanics*, Montreux, Switzerland, Vol. 1: 189–194.
12. Jeremic, M. L., 1981. Strength Properties of Coal Bearing Strata of the Plains Region, Western Canada, *Min. Met. Quat.* Yugoslavia, 28, 1: 64–76.
13. Shevyakov, L. D., 1941. Size and Deformability of Mine Pillars, *Izvestija ANSSSR*, 7, 8, 9, Moskow (in Russian).
14. Katkov, G. A. et al., 1979. Use of Optically Sensitive Materials to Model Brittle Failure of the Rocks Around Mine Openings, *Fiz. Tech. Prob. Raz. Polez. Iskop.*, 1: 92–95 (in Russian).

15. Trumbachev, C. & E. Melnikov, 1962. The Effect of the Angle of the Pillar Inclination on the Stress Distribution, *Technologia; Ekonomika Ugledobychi*, 3, Moskow (in Russian).

16. Jeremic, M. L., 1980. Philosophy of Hydraulic Mining, *CIM Bull.*, September: 81–84.

17. Fisecki, M. Y. & A. W. T. Grimley, 1977. Strata Control Studies at the Hydraulic Mine, *Strata Control Conference*, Banff, Canada, September.

18. Fisecki, M. Y., C. Chiang & W. V. Bannerman, 1980. Update in Hydraulic Mine Strata Mechanics Studies in Thick and Steep Seams of Rocky Mountains, *21st U.S. National Symp. on Rock Mech.*, University of Missouri-Rolla, May 27–30.

19. Jeremic, M. L., 1979. Coal Behavior in the Rocky Mountain Belt of Canada, *4th Inter. Congress of Rock Mech.*, Montreux, Switzerland, pp. 189–194.

20. Jeremic, M. L., 1980. Influence of Shear Deformation Structure in Coal on Selecting Methods of Mining, *Rock Mechanics*, Springer-Verlag, 13: 24–38.

21. Lutley, H. J., 1981. *Stability Analysis in Underground Hydraulic Mining*, M.Sc. Thesis, Dept. of Mineral Eng., The University of Alberta, Edmonton.

22. Baumgartner, P. & B. Stimpson, 1979. Technical Note – Development of a Tiltable Base Friction Frame for Kinematic Studies of Caving at Various Depths, *Int. J. Rock Mech., Min. Sci. & Geomech. Abstr.*, 16: 265–267.

23. Jeremic, M. L. & H. J. Lutley, 1980. Stress Analysis in Underground Extraction of Steeply Dipping Thick Coal Seams, *21st U.S. Symp. on Rock Mech.*, University of Missouri-Rolla, May 27–30.

24. Jeremic, M. L., 1980. Effect of Coal Seam Stability on Hydraulic Mine Design, *J. of Min. Met. Fuels.* India, May: 93–99.

25. Jeremic, M. L., 1980. Rupturing Criteria of Coal Bearing Strata, W. Canada, *Modern Geology*, 7: 191–199.

26. Hoek, E., 1974. Progressive Caving Induced by Mining an Inclined Ore Body, *Trans Inst. Min. and Met.*, London, October.

27. Jeremic, M. L., 1981. Strata Mechanics of Hydraulic Sublevel Coal Mining, *Int. J. Rock Mech., Min. Sci. & Geomech. Abstr.*, 19, 2: 135–142.

28. Lin, Y. L. & Y. Pang, 1977. Laws Governing the Manifestations of Rock Pressure, *Strata Control Conference*, Banff, Canada, September.

29. Jeremic, M. L., 1980. Coal Strengths in the Rocky Mountains, *World Coal*, 6, 9, September: 40–44.

30. Murashev, V. I. et al., 1973. Stress-Strain State of the Working Faces in Steep Coal Seams, *Fiz-Tekh. Probl. Raz. Polez. Iskop.*, 2: 11–16.

31. Ladanyi, B. & G. Archambault, 1972. Evaluation de la resistance au cisaillement d'un massif rocheux fragmente, *Proc. 24th Int. Geol. Cong.*, Montreal, Section 13, pp. 249–250.

32. Nascimento, V. & H. Teixeira, 1971. Mechanism of Internal Friction in Soils and Rock, *Proc. Symp. on Rock Fracture*, Nancy, France, Chapter II–3.

33. Hammett, R. D., 1974. *A Study of the Behaviour of Discontinuous Rock Masses*, Ph.D. Thesis, James Cook University of North Queensland, Australia.

34. Baumgartner, P., 1979. *The Effect of Joint Fabric on Rock Mass Caving*, M.Sc. Thesis, Dept. of Mineral Engineering, University of Alberta.

35. Jeremic, M. L., 1981. Strata Mechanics of Hydraulic Sublevel Coal Mining, *Int. J. Rock Mech., Min. Sci. & Geomech. Abstr.*, 19: 135–142.

36. Chirkov, Y. J. et al., 1975. Investigations of Displacement and Caving of Hanging Wall During Extraction of Thick Deposits of Krivbasa, *UGOL'*, February: 58–60 (in Russian).

37. Kalcov, A., 1965. Methodology of Investigation of Subsidence and Protection of Surface Structure, Tuzla Basin, *Archiv for Technology*, Tuzla, Vol. III, 2–3, pp. 77–88 (in Yugoslav).

38. Smailbegovic, F., 1963. *The Results of Measurements of Vertical and Horizontal Displacement with Precise Nivelman in Area of Tuzla City*, Ph.D. Thesis, The University of Zagreb (in Yugoslav).

39. Hoek, E., 1974. Progressive Caving Induced by Mining an Inclined Ore Body, *Trans. Inst. Min. & Met.*, London, October: 133–139.

CHAPTER 12

1. Hustrulid, W. A., 1976. A Review of Coal Pillar Strength Formulas, *Rock Mechanics*, Springer-Verlag, 8: 115–145.

2. Holland, C. T., 1964. The Strength of Coal in Mine Pillars, *Proceedings of the 6th Symp. on Rock Mechanics*, Univ. of Missouri, Rolla, pp. 450–466.

3. Bauschinger, J., 1876. *Mitteilungen aus dem Mechanisch-Technischem Laboratorium der K. Technischem Hochschule in Munchen*, Vol. 6.

4. Fadeev, A. D. & Grokholjskii, 1978. Mechanical Properties of Rocks at Deformations Level, *Fiz-Tech Prob. Raz. Pol. Iskop.*, January–February, 1: 21–27 (in Russian).

5. Jeremic, M. L., 1980. Characteristics of Western Canadian Coal Seams and Their Effect on Mine Design, *Min. Mag.*, London, December: 557–564.

6. Agoshkov, M. J., 1961. Influence of Geological-Structural Characteristics of the Strata on Dimension Ratio of the Pillars, *Izvestija ANSSSR*, 3, Moskow (in Russian).

7. Kunt'sch, M. F., 1978. Relationship Between a Dimension Ratio and Compressive Strength of Rock Samples, *UGOL'*, Moskow, March: 63–64 (in Russian).

8. Cruden, D. M., 1971. The Form of Creep Low for Rock Under Uniaxial Compression, *Int. J. of Rock Mech. and Min. Sci.*, 8: 105–126.

9. Rutter, E. H., 1972. On the Creep Testing of Rocks at Constant Stress and Constant Force, *Int. J. of Rock Mech. and Min. Sci.*, 9: 191–195.

10. Singh, D. P., 1975. A Study of Creep of Rock, *Int. J. of Rock Mech. and Min. Sci. & Geomech.*, 12: 271–276.

11. Brown, A., 1958. Ground Stress Investigations in Canadian Coal Mines, *Transaction AIME, Mining Engineering*, Augustus: 879–887.

12. Jeremic, M. L. & D. M. Cruden, 1979. Strength of Coal From Star-Key Mine Near Edmonton, Alberta, *CIM Bull.*, February: 94–99.

13. Barron, K., 1975. *An Injection Technique for Investigating the Integrity of Pillars and Ribs in Coal Mines*, Canmet Report ERP/MRL (CF) 76–86 (TR), Ottawa.

14. Brauner, G., 1964. Prognosis and Prevention of Rock Burst in Coal Mines at Ruhr Region, *Bergbau Arhiv*, December: 5 (in German).

15. Jeremic, M. L., 1979. Coal Behaviour in the Rocky Mountain Belt of Canada, *4th Int. Cong. on Rock Mech.*, Montreux, Switzerland, pp. 189–195.

16. Jeremic, M. L., 1980. Influence of Shear Deformation Structures in Coal on Selecting Method of Mining, *Rock Mech.*, Springer-Verlag, 13: 23–28.

17. Protodyakonow, M. M., 1958. *Investigations on the Brittleness and Viscosity of Coals*, Ugletkhizdat, Moskow.

18. Fairhurst, C., 1979. Rock Mechanics Studies of Proposed Underground Mining of Potash in Sergipe, Brazil, *4th Int. Congress on Rock Mech.*, Montreux, Switzerland, Vol. 1, pp. 131–137.

19. Barr, C. A., 1975. The Deformational Behaviour of Salt Rocks in Situ: Hypothesis vs. Measurements, *Bull. Int. Assoc. Engng. Geol.*, Drefeld, 12: 65–72.

20. Agoshkov, M. J., 1967. Influence of Geological Factors on the Pillar Sizes, *Izvestija ANSSSR*, 3 Moskow (in Russian).

21. Bieniawski, A. J., 1965. The Effect of Specimen Size on Compressive Strength of Coal, *Int. J. of Rock Mech. and Min. Sci.*, 5: 325–335.

22. Holland, C. T., 1964. The Strength of Coal in Mine Pillars, *Proceedings of the 6th Symp. on Rock Mechanics*, Univ. of Missouri, Rolla, pp. 450–466.

23. Salamon, M. D. G. & A. H. Munro, 1967. A Study of the Strength of Coal Pillars, *J. of the S. Afr. Inst. of Min. & Met.*, November: 185–187.

24. Jeremic, M. L., 1980. Coal Strength in the Rocky Mountains, *World Coal*, September: 40–44.

25. Jeremic, M. L., 1980. Characteristics of Western Canadian Coal Seams and Their Effect on Mine Design, *Min. Mag.*, London, December: 557–564.

26. Gramatikov, Y. A., 1977. Critical Size of Pillars Dangerous for Coal Outburst, *UGOL'*, September: 56–57 (in Russian).

27. Jeremic, M. L., 1980. Effect of Coal Seam Stability on Hydraulic Mine Design, *J. of Min. Met. Fuels*, May: 93–99.

28. Holland, C. T., 1958. Course of Occurrences of Coal Mine Bumps, *Min. Engng, Transaction AIME*, September: 994–1004B.

29. Bunting, D., 1911. Chamber Pillars in Deep Anthracite Mines, *Transactions AIME*,: 739–748.

30. Jeremic, M. L. & D. M. Cruden, 1979. The Strength of Coal from the Star-Key Mine, Near Edmonton, Alberta, *CIM Bull.*, April: 94–99.

31. Yemashchikov, V. S. & S. M. Shikinyanskii, 1978. Qualitative Evaluation of Limiting State of Rocks, *Fiz-Tekh. Prob. Raz. Polez. Iskop.*, September–October, 5: 28–36 (in Russian).
32. Hustrulid, W. A., 1976. A Review of Coal Pillar Strength Formulas, *Rock Mechanics*, 8: 115–145.
33. Jeremic, M. L., 1980. Influence of Shear Deformation Structures in Coal on Selecting Methods of Mining, *Rock Mechanics*, 13: 23–38.
34. Jeremic, M. L., 1979. Coal Behavior in the Rocky Mountain Belt of Canada, *4th Int. Cong. on Rock Mechanics*, Montreux, Switzerland, Vol. 1, pp. 189–194.
35. Protodyakonov, M. M., 1965. *Investigations on the Brittleness and Viscosity of Coals*, Ugletekizdat, Moscow (in Russian).
36. Cook, N. G. W. & M. Hood, 1978. The Stability of Underground Coal Mine Working, *Proceedings of the 1st Int. Symp. on Stability in Coal Mining*, Vancouver, Canada, pp. 135–147.
37. Holland, C. T., 1973. Mine Pillar Design, *SME Engineering Handbook*, AIME, New York, pp. 13–96 and 113–118.
38. Wilson, A. H. & D. P. Ashwin, 1972. Research Into the Determination of Pillar Sizes – Part I, A Hypothesis Concerning Pillar Stability – Part II, *The Min. Eng.*, June, 141: 409–430.
39. Ashley, G. H., 1930. Barrier Pillar Legislation in Pennsylvania, *Trans. AIME, Coal Div.*, 99: 76–96.
40. Kowalczyk, Z. & K. Trojanowski, 1964. Simplified Factors of Maximum Displacement of the Surface Caused by Mining Exploitation, *Przegl. Gorn.*, Katowice, 9: 117–124 (in Polish).
41. Kekuh, P. H., 1959. Analogy Method for Subsidence Investigations, *Mettal.*, 3: 53–59 (in Russian).
42. Brauner, G., 1973. *Subsidence due to Underground Mining, II Ground Movement and Mining Damages*, US Bureau of Mines, IC 8572, pp. 53.
43. Kalcov, A., 1965. Ground Surface Subsidence and Protection of Surface Structures, *Arch. for Technol.* 3, 2–3: 77–88 (in Yugoslav).
44. Jeremic, M. L., 1975. Subsidence Problems Caused by Solution Mining of the Rock Salt Deposits, *10th Canadian Rock Mech. Symp.*, Kingston, Vol. 1, September 2–4, pp. 203–223.
45. Kripakov, N. P., 1981. *Analysis of Pillar Stability on Steeply Pitching Seam Using the Finite Element Method*, USBM, RI 879, pp. 1–30.
46. Jeremic, M. L., 1969. Elastic Deformation of Rock Mass, *Min. Met. Bull.*, Belgrade, 7 & 8: 1010–1020 & 1205–1211 (in Yugoslav).
47. Jeremic, M. L., 1958. Concept of the Access to Coal Reserves of Raspotocje Coal Field, *Min. Met. Bull.*, 11, Belgrade (in Yugoslav).
48. Repko, A. A. & A. A. Chernysher, 1977. Deformation of the Shaft Pillars, *Fiz-Tek Probl. Raz. Pol. Iskop.*, 1: 11–17 (in Russian).
49. Schlieder, O., 1956. Vorausberechnung von Bodenbenvegungen, *Der Deutsche Steinkohlenbergbou*, B. and 2, Essen (in German).

Subject index

Abutment
  pillar 253, 477, 537
  stress 106, 176, 261, 324-
    325, 378, 382, 394, 425,
    472, 544
Accelerated deformation 221
Acoustic emission 112, 168
Air blast 477
Analogy models 15
Anchorage 44, 100, 101
Angles
  belt 340
  break 255, 257, 436, 493,
    541, 544
  dip 19, 245
  draw 293, 347
  failure 51, 198, 455
  friction 53, 55, 94, 203, 340,
    397, 455
  slide 94, 149, 210, 214
Anisotropy 39, 64, 454, 529
Anticline 48, 143
Anthracite coal 75, 522
Artificial
  pillar 420, 529
  reinforcement 217, 231
  support 31, 232, 360, 437,
    529
Arch 385, 436, 445
Average pillar stress 199, 245,
  254, 454, 457, 472

Banding 209
Barrier pillar 478, 530
B.C. coal 445, 463, 468, 478,
  485, 487
Beam deflection 251
Beams 103, 252, 377, 452
Bearing capacity 131-136, 378,
  418, 437, 455, 522, 527
Bedding 28, 42, 208, 212, 455,
  505
Benching 218, 411
Bituminous coal 68, 75, 194,
  498, 529
Blasting 186, 406, 409, 441
Block 46, 214, 248, 435, 483
Bolts 31, 44, 99, 107, 242, 445,
  505

Borehole 16, 70, 186, 468
Borehole mining 406
Brittle 70, 193, 258, 466, 497
Buckling 479
Bulking factor 338
Bumps 8, 111, 136, 164, 488

Canopy of powered support 315
Canadian plains region 21, 136,
  455, 494, 543
Carbon-dioxide 165
Carboniferous 61, 70, 73
Carbonate sediments 27
Caving 282, 288-293, 335, 351,
  393-396, 400, 430, 480, 482
Chevron fold 155
Clay 30-32, 543
Cleat 59, 83-91, 120, 477
Closure 177, 219-221, 421, 427,
  503
Coal
  bearing strata 16, 21
  black and dull 70
  clarain 68
  durain 70
  formation 59
  fusain 68
  petrography 66
  rank 74-80
  seam 25, 59-63, 354, 388
  vitrain 68
  zone 23
Coalification path 75
Coefficient
  expansion 119
  friction 179, 208, 498, 541
  poisson's 497
  safety 217
Cohesion 54, 208, 216, 386, 455
Coking coal 18, 73, 245, 522,
  526
Compaction 37, 65, 293, 389,
  436, 437
Competent rock 29, 97, 106, 540
Compressive strength
  uniaxial 192, 454, 466, 494,
    495
  triaxial 454
Computers 12, 518

Conglomerate 28, 377
Continuous
  excavation 235, 446
  miner 64, 122, 236, 241,
    442, 448
Conventional mining 406
Convergence 99, 150-155, 194,
  221, 379, 426, 506
Conveyor 235, 306, 406, 442,
  452, 453
Cracks 9, 45, 92, 433-434, 497
Creep 429, 493, 505-518
Critical
  arch 385
  length 11, 176, 433, 469
  pillar size 523
  span 433, 469
  stress 50, 147, 268, 386,
    422, 477, 499
  subsidence 256, 349
  width 256, 522, 538
Crosscut 99, 155, 236, 243, 303,
  443, 451
Crown pillar 405, 416-418
Crushing 313, 485, 503
Cost
  capital 238, 298, 320
  production 238, 304
Cutting machines 25, 305

Deflection 101, 251, 296, 428
Deformation 50, 125, 173, 194,
  218-229, 267, 388, 421, 427,
  477, 496, 504
Degradation 28, 124, 432, 515
Depillaring 236
Depth of mining 241, 268
Design principles 248
Destressing 106, 186, 472, 477,
  480, 487
Development 1, 25, 241, 407,
  409, 410, 442, 445
Deviation 291, 442
Dewatering 2, 171, 364
Diagenesis 21, 30, 36
Differential stress 224, 230
Dilatation 42, 437
Dip
  dip of the seam 19, 356, 405,

459, 529
dipping seam 19
flat seam 19
gently dipping seam 19
steeply dipping seam 19
vertical seam 19
Direct shear test 92
Displacement 97, 107, 128, 205-
210, 378, 379, 467, 482
Dolostone 28
Down-dip mining 301, 313, 405,
357
Drainage
gas 168, 445
water 169, 170
Drift 407, 442, 443, 446
Drill core 17, 499
Drilling 16, 186, 404, 406, 407,
409, 441
Dust 242, 451

Effective area 261
Effective stress 112
Elastic 65, 248, 377, 388, 490,
519
Elasto-brittle 27, 112, 468, 507,
522, 528
Elasto-plastic 12, 31, 153, 197,
203, 438, 466, 528
Endogenic 83
Energy 112, 179, 181-185, 498
Entry
cross entry 410
drift entry 243, 408
main entry 23, 448, 463
pillar entry 150, 502
sub-level entry 411, 443,
463
Environment 21, 27, 61
Equilibrium 4, 8, 283, 313, 314
Excavation 325, 341
Expansion 382
Extension 22, 43, 46, 175, 544,
546
Extraction 236, 240, 351-353,
375, 394, 408, 448
Exogenic 85

Face 133, 305, 306, 315, 324,
327-331, 425
Failure 6, 7, 8, 93, 103-109,
125, 155, 173, 197, 397,
477, 482, 498, 502
False roof 42, 106, 437
Fault 45, 54, 117-119, 213
Fill 364, 388, 389
Finite element program 276,
422, 463, 537
Fire 70, 164, 420, 530
Fissure 124, 349, 394, 426, 457,
489, 497
Flexure 107, 143, 252
Flow 166, 222, 393, 397, 399
Fluid 449, 505
Fold 47-50
Foundation 252
Fracturing 51, 52, 57, 130, 189,
290, 392

Friability 302
Friction 179, 339
Frozen stresses 140

Gas
emission 164, 166, 238, 448
outburst 17, 164, 173, 370
pressure 17, 164, 166
Geological
defects 45, 427, 481, 485
stresses 137
structures 17, 45, 472
Geostatic stresses 137
Gob 243, 304, 360, 382, 396
Gradient 19, 37, 447
Gravity
faults 51
loading 436
transportation 409, 445-446
Ground
control 18, 35, 42, 253, 460
relief 429
water 59, 168

Hardness 17, 426, 446, 449
Haulage 70, 247, 305, 447, 450
Hazard 164, 170, 185, 420
Heave 125, 149, 152, 429, 503
Hydraulic
fracturing 17, 29, 111, 185,
528
mining 48, 232, 404, 407,
445, 449
pressure 111
props 18, 240, 445, 452
Hydrodynamic stresses 163, 168
Hydromonitor 232, 445, 446,
478
Hydrostatic pressure 161, 366,
424, 456, 499, 539

Immediate roof 42, 106, 219,
382, 394, 437, 481
Inclination 291, 406, 437, 454,
456, 497, 534, 539, 544
Inclined drifts 451
Incompetent rock 136
Induced
cleat 85
deformation 109
stress 150, 328-329, 429,
465, 541
Inelastic 227, 388
In situ
stress 139, 459
test 133, 523
Instrumentation 17, 330, 389,
461, 505, 517
Integration 10, 11, 449
Interaction 277, 295, 333, 378,
388, 390, 420, 496, 544
Internal
friction 53, 203, 263, 397
structure 22, 25, 64
Isocline 48, 422
Isotropy 10, 64

Jet
energy 408, 449

impact force 446
pressure 408, 449
technology 445, 449
velocity 408
Joints 28, 45, 90, 92-96, 119-
121

Kinetic energy 182
Kuznetsk basin 526

Laminated 89, 100, 115
Layout 187, 217, 450, 530
Level 404, 407
Lignite 74, 75
Limestone 27
Limit angle 439, 544
Linear elastic 5, 284, 463, 497
Lithology 27, 68, 219, 437, 484
Load 310-313
Log 70
Long-pillar mining 242, 404
Longwall mining 26, 297, 337,
539

Major principal stress 52, 179,
280, 424, 470
Main roof 42, 106, 293, 437,
484
Mechanical
behaviour 3, 28, 36, 70, 80,
258
load 442
properties 3, 28, 36, 70, 80,
258
Microseismic 112
Mining methods 62, 297, 354,
404, 441
Minor principal stress 52, 179,
280, 424, 470
Modelling 12, 250, 258, 339,
466, 518, 537
Modulus of
elasticity 30, 41, 65, 98,
388, 390, 428, 497
rigidity 252
Mohr's circle 7, 174, 210-203,
397
Moisture 33, 430, 480
Monitor jet 186, 218, 232, 445-
446, 482
Mudstone 30, 377, 430, 543
Multiple
interaction 538
opening 242
seam mining 22, 188, 352

Oblique openings 119, 145, 153,
448
Open folds 47, 48
Orogenic movement 21, 37, 45,
511, 526
Outburst 17, 164, 173, 182-185,
505, 522
Overfolds 48
Overburden load 37, 76, 163,
245, 420, 457, 506
Overthrust 50, 76, 143

Panel 243, 256, 309, 406, 442,
448, 537

Partial extraction 407, 490, 536
Partings 22, 25, 209, 372
Penetration 386, 442
Permeability 28, 33, 166
Pillar
    design 205, 493, 521
    mini pillars 493
Pillars 233, 369, 373, 378, 420,
    468, 493, 504, 521, 530
Pinch out 23
Pipes 293, 447
Pitch 220, 158, 365, 412, 447
Planes of weakness 42, 92, 467,
    482
Porosity 28, 33, 389
Pothole 258, 290, 289
Powered support 135, 230, 315-
    323, 445, 452
Pressure
    arch 147, 452
    energy 142
    ram 268-270
Props 103, 323, 445, 451
Protodyakonov index 426
Production 243, 305, 407, 408,
    445, 449
Productivity 235, 238-244, 306,
    407, 443, 445, 449
Protective pillars 293, 530
Protective seam 188
Pumps 299, 407, 505

Raise 366, 411, 425, 446
Rapid extraction 164, 218, 291,
    441, 529
Recovery 65, 80, 426, 449, 538
Regulations 243
Reinforcement 97, 204, 217.
Relaxation 57, 159, 323, 544
Remote control 241, 232, 445
Resin bolt 100
Retreat mining 299-304, 521
Rib pillars 189, 344, 405, 413,
    416, 417
Rock
    beam 100, 107, 230, 251,
        429, 431, 488
    burst 17, 28, 111, 113, 142,
        394
    mechanics 8, 287
    pressure 35
    testing 36, 40, 41
Rocky mountain belt 37, 76, 85,
    88, 119, 214, 526
Rolls 63, 410
Roof
    bolting 100, 238, 443
    caving 105, 113, 477, 478
    displacement 98, 107
    fall 103, 147, 150, 386, 387,
        529
    sag 100, 150, 387, 427, 428,
        429
    span 42, 103, 109, 123, 187,
        426, 433, 481
    strata 97, 116, 393, 395, 467
    support 99-101, 315, 360,
        410, 451

Room-and-pillar mining 25, 233,
    413, 518
Run of mine coal 65

Sagging 151, 482, 504
Safety 241, 386, 477, 530, 531
Sandstone 22, 29, 122, 466
Screen 152, 451
Sealant 529
Seismic activity 112
Seismic effect 112
Semi-anthracite 497, 498, 506,
    515
Shaft 493, 530, 537, 539-545
Shaft mining 493
Shale 22, 29, 122, 377, 430, 543
Shearer 304, 305
Shear stress 38, 56, 436, 456,
    499
Shield mining 404-409
Shield support 318-321, 410,
    451
Shortwall mining 26, 241, 243
Siliceous sediments 28
Siltstone 22, 29, 122, 467, 543
Slab rock 10, 44, 191, 197, 503,
    514
Slaking 35
Slice mining 354
Slikensides 125, 130
Slips 55, 122, 428, 468
Spontaneous combustion 145
Stability analysis 454, 457, 539
Steel arch 360, 448, 479
Stope 218, 244, 404, 413, 463,
    476
Stowing 112, 287, 329, 364,
    405
Strain 81, 158, 181, 347, 514,
    533
Strain energy 112, 181, 498
Strata control 27, 62, 73, 106,
    529
Strata pressure 75, 111, 137, 382
Strength deterioration 43, 117,
    121, 481, 521
Stress concentration 173-176,
    259, 327-331, 394, 422, 476
Stresses 137-146, 163, 179, 326,
    423
Stress relaxation 174
Strike of seam 19, 356, 405
Structural stresses 141
Structure
    composite 63
    heterogenous 65, 66, 228
    homogenous 64, 537
    irregular 63
    uniform 62
Sub-bituminous coal 76, 189,
    194, 494, 511
Sub-coal strata 33, 125, 135
Subcritical width
Sub-level block 442, 470, 451,
    489
Sub-level mining 404, 408
Sub-level caving 441, 477
Subsidence 59, 285-292, 345-

351, 401, 403, 438, 489, 532
Superposition 276
Support 99, 101, 204-205, 240-
    244, 310-323, 407, 408, 443,
    445, 449
Swelling 30, 35, 127, 430, 480
Synclines 37, 45, 48, 143
Syngenetic 59, 61, 64

Tailgate 302, 306
Tangent modul of elasticity 497
Tectonic block 20, 51, 54
Tectonic stress 37, 52, 108, 139,
    145, 470, 540
Tensile strength 27, 58, 152,
    386, 485, 499
Tertiary creep 505, 510
Thermal stress 145
Thickness of seam
    medium thick 18, 25, 404,
        446
    thick 18, 25, 354, 448
    thin 18, 25
    very thick 18
    very thin 18
Thrust sheet 210
Tributary area loading 460
Tributary theory 245, 457
Trough subsidence 258, 288
Tunnel 20

Underground structure 3, 11,
    530
Undercut 422, 447, 451
Undulation 25, 26, 318
Unit weight 339, 459
Up-dip entries 313, 357, 405,
    457

Ventilation 235, 237-244, 407,
    445, 449
Violent pressure 111
Virgin stresses 224, 226, 539
Viscoelastic 541
Viscoplastic 12, 149, 225, 542
Viscous 158, 468, 482, 528
Voussoir beam 154

Washing plant 236
Water
    absorption 168
    flow 168, 407
    infusion 112
    jet 404, 407, 445
    pressure 168, 186, 449
Wearing capacity 35, 136
Weathering 28, 191, 494, 511,
    526, 189
Western Australian Collieries
    106, 194

Yieldable arch 443, 445, 449,
    453, 479
Yieldable pillar 413, 472
Yield deformation 160
Yield locus 397
Yield stress 159, 160

T - #0614 - 101024 - C0 - 254/178/31 - PB - 9789061915560 - Gloss Lamination